Biomedical Optical Imaging

Biomedical Optical Imaging

Edited by

James G. Fujimoto

and

Daniel L. Farkas

OXFORD

UNIVERSITY PRESS

2009

OXFORD
UNIVERSITY PRESS

Oxford University Press, Inc., publishes works that further
Oxford University's objective of excellence
in research, scholarship, and education.

Oxford New York
Auckland Cape Town Dar es Salaam Hong Kong Karachi
Kuala Lumpur Madrid Melbourne Mexico City Nairobi
New Delhi Shanghai Taipei Toronto

With offices in
Argentina Austria Brazil Chile Czech Republic France Greece
Guatemala Hungary Italy Japan Poland Portugal Singapore
South Korea Switzerland Thailand Turkey Ukraine Vietnam

Published by Oxford University Press, Inc.
198 Madison Avenue, New York, New York 10016

www.oup.com

Oxford is a registered trademark of Oxford University Press

Library of Congress Cataloging-in-Publication Data
Biomedical optical imaging / edited by James G. Fujimoto and Daniel L. Farkas.
    p. cm.
Includes bibliographical references and index.
ISBN: 978-0-19-515044-5
1. Imaging systems in medicine.   I. Fujimoto, James G. II. Farkas, Daniel L.
R857.O6B568 2008
616.07′54—dc22   2007041195

9 8 7 6 5 4 3 2 1

Printed in China
on acid-free paper

Preface

Biomedical optics is a rapidly emerging field that relies on advanced technologies, which give it high performance, and has important research, bio-industrial, and medical applications. This book provides an overview of biomedical optical imaging with contributions from leading international research groups who have pioneered many of its most important methods and applications.

Optical imaging—performed with the powerful human eye–brain combination—was probably the first method ever used for scientific investigation. In modern research, optical imaging is unique in its ability to span the realms of biology from microscopic to macroscopic, providing both structural and functional information and insights, with uses ranging from fundamental to clinical applications.

Microscopy is an icon of the sciences because of its history, versatility, and universality. Modern optical techniques such as confocal and multiphoton microscopy provide subcellular- resolution imaging in biological systems. The melding of this capability with exogenous chromophores can selectively enhance contrast for molecular targets as well as provide functional information on dynamic processes, such as nerve transduction. New methods integrate microscopy with other state-of-the-art technologies: nanoscopy, hyperspectral imaging, nonlinear excitation microscopy, fluorescence correlation spectroscopy, and optical coherence tomography can provide dynamic, molecular-scale, and three-dimensional visualization of important features and events in biological systems. Moving to the macroscopic scale, spectroscopic assessment and imaging methods based on properties of light and its interaction with matter, such as fluorescence, reflectance, scattering, polarization, and coherence can provide diagnostics of tissue pathology, including neoplastic changes. Techniques that use long-wavelength photon migration allow noninvasive exploration of processes that occur deep inside biological tissues and organs.

This book reviews the major thrust areas mentioned above, and will thus be suitable both as a reference book for beginning researchers in the field, whether they are in the realm of technology development or applications, and as a textbook for graduate courses on biomedical optical imaging. The field of biomedical imaging is very broad, and many excellent research groups are active in this area. Because this book is intended for a broader audience than the optics community, its contents focus mainly on reviewing technologies that are currently used in research, industry, or medicine, rather than those that are being developed.

The chapters begin with an introductory review of basic concepts to establish a foundation and to make the material more accessible to readers with backgrounds outside this field. The book's first section covers confocal, multiphoton, and spectral microscopy. Aside from classic light microscopy, these are the most widely used high-resolution biomedical imaging technologies to date. They have a significant base of commercially available instruments and are used in a wide range of application fields such as cellular and molecular biology, neurobiology, developmental biology, microbiology, pathology, and so forth. Confocal microscopy enables the imaging of both structure and function on a cellular level. The theoretical basis for optical imaging using confocal microscopy is well established and relies primarily upon classic optics. Multiphoton microscopy relies on nonlinear absorption of light, and provides deeper imaging than possible with confocal microscopy. By using near-infrared excitation, it reduces cell injury, enabling improved imaging of living cells and tissues. Spectral microscopy allows a quantitative analysis of cells and tissues based on their topologically resolved spectral signatures, yielding classification abilities, similar to those in satellite reconnaissance applications, to improve detection and diagnosis of abnormalities.

The use of exogenous fluorescent markers enables highly specific imaging functional processes and has been used broadly, especially in cell biology and neurobiology. Because fluorophores are a major adjunct to confocal, multiphoton, and spectral microscopy, they are covered next in the book, with chapters focusing on new probes and their use in monitoring messengerRNA and electrical events in cells.

The next section of the book covers newer methods of optical imaging, pushing the traditional limits of detection, resolution, and tissue penetration. Optical coherence tomography can provide cross-sectional and three-dimensional structural information of diagnostic relevance, imaging deeper than confocal or multiphoton microscopy. It represents an example of optical biopsy to image pathology in situ/in vivo and in real time. Nanoscopy and fluctuation correlation spectroscopy are advanced, laser-based methods that perform beyond the standard limits of spatial resolution and detection in optical imaging.

This book also covers medical applications of optical imaging, with emphasis on cancer research. Spectroscopic approaches have been especially interesting for imaging early neoplastic changes in tissue. Techniques such as autofluorescence, light scattering, and Raman have been explored, and several companies have developed spectroscopic diagnostic devices for clinical medicine.

Near-infrared light penetrates tissue well, and photon diffusion techniques can perform functional imaging of deep tissue. Their uses in cancer detection and neuroimaging are highlighted in the later chapters. The final chapters cover molecular imaging and optical reporter gene methods.

Thematically, the book covers the major areas of biomedical optical imaging by focusing on key technologies and applications, rather than attempting to be comprehensive. It surveys structural as well as functional imaging. Each imaging modality uses light in a specific way and is based on diverse contrast methods, delivery, and detection techniques, with different performance characteristics ranging from submicrometer resolution with limited penetration to millimeter resolution with centimeter penetration depths. Overall, the performance of optical imaging compares extremely well with that of other biomedical imaging technologies that have reached higher levels of maturity, commercial implementation, and clinical acceptance. Additionally, optics provides possibilities for probing function in real time, *in vivo*, and noninvasively with very high—indeed molecular—specificity and resolution. Thus, true molecular imaging is enabled for the first time by optical approaches. The advantages of complementary methods can be combined in a multimode setting, with further performance gains.

Currently, excellent multi-author books on confocal microscopy, multiphoton microscopy, and optical coherence tomography have been recently published or are in advanced stages of preparation. However, no book broadly covers all of biomedical optical imaging, a rapidly expanding area of research and commercial development, and we focused on collecting highlights of this field under one cover. *Biomedical Optical Imaging* has been developed so that it can serve as both a reference and as a text for graduate-level courses, and is intended to be useful for members of the optics community as well as the biology and medical communities. Our goal is to explain the uniqueness and potential of biomedical optical imaging. In our rapidly evolving field, it is not easy to keep up with developments, but it is certainly easy to get excited about possibilities—especially if one acquires a better knowledge of fundamentals and major trends, by throwing what Francis Bacon termed "the light of understanding" into depths and details of the living world that could previously only be imagined, not explored.

We wish to thank the leading groups and researchers who authored chapters in *Biomedical Optical Imaging*, as well as our colleagues and collaborators. We are particularly grateful to Ms. Dorothy Fleischer and Dr. Kevin Burton for their help with the editorial process. On behalf of all these contributors, we hope you find this book enlightening, useful, and stimulating.

Daniel L. Farkas
James G. Fujimoto

Contents

Contributors   xi

1. Confocal Microscopy   3
Tony Wilson

2. Spectral Optical Imaging in Biology and Medicine   29
Kevin Burton, Jihoon Jeong, Sebastian Wachsmann-Hogiu,
and Daniel L. Farkas

3. Multiphoton Microscopy in Neuroscience   73
Fritjof Helmchen, Samuel S.-H. Wang, and Winfried Denk

**4. Messenger RNA Imaging in Living Cells for Biomedical
Research**   102
Dahlene Fusco, Edouard Bertrand, and Robert H. Singer

5. Building New Fluorescent Probes   120
Alan S. Waggoner, Lauren A. Ernst, and Byron Ballou

6. Imaging Membrane Potential with Voltage-Sensitive Dyes   132
Dejan Vucinic, Efstratios Kosmidis, Chun X. Falk, Lawrence B. Cohen,
Leslie M. Loew, Maja Djurisic, and Dejan Zecevic

7. Biomedical Imaging Using Optical Coherence Tomography   161
James G. Fujimoto, Yu Chen, and Aaron Aguirre

8. Two-Photon Fluorescence Correlation Spectroscopy   196
Petra Schwille, Katrin Heinze, Petra Dittrich, and Elke Haustein

9. Nanoscopy: The Future of Optical Microscopy   237
Stefan W. Hell and Andreas Schönle

10. Fluorescence Imaging in Medical Diagnostics   265
Stefan Andersson-Engels, Katarina Svanberg, and Sune Svanberg

**11. Fluorescence and Spectroscopic Markers of Cervical
Neoplasia**   306
Ina Pavlova, Rebekah Drezek, Sung Chang, Dizem Arifler, Konstantin
Sokolov, Calum MacAulay, Michele Follen, and Rebecca Richards-Kortum

12. Quantitative Absorption and Scattering Spectra in Thick Tissues Using Broadband Diffuse Optical Spectroscopy   **330**

Dorota Jakubowski, Frederic Bevilacqua, Sean Merritt, Albert Cerussi, and Bruce J. Tromberg

13. Detection of Brain Activity by Near-Infrared Light   **356**

Enrico Gratton, Vlad Toronov, Ursula Wolf, and Martin Wolf

14. In Vivo Optical Imaging of Molecular Function Using Near-Infrared Fluorescent Probes   **374**

Vasilis Ntziachristos and Ralph Weissleder

15. Revealing the Subtleties of Disease and the Nuances of the Therapeutic Response with Optical Reporter Genes   **397**

Christopher H. Contag

Index   411

Contributors

Stefan Andersson-Engels
Department of Physics
Lund University
S-221 00 Lund
Sweden

Kevin Burton
Department of Surgery
Cedars-Sinai Medical Center
Los Angeles, CA 90048

Yu Chen
Fischell Department of
    Bioengineering
University of Maryland
College Park, MD 20742

Lawrence B. Cohen
Department of Cellular and Molecular
    Physiology
Yale University School of Medicine
New Haven, CT 06520–8026

Christopher H. Contag
Departments of Pediatrics, Microbiology
    and Immunology, and Radiology
Molecular Imaging Program
Stanford University School of Medicine
Stanford, CA 94305

Winfried Denk
Department of Biomedical Optics
Max-Planck Institute forMedical
    Research
69120 Heidelberg
Germany

Rebekah Drezek
Department of Bioengineering
Rice University
Houston, TX 77251

Daniel L. Farkas
Department of Surgery
Cedars-Sinai Medical Center
Los Angeles, CA 90048

Michele Follen
Gynecologic Oncology Center
University of Texas M. D. Anderson
    CancerCenter
Houston, TX 77230-1439

James G. Fujimoto
Research Laboratory of Electronics
Massachusetts Institute of Technology
Cambridge, MA 02139

Enrico Gratton
Department of Biomedical Engineering
University of California at Irvine
Irvine, CA 92697–2715

Stefan W. Hell
Department of NanoBiophotonics
Max-Planck-Institute forBiophysical
    Chemistry
37077 Gottingen
Germany

Fritjof Helmchen
Brain Research Institute
University of Zurich
8057 Zurich
Switzerland

Leslie M. Loew
R. D. Berlin CenterforCell   Analysis and
    Modeling
University of Connecticut Health Center
Farmington, CT 06030

Calum MacAulay
University of British Columbia
Department of Cancer Imaging
BC Cancer Agency Research Centre
Vancouver, BC
Canada V5Z 1L3

Vasilis Ntziachristos
Helmholtz CenterMunich
  GmbH
German Research Center for
  Environmental Health
Institute of Biological and Medical
  Imaging
85764 Neuherberg
Germany

Rebecca Richards-Kortum
Department of Bioengineering
Rice University
Houston, TX 77005

Andreas Schönle
Department for NanoBiophotonics
Max Planck Institute forBiophysical
  Chemistry
37077 Gottingen
Germany

Petra Schwille
Department of Biophysics
Biotechnologisches Zentrum der TU
  Dresden (Biotec)
01307 Dresden
Germany

Robert H. Singer
Department of Anatomy and Structural
  Biology
Albert Einstein College of Medicine
Bronx, NY 10461

Katarina Svanberg
Lund University Medical Laser
  Center
Division of Atomic Physics
Lund University
S-221 00 Lund
Sweden
and
Department of Oncology
Lund University Hospital
S-221 85 Lund
Sweden

Sune Svanberg
Department of Physics
Lund University
S-221 00 Lund
Sweden

Bruce J. Tromberg
University of California at Irvine
Beckman LaserInstitute and Medical
  Clinic
Irvine, CA 92612–3010

Alan S. Waggoner
MolecularBiosensorand   Imaging Center
Carnegie Mellon University
Pittsburgh, PA 15213

Samuel S.-H. Wang
Department of Molecular Biology
Princeton University
Princeton, NJ

Ralph Weissleder
CenterforMolecularImaging   Research
Department of Radiology
Massachusetts General Hospital
Harvard Medical School
Charlestown, MA 02129

Tony Wilson
University of Oxford
Engineering Science Department
Oxford, OX13PJ
United Kingdom

Biomedical Optical Imaging

1
Confocal Microscopy

TONY WILSON

It is probably fair to say that the development and wide commercial availability of the confocal microscope has been one of the most significant advances in light microscopy in the recent past. The main reason for the popularity of these instruments derives from their ability to permit the structure of thick specimens of biological tissue to be investigated in three dimensions. They achieve this important goal by resorting to a *scanning* approach togetherwith a novel (confocal) optical system.

The traditional wide-field conventional microscope is a parallel processing system that images the entire object field simultaneously. This is quite a severe requirement for the optical components. We can relax this design requirement if we no longer try to image the whole object field at once. The limit of this constraint relaxation is to require an image of only *one* point on the object at a time. In this case, all that we ask of the optics is that it provide a good image of one point. The price that we have to pay is that we must *scan* to build up an image of the entire field. The answerto the question as to whetherthis price is worth paying will, to some extent, depend on the application in question.

A typical arrangement of a scanning confocal optical microscope is shown in figure 1.1, in which the system is built around a host conventional microscope. The essential components are some form of mechanism for scanning the light beam (usually from a laser) relative to the specimen and appropriate photodetectors to collect the reflected or transmitted light (Wilson and Sheppard, 1984). Most of the early systems were analogue in nature, but it is now universal, thanks to the serial nature of the image formation, to use a computer both to drive the microscope and to collect, process, and display the image.

In the beam-scanning confocal configuration of figure 1.1, the scanning is typically achieved by using vibrating galvanometer-type mirrors or acousto-optic beam deflectors. The use of the latter gives the possibility of video-rate scanning, whereas vibrating mirrors are often relatively slow when imaging an extended region of the specimen, although significantly higherscanning speeds are achievable oversmallerscan regions. It should be noted that otherappr oaches to scanning may be implemented, such as specimen scanning and lens scanning. These methods, although not generally available commercially, do have an advantage in certain specialized applications (Wilson and Sheppard, 1984; Wilson, 1990). Because this chapter is necessarily limited in length, we recommend additional material that may be found in other sources (Wilson and Sheppard, 1984; Wilson, 1990; Pawley, 1995; Corle and Kino, 1996; Gu, 1996; Masters, 1996; Diaspro, 2002), where a detailed list of references may be found.

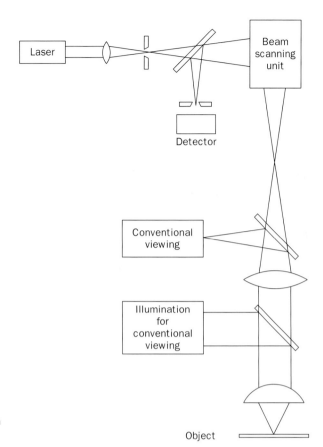

Figure 1.1 Schematic diagram of a confocal microscope.

IMAGE FORMATION IN SCANNING MICROSCOPES

We do not discuss the fine details of the optical properties of confocal systems because this is already widely available in the literature. The essence is shown in figure 1.1, where the confocal optical system consists simply of a point source of light that is then used to probe a single point on the specimen. The strength of the reflected or fluorescence light from the single point on the specimen is then measured via a point (pinhole) detector. The confocal—point source and point detector—optical *system* therefore merely produces an *image* of a single object point and hence some form of scanning is necessary to produce an image of an extended region of the specimen. However, the use of single-point illumination and single-point detection enables novel imaging capabilities that offera significant advantage over those possessed by conventional wide-field optical microscopes; these are enhanced lateral resolution and, perhaps more important, a unique depth discrimination or optical sectioning property. It is this latter property that leads to the ability to obtain three-dimensional images of volume specimens.

The improvement in lateral resolution may at first seem implausible. However, it can be explained simply by a principle given by Lukosz (1966), which states, in essence, that resolution can be increased at the expense of field of view. The field of view can then be increased by scanning. One way of taking advantage of Lukosz's principle is to place a very small aperture extremely close to the object. The resolution is now determined by the size of the aperture rather than the radiation. With the confocal microscope, we do not use a physical aperture in the focal plane, but rather use the back-projected image of a point

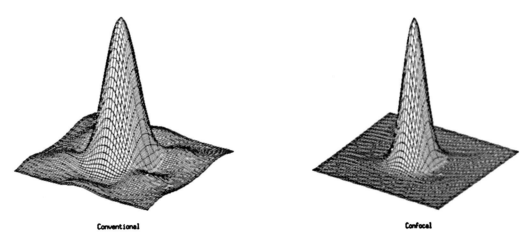

Conventional   Confocal

Figure 1.2 The point-spread functions of the conventional and confocal microscopes show the improvement in lateral resolution that may be obtained in the confocal case.

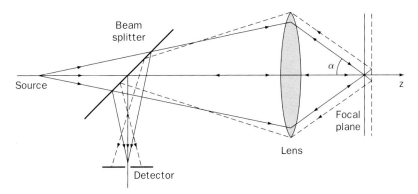

Figure 1.3 The origin of the optical sectioning or depth discrimination property of the confocal optical system.

detector in conjunction with the focused point source. Figure 1.2 indicates the improvement in lateral resolution that may be achieved.

The confocal principle, which was first described by Minsky (1961), was introduced in an attempt to obtain an image of a slice within a thick specimen that was free from the distracting presence of out-of-focus information from surrounding planes. The confocal optical systems fulfill this requirement, and their inherent optical sectioning or depth discrimination property has become the major motivation for using confocal microscopes and is the basis of many of the novel imaging modes of these instruments. The origin of the depth discrimination property may be understood very easily from figure 1.3, where we show a reflection-mode confocal microscope and consider the imaging of a specimen with a rough surface.

The solid lines show the optical path when an object feature lies in the focal plane of the lens. At a laterscan position, the object surface is supposed to be located in the plane of the vertical dashed line. In this case, simple ray tracing shows that the light reflected back to the detector pinhole arrives as a defocused blur, only the central portion of which is detected and contributes to the image. In this way, the system discriminates against features that do not lie within the focal region of the lens. A simple method to demonstrate this discrimination and to measure its strength is to axially scan a perfect reflector through the focal plane and measure the detected signal strength. Figure 1.4 shows a typical response.

These responses are frequently denoted by $V(z)$, by analogy with a similartechnique in scanning acoustic microscopy, although the correspondence is not perfect. A simple paraxial theory models this response as

$$I(u) = \left[\frac{\sin (u/2)}{u/2} \right]^2 \tag{1.1}$$

where u is a normalized axial coordinate that is related to the real axial distance, z, via

$$u = \frac{8\pi}{\lambda} nz \sin^2 (\alpha/2) \tag{1.2}$$

where λ is the wavelength and n sin α the numerical aperture. As a measure of the strength of the sectioning, we can choose the full width at half intensity of the $I(u)$ curves. Figure 1.5

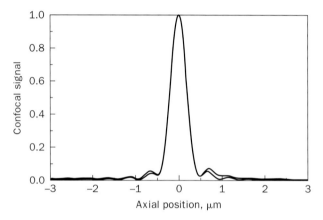

Figure 1.4 The variation in detected signal as a plane reflector is scanned axially through a focus. The measurement was taken with a 1.3 numerical aperture objective and 633 nm light.

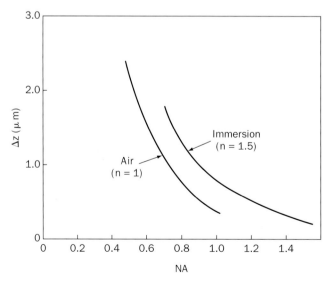

Figure 1.5 The optical sectioning width as a function of NA. The curves are for red light (0.6328-μm wavelength). Δz is the full width at the half-intensity points of the curves of $I(u)$ against u.

shows this value as a function of numerical aperture (NA) for the specific case of imaging with red light from a helium neon laser.

These curves were obtained using a high-aperture theory that is more reliable than equation 1.1 at the highest values of NA. Note that these numerical values refer to nonfluorescence imaging. The qualitative explanation of optical sectioning, of course, carries over to the fluorescence case, but the actual value of the optical sectioning strength is different in the fluorescence case.

APPLICATIONS OF DEPTH DISCRIMINATION

Because this property is one of the major reasons for the popularity of confocal microscopes, it is worthwhile, at this point, to review briefly some of the novel imaging techniques that have become available with confocal microscopy.

Figure 1.6 illustrates the essential effect. Figure 1.6A shows a conventional image of a planar microcircuit that has been deliberately mounted with its normal at an angle to the optic axis. We see that only one portion of the circuit, running diagonally, is in focus. Figure 1.6B shows the corresponding confocal image. Here, the discrimination against detail outside the focal plane is clear. The areas that were out of focus in view A have been rejected. Furthermore, the confocal image appears to be in focus throughout the visible band, which illustrates that the sectioning property is stronger than the depth of focus.

This suggests that if we try to image a thick, translucent specimen, we can arrange, by the choice of our focal position, to image detail exclusively from one specific region. In essence, we can section the specimen optically without having to resort to mechanical means. Figure 1.7 shows an idealized schematic of the process.

The portion of the beehive-shaped object that we see is determined by the focus position. In this way, it is possible to take a *through-focus* series and obtain data about the three-dimensional structure of the specimen. If we represent the volume image by $I(x, y, z)$, then by focussing at a position $z = z_1$, we obtain, ideally, the image $I(x, y, z_1)$. This, of course, is not strictly true in practice, because the optical section is not infinitely thin.

It is clear that the confocal microscope allows us to form high-resolution images with a depth of focus sufficiently small that all the detail that is imaged appears in focus. This suggests immediately that we can extend the depth of focus of the microscope by integrating the images taken at different focal settings *without* sacrificing the lateral resolution.

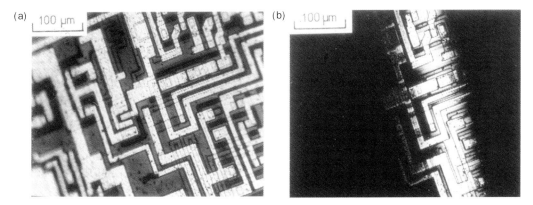

Figure 1.6 (A) Conventional scanning microscope image of a tilted microcircuit. The parts of the object outside the focal plane appear blurred. (B) Confocal image of the same microcircuit. Only the part of the specimen within the focal region is imaged strongly.

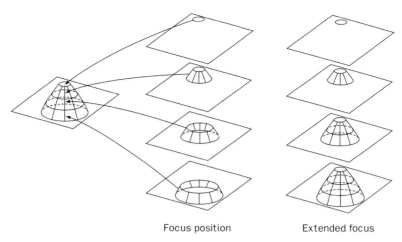

Focus position   Extended focus

Figure 1.7 An idealization of the optical sectioning property showing the ability to obtain a through-focus series of images that may then be used to reconstruct the original volume object at high resolution.

Mathematically, this *extended-focus* image is given by

$$I_{EF}(x, y) = \int I(x, y, z) \, dz \qquad (1.3)$$

As an alternative to the extended-focus method, we can form an *autofocus* image by scanning the object axially and, instead of integrating, selecting the focus at each picture point by recording the maximum in the detected signal. Mathematically, this might be written

$$I_{AF}(x, y) = I(x, y, z_{\max}) \qquad (1.4)$$

where $z_{\max}$ corresponds to the focus setting giving the maximum signal. The images obtained are somewhat similar to the extended focus method and, again, substantial increases in depth of focus may be obtained. We can go one step further in this case and turn the microscope into a noncontacting surface profilometer. Here we simply display $z_{\max}$.

It is clear by now that the confocal method gives us a convenient tool for studying three-dimensional structures in general. We essentially record the image as a series of slices and play it back in any desired fashion. Naturally, in practice it is not as simple as this, but we can, for example, display the data as an $x - z$ image rather than an $x - y$ image. This is somewhat similar to viewing the specimen from the side. As another example, we might choose to recombine the data as stereo pairs by introducing a slight lateral offset to each image slice as we add them up. If we do this twice, with an offset to the left in one case and the right in another, we obtain, very simply, stereo pairs. Mathematically, we create images of the form

$$\int I(x \pm \gamma z, y, z) \, dz \qquad (1.5)$$

where γ is a constant. In practice, it may not be necessary to introduce offsets in both directions to obtain an adequate stereo view.

All that we have said so far about these techniques has been by way of simplified introduction. In particular, we have not presented any fluorescence images. The key point is that, in both bright-field and fluorescence modes, the confocal principle permits the imaging of specimens in three dimensions. The situation is, of course, more involved than we have implied. A thorough knowledge of the image formation process, together with the

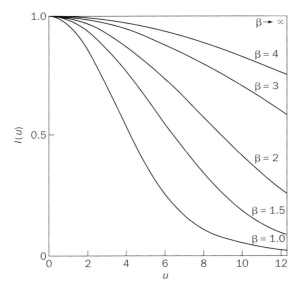

Figure 1.8 The detected signal as a perfect planar object is scanned axially through a focus for a variety of fluorescent wavelengths. Note that if we measure the sectioning by the half width of these curves, the strength of the sectioning is essentially proportional to β.

effects of lens aberrations and absorption, is necessary before accurate data manipulation can take place.

In conclusion to this section, it is important to emphasize that the confocal microscope does *not* produce *three-dimensional* images. The opposite is true. It essentially produces very high-quality *two-dimensional* images of a (thin) slice within a thick specimen. A three-dimensional rendering of the entire-volume specimen may then be generated by suitably combining a numberof these *two-dimensional* image slices from a through-focus series of images.

FLUORESCENCE MICROSCOPY

We now turn our attention to confocal fluorescence microscopy because this is the imaging mode that is most usually used in biological applications. Although we have introduced the confocal microscope in terms of bright-field imaging, the comments we have made concerning the origin of the optical sectioning and so forth carry over directly to the fluorescence case, although the numerical values describing the strength of the optical sectioning are, of course, different and we return to this point later.

If we assume that the fluorescence in the object destroys the coherence of the illuminating radiation and produces an incoherent fluorescent field proportional to the intensity of the incident radiation, $I(v, u)$, then we may write the effective intensity point-spread function, which describes image formation in the incoherent *confocal* fluorescence microscope, as

$$I(v, u), \quad I\left(\frac{v}{\beta} \frac{u}{\beta}\right) \tag{1.6}$$

where the optical coordinates u and v are defined relative to the *primary* radiation, and $\beta = \lambda_2/\lambda_1$ is the ratio of the fluorescence to the primary wavelength. We note that $v = (2\pi/\lambda)r\, n \sin\alpha$ where r denotes the actual radial distance.

This suggests that the imaging performance depends on the value of β. To illustrate this, figure 1.8 shows the variation in detected signal strength as a perfect fluorescent sheet through focus. This serves to characterize the strength of the optical sectioning in fluorescence microscopy in the same way that the mirror was used in the bright-field case.

Note that the half width of these curves is essentially proportional to β and so, for optimum sectioning, the wavelength ratio should be as close to unity as possible.

We have just discussed what we might call *one-photon fluorescence microscopy* in the sense that a fluorophore is excited by a single photon of a particular wavelength. It then returns to the ground state and emits a photon at the (slightly longer) fluorescence wavelength. It is this radiation that is detected via the confocal pinhole. Recently, however, much interest has centered on two-photon excitation fluorescence microscopy (Geoppert-Meyer, 1931; Denk et al., 1990). This process relies on the simultaneous absorption of two, longerwavelength photons, afterwhich a single fluorescence photon is emitted. The excitation wavelength is typically twice that used in the one-photon case. The beauty of the two-photon approach lies in the quadratic dependence of the fluorescence intensity on the intensity of the illumination. This leads to fluorescence emission that is always confined to the region of focus. In other words, the system has an inherent optical sectioning property. Otherbenefits of two-photon fluorescence overthe single-photon case include the use of red or infrared lasers to excite ultraviolet dyes, confinement of photobleaching to the focal region (the region of excitation), and the reduced effects of scattering and greater penetration. However, it should also be remembered that, compared with single-photon excitation, the fluorescence yield of many fluorescent dyes under two-photon excitation is relatively low (Xu and Webb, 1996a,b).

To make a theoretical comparison between the one- and two-photon modalities, let us assume that the *emission* wavelength, λ_{em}, is the same irrespective of the mode of excitation. Because the wavelength required for single-photon excitation is generally shorter than the emission wavelength, we may write it as $\gamma\lambda_{em}$ where $\gamma < 1$. Because two-photon excitation requires the simultaneous absorption of two photons of half the energy, let us assume that the excitation wavelength in this case may be written as $2\gamma\lambda_{em}$, which has been shown to be a reasonable approximation for many dyes (Xu et al., 1996a,b).

If we now introduce optical coordinates u and v normalized in terms of λ_{em}, we may write the effective point-spread functions in the one-photon confocal and two-photon cases as

$$I_{1p-conf} = I\left(\frac{v}{\gamma}, \frac{u}{\gamma}\right)I(v, u) \tag{1.7}$$

and

$$I_{2p} = I^2\left(\frac{v}{2\gamma}, \frac{u}{2\gamma}\right) \tag{1.8}$$

respectively. Note that although a pinhole is not usually used in two-photon microscopy, it is perfectly possible to include one if necessary. In this confocal two-photon geometry, the effective point spread function becomes

$$I_{2p-conf} = I^2\left(\frac{v}{2\gamma}, \frac{u}{2\gamma}\right)I(v, u) \tag{1.9}$$

If we now look at equations 1.7 and 1.8 in the $\gamma = 1$ limit, we find

$$I_{1p-conf} = I^2(v, u) \tag{1.10}$$

and

$$I_{2p} = I^2\left(\frac{v}{2}, \frac{u}{2}\right) \tag{1.11}$$

Thus we can see that because of the longerexcitation wavelength used in two-photon microscopy, the effective point-spread function is *twice* as large as that of the one-photon confocal in *both* the *lateral* and *axial* directions. The situation is somewhat improved in the confocal two-photon case, but it is worth remembering that the advantages of two-photon excitation microscopy are accompanied by a reduction in optical performance compared with the single-photon case.

From a practical point of view, it is worth noting that the two-photon approach has certain very important advantages over one-photon excitation in terms of image contrast when imaging through scattering media, apart from the greater depth of penetration afforded by the longerwavelength excitation. In a single-photon confocal case, it is quite possible that the desired fluorescence radiation from the focal plane may be scattered after generation in such a way that it is not detected through the confocal pinhole. Furthermore, because the fluorescence is generated throughout the entire focal volume, it is also possible that undesired fluorescence radiation that was not generated within the focal region may be scattered so as to be detected through the confocal pinhole. In either case, this leads to a reduction in image contrast. The situation is, however, completely different in the two-photon case. Here the fluorescence is generated *only* in the focal region and *not* throughout the focal volume. Furthermore, because all the fluorescence is detected via a large-area detector, with no pinhole involved, it is not so important if further scattering events take place. This leads to high-contrast images that are less sensitive to scattering. This is particularly important for specimens that are much more scattering at the fluorescence ($\lambda/2$) wavelength than the excitation (λ) wavelength.

OPTICAL ARCHITECTURES

In addition to the generic architecture of figure 1.1, a number of alternative implementations of confocal microscopes have been recently developed to reduce the alignment tolerances of the architecture of figure 1.1 as well as to increase image acquisition speed. Because great care is required to ensure that the illumination and detector pinholes lie in equivalent positions in the optical system, a reciprocal geometry confocal system has been developed that uses a single pinhole both to launch light into the microscope as well as to detect the returning confocal image signal. A development of this is to replace the physical pinholes with a single-mode optical fiber. However, these systems are essentially developments of the traditional confocal architecture. Let us now look at two recent attempts to develop new confocal architectures. The first system is based on the pinhole source–detector concept, whereas the second approach describes a simple modification to a conventional microscope to permit optically sectioned images to be obtained.

The Aperture Mask System

The question of system alignment and image acquisition speed have already been answered to a certain extent by Petrán et al. (1968), who originally developed the tandem scanning (confocal) microscope, which was later reconfigured into a one-sided disc embodiment (see, forexample, Corle and Kino, 1996). The main component of the system is a disc containing many pinholes. Each pinhole acts as both the illumination and detection pinhole, and thus the system acts rather like a large number of parallel, reciprocal geometry confocal microscopes, each imaging a specific point on the object. However, because it is important that no light from a particular confocal system enters an adjacent system (i.e., there should be no cross-talk between adjacent confocal systems), the pinholes in the disc must be placed far apart—typically 10 pinhole diameters apart—which has two immediate consequences. First, only a small amount (typically 1%) of the available light is used for imaging and, second, the wide spacing of the pinholes means that the object is only sparsely probed. To probe—and hence image—the whole object, it is necessary to arrange the pinhole apertures in a series of Archimedean spirals and to rotate the disc. These systems have been developed and are capable of producing high-quality images in real time with both TV rate and higher imaging speeds, without the need forlaserillumination.

To make greater use of the available light in this approach we must, inevitably, place the pinholes closer together. This, of course, leads to cross-talk between the neighboring confocal system; thus, a method must be devised to prevent this. To do this we replace the Nipkow disc of the tandem scanning microscope with an aperture mask consisting of many pinholes placed as close together as possible. This aperture mask has the property that any of its pinholes can be opened and closed independently of the others in any desired time sequence. This might be achieved, forexample, by using a liquid crystal spatial light modulator. Because we require there be no cross-talk between the many parallel confocal systems, it is necessary to use a sequence of openings and closings of each pinhole that is completely uncorrelated with the openings and closings of all the other pinholes. There are many such orthonormal sequences in the literature; however, they all require the use of both positive and negative numbers, and unfortunately we cannot have negative intensity of light! The pinhole is either open, which corresponds to one, or closed, which corresponds to zero. There is no position that can correspond to negative one. The way out of the dilemma is not to obtain the confocal signal directly. To use a particular orthonormal sequence, $b_i(t)$, of plus and minus ones, forthe ith pinhole, we must add a constant shift to the desired sequence to make a sequence of positive numbers that can be encoded in terms of pinhole opening and closing. We thus encode each of the pinhole openings and closings as $(1 + b_i(t))/2$, which will correspond to open (one) when $b_i(t) = 1$ and to close (zero) when $b_i(t) = -1$. The effect of adding the constant offset to the desired sequence is to produce a composite image that will be partly confocal as a result of the $b_i(t)$ terms and partly conventional as a result of the constant term. The method of operation is now clear. We first take an image with the pinholes encoded, as we have just discussed, and obtain a composite conventional plus confocal image. We then switch all the pinholes to the open state to obtain a conventional image. It is then a simple matterto subtract the two images in real time via computer to produce the confocal image.

Although this approach may be implemented using a liquid crystal spatial light modulator, it is cheaper and more simple merely to impress the correlation codes photolithographically on a disc and to rotate the disc so that the transmissivity at any picture point varies according to the desired orthonormal sequence. A blank sector may be used to provide the conventional image. If this approach is adopted, then all that is required is to replace the single-sided Nipkow disc of the tandem scanning microscope with a suitably encoded aperture disc (fig. 1.9) (Juškaitis et al., 1996; Wilson et al., 1997). Figure 1.10 shows

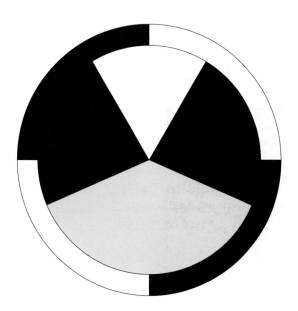

Figure 1.9 A typical aperture mask. When light passes through the encoded region of the disc, a composite conventional and confocal image is obtained. Light passing through the unobstructed sector provides a conventional image.

a through-focus series of images of a fly's eye, and figure 1.11 shows a three-dimensional representation of the fly's eye obtained with this white-light, real-time confocal system.

The Use of Structured Illumination to Achieve Optical Sectioning

The motivation for the development just described was to produce a light-efficient, real-time, three-dimensional imaging system. The approach was to start with the traditional confocal microscope design—point source and point detector—and to engineer a massively parallel, light-efficient three-dimensional imaging system (see also Liang et al., 1997; Hanley et al., 1999).

An alternative approach is to realize that the conventional light microscope already has many desirable properties: real-time image capture, standard illumination, ease of alignment, and so forth. However, it does not produce optically sectioned images in the sense usually understood in confocal microscopy. To see how this deficiency may be corrected via a simple modification of the illumination system, let us begin by looking at the theory of

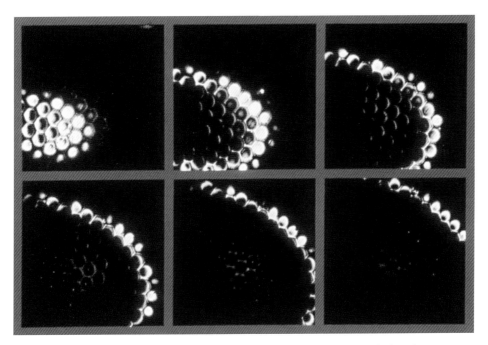

Figure 1.10 A through-focus series of images of a portion of a fly's eye. Each optical section was recorded in real time using standard nonlaser illumination.

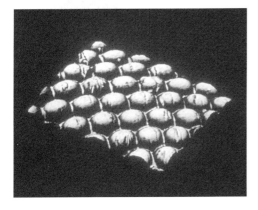

Figure 1.11 A computer-generated three-dimensional representation of the fly's eye.

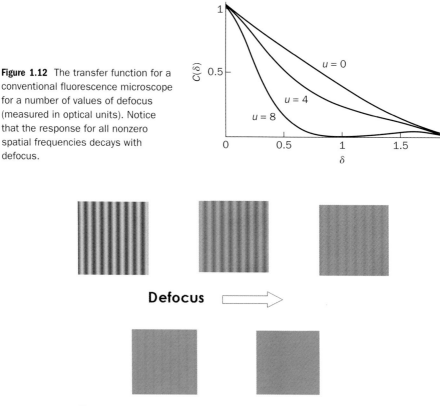

Figure 1.12 The transfer function for a conventional fluorescence microscope for a number of values of defocus (measured in optical units). Notice that the response for all nonzero spatial frequencies decays with defocus.

Figure 1.13 The image of a one-dimensional single spatial frequency bar pattern for varying degrees of defocus. Note that for sufficiently large values of defocus, the bar pattern is not imaged at all.

image formation in a conventional fluorescence microscope and start by asking in what way the image changes as the microscope is defocused. We know that in a confocal microscope the image signal from all object features attenuates with defocus and that this does not happen in a conventional microscope. However, when we look closely at the image formation process, we find that it is only the zero spatial frequency (constant) component that does not change with defocus (fig. 1.12); all otherspatial frequencies actually do attenuate to a greaterorlesserextent   with defocus.

Figure 1.13 shows the image of a single spatial frequency, one-dimensional bar pattern object for increasing degrees of defocus.

When the specimen is imaged in focus, a good image of the barpatter n is obtained. However, with increasing defocus, the image becomes progressively poorer and weaker until it eventually disappears, leaving a uniform gray level. This suggests a simple way to perform optical sectioning in a conventional microscope.

If we simply modify the illumination path of the microscope to project onto the object the image of a one-dimensional, single spatial frequency fringe pattern, then the image we see through the microscope will consist of a sharp image of those parts of the object, where the fringe pattern is in focus, together with an out-of-focus blurred image of the rest of the object. To obtain an optically sectioned image, it is necessary to remove the blurred out-of-focus portion as well as the fringe pattern from the in-focus optical section. There are many ways to do this. One of the simplest involves simple processing of three images taken at three different spatial positions of the fringe pattern. The out-of-focus regions remain fairly constant between these images and the relative spatial shift of the fringe pattern allows

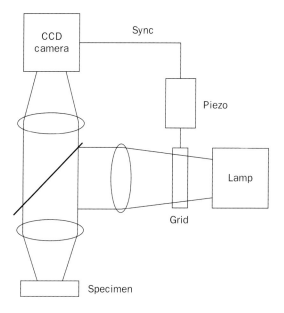

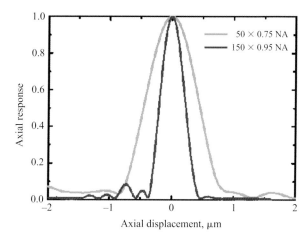

Figure 1.14 The optical system of the structured illumination microscope (top) together with experimentally obtained axial responses (bottom), which confirm the optical sectioning ability of the instrument.

the three images to be combined in such a way as to remove the fringes. This permits us to retrieve both an optically sectioned image as well as a conventional image in real time (Neil et al., 1997).

Because the approach involves processing three conventional microscope images, the image formation is fundamentally different from that of the confocal microscope. However, the depth discrimination or optical sectioning strength is very similar, and this approach, which requires very minimal modifications to the instrument, has been used to produce high-quality three-dimensional images of volume objects that are directly comparable with those obtained with confocal microscopes. A schematic of the optical arrangement is shown in figure 1.14A together with experimentally obtained axial responses in figure 1.14B. Note that these responses are substantially similar to those obtained in the true confocal case. As an example of the kind of images that can be obtained with this kind of microscope, figure 1.15 presents two images of a spiracle of a head louse. Figure 1.15A is an autofocus image of greatly extended depth of field constructed from a through-focus series of images. Figure 1.15B is a conventional image taken at midfocus. The dramatic increase in depth of field is clear when compared with a midfocus conventional image. Note that these images were taken using a standard microscope illuminator as a light source. Indeed, the system

(a)

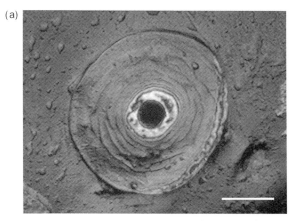

(b)

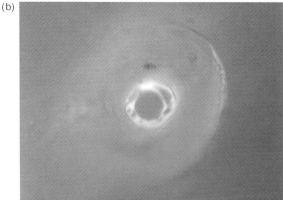

Figure 1.15 (A, B) Two images of the region around the spiracle of a head louse. View A is an autofocus image and view B shows a midfocus conventional image. The scale bar depicts a length of 10 μm.

is so light efficient that good-quality optical sections of transistor specimens have been obtained using simply a candle as light source!

Imaging using fluorescence light is also possible using this technique. However, an alternative approach that does not require a physical grid is possible if a laser is used as the light source. With this type of system, the laser illumination is split into two beams that are allowed to interfere at an angle in the fluorescent specimen. This has the effect of directly "writing" a one-dimensional fringe pattern in the specimen. Spatial shift of the fringe pattern is achieved by varying the phase of one of the interfering beams. As before, three images are taken from which both the optically sectioned image and the conventional image may be obtained (Neil et al., 1998a). The beauty of this approach is that no imaging optics are required at the illuminating wavelength, and system alignment is trivial.

ABERRATION CORRECTION

As we have seen, the most important feature of the confocal microscope is its optical sectioning property, which permits volume structures to be rendered in three dimensions via a suitable stack of through-focus images. However, to achieve the highest performance, it is necessary that the optical resolution be the same at all depths within the optical section. Unfortunately, there are fundamental reasons why this cannot be achieved in general. One is that high-aperture microscope objectives are often designed to give optimum performance when imaging features are located just below the coverglass. Another is more fundamental and is a result of specimen-induced refractive index mismatches. These could be caused by a variety of reasons, such as the use of an oil immersion objective to image into a watery specimen, or refractive index inhomogeneities within the specimen itself. In many cases, they cannot be removed by system design and a new approach must be taken. An *adaptive* method of correction in which the aberrations are measured using a wavefront sensor and are removed with an adaptive wavefront corrector, such as a deformable mirror, is required.

The key elements in an adaptive optics system are the wavefront sensor, which measures the (unwanted) aberration in an optical beam, together with a wavefront correction element that is able to introduce the appropriate amount of preaberration to cancel out the unwanted aberration and restore diffraction-limited performance. The existing techniques for the direct sensing of wavefronts have been mainly confined either to interferometric methods, which are naturally limited in application to use with coherent light sources, or to Hartmann-Shack wavefront sensors (Tyson, 1991; Hardy, 1998). The Hartmann-Shack sensor uses a zonal approach in which the wavefront is effectively broken down into a number of zones where the local gradient of the wavefront is measured. The overall wavefront shape is then inferred from this information. This method has been used extensively in adaptive optics for astronomy, but is not necessarily the most appropriate for microscopy.

A very powerful method of describing a particular wavefront aberration is in terms of a superposition of aberration modes, such as those described by the Zernike circle polynomials (Born and Wolf, 1975). These polynomials, which are two-dimensional functions, are orthogonal over the unit circle and have the advantage that real-world aberrations can often be described by only a few, low-order Zernike modes. We have shown in microscopy, for example, that the aberrations introduced when focusing on a biological specimen are dominated by a few lower order modes (Booth et al., 1998). As a further example, the same situation applies when focusing deep into the high-refractive index materials used in three-dimensional optical memory applications. In both cases it is necessary to use an adaptive feedback system to restore diffraction-limited performance at depth. It is therefore attractive to consider basing an adaptive system on Zernike polynomials, as it will be possible to simplify the overall system design because the Zernike modes of interest could be measured directly. Equally important, because only a small number of modes need be measured, the complexity of the wavefront sensor would be reduced.

The operation of the modal wavefront sensor is based upon the concept of *wavefront biasing*. This process involves the combination of certain amounts of the Zernike aberration mode being measured (the *bias mode*) with the input wavefront. Conceptually, this can be done by including appropriately shaped or etched glass plates in the optical beam path, although in practice other methods are used. First, the beam containing the input wavefront passes through a beam splitter to create two identical beams. In the first beam, a positive amount of the bias mode is added to the wavefront, which is then focused by a lens onto a pinhole. In the second beam, an equal but opposite negative amount of the bias mode is added to the wavefront. This is again focused onto a pinhole. Behind each pinhole lies a photodetectorthat measures the optical powerpassing through the pinhole. The output signal of the sensoris taken to be the difference between the two photodetectorsignals. Figure 1.16 illustrates the conceptual operation of the wavefront sensor.

The system presented in figure 1.16 is, of course, conceptual, and practical implementations differ. In one case, a computer-generated binary-phase hologram is used to allow several modes to be measured simultaneously (Neil et al., 2000a,b). A further implementation involves the sequential application of bias modes using an adaptive element such as a membrane or "bimorph" deformable mirror. Alternatively, a liquid crystal spatial light modulatormay be used.

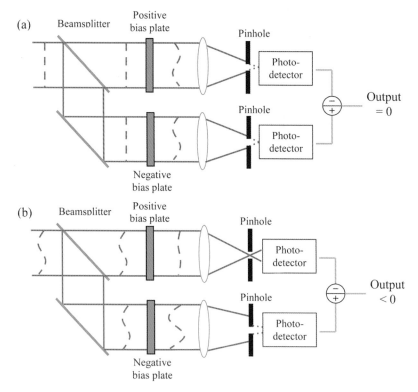

Figure 1.16 (A, B) Conceptual implementation of the wavefront sensor. In view A there is no aberration in the input wavefront, and the same amount of light is incident upon both photodetectors. The output signal is therefore zero. In view B there is a negative amount of the bias mode in the input wavefront. The aberration is therefore partially corrected by the positive bias plate and worsened by the negative bias plate. More light falls on the first detector than on the second. The output signal is therefore negative.

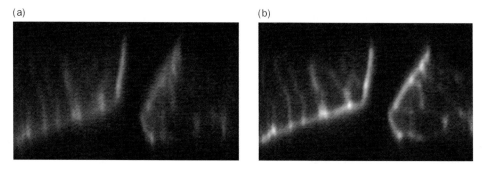

Figure 1.17 (A, B) Confocal microscope x–z scans of mouse intestine labeled with a fluorescent dye. The image in view A is without aberration correction. The image in view B was obtained with adaptive aberration correction using the sequentially implemented wavefront sensor. A membrane mirror was used as the biasing and correction element.

The practicability of this new sensor has been demonstrated in a confocal microscope (Booth et al., 2002a) and has been used to measure and correct aberrations in the imaging of biological specimens (Booth et al., 2002b) (fig. 1.17). It has also been combined with a commercial two-photon fluorescence microscope (Neil et al., 2000c).

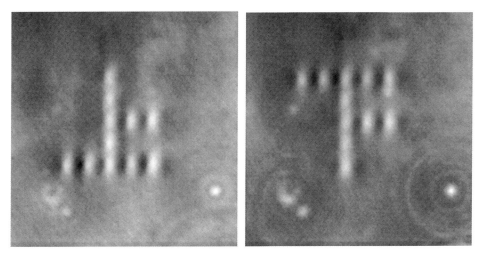

Figure 1.18 Phase contrast microscope images of two planes of bit data written at a depth of around 1 mm in a crystal of lithium niobate. The planes are approximately 20 μm apart. The data were written using an adaptive optics system incorporating the modal wavefront sensor.

The wavefront sensor was used in a system for the measurement and correction of aberrations for the writing of three-dimensional optical memory devices (Neil et al., 2002a). These devices are potential successors to current CD and DVD technology, and consist of many layers of data. The effects of aberrations particularly limit the data storage density of these devices; previously, data could only be written at depths up to around 100 μm. Our aberration correction system extended the feasible writing depth to 1 mm (fig. 1.18).

GENERATION OF OPTICAL FIELDS

A fundamental aspect of all optical imaging systems, regardless of their particular form, is that probe light is directed to be incident upon a specimen. The ensuing interaction gives rise to the optical emission that is used to form an image. The particular nature of the emission from which the image contrast originates is determined by the specific nature of the particular photophysical, photochemical, or photoelectrical interaction. Hence, optimum resolution and contrast can be achieved only if the optical field of the probe light is of the appropriate form. In many cases this will simply mean as small a spot of light as possible, and the challenge will be little more than correctly coupling light into the optical imaging system. In other cases it will be important to be able to control the amplitude, phase, and polarization of the optical field incident on the specimen. In this section we concentrate on controlling the polarization state of the light used to probe a specimen in an optical scanning microscope. In this case, a high-aperture objective lens produces a focused beam of light of the appropriate optical form both in terms of its shape and its polarization, and this focal distribution acts as the probe that is used to extract information from the specimen under study. To achieve this, it is necessary to control the optical field that is presented to the entrance pupil of the objective lens.

There are, however, an increasingly large number of instances when it is desirable to create arbitrary vectorial beams in which the polarization is spatially varying. Examples range from studies of singularities in optical fields (Swartzlander, 2001) to the tuning of the strength of the longitudinal component of the electrical field in the focal region of a microscope objective for single molecular experiments (Sick et al., 2000; Novotny et al., 2001). These applications require, for example, the use of radially and azimuthally

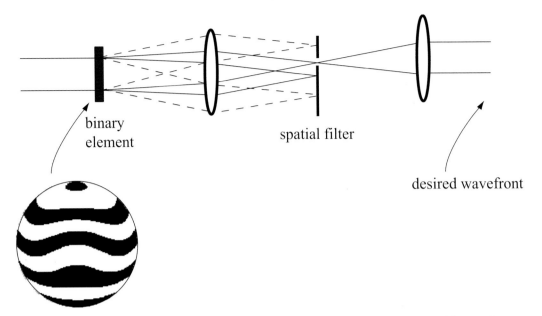

Figure 1.19 Schematic diagram of the 4-f system, together with the binary optical element, used for wavefront generation.

polarized beams that may be obtained by interferometric means (Tidwell et al., 1990, 1993; Youngsworth et al., 2000) or by the use of subwavelength gratings (Bomzon et al., 2002). In this section we also show how we can use radial polarization both to enhance the axial component of the electrical field in the focal plane and to generate a very finely focused spot of light (Quabis et al., 2000).

In the previous section we concentrated on the measurement of wavefronts, but said little about how the correction element or, more generally, the wavefront generation element was implemented. There are a number of approaches available, including the use of deformable mirrors. However, a very general approach, based on the use of computer-generated holograms, leads to a very versatile system that is able to generate beams of arbitrary phase, amplitude, and polarization. The basic idea is indicated in figure 1.19 in which a phase-only binary optical element (the pixels of which may be set to $+1$ or -1) is used in a 4-f optical system (Goodman, 1968).

In essence, we encode the binary phase-only element in such a way that one of the diffracted orders, which appears in the Fourier plane of the 4-f system, is proportional to the (inverse) Fourier transform of the desired output field. This diffracted order is then isolated using a spatial filter, and the action of the second lens is to transform this field to provide the required output. A number of encoding schemes have been described (Neil et al., 1998b, 2000d) to produce complex wavefronts and, in a further development, the system has been modified to permit the generation of arbitrary complex vector wavefronts (Neil et al., 2002b).

Generation and Focusing of Radially Polarized Electrical Fields

As we have indicated, there is increasing interest in using radially polarized light to illuminate a microscope objective fora numberof reasons. These include the enhancement of the axial component of the electrical field in the focal region as well as the narrowness of the radially symmetric focal distribution (Quabis, 2000). We shall therefore

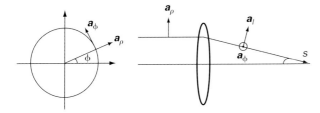

Figure 1.20 The focusing geometry.

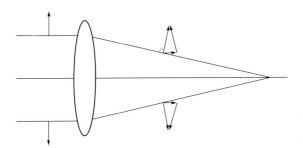

Figure 1.21 Schematic diagram of the electrical field components of a radially polarized beam when it is focused by an objective lens.

concentrate in the generation of such fields and their coupling into high-NA microscope objective lenses.

We begin by recalling that when describing focusing by a (high NA) objective lens, it is usual to decompose the incident field, E_i, into its radial and azimuthal components (fig. 1.20).

Thus,

$$E_i = \alpha_\rho \, a_\rho + \beta_\phi \, a_\phi \tag{1.12}$$

where α_ρ and β_ϕ are the strengths of the radial and azimuthal components, respectively.

The vectors $a_\rho = (\cos\phi, \sin\phi)$ and $a_\varphi = (-\sin\phi, \cos\phi)$ represent unit vectors in the radial and azimuthal directions. The field in the focal region may now be written using the vectorial form of the Debye integral (Sommerfeld, 1954; Richards and Wolf, 1959) as

$$E \sim \int_0^{2\pi} \int_0^{\alpha} \cos^{\frac{1}{2}}\theta \, \sin\theta \left\{ \alpha_\rho \begin{pmatrix} \cos\theta\,\cos\phi \\ \cos\theta\,\sin\phi \\ \sin\theta \end{pmatrix} + \beta_\phi \begin{pmatrix} -\sin\phi \\ \cos\phi \\ 0 \end{pmatrix} \right\} \tag{1.13}$$

$$\exp\left(-j\,kr_p \, \sin\theta \, \cos\left(\phi - \phi_p\right)\right) \exp\left(j\,kz \, \cos\theta\right) \, d\theta d\phi$$

where $\left(r_p, \phi_p, z\right)$ denotes a point in the focal region $k = 2\pi n/\lambda$, where λ is the wavelength and n is the refractive index. Furthermore, α is the angular limit of integration determined by the NA ($n \sin\alpha$) of the objective lens. We note that for the case of linear input $E_i = i_x$, say, which gives $\alpha_\rho = \cos\phi$ and $\beta_\phi = -\sin\phi$ and leads to the classic result of Richards and Wolf (1959) when these expressions are substituted into equation 1.13.

We shall now specialize to the case of the focusing action of an objective lens when radially polarized light is incident (fig. 1.21).

The incident beam is refracted by the lens toward the focus and, as can be seen, the refraction leads to the presence of an axially polarized component in the focal field. The relative strength of the radial and axial polarization components of the focal field is determined by the NA. It is to be expected that the strength of the axial component will be enhanced at highervalues of NA. If we now set $\alpha_\rho = 1$ and $\beta_\phi = 0$, then it is straightforward to evaluate the integrals of equation 1.13 to reveal a field in the focal region described by

$$E \sim K_{11}(r,\alpha,z) \begin{pmatrix} \cos\phi \\ \sin\phi \\ 0 \end{pmatrix} + i K_0(r,\alpha,z) \begin{pmatrix} 0 \\ 0 \\ 1 \end{pmatrix} \qquad (1.14)$$

where K_{11} and K_0 are defined as the following integrals:

$$K_{11}(r,\alpha,z) = \int_0^\alpha \cos^{\frac{1}{2}}\theta \, \sin\theta \, \cos\theta \, J_1(kr\sin\theta) \, e^{ikz\cos\theta} \, d\theta \qquad (1.15)$$

and

$$K_0(r,\alpha,z) = \int_0^\alpha \cos^{\frac{1}{2}}\theta \, \sin^2\theta \, J_0(kr\sin\theta) \, e^{ikz\cos\theta} \, d\theta \qquad (1.16)$$

where r,ϕ,z are coordinates in the focal region, k is the wave number, and α is the semi-aperture of the objective lens, related to the NA via $n\sin\alpha$, where n denote the refractive index.

Figure 1.22 shows the relative strengths of the radial and axial polarization components together with the normalized intensity distributions in the focal region. However, even at an NA of unity, it is evident that a significant lateral component is still present.

To suppress the lateral component of electrical field further, annular apodization may be used. Figure 1.23 illustrates the process as the degree of annular apodization is increased.

Indeed, for the limiting case of an infinitely narrow annulus, the electrical field in the focal region becomes

$$E \sim \cot\alpha \, J_1(v) \begin{pmatrix} \cos\phi \\ \sin\phi \\ 1 \end{pmatrix} + i J_0(v) \begin{pmatrix} 0 \\ 0 \\ 1 \end{pmatrix} \qquad (1.17)$$

where we have introduced the normalized coordinate $v = kr\sin\alpha$ and, as we see, the field becomes essentially axial at high aperture.

To demonstrate the ability of our wavefront generator to accurately produce radial polarized light, we elect to place a subresolution scatterer in the focal region of a high NA objective lens and to examine the form of the field scattered back through the objective lens. It is clear from figure 1.21 and the mathematical expressions that the polarization of the electrical field on-axis at the focal point is purely axial. We assume that the scatter is sufficiently small that it radiates as an electrical dipole (Wilson et al., 1997), with a dipole moment that is proportional to the electrical field of the incident light (fig. 1.24).

The far-field radiation pattern, from the (dipole) scatterer, may be written as $E = -r_\wedge (r_\wedge p)$, where p is the dipole moment. We note, of course, that $E \cdot r = 0$ (i.e., no component along the propagation direction r). It is then straightforward to refract this field by the objective lens to find the field, E_2, in the back focal plane. Forourspecial case in which the dipole radiates as p_z only, it is a simple matterto write the field in the back focal plane (fig. 1.24, dashed line) as

$$E_2 \sim \frac{\sin\theta}{\sqrt{\cos\theta}} \begin{pmatrix} \sin\phi \\ \cos\phi \end{pmatrix} \qquad (1.18)$$

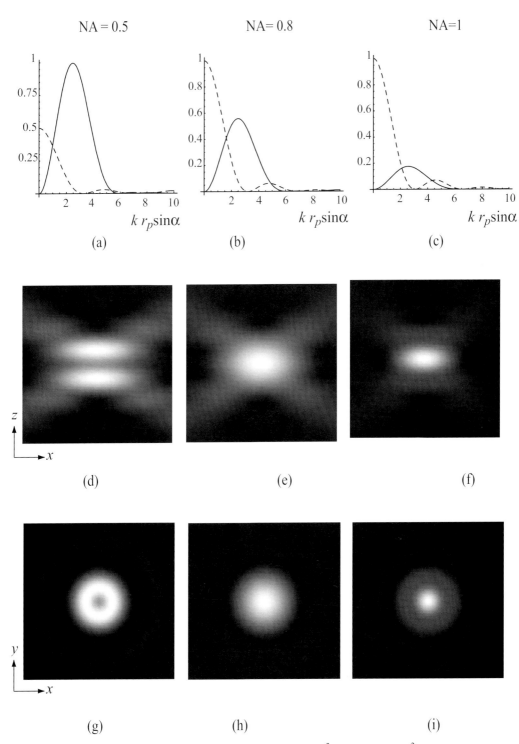

Figure 1.22 (A–C) Theoretical simulations of the relative strengths of K_{11}^2 (solid line) and K_0^2 (dashed line) at different NAs: 0.5 (A), 0.8 (B), and 1 (C). (D–I) Normalized intensity distributions at the focus of the following aperture sizes: (D, G) NA = 0.5, (E, H) NA = 0.8, and (F, I), NA = 1 when the objective lens is illuminated with radially polarized light.

$$\varepsilon = 0 \qquad\qquad \varepsilon = 0.8 \qquad\qquad \varepsilon = 0.99$$

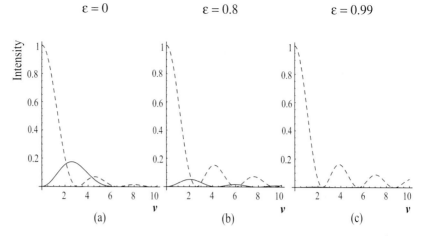

(a)   (b)   (c)

Figure 1.23 (A–C) The effect of annular apodization of a radially polarized beam on K_0^2 (dashed line) and K_{11}^2 (solid line) when focused by an objective lens of NA = 1 for no apodization (A), 80% apodization (B), and 99% apodization (C).

Figure 1.24 The use of radially polarized light results in a pure axially polarized electrical field at the geometric focus. A point scatterer located at the focus radiates as a dipole with a far-field pattern as shown with the dashed lines.

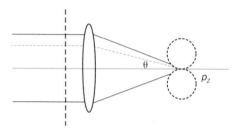

where the $\sqrt{\cos\theta}$ factor takes account of the change in the direction of the energy stream (Richards and Wolf, 1959). For objective lenses obeying the sine condition, we may introduce a radial coordinate in the pupil plane: $\rho = \sin\theta/\sin\alpha$. Figure 1.25 shows the intensity in the back focal plane when a 100-nm gold bead is placed on-axis in the focal plane and the objective lens is illuminated with a radially polarized field. Good agreement between theory and experiment is obtained (Wilson et al., 2003) and, as expected, zero intensity is found in the center of the pupil.

Having shown that radial polarization can be effectively coupled into a high-aperture objective, it is natural to try to use it for imaging, particularly because the focal spot can be made very narrow if annular illumination is also introduced (Quabis et al., 2000). However, care needs to be taken to understand the physics of the image formation process. If we attempt to image a small scattering bead in a confocal microscope, for example, we will obtain ringlike images rather than a full point: $I\left(r_p = 0\right) = 0$ (fig 1.26).

This is to be expected, because the intensity in the confocal image of a point scatter is proportional to the modulus square of the average field image, despite the very tightly focused spot illuminating the scatterer. The origin of the contrast lies in the polarization of the field in the spot. On-axis, this is purely axial, producing the radiation pattern (fig. 1.24) that inevitably leads to zero intensity in the image.

It might be thought, therefore, that radially polarized illumination is not useful in confocal microscopy, but we must remember that these results are for point scatters and that much confocal microscopy is concerned with fluorescence imaging, where the image formation process is completely different. If we assume that the fluorescent molecule radiates as a randomly oriented dipole and we average over all possible orientations, the situation

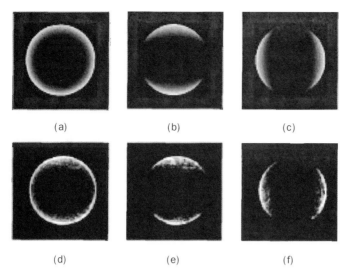

(a)　　　　　　(b)　　　　　　(c)

(d)　　　　　　(e)　　　　　　(f)

Figure 1.25 (A, D) Theoretical and experimental backfocal plane intensities. The vectorial nature of the field is revealed with the aid of horizontal (B, E) and vertical (C, F) linear poles, respectively. The objective used was a 63×, 1.4-NA oil immersion objective, and a 0.532-μm laser was used as the light source.

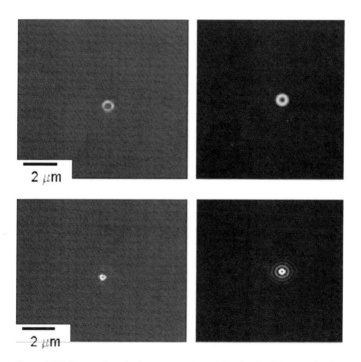

Figure 1.26 Comparison between experimentally obtained images (top) and theory (bottom) of a spherical gold bead in a confocal microscope with radially polarized illumination. The upper images correspond to no annular apodization whereas the lower images show the effect of 75% annular apodization.

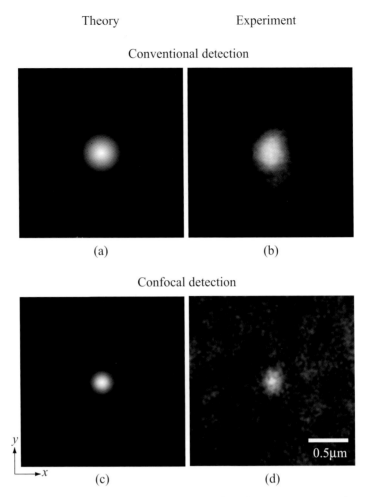

Figure 1.27 (A–D) Comparing the images of a 20-nm fluorescent quantum dot with the corresponding theoretical simulations for conventional microscopy (A, B) and confocal microscopy with radially polarized illumination (C, D).

is different. Figure 1.27 shows the improvement in resolution that is obtained in this case by using radially polarized light.

CONCLUSION

We have examined the origin of the optical sectioning property in the confocal microscope to introduce the range of imaging modes to which this unique form of microscope leads. A range of optical architectures has also been described. By far the most universal is that shown in figure 1.1, where a confocal module is integrated around a conventional optical microscope. Other, more recent, real-time implementations were also described. A number of practical aspects of confocal microscopy were not discussed because they are readily available elsewhere, such as advice on the correct choice of detector pinhole size, or because they are still the focus of active research, such as the development of new contrast mechanisms for achieving enhanced three-dimensional resolution including stimulated emission depletion methods known as STED (Hell and Wichman, 1994) or4-Pi (Hell and Stelzer, 1992), which are still under development but offer great promise.

We further looked at a number of approaches by which the point-spread function of a microscope objective may be modified. In the first example, the modification of the point-spread function removed an unwanted specimen or a system-induced aberration, and an adaptive system was developed. In the last example, a more general system was described in which a binary optical element-based system was used to generate arbitrary vector wave-fronts. Radially polarized beams were considered as a particular example, and images in confocal microscopes were presented.

ACKNOWLEDGMENTS

The contributions, in alphabetical order, of Martin Booth, Rimas Juškaitis, Farnaz Massoumian, and Mark Neil were invaluable in the development of this work.

REFERENCES

Booth MJ, Neil MAA, and Wilson T. (1988). Aberration correction for confocal imaging in refractive-index-mismatched media. *J. Microscopy.*, **192**, 90.

Booth MJ, Neil MAA, Juskaitis R, and Wilson T. (2002a). Adaptive aberration correction in a confocal microscope. *Proc. Natl. Acad. Sci USA*, **99**, 5788.

Booth MJ, Neil MAA, and Wilson T. (2002b). New modal wave-front sensor: Application to adaptive confocal fluorescence microscopy and two-photon excitation fluorescence microscopy. *J. Opt. Soc. Am.*, **19**, 2112.

Bomzon Z, BienerG, KleinerV, and Hasman E. (2002). Radially and azimuthally polarized beams generated by space-variant dielectric subwavelength gratings.*Opt. Lett.*, **27**, 285.

Born M and Wolf E. (1975). *Principles ofOptics.* Pergamon, Oxford.

Corle TR and Kino GK. (1996). *Confocal Scanning Optical Microscopy and Related Imaging Systems.* Academic Press, New York.

Denk W, Strikler JH, and Webb WW. (1990). 2-photon laser scanning fluorescence microscopy. *Science,* **248**, 73.

Diaspro A. (2002). *Confocal and Two-Photon Microscopy.* Wiley-Liss, New York.

Goodman JW. (1968). *Introduction to Fourier Optics.* McGraw Hill, New York.

Goeppert-Meyer M. (1931). Two-quantum processes. *Ann. Phys.*, **9**, 273.

Gu M. (1996). *Principles ofThree-Dimensional Imaging in Confocal Microscopes.* World Scientific, Singapore.

Hanley QS ,Verveer PJ, Gemkow MJ, Arndt-Jovin D, and Jovin TM. (1999). An optical sectioning programmable array microscope implemented with a digital micromirror device. *J. Microsc.*, **196**, 317.

Hardy JW. (1998). *Adaptive Optics for Astronomical Telescopes.* Oxford University Press, Oxford, UK.

Hell S and Stelzer EHK. (1992). Fundamental improvement of resolution with a 4pi-confocal fluorescence microscope using 2-photon excitation. *Opt. Commun.*, **93**, 277.

Hell S and Wichmann J. (1994). Breaking the diffraction resolution limit by stimulated-emission-stimulated-emission-depletion fluorescence microscopy. *Opt. Lett.*, **19**, 870.

Juškaitis R, Wilson T, Neil MAA, and Kozubek M. (1996). Efficient real-time confocal microscopy with white light sources. *Nature,* **383**, 804.

Liang MH, Stehr RL, and Krause AW. (1997). Confocal pattern period in multiple-aperture confocal imaging systems with coherent illumination. *Opt. Lett.*, **22**, 751.

Lukosz W. (1966). Optical systems with resolving powers exceeding the classical limit. *J. Opt. Soc. Am.*, **56**, 1463.

Masters BR. (1996). *Confocal Microscopy.* SPIE, Bellingham, Washington.

Minsky M. (1961). Microscopy apparatus. U.S. patent 3,013,467, December 19, 1961. (Filed November 7, 1957.)

Neil MAA, Juskaitis R, and Wilson T. (1997). Method of obtaining optical sectioning by using structured light in a conventional microscope. *Opt. Lett.*, **22**, 1905.

Neil MAA, Booth MJ, and Wilson T. (1998a). Dynamic wave-front generation for the characterization and testing of optical systems. *Opt. Lett.*, **23**, 1849.

Neil MAA, Booth MJ, and Wilson T. (2000a). New modal wave-front sensor: A theoretical analysis. *J. Opt. Soc. Am.,* **17**, 1098.

Neil MAA, Booth MJ, and Wilson T. (2000b). Closed-loop aberration correction by use of a modal Zernike wave-front sensor. *Opt. Lett.,* **25**, 1083.

Neil MAA, Juškaitis R, Booth MJ, Wilson T, Tanaka T, and Kawata SJ. (2000c). Adaptive aberration correction in a two-photon microscope. *J. Microscopy,* **200**, 105.

Neil MAA, Juškaitis R, Booth MJ, Wilson T, Tanaka T, and Kawata S. (2002a). Active aberration correction for the writing of three-dimensional optical memory devices. *Appl. Opt.,* **41**, 1374.

Neil MAA, Juskaitis R, and Wilson T. (1998b). Real time 3D fluorescence microscopy by two beam interference illumination. *Opt. Commun.,* **153**, 1.

Neil MAA, Wilson T, and Juškaitis R. (2000d). A wavefront generator for complex pupil function synthesis and point spread function engineering. *J. Microscopy,* **197**, 219.

Neil MAA, Massoumian F, Juškaitis R, and Wilson T. (2002b). Method for the generation of arbitrary complex vector wave fronts. *Opt. Lett.,* **27**, 1929.

Novotny L, Beverluis MR, Youngworth KS, and Brown TG. (2001). Longitudinal field modes probed by single molecules. *Phys. Rev. Lett.,* **86**, 5251.

Pawley JB (ed.). (1995). *Handbook ofBiological Confocal Microscopy.* Plenum Press, New York.

Petrán M, Hadravsky M, Egger MD, and Galambos R. (1968). Tandem-scanning reflected-light microscope. *J Opt. Soc. Am.,* **58**, 661.

Quabis S, Dorn R, Eberler M, Gloeckl O, and Leuchs G. (2000). Focusing light to a tighter spot. *Opt. Commun.,* **179**, 1.

Richards B and Wolf E. (1959). *Proc. R. Soc. (Lond) A,* **253**, 358.

Sick B, Hecht B, and Novotny L. (2000). Orientational imaging of single molecules by annular illumination. *Phys. Rev. Lett.,* **85**, 4482.

Sommerfeld A. (1954). *Optics.* Academic Press, London, UK.

Swartzlander GA Jr. (2001). Peering into darkness with a vortex spatial filter. *Opt. Lett.,* **26**, 497.

Tidwell SC, Ford DH, and Kimura WD. (1990). Generating radially polarized beams interferometrically. *Appl. Opt.,* **29**, 2234.

Tidwell SC, Kim GH, and Kimura WD. (1993). Efficient radially polarized laser-beam generation with a double interferometer. *Appl. Opt.,* **32**, 5222.

Tyson RK. (1991). *Principles ofAdaptive Optics.* Academic Press, London, UK.

Wilson T (ed.). (1990). *Confocal Microscopy.* Academic Press, London, UK.

Wilson T, Juškaitis R, and Higdon PD. (1997). The imaging of dielectric point scatterers in conventional and confocal polarisation microscopes. *Opt. Commun.,* **141**, 298–313.

Wilson T, Massournian F, and Juškaitis R. (2003). Generation and focusing of radially polarized electric fields. *Opt. Eng.,* **42**, 3088.

Wilson T and Sheppard CJR. (1984). *Theory and Practice ofScanning Optical Microscopy.* Academic Press, London, UK.

Youngworth KS and Brown TG. (2000). Inhomogenous polarization in scanning optical microscopy. *Proc. SPIE,* **3919**, 75.

Xu C and Webb WW. (1996a). Measurement of two-photon excitation cross sections of molecular fluorophores with data from 690 to 1050 nm. *J. Opt. Soc. Am.,* **B13**, 481.

Xu C, Williams Zipfel W, and Webb WW. (1996b). Multiphoton excitation cross-sections of molecular fluorophores. *Bioimaging.* **4**, 198.

2

Spectral Optical Imaging in Biology and Medicine

KEVIN BURTON, JIHOON JEONG,
SEBASTIAN WACHSMANN-HOGIU,
AND DANIEL L. FARKAS

The utility of light as an investigational tool for biological research and medical diagnosis is largely based on its noninvasive nature, and its great spatiotemporal range and resolution. Additionally, handling optical images is very intuitive, as perception of the world around us relies heavily on our eyes and brains.

In the quest for better quantitating biological structure, understanding function, and applying both of these to the detection and treatment of disease, the mere extension of our sight is no longer sufficient. Light proves itself highly versatile, because its additional properties—usable for imaging only by deployment of relatively advanced technologies—allow the creation of optical images that carry valuable information (Farkas, 2003). Thus, in addition to intensity, features that can be used to generate contrast (Farkas et al., 1997) and, hence, images include wavelength (Farkas, 2001), polarization, coherence, lifetime and nonlinear effects, and this book contains illustrations of methods and applications in this respect.

Wavelength is one of the basic properties of light, and one that humans have used and appreciated for a long time in their everyday lives (and decision making). Spectral optical imaging is a relatively new field in which the advantages of optical spectroscopy as an analytical tool are combined with the power of object visualization as obtained by optical imaging. Applications in industry (machine vision) and remote sensing (including satellite reconnaissance) are relatively well-known, but biological and medical applications lagged surprisingly behind until a recent surge in interest within the past decade.

We review here the field of spectral optical imaging, with emphasis on basic concepts, and aim for a balance between acquisition methods, analysis and display, and applications, both biological and medical.

SPECTRAL SELECTION METHODS FOR OPTICAL IMAGING

Optical spectroscopy transforms an incoming time domain optical waveform into a frequency (or wavelength) spectrum. In spectral imaging, each pixel of the image contains spectral information, which is added as a third dimension to the two-dimensional spatial image, generating a three-dimensional data cube.

As a matter of terminology, we refer to *spectral imaging* when this third dimension is a well-sampled spectral domain covering the visible range (usually 20–50 wavelength bands, equispaced) (Levenson et al., 1999). When such imaging extends to otherspectr al domains beyond the visible realm (ultraviolet, infrared, and so forth), the term, derived from remote sensing, is *hyperspectral imaging*. Last, *multispectral imaging* should probably be reserved for imaging that simultaneously uses two or more different spectroscopy methods in the imaging mode (e.g., wavelength and fluorescence lifetime) (see Wachsmann-Hogiu and Farkas, 2008). The result of imaging with a couple of color filters should not be termed *spectral imaging*, much less *multispectral,* but unfortunately these terms have been thus misused, especially in the biological literature.

There are a number of different techniques that permit the acquisition of spectrally resolved image sets, corresponding to spectral discrimination devices that can be inserted eitherin the illumination orin the imaging path of the light. In remote sensing, the illumination source is the sun, with spectral qualities that are constant; consequently, all spectral discrimination is effected on the returned light. In contrast, in "nonremote" sensing, the illumination arm can be varied spectrally, providing information, in reflection or absorbance, very similar to what can be obtained by spectrally filtering the remitted light. Thus, with a tunable light source, the illuminating light is scanned continuously or discontinuously through a number of wavelengths. The light sources can be tuned either using diffraction gratings, as in most monochromators, or tunable filters, such as acousto-optic tunable filters (AOTFs) or liquid crystal tunable filters (LCTFs), in case there is need for greater speed in switching wavelengths than can be achieved with a monochromator, which must physically shift the grating position. Tunable filters, as their name implies, can be tuned to permit the transmission of narrow, preselected bands of light, and these can be rapidly and randomly switched. A "continuous" spectrum can be obtained by collecting a grayscale image at each contiguous spectral band. Thus, these devices are *band sequential* in operation. It should be noted that only a portion of the photons emerging from the imaged specimen will be passed to the collector, because the filter is only transparent to a narrow wavelength region at any one time.

A whole data cube may contain up to 10^9 pixels (10^3 pixels perdimension), making recording, storing, and data analysis a technical and computational challenge. To increase the efficiency of recording a spectral cube, multiplexing is often required. This can be realized in three different ways. The first way is to detect simultaneously all the desired wavelengths in the experiment by using a dispersive element equipped with one or more slits, and performing scanning in spatial coordinates. This approach is applied in point-scanning spectrometers, pushbroom imaging spectrometers, moving slit spectrometers, and more recently in Hadamard spectrometers. A second way to obtain signal multiplexing is to detect a full spatial image via a two-dimensional detector (most usually a charge-coupled device [CCD]), and perform scanning in the wavelength domain. The most common wavelength-scanning techniques using this approach are filter wheels, Fourier transform imaging spectrometers (FTS), volume holographic spectral imagers, and techniques based on AOTFs and LCTFs. The third way is to collect all spatial and spectral information simultaneously in a snapshot. These are the so-called *nonscanning techniques,* and one prominent example is the computed tomography imaging spectrometer (CTIS).

In the following pages, we survey the most interesting and frequently used modalities to record spectral images, discussing some of their advantages and disadvantages for biomedical applications.

Slit Spectrometers

Slit spectrometers use one or multiple entrance slits that, in combination with a dispersive element (prism or grating), project a spectrum on a CCD detector. Spatial information is then obtained by spatial scanning.

Single-Slit Spectrometers

A single fixed slit is used in point-scanning spectrometers, which are commonly attached to confocal microscopes. These spectrometers (fig. 2.1) consist of a diffractive element (usually grating or prism) and a linear detector array, allowing both diffraction-limited spatial resolution (by x–y spatial scanning of a focused laser beam) and high spectral resolution (limited to the Rayleigh criterion of resolution for a specific grating/prism).

An improvement to the single-slit spectrometer is the implementation of line scanning. The object is wide-field illuminated, and scanning of eitherthe specimen (*pushbroom imaging spectrometer*) orthe slit (*moving slit spectrometer*) is performed to obtain the second spatial dimension. A two-dimensional detector is required for these arrangements (fig. 2.2).

Multislit Spectrometers

Although single-slit spectrometers offer high spectral and spatial resolution, they have a relatively low throughput. This is especially critical in low-light-level applications like autofluorescence measurements or Raman imaging. Increased light throughput and a higher signal-to-noise ratio can be realized by replacing the single entrance slit with multiple slits or, in general, with a mask (fig. 2.3). A more general implementation is by using special masks, called *Hadamard masks*, which are made up of a pattern of either transmissive and

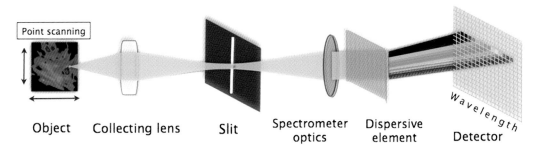

Figure 2.1 Schematic of the point-scanning multispectral imaging technique. Spectral (λ) information from a single point of the object is obtained using a single-slit spectrometer with a dispersive element (prism, grating) and a linear array or CCD detector. A complete image (a spectral data cube) is obtained by x–y scanning of the object or excitation light.

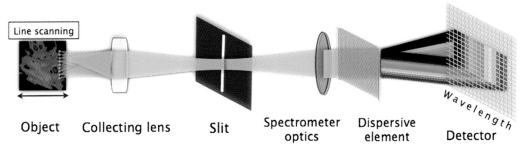

Figure 2.2 Schematics of the line-scanning multispectral imaging technique. Spectral (λ) information from a line-illuminated object is obtained using a single-slit spectrometer with a dispersive element (prism, grating) and a linear array or CCD detector. A complete image (a spectral data cube) is obtained by scanning the object (or the excitation light) in the direction perpendicular to the excitation line. A similar approach uses full-field illumination of the object and slit scanning.

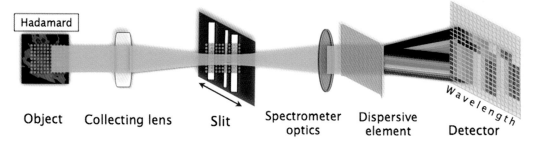

Object   Collecting lens   Slit   Spectrometer   Dispersive   Detector
                                                optics   element

Figure 2.3 Schematics of a multislit Hadamard spectrometer. The object is full-field illuminated and spectral (λ) information is obtained using a multiple-slit (Hadamard mask) spectrometer with a dispersive element (prism, grating) and a linear array or CCD detector. A complete image (a spectral data cube) is obtained by scanning the slit (varying the mask) and performing a Hadamard transform.

reflective or transmissive and opaque slits of various widths constructed from a Hadamard matrix. The spectral information is extracted by performing a Hadamard transform.

The Hadamard transform is an example of a generalized class of Fourier transforms and consists of a projection onto a set of square waves called *Walsh functions* (in comparison, the Fourier transform consists of a projection onto a set of orthogonal sinusoidal waveforms). The Hadamard transform is traditionally used in many signal processing and data compression algorithms, and, more recently, in spectral imaging (Treado and Morris, 1988, 1990; Goelman, 1994; Macgregor and Young, 1997; Hanley et al., 1998, 1999, 2000; Tang et al., 2002; Wuttig, 2005). For multispectral imaging, the most common use of Hadamard methods is by modulating a set of parallel spectrograph entrance slits by a spatial light modulatoraccor ding to **S** matrix encoding (Hanley et al., 2002).

Although Hadamard multiplexing has a major advantage in measurements in which the noise is detector limited, it may create masking effects and artifacts that need to be accounted for(Hanley and Jovin, 2001).

Band-Sequential Spectrometers

Band-sequential spectrometers record grayscale images by scanning the illuminating or emission light continuously ordiscontinuously through a numberof wavelengths. The result is always an image stack that constitutes a spectral cube and can be further analyzed for relevant information. The most common methods for wavelength scanning are described in the following subsections and are presented schematically in figure 2.4.

Filter Wheels

Although slit spectrometers measure the whole spectrum at once and scan the spatial domain, band-sequential spectrometers record the whole image at once and scan the spectral domain. The most basic approach to wavelength-scanning multispectral imaging (Waggoner et al., 2001) is the use of a set of narrow bandpass filters mounted on a spinning disc (*filter wheel*). As the disc rotates in synchrony with the frame rate of the CCD detector, different filters perpendicularly cross the optical path, leading to wavelength selection. Although not providing spectral imaging in the strict sense of more or less continuous, high-resolution spectral content, filter changers represent the main competition to all more advanced spectral imaging devices. With suitable selection of dyes and filter combinations, as many as five to eight different fluorescent dyes (all emitting in the visible region) can be imaged simultaneously (DeBiasio et al., 1987). Cross-talk correction is required to compensate for emission from one dye entering into another dye's main collection channel, but such algorithms have been well worked out (Galbraith et al., 1989, 1991; Morrison et al., 1998).

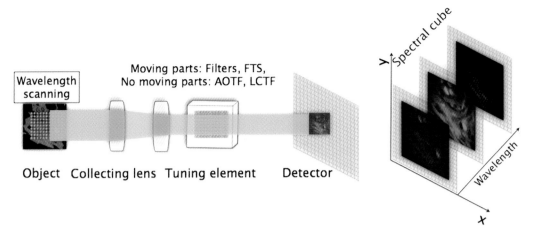

Figure 2.4 Schematics of the band-sequential technique. The object is full-field illuminated, and the collected light is sent through an imaging wavelength selection device (filter wheel, FTS, AOTF, LCTF, interferometer, computer generated hologram). For all these methods, a spectral data cube is generated by scanning in the wavelength dimension.

An advantage of this approach is its basic simplicity and relatively low cost. Of all the technologies discussed here, fixed-wavelength filter-based systems are the most likely to find theirway into turnkey systems (including clinical). Potential disadvantages include their inflexibility: New dyes may be difficult to add to the repertoire, and there remains the possibility of image shift associated with each change of filter, although with careful engineering this can be minimized or, if necessary, corrected for ex post facto. Such systems are not well suited for spectral analysis of complex samples with signals that are not made up of combinations of well-characterized fluorophores or chromophores. The spectral resolution is determined by the bandwidth and number of the filters used. For example, a spectral resolution of 5 nm over a spectral range of 200 nm requires 40 filters.

Fourier Transform Imaging Spectrometers

Fourier transform interference spectroscopy is the technique most widely used for generating spectral information in the infrared region (Buican, 1990; Lewis et al., 1995; Jackson and Mantsch, 1997). The method takes advantage of the principle that when light is allowed to interfere with itself at a number of optical path lengths, the resulting interferogram reflects its spectral constitution. Monochromatic light generates a pure cosine wave, and multispectral light, which consists of a mixture of wavelengths, will generate an interferogram that can be modeled as a sum of their respective cosine waves. An inverse Fourier transform of such an interferogram will regenerate the presence and intensities of all the contributing wavelengths (Bell and Sanderson, 1972). A Michelson interferometer can be used, but a far more stable design has been found in a Sagnac type of interferometer, in which the two beams travel the same path, but in opposite directions (Malik et al., 1996a,b). In this way, a method traditionally applied in infrared can be used in the visible part of the spectrum, where smaller wavelengths imply higher sensitivity to vibrations. By rotating the beam or the whole interferometer at very small angles, an optical path difference (OPD) is created (within the beam splitter) that depends on the incidence angle of light on the beam splitter. This rotation causes the interference fringes to travel across the detector onto the image plane. In the implementation developed by Applied Spectral Imaging (Carlsbad, California), the Sagnac interferometer is placed on the emission side of the microscope (Garini et al., 1999). The beam coming from the collecting objective is first collimated and then divided into two coherent beams. The beams are then recombined and produce an interference pattern, which depends on the wavelengths of light and the OPD introduced between the two beams. A two-dimensional detector measures the intensity of this interference pattern as a function of OPD, generating an interferogram, which has discrete values. The subsequent fast Fourier transformation of the interferogram recovers the spectrum.

Recently, Heintzmann et al. (2004) demonstrated the possibility of double-pass Fourier Transform Imaging Spectroscopy (FTIS) , in which both the excitation and emission are spectrally modulated, allowing one to obtain an excitation as well as an emission spectrum of a fluorescent sample with a single sweep of the interferometer. In addition, optical sectioning can be achieved at excitation wavelengths as a result of pattern excitation through the Sagnac interferometer.

Another approach, based on a scanning birefringent interferometer (Harvey and Fletcher-Holmes, 2004), uses two Wollatson prisms of equal and opposite splitting angles located close to the exit pupil of the microscope to address the problem of sensitivity to vibrations of FTISs. The light coming from the sample is first linearly polarized at 45 deg to the optic axes of the prisms and is then split by the first prism into two orthogonally components of equal amplitude. The second prism refracts the two components so that they propagate in parallel, and a lens focuses them in the same spot on the detector, allowing them to interfere. Because the two components are orthogonally polarized, an OPD is introduced, which can be modulated by the translation of the second Wollatson prism. Although the main advantages are the improved robustness and the large spectral range (can be operated at wavelengths between 200 nm and 14 μm), this approach has lower optical efficiency compared with the Michelson or Sagnac interferometers (maximum, 50% in polarized light and 25% in nonpolarized light).

There are advantages and drawbacks to Fourier transform techniques. As with tunable filters, and in contrast to fixed interference filters, wavelength ranges can be flexibly tailored to the exact spectral properties of the image. Compared with slit-based devices, Fourier transform spectroscopy enjoys a throughput advantage, because it does not require a narrow-slit aperture that would reduce the signal-to-noise ratio. The so-called *Fellgett advantage* refers to the fact that the interferometer can collect all the emitted or transmitted photons simultaneously (subject, of course, to normal losses), rather than collecting photons in a single, narrow, wavelength band at any one time, as with the band-sequential devices. If photobleaching of labile fluorescent labels occurs during imaging, the Fellgett advantage could be significant. On the otherhand, unlike a band-sequential device, an FTIS is unable to alter its sensitivity variably (or imaging time) in different spectral regions to compensate forchanges in detectorsensitivity orlight flux (Farkas et al., 1998a). The device also requires high stability and suffers from phase errors, aliasing, and incomplete modulation. Data analysis methods have evolved to include pre- and posttransform procedures to reduce artifacts, including apodizing, phase correction, zero filling, spectral averaging, and frequency-selective filtering.

Liquid Crystal Tunable Filters

Liquid crystal tunable filters select the wavelengths they transmit based on the same principles as fixed-wavelength interference filters, but with faster (approximately tens of milliseconds) electronically controlled tuning. They have the advantage of no moving parts and random wavelength access within their spectral range. Liquid crystal tunable filters consist of a number of liquid crystal layers, each of which passes a number of different frequencies. Stacking them results in a single, dominant transmission band, along with much smallerside bands; thus, m + 1 polarizers are separated by m layers of liquid crystals (m is typically 3–10), sandwiched between birefringent crystals. The OPD in birefringent crystals is dependent upon crystal thickness and the refractive index difference between the ordinary and extraordinary light rays produced at the wavelength λ of incident illumination.

Transmission of light through the crystal is dependent upon the phase delay created by the difference in propagation speed between the extraordinary and ordinary rays. Crystals are often selected for a binary sequence of retardation so that transmission is maximum at the wavelength determined by the thickest crystal retarder. Other stages in the filter serve to block the transmission of unwanted wavelengths. All this yields versatile and relatively fast wavelength selection, without much image distortion orshift. Each filterassembly can span approximately one octave of wavelength (e.g., 400–800 nm) and their useful range can extend into the near infrared. They can be introduced either into the illumination (excitation) or emission pathways, or both. Throughput is a problem, in that half the light corresponding to one polarization state is lost automatically, and peak transmission of the other half probably does not exceed 40% at best. Out-of-band rejection is not sufficient to prevent excitation light from leaking into the emission channel without the use of a dichroic mirror or cross-polarization (Morris et al., 1996). A small controller box in addition to a PC is needed to drive an LCTF assembly.

Acousto-optic Tunable Filters

Acousto-optic tunable filters provide electronically controllable, solid-state wavelength tunability of light from the ultraviolet to the near infrared, with random access, bandpass variability, and high throughput. They function based on light–sound (photon–phonon) interactions in a TeO_2 crystal, and are, therefore, tunable at speeds limited by the speed of sound propagation in the crystal. This yields typical wavelength switching times in the tens of microseconds, making them among the fastest wavelength switching devices available. These features have led to the use of AOTFs in a wide variety of spectroscopic applications. Interest in AOTFs for multispectral biological and biomedical imaging has, however, been more recent. After remote sensing-type imaging applications for ground-based and planetary targets were demonstrated in the 1970s and '80s, several groups have reported breadboard AOTF imaging demonstrations (Suhre et al., 1992; Cui et al., 1993), and their use in biologically relevant experiments such as fluorescence (Morris et al., 1994) and Raman (Treado et al., 1992) microscopy of biological samples.

In an AOTF, filtered light of narrow spectral bandwidth is angularly deflected away from the incident beam at the output of the crystal. The central wavelength of this filtered beam is determined by the acoustic frequency of the AOTF; this wavelength can be changed within approximately 25 µs to any otherwavelength. High-end AOTFs involve additional optics compared with LCTFs, and require higher power and more involved electronics. The payoff is much fasterswitching and the ability to vary not only the wavelength, but also the bandwidth and the intensity of the transmitted light. Thus, experiments involving luminescence lifetimes or very rapid acquisition of multiple wavelengths are possible using this technology. Imaging applications have been, however, impeded for many years by the facts that usually only one polarization state is available, there is some image shift when wavelengths are changed, intrinsic image blur is present, and out-of-band rejection is no greater than 10^{-2} to 10^{-3}. Approaches yielding solutions to some of these problems have been described previously (Wachman et al., 1997). Improvements include transducer apodization (in the emission path) and the use of two AOTFs in tandem (in the excitation path) to improve out-of-band rejection further (Farkas et al., 1998b). *Apodization* refers to a technique in which the transduceris sectioned into a numberof discrete slices with the relative amplitudes of the electronic signal used to drive each arranged to generate an acoustic spatial profile within the crystal that optimizes the shape of the filter spectral bandpass. With proper apodization (11 electronically isolated, individually driven slices with a slice-to-slice separation of $\approx$50 µm), Wachman et al. (unpublished) obtained a more than 10-dB decrease in the out-of-band light between the primary side lobes, with even greater decreases obtainable in specific regions. In addition, they developed an AOTF illumination system that can be fibercoupled to any microscope (orotheroptical system) to

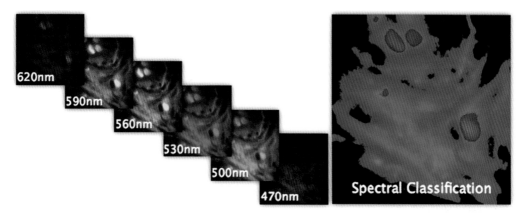

Figure 2.5 Multispectral images of two-photon excited fluorescence. Fluorescence of breast tissue (H&E stained) was measured by wide-field two-photon excitation (100 fs, 800 nm, 200 mW, 80-MHz repetition rate, 60-μm diameter field of view) and spectral/lifetime detection. Spectral selection at a 8-nm bandwidth and a 5-nm step size between 460 and 650 nm was realized by using a 2-cm imaging AOTF coupled to a Hamamatsu ORCA-ER CCD camera. Images recorded at 470, 500, 530, 560, 590, and 620 nm are shown, together with the spectral classification performed using linear discriminant analysis. The objects classified as (pseudocolor) red are red blood cells.

provide speed and spectral versatility for excitation as well as detection. This source consists of a 500-W short-arc xenon lamp followed by a pair of identical AOTFs placed in series, with the output of the second AOTF coupled into a fiber. This crystal configuration enables both polarizations to be filtered twice and recombined with a minimum of extra optics. The double filtering greatly reduces both the out-of-band light levels ($<10^{-4}$) relative to the peak. This system has been tested for its suitability for both bright-field and fluorescence applications. Recent developments include the use of a supercontinuum laser light source that can provide, because of the short duration of the laser pulses (several picoseconds), not only wavelength selection but also multiphoton excitation of the sample and time-resolved measurements (like Fluorescence Lifetime Imaging Microscopy [FLIM], for example).

For advanced spectral imaging applications involving AOTFs, we have built a microscopy workstation, similar in concept to the one we described previously (Shonat et al., 1997; Wachman et al., 1997), in which both excitation and emission wavelength selection is achieved by AOTFs, but uses fast lasers typical for nonlinear applications as excitation sources. This imparts great flexibility to the experimental capabilities, allows for true multispectral imaging, and could be used in a number of new nonlinear imaging experiments. One such application is illustrated in figure 2.5, with a femtosecond laser inducing wide-field two-photon excitation of a hematoxylin and eosin (H&E)-stained tissue specimen. The two-photon excited fluorescence is separated into spectral components by an AOTF-based imager and is recorded with a two-dimensional, cooled CCD camera. Spectral analysis and classification are performed, and are presented in figure 2.5 (see also Wachsmann-Hogiu and Farkas, 2007). A further advance includes the use of a broadband (*white light*) laser as the light source, in a setting that enables multispectral imaging amenable to deployment in a surgical (operating room) setting (Nowatzyk, Wachsmann-Hogiu, and Farkas, in preparation).

Volume Holographic Spectral Imaging

Volume holograms are a new type of wavelength selective element for spectral imaging consisting of a three-dimensional periodic phase or absorption perturbation throughout the entire volume of the element. This three-dimensional pattern gives high diffraction

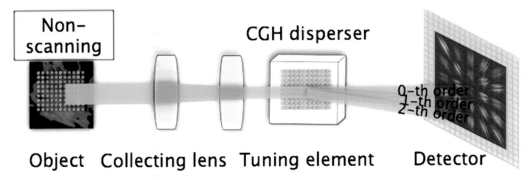

Figure 2.6 Schematics of a single-shot spectrometer. The object is full-field illuminated, and the collected light is sent through a computer-generated hologram (CGH) disperser that is used to distribute various diffraction orders of the primary scene over a large two-dimensional CCD array.

efficiencies (close to 100% at a single wavelength), high angle and spectral selectivity, and the ability of multiplexing many holograms in the same volume (Liu et al., 2004). Typically, a volume hologram is created by recording the interference pattern of two mutually coherent light beams. The spectral coverage ranges from 400 to 800 nm, which is limited by the mechanical constraint on the incident angle on the hologram.

The spatial resolution is different along x and y directions. The x resolution (which is degraded by diffraction) is given by the angle Bragg selectivity of the hologram multiplied by the focal length of the collimating lens (Sinha and Barbastathis, 2004). For a 10-mm focal length, the resolution along the x-axis is approximately 3 μm. Because there is no angle Bragg selectivity along the y-axis, the spatial resolution in this direction is diffraction limited by the imaging optics. Although the scanning speed, as determined by the rotation stage, is relatively high (approximately 1 nm/1 ms at 550 nm), the acquisition speed is limited by the low throughput. This is, in turn, determined by the narrow bandwidth and polarization sensitivity of the hologram (Liu et al., 2002).

An intriguing feature of volume holograms is that they have the potential of recording single-shot spectral images without postcapture computation. This goal could, in principle, be realized by multiplexing several tens of holograms in the same material, whereas each hologram diffracts light in a narrow spectral band to a distinct direction so that image slices at different wavelengths can be projected onto different areas of the same detector.

Single-Shot Spectrometers

Computed Tomography Imaging Spectrometer

Spectral imaging can be accomplished using a number of techniques that disperse the spectral information onto detectors either in sequential mode or, more recently, simultaneously. The methods presented earlier are all based on spatial/wavelength scanning across the respective dimension. Even though fast scanners exist, limitations imposed by the low number of photons detected in many applications makes spectral imaging performed in scanning mode time costly. Many fast biological processes cannot, therefore, be observed with scanning devices. To record the spectral and spatial information on a CCD chip simultaneously, CTIS has been developed (Volin et al., 1998; Ford et al., 2001; Volin et al., 2001). In its most common design, a computer-generated hologram disperser is used to distribute various diffraction orders of the primary scene over a large two-dimensional CCD array (fig. 2.6). The position and intensity of the resulting multiple-image mosaic reflect the spectral content of the original scene; the spectral content of the image is reconstructed using an iterative multiplicative algebraic reconstruction algorithm. The advantage of this

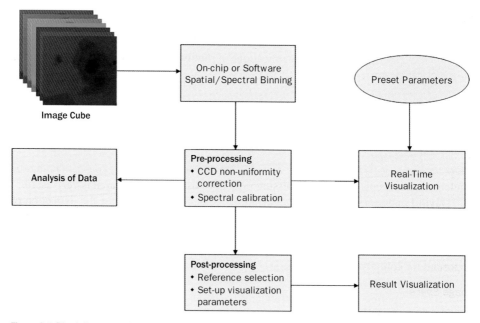

Figure 2.7 Blockdiagram of general computational processing for spectral optical imaging.

approach is that spatial and spectral information can be acquired with one image exposure, without any necessity forscanning in eitherthe spatial orwavelength domains. Therefore, acquisition times as short as 50 ms (for bright samples) and 2 s (fordim samples) can be used. The requirement is, however, a much larger CCD array size, because the CCD must capture not only the primary image but all of its higher order diffraction images. Considering the high number of diffracting orders measured (typically tens, which should have minimum overlap), and the size of current CCD detectors, the number of spatial resolution elements fora given image is rathersmall: approximately 200 × 200 pixels (Ford et al., 2001). In addition, the reconstruction of the spectral date cube from the dispersed images is computationally intense.

SPECTRAL DATA PROCESSING AND VISUALIZATION METHODS

From a technical perspective, spectral optical imaging is largely composed of acquisition optics and computational processing to visualize and analyze image "cubes" (x-, y-axes forspatial two-dimensional dimension and z-axis forthe spectral dimension). During the past several years, various technologies have been developed for acquiring spectral images (acquisition optics) that are suitable for use in microscopy, endoscopy, and organ imaging as explained in previous sections of this chapter.

In comparison with acquisition optics, there has been relatively little research on computational processing and this has mostly focused on analysis of three-dimensional image cubes forobject detection orclassification (Landgrebe, 1998). This computational processing is applied mainly to visualization and analysis of acquired image cubes, and management of spectral signatures contained within them. Although imaging parameters and visualization tactics for different applications may vary and often depend on methods of acquisition and chemical and structural characteristics of the specimen, the general approach to computational processing is relatively constant. Figure 2.7 is a block diagram of the computational steps.

Three Views of Spectral Data

Research in spectral image data processing has historically been closely tied to developments in earth science and military applications of satellite and aircraft imaging. One of the most innovative events was the launch, in 1989, of AVIRIS (Airborne Visible Infrared Imaging Spectrometer), an aircraft-based device that could acquire 224 spectral bands. Geologists used these spectral data to identify objects and materials, and to produce information about land cover.

There has been a developing tradition of research on computational algorithms applied to spectral data on earth science. Although the requirements for biomedical applications of spectral imaging are quite different from earth science and geology, understanding the fundamental concepts and data processing principles used in that research is very useful.

David Landgrebe (Purdue University) has pioneered principles for the extraction of information from spectral data (Kerekes and Landgrebe, 1989; Lee and Landgrebe, 1993; Landgrebe, 1998). Among these, understanding the concept of three views of the spectral data in terms of a data representation scheme or *representation space* is fundamentally important for computational processing of spectral data.

Image Space

Image space is a natural concept in the sense that the human vision system can be considered a wide "channel" into the human brain. It displays the data samples in relation to one another spatially and provides a "picture" of the scene for the human viewer. However, this space cannot represent the information in the spectral data cube unless appropriate image visualization methods are used—in a mathematical sense, mapping the spectral image cube onto image space through various visualization schemes (an explanation of this follows later in the chapter). Image space can be represented as a two-dimensional array (black-and-white orgr ayscale image, x- and y-axes forthe spatial dimension and signal intensity) or a three-dimensional array (color image, x- and y-axes for the spatial dimension, and z-axis for color space model [e.g., red, green, blue {RGB} and hue, saturation, and brightness {HSB}]). It serves to provide an overview of the data.

Spectral Space

Spectral space refers to the spectrum of each pixel. Intensity response as a function of wavelength can provide the analyst with spectral information that is often directly interpretable. Especially when a high degree of spectral detail is present, characteristics of a given pixel response can be related to physical properties of the contents of the pixel area, and it is called *spectral signature*. Whetheror iginating in one orsever al optical phenomena (e.g., absorption, reflection, scattering, emission, or any combination thereof), if such a signature is reproducible, it is considered intrinsic to and therefore characteristic of the subdomain represented by the pixel. It usually allows for an objective segmentation of the imaged object, based on these spectral signatures.

Feature Space

If one samples at two different wavelengths (x and y), the resulting values can be plotted as shown at the right of figure 2.8. If one samples at more values for different wavelengths, for example, 10 values, the point representing each spectral response would then be a point in 10-dimensional space. This hyperspace is called *feature space*. This space has been used in earth sciences for detecting correct end-member pixels using multidimensional exploration tools included in commercial or noncommercial software such as ENVI by ITT or MultiSpec by Purdue University.

Each of these three data spaces has its advantages and limitations. Image space shows the relationship of spectral response to its geographic position, and it provides a way to associate each pixel with a location on the ground. It also provides some additional information

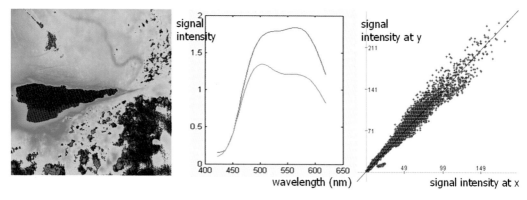

Figure 2.8 Three views of spectral data: image space (left), spectral space (middle), and feature space (right).

useful in analysis. Spectral space and feature space are convenient for computational processing. Figure 2.8 shows typical representations of these three spaces (image space [left], spectral space [middle], feature space [right]). Most computational processing methods for spectral imaging presented in this book deal with the mapping between spectral space, feature space, and image space.

Spectral Signature and Spectral Similarity Measures

The term *spectral signature* in this book is anothername fora plot of the variations in absorbed, reflected, or emitted light intensity as a function of wavelength, time, and other possible scale units foreach pixel in the image fora given imaging mode. These signatures are useful for identifying and separating tissues or objects of interest, and we can then interpret these signatures to represent tissue characteristics.

Spectral imaging is designed to measure the signal collected from an object as a function of two spatial dimensions and one spectral dimension. The resulting three-dimensional data set is often referred to as an *object cube* or *data cube*. In single spectral imaging data, a pixel in a scene is usually represented by a column vector in which each pixel contains specific spectral information provided by its spectral signature. For spectral matching, it is crucial to measure the spectral similarity using calculation algorithms to evaluate the difference between a reference signature chosen by the user (supervised) or detected automatically (unsupervised) and the target pixel spectral signature in the image.

In this section we introduce the most popular algorithms for calculating measures of similarity in spectral signatures. All these algorithms can be implemented easily in computer programming languages. More advanced and complex algorithms are primarily in remote sensing and geologic science fields, but are beyond the scope of this book.

A pixel is usually represented by a column vector and we define the pixel vector as p and a reference vector as r. The numberof spectral channels is N. In otherwor ds, N is the dimensionality of the image cube. To be useful in general measures of similarity, each algorithm was modified so the resulting value would be between zero and one to indicate the range between no difference from the reference signature and the maximum difference for the entire image cube, respectively.

Euclidean Distance Measure

The most popular and simple spectral similarity measure is based on Euclidean distance between two spectra (Chang, 2000). In statistical analysis and signal processing, the metric of this distance is frequently used for measuring the separation or closeness of data points. The Euclidian distance is defined as following two different measures: the root sum of square error (RSSE) and the sum of area difference (SAD):

$$RSSE_{orig} = \sqrt{\sum_{i=1}^{N} (p_i - r_i)^2}$$

$$SAD_{orig} = \sum_{i=1}^{N} |(p_i - r_i) \cdot \lambda_{inc}|$$

where N represents number of spectra, r_i and p_i are the mean intensity of the i_{th} spectrum of the reference signature and the sample pixel signature, and λ_{inc} represents the increment of the spectra. These two Euclidean distances have similar characteristics, but RSSE does not consider the possibilities of the different steps in the spectral axis whereas SAD does.

For a logical comparison, these could be modified to scale the distances between zero and one:

$$SAD = \left(SAD_{orig} - m \right) / (M - m)$$

$$RSSE = \left(RSSE_{orig} - m \right) / (M - m)$$

where m and M are the minimum and maximum of RSSE or SAD values, respectively.

Correlation Measure

Correlation, also called *correlation coefficient,* indicates the strength and direction of a linear relationship between two random variables. A number of different coefficients are used for different situations. The correlation is one in the case of a perfect linear relationship, and zero if the variables are independent.

Spectral Correlation Similarity

Spectral correlation similarity (SCS) uses the Pearson correlation coefficient as a similarity measure. The Pearson coefficient can be measured using the following equation:

$$SCS = \frac{1}{N-1} \cdot \left(\frac{\sum_{i=1}^{N} \left(r_i - \mu_{ref} \right) \left(p_i - \mu_{samp} \right)}{\sigma_{ref} \sigma_{samp}} \right)$$

where μ_{ref} and σ_{ref} represent the mean and standard deviation of the reference signature vector, and μ_{samp} and σ_{samp} represent those of the sample pixel vector.

Spectral Similarity Value

The spectral similarity value (SSV) is a combined measure of the correlation similarity and the Euclidian distance (RSSE). It can be formulated as

$$SSV_{orig} = \sqrt{\left\{ \left(RSSE_{orig} - m \right) / (M - m) \right\}^2 + \left(1 - SCS^2 \right)}$$

Identical vectors have identical magnitudes and directions. For a spectrum considered as a vector, the magnitude corresponds to the average spectral reflectance (brightness) and the direction corresponds to the spectral shape. Both dimensions of vector identity must be quantified when determining the similarity (or *closeness*) between two spectra. Euclidean distance primarily measures the signal difference between two vectors. Correlation compares the shapes of two spectra. By definition, the SSV combines signal difference and shape similarity (Homayouni and Roux, 2003). It has a minimum of zero and a maximum

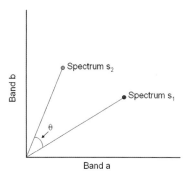

Figure 2.9 Spectral angle in the scatterplot (see text).

of the square root of two. In other word, smaller SSV indicates spectra that are more similar (Granahan and Sweet, 2001).

For a logical comparison, SSV could also be modified to scale between zero and one:

$$SSV = SSV_{orig} \Big/ \sqrt{2}$$

Spectral Angle Measure

The spectral angle measure (SAM) has been widely used in hyperspectral image analysis to measure spectral similarity between substance signatures for material identification. The SAM calculates the angle between two spectra and uses it as a measure of discrimination. Assume a two-band spectral remote sensing system. Each two-point spectrum is a point in band *b* versus band *a* space (fig. 2.9). The angle between the two lines connecting each spectrum (point) to the origin is the angular separation of the two spectra.

The smaller angle means more similarity between the sample pixel and reference spectra. The SAM can be modified to rescale the values of measure to [0, 1] (Schwarz and Staenz, 2001) to facilitate comparison with other measures. The angle is calculated using the following equation:

$$SAM = 2\cos^{-1}\left(\frac{\sum\limits_{i=1}^{N} r_i p_i}{\sqrt{\sum\limits_{i=1}^{N} r_i^2}\sqrt{\sum\limits_{i=1}^{N} p_i^2}}\right)\Big/ \pi$$

Visualization of Spectral Data

Visualization in image space is usually the ultimate goal in biomedical applications of spectral optical imaging. Although the practical importance is enormous, there has been very little research on visualization methods for spectral imaging.

There are three different strategies to visualize spectral signatures as an image (Jeong et al., 2005). The most popularone is *classification imaging*, during which we classify pixels into several groups after matching their spectral signature, and each group is displayed in specified (pseudo) color. Various kinds of classification algorithms can be used, and this method is very useful to segment tissue images objectively for computerized detection. Currently, only variations on the classification method are used to visualize spectral signatures.

A second visualization strategy formulates the image based on the calculation of various intensity functions depending on the spectral imaging modes. This can be called *quantitative imaging*. This imaging strategy shows us very fine image detail with flexible modulation capability, although colors in the image are not directly related to tissue identification.

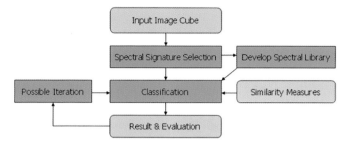

Figure 2.10 General process of classification imaging.

Typical examples include narrowband imaging with custom color bars to visualize certain ranges of spectra typically used in Raman spectral imaging, and conventional FLIM measurement of the average fluorescence decay time for each pixel yielding a map (or series of maps) of lifetime data.

The final strategy is a hybrid of these first two strategies, called *classification–quantitative hybrid imaging*. It classifies pixels into several groups using the first strategy, and then calculates intensities for that group using the second strategy. This yields a powerful approach to visualizing data.

Classification Imaging Methods

Classification imaging is the most common technique in spectral image analysis. In spectral image classification, multiple spectral bands (and sometimes ancillary information such as slope/aspect) are used to produce a thematic classification of an image.

Generally speaking, there are two types of classification schemes—supervised (Jimenez and Landgrebe, 1998) and unsupervised—both of which have been in common use in remote sensing for many years. Supervised and unsupervised classification rely on classic set theory in assigning pixels into discrete classes based on a set of fixed decision boundaries.

The analysis and imaging process begins with a general overview of the data set to be analyzed. The intent is to create a list of classes that is suitably exhaustive and that includes the classes of user interest. To the extent possible at this point, and from such an image presentation, consideration should be given to picking classes in such a way as to provide for a set that is separable from a spectral standpoint. Afterward, or as a part of listing the desired classes, the spectral description of the classes must be designated. Each signature can be drawn about individual pixels by observing a spectral space representation of a pixel. The use of *imaging spectroscopy* characteristics—when specific spectral bands of an individual reference signature are used to identify specific tissue—are an example of this.

Using the various spectral similarity measurement algorithms outlined earlier, the similarity between a reference signature and each pixel signature in the image is calculated and the pixel placed into the specified class and represented as a specific color. The last step is to evaluate of the result and, if needed, iterate the classification (fig. 2.10). Classification is very useful to segment the tissues objectively for computerized detection, although it suffers from the all-or-nothing choice of a given color.

Quantitative Imaging Methods

For quantitative imaging, the image is formulated based on calculation of various intensity functions depending on the spectral imaging modes. Therefore, there are no general rules or processes, but it does not require user intervention in the process—meaning, no heuristic color information is considered. This imaging method attaches importance to visualizing signal intensities from a given imaging mode.

Classification–Quantitative Hybrid Imaging Methods

Classification–quantitative hybrid imaging has, as the name implies, both classification and quantitative imaging components. Mathematically, classification imaging and hybrid imaging methods are based on spectral signature measures as a fundamental function. Spectral similarity measurement values are used for projection of spectral column vectors to specified image space column vectors.

The projection process starts from the calculation of spectral similarity measurements based on the reference signatures and can be modulated by various control functions or algorithms according to the nature of the specified image space or visualization strategies. Therefore, understanding the characteristics of the image space model is very important.

Image Space Models

Image space can be reinterpreted as an abstract mathematical model describing the way pixels can be represented as a series of numbers (in black and white or as a grayscale image) ormultiple series of numbers, typically as three orfourvalues (in a colorimage). When this mathematical model is associated with a precise description of how the components are to be interpreted (viewing conditions and so forth), the resulting set of pixels can be regarded as an image space.

Because any image space defines pixels as a function of the absolute reference frame, image spaces allow reproducible representations of pixels. For black-and-white and grayscale images, the image space model is simple. Each pixel has a certain number for the on and off or signal intensity value; therefore, there is just one-to-one mapping of the scalar value. Forcolorimages, there are many colorspace schemes, including several available commercially, but the most popularfordigital imaging are RGB and HSB.

Red, Green, Blue Color Space   A medium that transmits light uses additive color mixing with primary colors of red, green, and blue, each of which stimulates one of the three types of the eye's color receptors with as little stimulation as possible of the other two. This is called *RGB color space.* Mixtures of light of these primary colors cover a large part of the human color space and thus produce a large part of human color experiences.

Red, green, and blue spaces are generally specified by defining three primary colors and a white point. Unfortunately, there is no exact consensus regarding which loci in the chromaticity diagram the red, green, and blue colors should have, so the same RGB values can give rise to slightly different colors on different screens (Süstrunk et al., 1999).

Red, green, blue is the optimal colorspace forelectr onic devices. Each pixel on the screen can be represented in the computer or interface hardware as the sum of red, green, and blue values. These are converted into intensities that are then used for display. Typical display adapters currently use up to 24 bits of information for each pixel. This is commonly specified using three integers between 0 and 255, each representing red, green, and blue intensities in that order (fig. 2.11).

Hue, Saturation, Brightness Color Space   The HSB orhue, saturation, value model defines a color space in terms of three constituent components. Hue is the color type, such as red, blue, or yellow. It ranges from 0 to 360 deg, but can be rescaled to 0 to 1 or 0 to 255 for a byte representation. Saturation is the "vibrancy" of the color. The lower the saturation of a color, the more "grayness" is present and the more faded the color will appear. Brightness is an attribute of visual perception in which a source appears to emit a given amount of light.

The HSB model is commonly used in computer graphics applications. There are several visualization methods of the HSB model. Figure 2.12 shows one of these color space

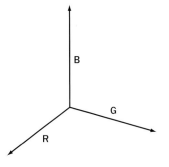

Figure 2.11 Red (R), green (G), blue (B) color space.

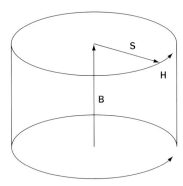

Figure 2.12 A cylindrical representation of the HSB model.

visualizations as a cylindrical object. The hue varies along the outer circumference of a cylinder, with saturation varying with distance from the center of a circular cross-section. Brightness varies from top to bottom.

The HSB model is similar to the way humans tend to perceive color. Red, green, blue is an additive model that defines color in terms of the combination of primaries, whereas HSB encapsulates information about a color in terms that are more familiar to humans.

Hybrid Imaging Methods: Examples

There are two different approaches for hybrid imaging with respect to the steps that follow quantification: either classification occurs first followed by quantification or classification and quantification occursimultaneously.

The multipeak contrast method (MPCM) is a visualization technique that first classifies target sample pixels into predefined classes and then quantifies relative closeness based on a spectral similarity measurement algorithm. The alternative multichannel fusion method (MCFM) is a "fuzzy classification" imaging technique in which classification and quantification are performed together.

Multipeak Contrast Method   The first step of the MPCM is exactly the same as with classification imaging methods. Using specified spectral similarity measurement algorithms, similarity values between a reference signature and each pixel signature in the image are calculated and the is pixel assigned to the specified class, which then determines the color of the pixel in the segmented image.

The second step is to calculate the degree of relative fitness between all selected reference signatures using the following equation. This calculated value, F_r, is used to represent quantification. Several imaging parameters can be modulated using this value, such as

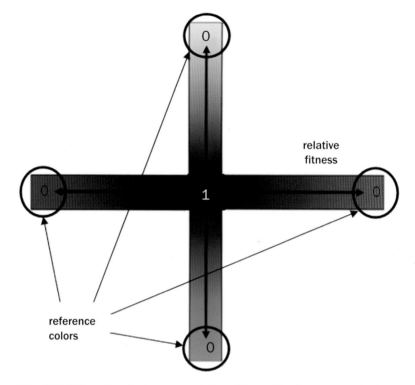

Figure 2.13 Schematic color bar representation of the multipeakcontrast method (four-parameter peaks model).

brightness, saturation, transparency, or height for a three-dimensional representation:

$$F_r = S_m\left(r_f, p\right) \Big/ \frac{\sum\limits_{i=1}^{N} S_m(r_i, p)}{N}$$

where S_m is a function of spectral similarity with range [0, 1 (zero means perfect fitness and one means no fitness)], r_f is the reference signature determined by the best fit, r_i refers to all reference signatures, p is the pixel signature, and N is the number of reference signatures.

The F_r value also has a range of [0, 1], where zero means perfect fitness of the pixel signature p with the signature r_f and one means that all the similarity measure values are the same so that this pixel cannot be classified into a colorclass. The F_r value can be used to modulate any imaging parameter, such as brightness, saturation, and transparency. For example, if F_r is used to control brightness, the pixel will be painted with its specified color when the value is zero and it will be painted with black if the value is one. Therefore, the transition line between the classes in the image is black and the transition area has dark gradations (fig. 2.13).

Because the MPCM uses determined colors of specified classes only, it has greater contrast between the classes, and the pixels between the classes are painted with dark gradations marking transition boundaries. However, it cannot display mixtures of classes. Therefore, this method is similar to classification imaging with a consideration of degree of fitness.

Multichannel Fusion Method   The MCFM is a kind of fuzzy-set classification imaging in which classification and quantification are performed together. Fuzzy-set theory attempts

(a)

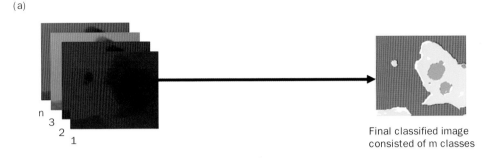

Final classified image
consisted of m classes

(b)

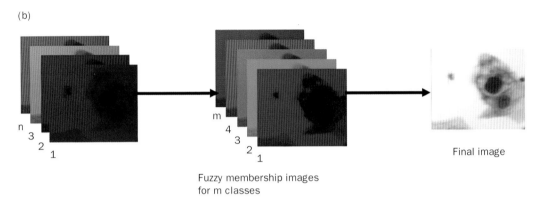

Fuzzy membership images
for m classes

Final image

Figure 2.14 (A, B) Hard classification imaging (A) versus soft fuzzy-set classification imaging (B).

to improve classification accuracy by taking into account graded boundaries that are typical of the real world. Figure 2.14 attempts to describe the differences between conventional hard classification and fuzzy-set classification.

In hard classification imaging, hard partitioning of an image cube and assignment of each pixel to one of m classes is performed during classification. On the other hand, fuzzy-set classification performs a fuzzy partition of an image cube wherein each pixel has a membership grade value (from zero to one) for m classes using classification processes and produces the resulting image from the fuzzy membership images using additional reconstruction algorithms.

The MCFM calculates the whole image cube with all m classes using the specified spectral similarity algorithm, and this value is mapped with each pixel color intensity into fuzzy membership images (channels). Therefore, if we have m classes (reference signatures), it creates m image channels. The final image is formulated from a fusion of these channels, which gives rise to the name *multichannel fusion method.*

Figure 2.15 shows the overall projection processes from spectral space into image space for general classification imaging and these two different hybrid imaging methods.

Comparison of Quantitative Imaging with Hybrid Imaging

Figure 2.16 is a good example of the difference between quantification imaging (multiband overlay), MPCM, and MCFM hybrid imaging. The images shown in figure 2.16A through C are quantitative overlays. As the number of bands increases, a given pixel rapidly changes color toward white and becomes indistinct as a result of loss of the original assigned colors. The images shown in figure 2.16D through F utilize the classification–quantitative hybrid imaging strategy to mitigate this effect.

A pixel spectra column

Classification Imaging

$$\begin{bmatrix} 75 \\ 60 \\ 54 \\ 38 \\ 49 \\ 98 \\ \cdots \end{bmatrix}$$

Spectral
Similarity Measure →

Multi-Channel Fusion

$$\begin{bmatrix} 0.3 \\ 0.1 \\ 0.6 \end{bmatrix}$$

**Reference similarity
result column
(3 references for example)**

x Brightness (F_r)
Multi-Peak Contrast

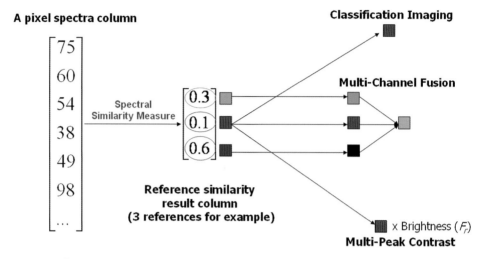

Figure 2.15 Projection processes overview of traditional classification, multipeakcontrast, and multichannel fusion hybrid imaging methods.

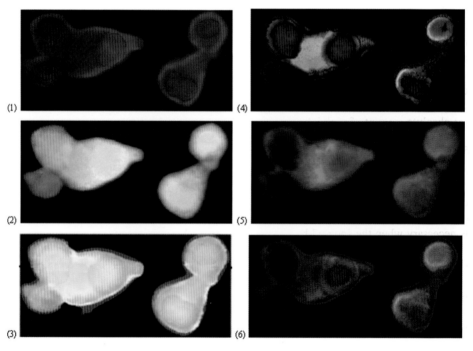

(1)

(4)

(2)

(5)

(3)

(6)

Figure 2.16 (A–F) Color images created by various hybrid color representation methods for six-parameter intracellular imaging: two-band overlay (A), three-band overlay (B), six-band overlay, (C), various hybrid color representations (with six colors, obtained using different algorithms) (D–F).

Figure 2.16D is generated by using MPCM imaging in which each pixel is allocated into the color having the strongest class/reference signature signal, and the relative contribution is calculated from the fitness, which is then indicated by brightness. This algorithm is useful to emphasize the contrast between different spectral signatures without losing their own color hue and saturation values. Figure 2.16E and F are formed by MCFM, which produces fuzzy-set color channel images for each reference spectral signature based on the spectral similarity measures. These channels are fused to form the resulting image. Figure 2.16E and F differs in the specificity value used to amplify the difference of the signatures and the calculation methods used to determine the degree of spectral similarity. The MCFM has an advantage overthe MPCM in that it allows a mixture of colorcomponents based on fuzzy membership values, and thus a more natural image is formed.

Spectral Unmixing

In a spectral image cube, a pixel spectral signature can be a mixture of more than one distinct spectral signature. *Spectral unmixing* is the procedure by which the measured spectrum of a mixed pixel is decomposed into a collection of constituent spectra, or *end members*, and a set of corresponding fractions, or *abundances*, that indicate the proportion of each end member present in the pixel (Keshava and Mustard, 2002). The term *end member* is approximately equivalent to *spectral signature* in spectral space, but it has a more general usage because it also can be used for feature space representation. It originated in mineralogy, where it refers to the mineral that is at the extreme end of a mineral series. Because spectral data analysis was originally developed for remote sensing applications, it is also used for indicating the pixel or region in image space representing a certain object or substance. *End member* substitutes for *spectral signature* when we need to considerfeatur e space analysis as well as spectral space.

Some types of images in biomedical sciences, such as those generated by immunofluorescence or in situ hybridization procedures, contain multiple, distinct spectral signals that may coexist in a single pixel to form the signal. In fluorescence, because of the additive nature of the light signal, the observed spectrum is a linear mixture of the component spectra, weighted by the amount of each probe. A linear combination algorithm can be used to unmix the summed signal arising from the pure spectral components to recover the weighting coefficients. Given an appropriate set of standards, the algorithm can quantitate the absolute amount of each label present (Farkas et al., 1997; 1998).

Steps in Unmixing

Algorithms for spectral unmixing use a variety of different mathematical techniques to estimate end members and abundances. Generally, we can decompose the complete end-to-end unmixing problem in a sequence of three consecutive steps: dimension reduction, end-member determination, and inversion (Keshava, 2003). The dimension reduction step is necessary when the spectral band number is too high (e.g., most remote sensing hyperspectral data sets), but we can ignore this step because the spectral band number is not that high in biomedical spectral imaging applications.

In the end-member determination step, we estimate the set of distinct spectra (end members) that constitute the mixed pixels in the scene; during the inversion step we generate abundance planes that allow us to estimate the fractional abundances for each mixed pixel from its spectrum and the end-member spectra.

Linear versus Nonlinear Mixing Models

Spectral mixing models attempt to represent the underlying physics that are the foundation of spectral signatures, and unmixing algorithms use these models to perform the inverse operation, attempting to recover the end members and their associated fractional abundances from the mixed-pixel spectrum. There are two categories of mixing models: the linear mixing model and the nonlinearmixing model.

Linear Mixing Model   In linear mixing model, if the pixel area is divided proportionally according to the fractional abundances of the constituent end members, a linear relationship exists between the fractional abundance of the end members and the spectrum of the pixel. If we have K spectral bands, and we denote the ith end-memberspectr um as s_i and the abundance of the ith end member as a_i, the observed spectral signature x for any pixel in the scene can be expressed as

$$x = \sum_{i=1}^{N} a_i s_i + e = Sa + e$$

where N is the number of end members, S is the matrix of end members, and e is an error term accounting for additive noise. This can easily be extended for multiple pixel spectra. If we define a matrix $\mathbf{X}$ having a corresponding abundance matrix $\mathbf{A}$ and noise matrix $\mathbf{E}$, the linear equation can be expressed as

$$\mathbf{X} = \mathbf{SA} + \mathbf{E}$$

Nonlinear Mixing Model   If the mixture of end members in the pixel is an aggregate spectrum of reflected, scattered, and/or absorbed photon properties, it may not have the linear proportions of the end-member spectral signature. To explain this phenomenon, various nonlinear mixing models have been proposed (Verstraete et al., 1990; Iaquinta et al., 1997; Asner et al., 2000). Although there are obvious advantages of using a nonlinear approach, it has not been widely applied to biomedical applications because of (1) the complexity of the model and (2) the recovery of mixture parameters requires knowledge of too many parameters.

Linear Unmixing Algorithm

There are two different strategies to determine end members. In a linear mixture model (LMM), this is the step when the matrix $\mathbf{S}$ is determined. The first strategy is to determine end members using a process in the image space or using predetermined spectral signatures in case of known fluorescence spectra. The second strategy is to determine end members using automated detection algorithms based on machine learning statistics.

After determining the matrix $\mathbf{S}$ using either strategy, inversion of the matrix based on minimizing the squared error is the simplest form of the algorithm. Variations on the least squares concept have been adopted to reflect the unique circumstances associated with spectral data. Starting with the LMM, and the assumption of no additive noise, the unconstrained least squares solution for a is

$$\hat{a}^U = (S^T S)^{-1} S^T x$$

This unconstrained estimate for a minimizes $|x - S\hat{a}^U|^2$ (Strang, 1988). This form requires no estimate of the additive noise and exists when there are more bands than end members, and when S has a full column rank (most of the spectral data will fulfill these requirements).

There are several other algorithms for the inversion processes, but a more detailed explanation is beyond our scope here. For more detailed information on spectral unmixing algorithms, see Keshava and Mustard (2002).

Dimensionality Reduction

Frequently, spectral data analysis is either impossible or inefficient when all wavelengths are included in the data set. Because there is a great deal of covariance in typical data sets (i.e., the intensity at one wavelength predicts with high confidence the intensity at neighboring wavelengths), the number of dimensions needed to express the actual information content in a data set is often far less than the number of dimensions in the data set itself. Not surprisingly, the familiar trade-off for less burdensome computation is decreased accuracy incurred by discarding information. Therefore, this step is optional and is only invoked by some algorithms to reduce the computational load of subsequent steps. In earth science

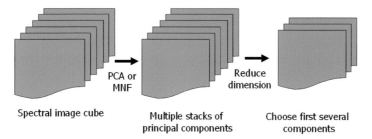

Figure 2.17 Dimensionality reduction process. MFN, minimum noise fraction.

applications, dimensionality reduction is very important because of the high dimensionality of such data (several hundred spectral bands); in biomedical optical spectral imaging, with tens of bands, this is not an issue.

Principal Component Analysis

Principal component analysis (PCA) is one of a family of statistical tools that can identify the most informative combinations of wavelengths (by rotating the basis vectors of the original data set), and can segregate signal from noise (with some major limitations). Typically, the dimensionality of a 20- to 30-wavelength image cube of a standard histology sample can be reduced to three or four dimensions (which are composed of linear combinations of many of the original wavelengths) while preserving most of the spectral information.

Principal component analysis identifies orthogonal axes for dimension reduction by performing an eigenvalue decomposition of the sample covariance matrix of the data. If we have k original variables $\{x_1, x_2, \ldots, x_k\}$, these can produce k new variables $\{y_1, y_2, \ldots, y_k\}$ (called *principal components*) through the following linear transformation that projects each image pixel's spectrum to a new set of orthogonal coordinate axes (Hastie et al., 2001):

$$y_1 = a_{11}x_1 + a_{12}x_2 + \cdots + a_{1k}x_k$$
$$y_2 = a_{21}x_1 + a_{22}x_2 + \cdots + a_{2k}x_k$$
$$y_k = a_{k1}x_1 + a_{k2}x_2 + \cdots + a_{kk}x_k$$

where y_k's are uncorrelated (orthogonal) vectors. The first principal component, y_1, explains as much as possible of the original variance in the data set; y_2, the second principal component, explains as much as possible of the remaining variance, and so on. Therefore, the first several vectors explain most of the variances in whole dimensions. $\{a_{k1}, a_{k2}, \ldots, a_{kk}\}$ is the k_{th} eigenvector of the correlation/covariance matrix, and coefficients of the k_{th} principal component. Dimensionality reduction is performed after a PCA transform to choose the first several principal components that explain more than a threshold percentage (e.g., 99%) variance (fig. 2.17).

Minimum Noise Fraction Transform

Principal component analysis works independently of any estimates of the noise in the signal. Therefore, if bands in a spectral image have differing amounts of noise, standard principal components derived from them may not show the usual trend of steadily increasing noise with increasing component number. The minimum noise fraction (MNF) transform is a modified version of the PCA that orders the output components by decreasing the signal-to-noise ratio (Green et al., 1988).

The MNF procedure first estimates the noise in each image band using the spatial variations in brightness values. It then applies two successive PCAs. The first uses the noise estimates to transform the data set to a coordinate system in which the noise is uncorrelated and is equal in each component. Then a standard PCA is applied to the noise-adjusted data, with output components ordered by decreasing variance. This procedure produces a component set in which noise levels increase uniformly with increasing component number. The low-order components should contain most of the image information and little image noise.

Automated End-member Determination

End-member determination is a very important step for the analysis and visualization of spectral data. Typically in biomedical applications of optical spectral imaging, one chooses end members from the image space; if performed using a spectral space representation, the process is called *reference spectral signature selection.* This is an empirical estimate of end members from an image space. This approach utilizes characteristics of end members obtained from empirical estimates through observation and physical intuition. Visualization of spectral data based on this approach is regarded as *supervised end-member determination* because it requires a priori knowledge of the properties of interest. Typically, the user will know the location of a number of pixels or regions representing these properties.

In contrast, if the end members must be determined with no or very limited a priori information, an automated method of determining these essential components is needed. This is also called *unsupervised end-member determination.* In unsupervised end-member determination, the computer attempts to group pixels with similar spectral characteristics based on the statistical properties of the image.

Spectral data can be expressed as points in hyperspace. This space is also called *feature space,* as explained previously. Spectrally similar pixels will cluster together, and some algorithms can be used to identify such clusters, which might represent meaningful bases forautomated end-memberdeter mination (Landgrebe, 1999).

Clustering Algorithms

Clustering is a data mining algorithm for unsupervised learning or indirect knowledge discovery and it is most widely used for unsupervised automated end-member determination algorithms. Most of the data mining methods develop models that predict how to classify new data from classified training data sets. But neither classified training data sets nor discrimination between independent and dependent variables are needed in the clustering algorithms. Instead, assuming similar data records will act similarly, those data will be found as a same group called a *cluster* (Hastie et al., 2001; Kang et al., 2004).

Clustering algorithms are useful when the only information available is a set of mixed pixel spectra, X, from a scene. Several variations on traditional clustering algorithms, such as K-means clustering, have attempted the resulting centroids to serve as estimates of end members. Similarly, as an extension of classification algorithms that assign the class label of the nearest centroid to pixels, abundance estimates that implicitly observe the nonnegativity and full additivity conditions are derived from the relative proximity of a pixel to each centroid. This is called *fuzzy C-means algorithms* (Bezdek et al., 1984), and it was used for various biomedical applications using LCTF-based spectroscopy (Mansfield et al., 1997).

A conventional K-means clustering algorithm can classify clusters by minimizing Euclidean distances between the points in the hyperspace of feature space (fig. 2.18A). This algorithm also can be modified for the use of spectral space if the clustering algorithm classifies clusters that maximize consistency between spectral signatures of the spectral data instead of Euclidean distances between the points in the feature space. It is possible to use any of various spectral similarity measures explained previously in this chapter—especially, RSSE is mathematically the same as the Euclidean distance between the points in the hyperspace of the feature space because it calculates the root square sum of all the band axes.

Anotherpossible modification of the conventional K-means algorithm is the usage of the threshold similarity index and minimum share of the cluster to determine optimal

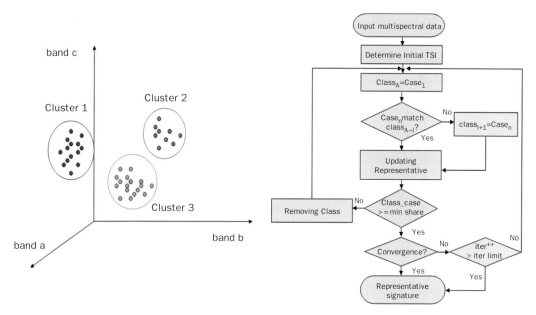

Figure 2.18 (left) Conventional K-means algorithm classifying points in the feature space (K = 3 in three-band feature space, for example). (right) Flow chart of the automated end-member determination process using a modified K-means algorithm. TSI, threshold similarity index.

cluster numbers and to remove the outlier signatures, whereas the conventional algorithm chooses clusters based on the predetermined number of clusters (K), which possibly leads to unfavorable results if the data characteristics are unknown. The median (K-means) values of the cluster members are chosen as the cluster representatives. Partitioning spectral data into initial clusters, finding the centroid for each collection, repartitioning into K clusters from the results, and refinding the centroid are performed repeatedly until relative changes of the total distortion are smaller than threshold values given. The results from those classifiers are reported as detected end members. The flow chart in figure 2.18B shows this process in detail (Jeong et al., 2007).

Support Vector Machine

Currently, many algorithms for automatic determination and classification of end members in feature space of a spectral data cube are less than ideal. In addition to clustering algorithms, support vector machines are one of the most promising approaches. Support vector machines (SVMs) are "learning machines" that view classification as a quadratic optimization problem. Support vector machines classify data by determining a set of support vectors that outline a hyperplane in feature space (Gunn, 1998; Perkins et al., 2001). The goal of the SVM is to find the hyperplane that maximizes the margin—or separation—between the hyperplane and the closest data points. The larger the margin, the more separation that occurs between the classes and therefore the greater ability to generalize. Data points that exist at the margin are called *support vectors* and they serve to define the optimal decision surface (Fletcher and Kong, 2003). Support vector machine-based algorithms have been applied to biomedical spectral imaging in several areas, including cervical cancer cell detection (Zhang and Liu, 2004), colon tissue classification (Rajpoot and Rajpoot, 2004), and white blood cell detection (Zhang et al., 2005).

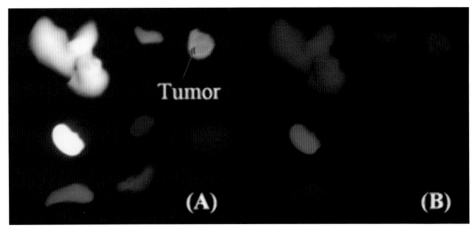

Figure 2.19 The classification of ex vivo specimens by multispectral analysis. (A) The raw image. (B) The classified image.

Illustrations of Spectral Image Processing Using in Vivo Data

Spectral Subtraction

Spectral subtraction is a process by which a known confounding fluorophore's (KCF's) signal is "subtracted" from the spectra in an attempt to isolate the signal of interest in the image (Farkas et al., 1997, 1998a). The first step in the spectral subtraction process is to estimate the amplitude (α) of the KCF foreach pixel in the image; thus we assume that every pixel can be expressed as [(signal of interest) $+ \alpha$]. Ideally, α would be estimated using an independent optical method, such as FLIM for collagen autofluorescence. Spectral subtraction can, if necessary, be performed using *spectral unmixing* to infer α. The inferred value of α, which will be referred to as α', is constrained to be no larger than that obtained from a sample known to generate a large KCF signal of amplitude α' max. In figure 2.19, the average signal from mouse brain was used, because its signal is dominated by our KCF (in this example, the autofluorescence of collagen). The second step in the spectral subtraction process is to remove the KCF signal from each pixel in the image, leaving only the signal of interest. Removal of the KCF signal from the spectral signal does not preclude any subsequent spectral analysis, and usually results in the KCF signal being mitigated or removed entirely from the image. In figure 2.19, α' was generated on a pixel-by-pixel basis and then subtracted from the image in figure 2.19A to yield the spectrally unmixed equivalent in figure 2.19B.

Spectral Signature Display Algorithm

In our study of fresh breast cancer tissue ex vivo, we successfully discriminated between normal and cancer tissue and visualized the border between them with the aim of assisting surgeons. This success resulted from a clear difference between normal and cancer spectral signatures as shown in figure 2.20 (Chung et al., 2006a).

We applied a similar strategy to discriminating important anatomic structures in the necks of rats in vivo during parathyroid surgery. Because of the numerous small nerves and other important structures within the neck that reside around the parathyroid glands, this operation can be technically challenging and risky, and it is quite difficult to discriminate these structures by the naked eye (fig. 2.21A). Although the spectrally segmented images displayed some important structures, the images were nevertheless disappointing because demarcation of parathyroids was very poor (fig. 2.21B).

A spectral signature analysis using MATLAB showed that there were very few differences between thyroid, parathyroid, and implanted parathyroid. However the very small

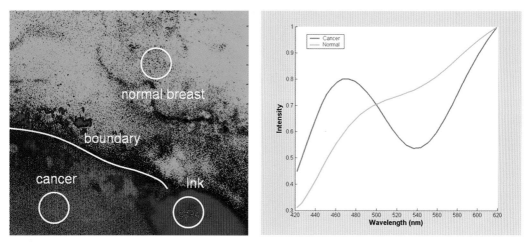

Figure 2.20 Spectral signature imaging (left) and signature difference (right) between normal and cancerous breast tissue.

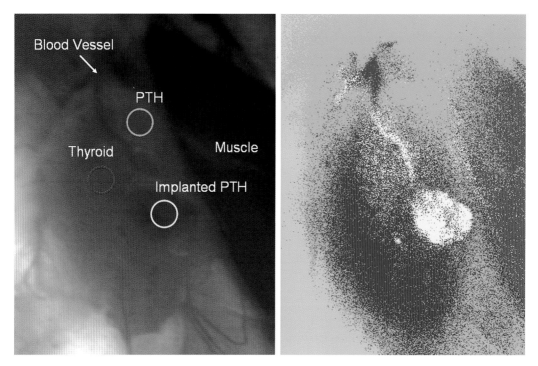

Figure 2.21 In vivo dissected rat neckimage seen with the naked eye (left) and its simple spectral classification image (right). PTH, parathyroid. See figure 2.24 for a better classification.

standard deviations for each signature indicated that the spectral signatures for each tissue could be discriminated based on amplification of the differences (fig. 2.22). We therefore introduced the concept of *signature specificity degree* and *signature contribution*. Signature specificity degree controls the brightness value corresponding to spectral signature similarity. A "high" degree indicates that the brightness of a pixel will decrease sharply when the signature difference increases from a specified/detected signature. The concept of signature contribution determines the color of the pixel when multiple signatures could be

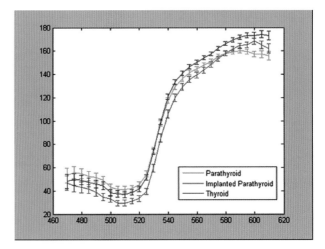

Figure 2.22 Spectral signatures differences between parathyroid, implanted parathyroid, and thyroid.

considered for the pixel signature. These two concepts can be implemented using various methods. Figure 2.23 shows a conceptual diagram of this approach.

The MCFM is one approach to implementing these concepts to produce multicolor channel images based on signature similarity and specificity degree. An image is produced by fusion of the colorchannels and we expect that this approach will facilitate segmentation of more complex tissue structures, resulting in fluorescencelike images. Using this new approach, we were much more successful in spectrally segmenting the images showing parathyroid and surrounding structures (fig. 2.24).

SELECTED BIOLOGICAL APPLICATIONS OF SPECTRAL IMAGING

The preceding sections of this chapter describe ways in which spectral techniques can enhance a wide variety of optical imaging methods applicable to biological systems, extending from "molecular" imaging to entire ecosystems imaged by satellite. In this section we summarize recent applications of spectral methods to optical imaging at the cellular and tissue levels. Initially we discuss applications to a variety of biological systems, and then follow with an overview of how, armed with an understanding gained from basic science studies, biomedical and clinical researchers have increasingly targeted spectral methods to human cells and tissues in vitro and in vivo with the aim of improving clinical diagnosis and treatment (see also Andersson-Engles et al., this volume; Pavlova et al., this volume; Wax et al., this volume). Also refer to recent reviews (McNeil and Ried, 2000; Dickinson et al., 2001; Farkas and Becker, 2001; Hiraoka et al., 2002; Kollias and Stamatas, 2002; Shafer-Peltier et al., 2002; Chenery and Bowring, 2003; Grow et al., 2003; Berg, 2004; Zimmermann et al., 2003; Ecker & Steiner, 2004). Although no review can include all relevant applications, we hope that this survey will assist you in evaluating how these technologies may apply to yourown biomedical optical imaging activities.

Bacteria, plant, and animal cells have helped to reveal some of the molecular and cellular entities responsible for producing spectral "signatures" that reveal their presence (see also Pavlova et al., this volume, fora discussion of animal cells). Spectral imaging techniques have recently been used to increase the sensitivity of bacterial detection using

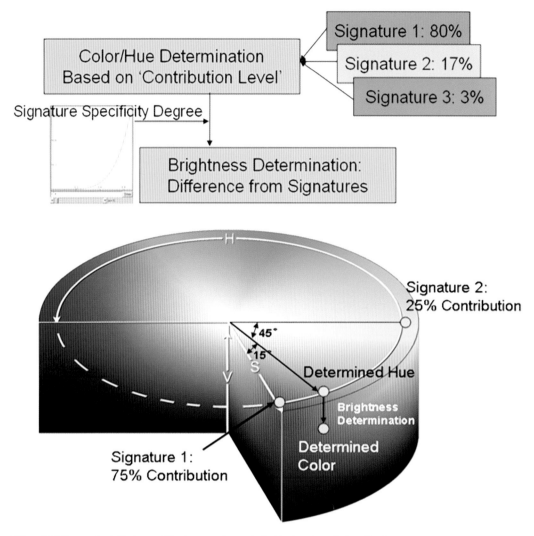

Figure 2.23 Conceptual diagram of the improved spectral signature imaging method.

autofluorescence and extrinsic fluorescent probes. Miyazawa et al. (2005) have used near-infrared imaging and spectral selection via an AOTF to distinguish four species of bacteria in separate single colonies in vitro and conclude that this approach could reduce assay times for detecting microbial contamination in food processing plants. Sunamura et al. (2003) used spectral microscopy to distinguish bacteria from other autofluorescent particles in sediments, allowing for semiautomatic quantitation of bacterial density. Autofluorescence is especially strong in many plant cells as a result of endogenous chromophores, and spectral imaging has been used to separate contributions from mitochondria, cell walls, and chloroplasts, as well as fluorescence from exogenous fluorescent proteins and background arising from fixatives (Berg, 2004). Spectral separation of absorbing stains that bind to nucleic acids, protein, and starch has been used to distinguish sterile from fertile pollen grains in rice (Hu et al., 2005). In water plant leaves, spectral images have been obtained using a Sagnac interferometer attached to a scanning microscope coupled to a pulsed infrared laser that provided for multiphoton excitation of fluorescence, second harmonic generation, and fluorescence lifetime imaging (Ulrich et al., 2004).

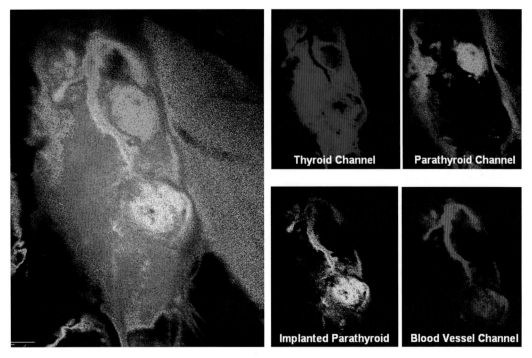

Figure 2.24 In vivo dissected rat neckspectral signature image based on the spectral angle similarity algorithm and the MCFM (specificity degree $= 20$). (Compare to figure 2.21.)

Most research using spectral microscopy involves studies of animal cells, either isolated in culture or within tissues in vitro and in vivo. Sánchez-Armáss et al. (2006) used spectral imaging of fluorescence from the pH-sensitive dye SNARF to distinguish protonated and deprotonated forms of the dye, the fluorescence emission spectra of which vary with cellular microdomains in glial cells. Sharonov et al. (2005) assessed proapoptotic activity in murine monocytes using five fluorescent probes to monitor mitochondrial potential and membrane structure and integrity. Spectral imaging allowed differentiation of structural and functional features of the cells, as well as separation of fluorescence of free rhodamine from rhodamine conjugated to mutant cytochrome c loaded into the cells by electroporation. Several applications of multicolor and spectral imaging at the single-cell level within tissues were discussed in a recent review by Ecker and Steiner (2004), including characterization of infiltrating leukocytes, quantification of apoptosis and proliferation, and simultaneous quantification of the relative concentrations of multiple tumor-suppressing proteins. Spectral fluorescence, absorbance, and reflectance methods have also been applied to tissues, primarily in mice and rats, and often without added optical contrast agents (see Chung et al., 2006b, and Leavesley et al., 2005, for recent surveys). High spectral resolution measurements have been made using an AOTF or liquid crystal filter, followed by computational separation of signals and classification of features recorded in digital images. Images were acquired on whole animals, eye and skin, and lymph nodes, and the implications of these results forhuman tissues were discussed (see also the section laterin this chapteron clinical studies).

The following sections describe additional research on animal cells and tissues using two spectrally based approaches that have recently seen a rapid expansion in their use. The first is *spectral unmixing,* which computationally separates fluorescence from multiple sources with significant overlap in their emission spectra. The second is *Raman scattering,* a spectroscopic technique that, when applied to imaging, provides for structural specificity with spatial resolution at the subcellular level without the need for extrinsic probes. We then considerclinical applications of spectral imaging.

Spectral Unmixing in Fluorescence Imaging

A very active area of biomedical optical imaging for which spectral techniques have become critical is fluorescence microscopy in which contributions from multiple probes with overlapping emission spectra must separated, not only from each other, but from nonspecific autofluorescence arising from endogenous cellular structures, fixatives, and optical components. The development of multicolor variants of fluorescent proteins has greatly facilitated monitoring the expression and spatial location of multiple subcellular targets in live cells, but this also has increased the challenge of distinguishing between them because their colors often overlap. Improvements in illumination sources and spectral selection methods in commercial microscopes have contributed to increasing application of spectral imaging techniques in these studies (reviewed by Dickinson et al., 2001; Farkas, 2001; and Leavesley et al., 2005). Leavesley et al. (2005) demonstrate how spectral unmixing improves objectivity in the interpretation of biopsy specimens treated with traditional histologic stains by facilitating classification of phenotypic features. They also describe applications to whole-animal macroscopic fluorescence imaging and the utility of evaluating fluorescence spectra in retinal imaging and near-infrared imaging of skin. Huth et al. (2004) have studied internalization of liposomes by COS-7 cells as part of a drug-targeting protocol. The liposomes were made more hydrophilic by labeling with fluorescein isothiocyanate (FITC)–dextran or more lipophilic by labeling with rhodamine bound to phosphatidyl ethanolamine (Rh-PE), and three additional fluorophores were used to label intracellular compartments. A range of spectra were present in the microscope images depending on which liposomes were targeted to which regions of the cell, which could also overlap spatially. Spectral unmixing quantitatively distinguished liposome targeting to different compartments. Timlin et al. (2005) analyzed fluorescence signals from printed DNA arrays obtained from commercial sources or custom made in laboratories. They were able to distinguish complementary DNA (cDNA) labeled by green and red fluorophores (Cy 3 and Cy 5) from background fluorescence emitted by the glass substrates as well as a contaminant frequently present in printed arrays, thus overcoming several artifacts that currently plague microarray data used to monitor gene expression. Mansfield et al. (2005) describe the unmixing of quantum dot fluorescence and autofluorescence.

Among the advantages of using spectral imaging technology to distinguish fluorescence signals computationally is its flexibility, which is farsuper iorto colorfilter s because it can be used immediately with new probes as they become available (see Farkas et al., 1996; and Waggoner et al., this volume). A case in point is the rapidly expanding library of fluorescent proteins that often exhibit significant spectral overlap in their excitation and emission spectra (see reviews by Dickinson et al., 2001; Hiraoka et al., 2002; Zimmermann et al., 2003; Zimmerman, 2005) but can be distinguished from other probes and from each otherby linearunmixing of spectral images (Dickenson et al., 2001; Lansford et al., 2001; Conchello and Lichtman, 2005). As an example, FITC and green fluorescent protein (GFP) are separated by only 7 nm at their peak emission intensity, but it has been shown that a GFP–histone fusion protein can be distinguished from FITC-labeled microtubules by linear unmixing algorithms (Haraguchi et al., 2002). Kramer-Hämmerle et al. (2005) used fluorescence microscopy to investigate the interaction of a GFP–fusion protein with a human immunodeficiency virus (HIV) Rev–cyan fluorescent protein (CFP) fusion. The significant spectral overlap of these two green/blue-green fluorescent tags required them to be unmixed in images to demonstrate theircolocalization at the subcellularlevel. Nadrigny et al. (2006) separated autofluorescence from GFP and yellow fluorescent protein (YFP) fluorescence originating from secretory vesicles in mouse cortical astrocytes.

Much effort in this area has gone toward unmixing donor and acceptor fluorescence emission in experiments utilizing resonance energy transfer to identify protein interactions within cells (Haraguchi et al., 2002; Zimmermann et al., 2002; Ecker et al., 2004; Nishi et al., 2004; Zhang et al., 2004; De and Gambhir, 2005; Raicu et al., 2005; Thaler et al., 2005). Spectral unmixing has been used to distinguish the fluorescence of CFPs and YFPs bound to glucocorticoid and mineralocorticoid receptors, respectively, in COS-1 cells and hippocampal neurons (Nishi et al., 2004). Resonance energy transfer measurements showed that the receptors form heterodimers in the cell nuclei, but not in the cytoplasm, and therefore they are not translocated into the nucleus as heterodimers. Zhang et al. (2004) studied mitochondrial caspase activity during apoptosis by monitoring fluorescence resonance energy transfer (FRET) between CFP and YFP. The CFP and YFP were conjugated to opposite ends of a short (< 10-nm) linkerthat was a substrate forcaspase, so that FRET between the two fluorescent proteins was eliminated when the linker was cleaved and the CFP and YFP diffused away.

Fluorescent proteins and spectral unmixing are also becoming critical tools for biomedical optical imaging, as we discuss in more detail later in the section on clinical applications. We first survey applications of another technique with potential applications in vivo—Raman imaging—which uses intrinsic signals from the cells and tissues themselves.

Raman Spectral Imaging

A relatively new imaging technique that evolved from spectroscopy is Raman spectral imaging (discussed earlier in this chapter). The spectra are specific to different classes of biomolecules and hence the method does not require exogenous contrast agents. This has two advantages: (1) several classes of molecules and subcellular compartments can be localized with optical resolution and (2) it is promising for use with tissues in vivo (Kollias and Stamatas, 2002; Shafer-Peltier et al., 2002; Chenery and Bowring, 2003). The carbon chains in fatty acids generate specific Raman signatures that have been used to image lipid bodies and vesicles enclosed by lipid bilayers. Otto and colleagues (Manen et al., 2005) have recently studied phagosomes in leukocytes and macrophages. By selecting the spectra of arachidonate in confocal Raman images, they found it localized to lipid bodies that associate with latex beads phagocytosed by the cells. Arachidonate has been shown to be involved in NADPH oxidase activation, and its identification in lipid bodies suggested that these structures serve as reservoirs for localized activation of this important leukocyte function. Several types of phospholipid-containing organelles were imaged in yeast (*Saccharomyces pombe*) by Hamaguchi and colleagues (Huang et al., 2005), with the Raman spectrum of phosphatidylcholine closely matching that of mitochondria. The identity of mitochondria and nuclei were confirmed by fluorescence of GFP fused to proteins used as markers for these organelles (the blue excitation of the GFP did not overlap the red light used for the Raman imaging). An interesting observation was that one Raman spectral peak from an unidentified molecule disappeared when the cells were treated with a respiration inhibitor that targets mitochondria (potassium cyanide). This signal therefore distinguished live from dead cells and prompted the authors to refer to this band as "the Raman spectroscopic signature of life" (Huang et al., 2005, 10010). Raman spectral imaging has also been used to study the spatial and temporal behaviorof otherbiomolecules, including proteins and DNA (Uzunbajakava et al., 2003a,b), and changes in collagen in murine calvaria in response to fibroblast growth factor 2 (Crane et al., 2004). Raman imaging has even been applied to cells in suspension by using a "holographic" optical trap to control the position of cells while preventing rotation that would occur around the optical axis of a single-beam trap (Creely et al., 2005). The distribution of proteins and lipids was studied in Jurkat cells (cancer cells derived from human T-cell leukemia).

Coherent Anti-Stokes Raman Spectroscopy (CARS) is a sensitive variant of Raman spectroscopy and has been used to image a variety of molecules, including water, DNA, proteins, and aliphatic compounds (carbon–hydrogen bonds in open carbon chains) (Potma et al., 2001; Cheng et al., 2002, 2003; Wang et al., 2005). Wiersma and colleagues (Potma et al., 2001) imaged intracellular hydrodynamics in single cells of the slime mold *Dictyostelium* to calculate the concentration and mobility of water throughout the cell and to model water diffusion and membrane permeability constants. It was found that water mobility was greatly reduced near the plasma membrane. Xie, Cheng, and colleagues (Cheng et al., 2003; Wang et al., 2005) have imaged phospholipid bilayer structures and their associated waterlayer s, including multilamellarvesicles in vitro and nodes of Ranvierin axons of spinal cord white matter obtained from guinea pigs. At the same time, the ordering of water molecules between, and of lipids within, the bilayers in the myelin sheaths was measured.

CLINICAL APPLICATIONS OF SPECTRAL IMAGING

The groundwork laid by use of spectral imaging approaches in the basic biological sciences has allowed theirapplication to clinical problems (forr eviews see McNeil and Ried, 2000; Farkas and Becker, 2001; Kollias and Stamatas, 2002; Levsky and Singer, 2003). We summarize applications to pathology and genetic analyses, as well as noninvasive imaging in vivo using intrinsic optical signals from tissues. These approaches are targeting a number of clinical conditions, but none more so than cancer, and we conclude with examples of spectral imaging methods applied to cancer diagnosis and treatment.

Pathology of Stained Tissue

We have advocated the use of spectral imaging in cyto- and histopathology for a number of years (Farkas et al., 1997; Levenson and Farkas, 1997). Sectioned tissue specimens stained with colored dyes naturally lend themselves to spectral analysis as a quantitative supplement to subjective interpretation, which is subject to considerable variation among observers. As an example, thymidylate synthase expression has recently been quantified in spectral images of immunohistochemically stained sections of cancerous and normal rectal tissue by comparing it with manual visual grading (Atkin et al., 2005). Visual grading correlated well with quantitative expression of the enzyme, suggesting that the latter can provide a more precise and reliable method of grading those specimens. Angeletti et al. (2005) applied spectral imaging and classification techniques to bladder cells, both normal and malignant, stained with Papanicolaou stain. The automated analysis could differentiate malignant from benign cells, and in addition showed significantly greater sensitivity and specificity compared with morphological examination by cytopathologists. Improvement was also observed in efficiency of predicting follow-up results. Katzilakis et al. (2004) studied stained bone marrow smears in which morphological differences between normal lymphocytes and lymphoblasts of children with acute lymphoblastic leukemia are not usually visible. They found that the absorption spectra of stained normal lymphocytes was significantly different from lymphoblasts from acute lymphoblastic leukemia, thus allowing them to be classified and clearly identified by pseudocolor in microscope images.

Fluorescent DNA Probes

During the past few years there have been increasingly large numbers of studies of chromosomes or genes using multiple DNA-hybridizing fluorescent probes, and unique identification of these signals has been made possible by spectral optical imaging. Spectral karyotyping constitutes a major application, in which complete sets of chromosomes with bound fluorescent probes are imaged either in vitro or in situ (McNeil and Ried, 2000; Rehen et al., 2001; Levsky and Singer, 2003; Bérgamo et al., 2005; Robertson et al., 2005; Tran et al., 2005; formessengerRNA [mRNA] probes, see Fusco et al., this volume). In one study which used both spectral karyotyping (SKY) and fluorescence in situ hybridization, aneuploidy was assessed in the embryonic and adult nervous systems of mice (Rehen et al., 2001). A surprising $\approx$33% of embryonic neuroblasts were found to be aneuploid, whereas the proportion of adult neurons exhibiting aneuploidy was much reduced ($\approx$1%) in the X and Y chromosomes. Spectral karyotyping was also used to demonstrate a normal karyotype in human submandibular gland cells in culture that are epithelial in nature and were intended to act as autologous replacements for salivary cells in clinical applications (Tran et al., 2005). Needless to say, an important application of spectral karyotyping is in characterizing chromosomal aberrations associated with cancer. In recent studies it was found that gains/losses of chromosomes correlated with survival in head-and-neck cancer patients (Bérgamo et al., 2005), and that chromosomal abnormalities become more common in sarcomas with passage number in culture (Robertson et al., 2005). As discussed earlier, spectral imaging has also been used to separate background fluorescence from signals arising from multiple dyes used to label cDNA in microarrays (Timlin et al., 2005).

Intrinsic Optical Signals

A frequent requirement of biomedical optical imaging is for tissue to be monitored in vivo without adding contrast agents such as dyes. Acquisition of high-resolution spectra can, in many cases, expand existing optical techniques to meet this requirement. A recent effort combines hyperspectral imaging with selection of polarization in a compact and portable device that was used to image human hands in vivo (Gupta, 2005). Using an AOTF and CCD camera optimized for the chosen range of wavelengths (400–880 nm), it was shown that in addition to the increased transmission of light through the hand at longer wavelengths, the polarization dependence of transmission also depended on wavelength. These effects depended on which finger was imaged and the location on the finger, suggesting that each has a distinctive optical signature dependent on wavelength and polarization. The advantage of using optical imaging is that it (1) eliminates the need forbiopsy and (2) quantitates structural features, thus avoiding the subjective assessments that can vary considerably among trained observers.

Blood Imaging

Oxygen saturation of hemoglobin can be measured by spectral analysis of reflected absorbance of blood in vivo (Shonat et al., 1997; 1998), and similar approaches have been used recently, including imaging of ulcers and other wounds in humans (Martinez, 2002), and development of hypoxia in tumor vasculature (Sorg et al., 2005). *Orthogonal polarization spectral imaging* (Sorg et al., 2005; Zuurbier et al., 2005) is another optical technique commonly used to study blood and vasculature, but only a single wavelength is chosen forobser vation, and hence *spectral* is arguably a misnomerforthis technique.

Spectral Endoscopy

An optical imaging method commonly used in the clinic is endoscopy, and spectral imaging hardware and software are now being coupled to endoscopes for medical applications during which only the tissue's endogenous chromophores or other features provide optical signals. Gono et al. (2003, 2004) assembled a multicolorimaging endoscope using sets of narrowband color filters to image reflected/scattered light from several human epithelial tissues, including the back of the tongue, the gastrointestinal tract, and the esophagus. They found that by choosing appropriate spectral ranges they were able to enhance contrast of capillaries and crypts in the intestinal mucosa. Vo-Dinh et al. (2004) have recently developed an AOTF-based hyperspectral imaging endoscope to monitor fluorescence from tissue surfaces. The system was tested on sheep brain and mouse kidney coated with various dyes, the fluorescence of which accentuated surface features of the organs. In chicken liver, autofluorescence exhibited a strong peak at $\approx$630 nm (porphyrin), although the intensity varied widely over the surface of the organ. The autofluorescence was suggested to reflect synthesis of heme and porphyrin-related compounds, and hence may be diagnostic of liver function.

An interesting application of spectral imaging endoscopy is the assessment of tissue health based on spectrally detected Mie scattering (Lindsley et al., 2004). The mere enlargement of nuclei, as it occurs in dysplasia, followed by cancerization, can be located in the images, because the spectrum of scattered light depends on the average scatterer size, and the main scatterers in tissue are the nuclei. This should help with driving biopsies and, ultimately, with early diagnosis (discussed later).

Cancer

A wide array of spectral imaging approaches have been targeted toward diagnosis and treatment of cancer, with studies being conducted at the cellular (in vitro), tissue (ex vivo and in vivo), animal model (preclinical), and human (clinical) levels. We bring discussion of these applications together here to illustrate how the technological approaches discussed earlier are broadening our knowledge of the family of diseases that is cancer.

Exogenous Markers

Stains and fluorophores that provide optical contrast are critical to traditional pathology of tissue sections as well as more recent molecular imaging approaches at the cellular and preclinical levels. As discussed earlier, quantitative analysis of the full optical spectra of stained thin sections (H&E, immuno- and histochemistry, and so forth) has been shown to distinguish cells and tissues more reliably than subjective visual assessment of color. The relevance to cancer diagnosis seems clear, as these studies addressed several types of cancer, including rectal, bladder, and blood (Katzilakis et al., 2004; Angeletti et al., 2005; Atkin et al., 2005).

Cancers are notable for the variety of cellular and molecular changes that distinguish them from normal cells. Detailed characterization at these spatial scales is required to design efficiently approaches to diagnosis and treatment, and figure 2.25 illustrates how spectral microscopy has been used to visualize simultaneously intracellular distributions of six molecularcomponents known to be associated with cancercell and tumorgr owth. Figure 2.25 shows simultaneous imaging of six intracellular components: (A) DAPI DNA, (B) FITC her2/neu, (C) Cy3 Rb, (D) Cy3.5 NFκB, (E) Cy5 Ras, (F) Cy5.5 p53. Views A through F in figure 2.25 show the individual distribution of these entities in the five cells that are within the field of view, and the middle of the figure shows an overlay of all these images. Simultaneous quantitation (such as intensity profiles along the white line shown) is evidenced by the intensity profiles in the bottom of figure 2.25. Figure 2.26 shows a nucleus labeled with DAPI (blue), her-2/neu with FITC (cyan), vascular endothelial growth factor with Cy3 (green), cyclin D1 with Cy3.5 (yellow), ras with Cy5 (magenta), c-myc with Cy5.5 (red), and p53 with Cy7 (orange). Pseudocolored images representing the amount of each of the seven labeled proteins are shown (fig. 2.26A–G). Differences in the spatial distribution of the fluorescence between the various images correlate with expected differences in the spatial location of the various proteins within the cells.

Fluorescence spectral microscopy has been used by Siboni et al. (2001) to visualize localization of an environment-sensitive dye (merocyanine 540) at the subcellular level, and how this changed in the typically acidic conditions of tumors. The emission spectrum of the dye depends on the polarity of the surroundings (lipophilic vs. hydrophilic), so that

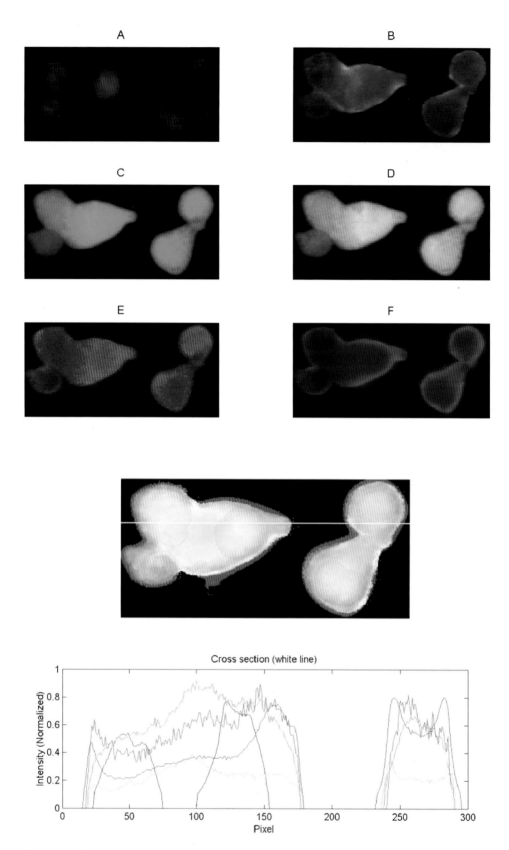

Figure 2.25 Six-parameter intracellular imaging and quantitation of cancer-relevant cellular components in human breast cancer cells.

A   B   C   D

E   F   G   Merged

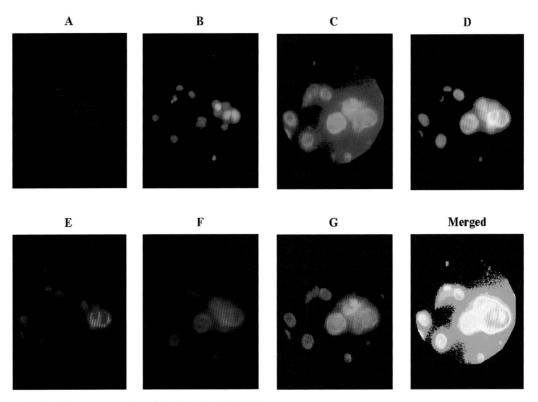

Figure 2.26 Seven-parameter cellular imaging with AOTFs.

detailed spectral analysis revealed changes in polarity in different cellular compartments in murine and human colon carcinoma cells and how these varied with pH corresponding to normal and tumor tissues. These effects correlated with the proliferative state of cancer cells in tumors. Cancer cells are also characterized by chromosomal aberrations, and these have been characterized by Bérgamo et al. (2005) in head-and-neck cancers by the spectral karyotyping approaches discussed earlier.

Spectral unmixing is becoming increasingly important in whole-animal imaging for distinguishing cancer cells from surrounding tissues and from background autofluorescence. Hoffman and colleagues have used fluorescent proteins expressed in mice to visualize tumor growth and angiogenesis in human pancreatic, breast, and skin tumors grown orthotopically in mice (reviewed in Hoffman, 2004). Different fluorescent proteins were expressed in the human and mouse cells, allowing visualization of tumor growth separately from infiltration of host cells into the tumor, including endothelial cells during angiogenesis. Using an approach we proposed (Farkas et al., 1997), the fluorescence emission of the fluorescent proteins was unmixed from each other as well as from autofluorescence to allow visualization of the different tissues in the living mice. In another study also using mouse models, spectral fluorescence images were acquired from xenografted prostate tumors grown subcutaneously (Gao et al., 2004). Quantum dots were targeted to the cancer cells and autofluorescence was removed computationally, allowing the tumors to be better visualized in cases when the signal was weak as a result of a low concentration or great depth within the tissue.

The depth of a tumor within a tissue is often as important to prognosis as its location in the other two dimensions (Farkas et al., 1995). One approach is to limit the depth viewed in an image by illuminating with linearly polarized light while rejecting depolarized light in the image, because the latter arises from relatively deep layers within the tissue. This depth selection can be combined with topical application of chromophores in a variant on traditional staining of excised tissue. As a demonstration of this approach, discarded tumor material from excised skin has been dyed and viewed using spectral polarized light imaging (Yaroslavsky et al., 2003). It was shown that the dye enhanced tumor margins sufficiently well to make them nearly as identifiable as in frozen H&E sections. In addition, spectral techniques allowed blood vessels to be distinguished from the surrounding tissue, and the depth of the tumors to be localized to within ≈ 150 μm of the skin surface. This approach may be useful clinically because it is sufficiently rapid to be used intraoperatively and is less expensive than the usual preparative procedures for tissue sectioning. In another approach to measuring depth, Chaudhari et al. (2005) have constructed "tomographic" images of brain tumors expressing luciferase in mice by a clever analysis of bioluminescence spectra. They took advantage of the fact that biological tissue preferentially absorbs the shorter wavelengths of the luciferase emission, so that it becomes more red when a tumor is located more deeply within a tissue. By observing red shifts in bioluminescence spectra and taking the tissue optical properties into account, they were able to estimate the depths of tumors, and these values showed reasonable correspondence with those taken from magnetic resonance images (see Ntziachristos and Weissleder, this volume, for other approaches to tomography).

Endogenous Markers

Currently, the most feasible clinical applications of spectral techniques make use of intrinsic signals arising from molecules that naturally occur within cells and tissues (Chung et al., 2005). These methods are generally sensitive to many cell types and extracellular matrix molecules within a tumor, which can often be as diagnostic of tumors and their margins as the cancer cells themselves. Steller et al. (2006) have recently used infrared spectral imaging microscopy to study thin sections of unstained cervix uteri encompassing normal, precancerous, and cancerous tissue. The spectra of the images were analyzed by cluster analysis (described earlier in this chapter), and these spectral "signatures" correlated strongly with morphological features in adjacent stained sections. Dysplastic lesions and squamous cell carcinoma could be distinguished from each other and from the basal layer. In addition, cancer cells could be distinguished from other tissues characteristic of tumors, including stroma, epithelium, inflammation, and blood vessels (see Pavlova et al., this volume).

For applications in vivo, tissues have to be accessible to optical probes and therefore are generally of epithelial or endothelial origin. An obvious example is the skin, and the applicability of spectral imaging to melanoma has been explored (reviewed by Farkas and Becker, 2001). A recent clinical study used spectral imaging in vivo and neural network-based classification to identify cutaneous pigmented lesions as eithermelanoma ornonmelanoma (Tomatis et al., 2005). The analysis was able discriminate well between the two categories as judged by comparison with histologic diagnosis after excision of the lesions.

The general applicability of spectral imaging to diagnosis and treatment of cancer without the use of added contrast agents has recently been assessed by Chung et al. (2006b). They summarized the use of reflectance, one- and two-photon excitation of autofluorescence, and Mie scattering to image live cancer cells in vitro and in tissues that were either fixed orfr esh ex vivo orin vivo. The potential of these approaches to detect and localize cancer tissue intrasurgically is quite exciting.

FUTURE WORK

Biophotonics constitutes an important, fast-evolving, and promising area of investigation. It appears that spectral imaging, relying on advances in domains such as laser physics, satellite reconnaissance, and nanotechnology, can bring a major contribution to the general field of biophotonics, and particularly to its translational applications in medicine and surgery.

We believe that the portability and relatively low cost of spectral imaging methodologies, coupled with the high specificity of the information derived from the spectral approaches, constitute a set of advantages that makes theiradoption appealing. However, the *likelihood* of such adoption into clinical practice will also depend on the usefulness, sophistication, and real-time availability of any analysis and visualization of the spectral imaging data, and efforts are necessary in this direction. Moreover, the ability of spectral/multispectral imaging to generate quantitative maps of tissue based on intrinsic signatures at the loci of interest constitutes probably one of our best hopes for achieving the high-quality intrasurgical imaging—not necessitating contrast agents—that could not only improve diagnosis, but also more closely couple it to intervention and its assessment. Advances in optics, hardware, and software are all necessary to achieve this worthy goal, as is a careful reconsideration of the role that *advanced* biophotonics should play in the clinic: Although photomechanical, photothermal, and photochemical laser effects have been used successfully in an increasing range of surgical treatments, mostly to remove or otherwise "fix" abnormalities, photonic *imaging,* best suited for detecting and characterizing such abnormalities, has been lagging in clinical acceptance, despite clear technological advances in laboratory research and preclinical trials, and an equally clear need for the kind of resolution (spatial, temporal, spectral, and so on) and specificity that optical imaging alone can offer.

ACKNOWLEDGMENTS

We thank our colleagues and particularly our collaborators, Drs. Erik Lindsley, Julia Ljubimova, Andreas Nowatzyk, Stanley E. Shackney, and Elliot Wachman, for permission to reproduce results from our shared work and for valuable ideas and discussions. Support from the National Science Foundation, the National Institutes of Health, and the U.S. Navy Bureau of Medicine and Surgery, and from corporate partners is much appreciated.

REFERENCES

Angeletti, C., N. R. Harvey, V. Khomitch, A. H. Fischer, R. M. Levenson, and D. L. Rimm. Detection of malignancy in cytology specimens using spectral–spatial analysis. *Lab. Invest.* 85:1555–1564, 2005.

Asner, G. P., A. R. Townsend, and B. H. Braswell. Satellite observation of El Nino effects on Amazon forest phenology and productivity. *Geophys. Res. Lett.* 27:981–984, 2000.

Atkin, G., P. R. Barbera, B. Vojnovica, F. M. Daleya, R. Glynne-Jones, and G. D. Wilson. Correlation of spectral imaging and visual grading for the quantification of thymidylate synthase protein expression in rectal cancer. *Hum. Pathol.* 36:1302–1308, 2005.

Bell, E. E., and R. Sanderson B. Spectral errors resulting from random sampling position errors in Fourier-transform spectroscopy. *Appl. Opt.* 11:688, 1972.

Berg, R. H. Evaluation of spectral imaging for plant cell analysis. *J. Microsc.* 214:174–181, 2004.

Bérgamo, N. A., L. C. da Silva Veiga, P. P. dos Reis, I. N. Nishimoto, J. Magrin, L. Paulo Kowalski, J. A. Squire, and S. R. Rogatto. Classic and molecular cytogenetic analyses reveal chromosomal gains and losses correlated with survival in head and neck cancer patients. *Clin. Can. Res.* 11:621–631, 2005.

Bezdek, J. C., R. Ehrlich, and W. Full. FCM: The fuzzy C-means clustering-algorithm. *Comp. Geosci.* 10:191–203, 1984.

Buican, T. N. Automated cell-separation techniques based on optical trapping. *Abstr. Pap. Am. Chem. S.* 199:42-IAEC, 1990.

Chang, C. I. An information–theoretic approach to spectral variability, similarity, and discrimination for hyperspectral image analysis. *IEEE Trans. Inform. Theory* 46:1927–1932, 2000.

Chaudhari, A. J., F. Darvas, J. R. Bading, R. A. Moats, P. S. Conti, D. J. Smith, S. R. Cherry, and R. M. Leahy. Hyperspectral and multispectral bioluminescence optical tomography for small animal imaging. *Phys. Med. Biol.* 50:5421–5441, 2005.

Chenery, D., and H. Bowring. Infrared and Raman spectroscopic imaging in biosciences. *Spectrosc. Eur.* 15:8–14, 2003.

Cheng, J. X., Y. K. Jia, G. Zheng, and X. S. Xie. Laser-scanning coherent anti-Stokes Raman scattering microscopy and applications to cell biology. *Biophys. J.* 83:502–509, 2002.

Cheng, J. X., S. Pautot, D. A. Weitz, and X. S. Xie. Ordering of water molecules between phospholipid bilayers visualized by coherent anti-Stokes Raman scattering microscopy. *Proc. Natl. Acad. Sci. USA* 100:9826–9830, 2003.

Chung, A., M. Gaon, J. Jeong, S. Karlan, E. Lindsley, S. Wachsmann-Hogiu, Y. Xiong, T. Zhao, and D. L. Farkas. Spectral imaging detects breast cancer in fresh unstained specimens. *Proc. SPIE* 6088: 6008806-1-7, 2006a.

Chung, A., S. Karlan, E. Lindsley, S. Wachsmann-Hogiu, and D. L. Farkas. In vivo cytometry: A spectrum of possibilities. *Cytometry* 69A:142–146, 2006b.

Chung, A., S. Wachsmann-Hogiu, T. Zhao, Y. Xiong, A. Joseph, and D. L. Farkas. Advanced optical imaging requiring no contrast agents: A new armamentarium for medicine and surgery. *Curr. Surg.* 62:365–370, 2005.

Conchello, J.- A., and J. W. Lichtman. Optical sectioning microscopy. *Nat. Meth.* 2:920–931, 2005.

Crane, N. J., G. G. Yu, M. A. Ignelzi Jr., and M. D. Morris. Study of localization of response to fibroblast growth factor-2 in murine calvaria using Raman spectroscopic imaging. *Proc. SPIE* 5321:242–249, 2004.

Creely, C. M., G. Volpe, G. P. Singh, M. Soler, and D. V. Petrov. Raman imaging of floating cells. *Opt. Exp.* 13(16):6105–6110, 2005.

Cui, Y., D. J. Cui, and J. H. Tang. Study on the characteristics of an imaging spectrum system by means of an acoustooptic tunable filter. *Opt. Eng.* 32:2899–2903, 1993.

De, A., and S. S. Gambhir. Noninvasive imaging of protein–protein interactions from live cells and living subjects using bioluminescence resonance energy transfer. *FASEB J.* 19:2017–2019, 2005.

DeBiasio, R., G. R. Bright, L. A. Ernst, A. S. Waggoner, and D. L. Taylor. 5-Parameter fluorescence imaging: Wound-healing of living Swiss 3T3 cells. *J. Cell Biol.* 105:1613–1622, 1987.

Dickinson, M. E., G. Bearman, S. Tille, R. Lansford, and S. E. Fraser. Multi-spectral imaging and linear unmixing add a whole new dimension to laser scanning fluorescence microscopy. *Biotechniques* 31:1272–1278, 2001.

Ecker, R. C., R. de Martin, G. E. Steiner, and J. A. Schmid. Application of spectral imaging microscopy in cytomics and fluorescence resonance energy transfer (FRET) analysis. *Cytometry* 59A:172–181, 2004.

Ecker, R. C., and G. E. Steiner. Microscopy-based multicolor tissue cytometry at the single-cell level. *Cytometry* 59A: 182–190, 2004.

Farkas, D. L. Spectral microscopy for quantitative cell and tissue imaging. In *Methods in Cellular Imaging* (A. Periasamy, ed.), pp. 345–361. Oxford University Press, New York. 2001.

Farkas, D. L. Invention and commercialization in optical bioimaging. *Nat. Biotechnol.* 21:1269–1271, 2003.

Farkas, D. L., B. T. Ballou, G. W. Fisher, and D. Fishman. Microscopic and mesoscopic spectral bio-imaging. *Proc. SPIE* 2678:200–209, 1996.

Farkas, D. L., B. T. Ballou, G. W. Fisher, C. Lau, W. Niu, E. S. Wachman, and R. M. Levenson. Optical image acquisition, analysis and processing for biomedical applications. *Springer Lect. Notes Comp. Sci.*, 1311:663–671, 1997.

Farkas, D. L., B. T. Ballou, G. W. Fisher, and D. L. Taylor. From *in vitro* to *in vivo* by dynamic multiwavelength imaging. *Proc. SPIE* 2386:138–149, 1995.

Farkas, D. L., and D. Becker. Applications of spectral imaging: Detection and analysis of human melanoma and its precursors. *Pigment Cell Res.* 14:2–8, 2001.

Farkas, D. L., R. M. Levenson, B. C. Ballou, C. Du, C. Lau, E. S. Wachman, and G. W. Fisher. Non-invasive image acquisition and advanced processing in optical bioimaging. *Comput. Med. Imaging Graphics* 22:89–102, 1998a.

Farkas, D. L., W. Niu, and E. S. Wachman. The acousto-optic tunable filter microscope: Recent advances. *Biophys. J.* 74:A227, 1998b.

Fletcher, J. T., and S. G. Kong. Principal component analysis for poultry tumor inspection using hyperspectral fluorescence imaging. *Proc. Intl. Joint Conf. Neural Networks* 1:149–153, 2003.

Ford, B. K., C. E. Volin, S. M. Murphy, R. M. Lynch, and M. R. Descour. Computed tomography-based spectral imaging for fluorescence microscopy. *Biophys. J.* 80:986–993, 2001.

Galbraith, W., K. W. Ryan, N. Gliksman, D. L. Taylor, and A. S. Waggoner. Multiple spectral parameter imaging in quantitative fluorescence microscopy. 1. Quantitation of bead standards. *Comput. Med. Imaging Graphics* 13:47–60, 1989.

Galbraith, W., M. C. E. Wagner, J. Chao, M. Abaza, L. A. Ernst, M. A. Nederlof, R. J. Hartsock, D. L. Taylor, and A. S. Waggoner. Imaging cytometry by multiparameter fluorescence. *Cytometry* 12:579–596, 1991.

Gao, X., Y. Cui, R. M. Levenson, L. W. K. Chung, and S. Nie. In vivo cancer targeting and imaging with semiconductor quantum dots. *Nat. Biotech.* 22:969–976, 2004.

Garini, Y., A. Gil, I. Bar-Am, D. Cabib, and N. Katzir. Signal to noise analysis of multiple color fluorescence imaging microscopy. *Cytometry* 35:214–226, 1999.

Goelman, G. Fast Hadamard spectroscopic imaging techniques. *J. Magn. Res. Ser. B* 104:212–218, 1994.

Gono, K., T. Obi, M. Yamaguchi, N. Ohyama, H. Machida, Y. Sano, S. Yoshida, Y. Hamamoto, and T. Endo. Appearance of enhanced tissue features in narrow-band endoscopic imaging. *J. Biomed. Opt.* 9:568–577, 2004.

Gono, K., K. Yamazaki, N. Doguchi, T. Nonami, T. Obi, M. Yamaguchi, N. Ohyama, H. Machida, Y. Sano, S. Yoshida, Y. Hamamoto, and T. Endo. Endoscopic observation of tissue by narrowband illumination. *Opt. Rev.* 10:211–215, 2003.

Granahan, J. C., and J. N. Sweet. An evaluation of atmospheric correction techniques using the spectral similarity scale. *IEEE 2001 International Geoscience and Remote Sensing Symposium* 5:2022–2024, 2001.

Green, A. A., M. Berman, P. Switzer, and M. D. Craig. A transformation for ordering multispectral data in terms of image quality with implications fornoise removal. *IEEE T. Geosci. Remote* 26:65–74, 1988.

Grow, A. E., L. L. Wood, J. L. Claycomb, and P. A. Thompson. New biochip technology for label-free detection of pathogens and theirtoxins. *J. Microbiol. Meth.* 53:221–233, 2003.

Gunn, S. R. *Support Vector Machines for Classification and Regression.* Technical report, ISIS, Department of Electronics and ComputerScience, University of Southhampton, 1998.

Gupta, N. Acousto-optic-tunable-filter-based spectropolarimetric imagers for medical diagnostic applications: Instrument design point of view. *J. Biomed. Opt.* 10:151802-1-6, 2005.

Hanley, Q. S., D. J. Arndt-Jovin, and T. M. Jovin. Spectrally resolved fluorescence lifetime imaging microscopy. *Appl. Spectrosc.* 56:155–166, 2002.

Hanley, Q. S., and T. M. Jovin. Highly multiplexed optically sectioned spectroscopic imaging in a programmable array microscope. *Appl. Spectrosc.* 55:1115–1123, 2001.

Hanley, Q. S., P. J. Verveer, D. J. Arndt-Jovin, and T. M. Jovin. Three-dimensional spectral imaging by Hadamard transform spectroscopy in a programmable array microscope. *J. Microsc. (Oxford)* 197:5–14, 2000.

Hanley, Q. S., P. J. Verveer, and T. M. Jovin. Optical sectioning fluorescence spectroscopy in a programmable array microscope. *Appl. Spectrosc.* 52:783–789, 1998.

Hanley, Q. S., P. J. Verveer, and T. M. Jovin. Spectral imaging in a programmable array microscope by Hadamard transform fluorescence spectroscopy. *Appl. Spectrosc.* 53:1–10, 1999.

Haraguchi, T., T. Shimi, T. Koujin, N. Hashiguchi, and Y. Hiraoka. Spectral imaging fluorescence microscopy. *Genes Cells* 7:881–887, 2002.

Harvey, A. R., and D. W. Fletcher-Holmes. Birefringent Fourier-transform imaging spectrometer. *Opt. Exp.* 12:5368–5374, 2004.

Hastie, T., R. Tibshirani, and J. Friedman. *The Elements ofStatistical Learning: Data Mining, Inference, and Prediction.* Springer, New York, 2001.

Heintzmann, R., K. A. Lidke, and T. M. Jovin. Double-pass Fourier transform imaging spectroscopy. *Opt. Exp.* 12:753–763, 2004.

Hiraoka, Y., T. Shimi, and T. Haraguchi. Multispectral imaging fluorescence microscopy for living cells. *Cell Struct. Funct.* 27:367–374, 2002.

Hoffman, R. M. Imaging tumor angiogenesis with fluorescent proteins. *APMIS* 112:441–449, 2004.

Homayouni, S., and M. Roux. *Material Mapping from Hyperspectral Images using Spectral Matching in Urban Area.* Presented at the IEEE Workshop in honor of Prof. Landgrebe, Washington, D.C., USA, 2003.

Hu, Y., Q. Wu, S. Liu, L. Wei, X. Chen, Z. Yan, J. Yu, L. Zeng, and Y. Ding. Study of rice pollen grains by multispectral imaging microscopy. *Microsc. Res. Tech.* 68(6):335–346, 2005.

Huang, Y.- S., T. Karashima, M. Yamamoto, and H. Hamaguchi. Molecular-level investigation of the structure, transformation, and bioactivity of single living fission yeast cells by time- and space-resolved Raman spectroscopy. *Biochemistry* 44:10009–10019, 2005.

Huth, U., A. Wieschollek, Y. Garini, R. Schubert, and R. Peschka-Su. Fourier transformed spectral bio-imaging for studying the intracellular fate of liposomes. *Cytometry* 57A:10–21, 2004.

Iaquinta, J., B. Pinty, and J. L. Privette. Inversion of a physically based bidirectional reflectance model of vegetation. *IEEE Transactions on Geoscience and Remote Sensing* 35:687–698, 1997.

Jackson, M., and H. H. Mantsch. The medical challenge to infrared spectroscopy. *J. Mol. Struct.* 408:105–111, 1997.

Jeong, J., P. K. Frykman, M. Gaon, A. P. Chung, E. H. Lindsley, J. Y. Hwang, and D. L. Farkas. Intelligent spectral signature bio-imaging for surgical applications. *Proc. SPIE* 6441:57, 2007.

Jeong, J., T. Zhao, Y. Xiong, and D. L. Farkas. Designing color representation methods for multi-spectral signatures in bioimaging. In *9th Annual Proceedings ofFred S. Grodins Graduate Research Symposium* (M. Khoo, ed.), pp. 69–70. University of Southern California, Los Angeles, 2005.

Jimenez, L. O., and D. A. Landgrebe. Supervised classification in high-dimensional space: Geometrical, statistical, and asymptotical properties of multivariate data. *IEEE Transactions on Systems Man and Cybernetics Part C—Applications and Reviews* 28:39–54, 1998.

Kang, H., S. Yoo, K. Choi, and J. Jeong. Automatic clusterdetection of lumbarspine 3D CT images. *Proc. ofthe 18th International Congress and Exhibition. Computer Assisted Radiology and Surgery. ICS* 1268:1331, 2004.

Katzilakis, N., E. Stiakaki, A. Papadakis, H. Dimitriou, E. Stathopoulos, E.- A. Markaki, C. Balas, and M. Kalmanti. Spectral characteristics of acute lymphoblastic leukemia in childhood. *Leukemia Res.* 28:1159–1164, 2004.

Kerekes, J. P., and D. A. Landgrebe. Simulation of remote-sensing systems. *IEEE T. Geosci. Remote* 27:762–771, 1989.

Keshava, N. A survey of spectral u algorithms. *Lincoln Lab. J.* 14:55–78, 2003.

Keshava, N., and J. F. Mustard. Spectral unmixing. *IEEE Signal Proc. Mag.* 19:44–57, 2002.

Kollias, N., and G. N. Stamatas. Optical non-invasive approaches to diagnosis of skin diseases. *J. Invest. Dermatol.* 7:64–75, 2002.

Kramer-Hämmerle, S., F. Ceccherini-Silberstein, C. Bickel, H. Wolff, M. Vincendeau, T. Werner, V. Erfle, and R. Brack-Werner. Identification of a novel Rev-interacting cellular protein. *BMC Cell Biol.* 6:1–22, 2005.

Landgrebe, D. *Multispectral Data Analysis: A Signal Theory Perspective.* Online. http://dynamo.ecn.purdue.edu/~biehl/MultiSpec/documentation.html. 1998.

Landgrebe, D. A. Information extraction principles and methods for multispectral and hyperspectral image data. In *Information Processing for Remote Sensing* (C. H. Chen, ed.), pp. 3–37. RiverEdge: World Scientific, 1999.

Lansford, R., G. Bearman, and S. Fraser. Resolution of multiple green fluorescent protein color variants and dyes using two-photon microscopy and imaging spectroscopy. *J. Biomed. Opt.* 6:311–318, 2001.

Leavesley, S., W. Ahmed, B. Bayraktar, B. Rajwa, J. Sturgis, and J. P. Robinson. Multispectral imaging analysis: Spectral deconvolution and applications in biology. *Proc. SPIE* 5699:121–128, 2005.

Lee, C., and D. A. Landgrebe. Analyzing high-dimensional multispectral data. *IEEE Transactions on Geoscience and Remote Sensing* 31:792–800, 1993.

Levenson, R. M., E. M. Balestreire, and D. L. Farkas. Spectral imaging: Prospects for pathology. In *Applications of Optical Engineering to the Study ofCellular Pathology* (E. Kohen, ed.), pp. 133–149. Trivandrum, Signal Research Publishers, 1999.

Levenson, R. M., and D. L. Farkas, Digital spectral imaging for histopathology and cytopathology, *Proc. SPIE* 2983: 123–135, 1997.

Levsky, J. M., and R. H. Singer. Fluorescence in situ hybridization: Past, present and future. *Cell Sci.* 116:2833–2838, 2003.

Lewis, E. N., P. J. Treado, R. C. Reeder, G. M. Story, A. E. Dowrey, C. Marcott, and I. W. Levin. Fourier-transform spectroscopic imaging using an infrared focal-plane array detector. *Anal. Chem.* 67:3377–3381, 1995.

Lindsley, E., E. S. Wachman, and D. L. Farkas. The hyperspectral imaging endoscope: A new tool for in vivo cancer detection. *Prog. Biomed. Opt. Imaging* 5:75–82, 2004.

Liu, W. H., G. Barbastathis, and D. Psaltis. Volume holographic hyperspectral imaging. *Appl. Opt.* 43:3581–3599, 2004.

Liu, W. H., D. Psaltis, and G. Barbastathis. Real-time spectral imaging in three spatial dimensions. *Opt. Lett.* 27:854–856, 2002.

Macgregor, A. E., and R. I. Young. Hadamard transforms of images by use of inexpensive liquid-crystal spatial light modulators. *Appl. Opt.* 36:1726–1729, 1997.

Malik, Z., D. Cabib, R. A. Buckwald, A. Talmi, Y. Garini, and S. G. Lipson. Fourier transform multipixel spectroscopy forquantitative cytology. *J. Microsc. (Oxford)* 182:133–140, 1996a.

Malik, Z., M. Dishi, and Y. Garini. Fourier transform multipixel spectroscopy and spectral imaging of protoporphyrin in single melanoma cells. *Photochem. Photobiol.* 63:608–614, 1996b.

Manen, H.- J. V., Y. M. Kraan, D. Roos, and C. Otto. Single-cell Raman and fluorescence microscopy reveals the association of lipid bodies with phagosomes in leukocytes. *Proc. Natl. Acad. Sci. USA* 102(29):10159–10164, 2005.

Mansfield, J. R., K. W. Gossage, C. C. Hoyt, and R. M. Levenson. Autofluorescence removal, multiplexing, and automated analysis methods forin-vivo fluorescence imaging. *J. Biomed. Opt.* 10:041207-1–9, 2005.

Mansfield, J. R., M. G. Sowa, G. B. Scarth, R. L. Somorjai, and H. H. Mantsch. Fuzzy C-means clustering and principal component analysis of time series from near-infrared imaging of forearm ischemia. *Comput. Med. Imaging Graphics* 21:299–308, 1997.

Martinez, L. *A Non-invasive Spectral Reflectance Method for Mapping Blood Oxygen Saturation in Wounds.* Presented at the 31st Applied Image Pattern Recognition Workshop, San Jose, 2002.

McNeil, N., and T. Ried. Novel molecular cytogenetic techniques for identifying complex chromosomal rearrangements: Technology and applications in molecularmedicine. *Exp. Rev. Mol. Med.* 2:1–14, 2000.

Miyazawa, K., K. Kobayashi, S. Nakauchi, and A. Hiraishi. In situ detection and identification of microorganisms at single colony resolution using spectral imaging technique. *LNCS* 3540:419–428, 2005.

Morris, H. R., C. C. Hoyt, P. Miller, and P. J. Treado. Liquid crystal tunable filter Raman chemical imaging. *Appl. Spectrosc.* 50:805–811, 1996.

Morris, H. R., C. C. Hoyt, and P. J. Treado. Imaging spectrometers for fluorescence and Raman microscopy: Acoustooptic and liquid-crystal tunable filters. *Appl. Spectrosc.* 48:857–866, 1994.

Morrison, T. B., J. J. Weis, and C. T. Wittwer. Quantification of low-copy transcripts by continuous SYBR (R) green I monitoring during amplification. *Biotechniques* 24:954, 1998.

Nadrigny, F., I. Rivals, P. G. Hirrlinger, A. Koulakoff, L. Personnaz, M. Vernet, M. Allioux, M. Chaumeil, N. Ropert, C. Giaume, F. Kirchhoff, and M. Oheim. Detecting fluorescent protein expression and co-localisation on single secretory vesicles with linear spectral unmixing. *Eur. Biophys. J.* 35:533–547, 2006.

Nishi, M., M. Tanaka, K. Matsuda, M. Sunaguchi, and M. Kawata. Visualization of glucocorticoid receptor and mineralocorticoid receptor interactions in living cells with GFP-based fluorescence resonance energy transfer. *J. Neurosci.* 24:4918–4927, 2004.

Perkins, S., N. R. Harvey, S. P. Brumby, and K. Lacker. Support vector machines for broad area feature extraction in remotely sensed images. *Proc. SPIE* 4381:268–295, 2001.

Potma, E. O., W. P. D. Boeij, P. J. M. V. Haastert, and D. A. Wiersma. Real-time visualization of intracellular hydrodynamics in single living cells. *Proc. Natl. Acad. Sci. USA* 98:1577–1582, 2001.

Raicu, V., D. B. Jansma, R. J. D. Miller, and J. D. Friesen. Protein interaction quantified in vivo by spectrally resolved fluorescence resonance energy transfer. *Biochem. J.* 385:265–277, 2005.

Rajpoot, K., and N. Rajpoot. SVM optimization for hyperspectral colon tissue cell classification. *Medical Image Computing and Computer-Assisted Intervention—MICCAI, Pt 2, Proceedings* 3217:829–837, 2004.

Rehen, S. K., M. J. McConnell, D. Kaushal, M. A. Kingsbury, A. H. Yang, and J. Chun. Chromosomal variation in neurons of the developing and adult mammalian nervous system. *Proc. Natl. Acad. Sci. USA* 98:13361–13366, 2001.

Robertson, S. A., J. Schoumans, B. D. Looyenga, J. A. Yuhas, C. R. Zylstra, J. M. Koeman, P. J. Swiatek, B. T. Teh, and B. O. Williams. Spectral karyotyping of sarcomas and fibroblasts derived from Ink4a/Arf-deficient mice reveals chromosomal instability in vitro. *Int. J. Oncol.* 26:629–634, 2005.

Sánchez-Armáss, S., S. R. Sennoune, D. Maiti, F. Ortega, and R. Martínez-Zaguilán. Spectral imaging microscopy demonstrates cytoplasmic pH oscillations in glial cells. *Am. J. Physiol. Cell Physiol.* 290:C524–C538, 2006.

Schwarz J., and K. Staenz. Adaptive threshold for spectral matching of hyperspectral data. *Can. J. Rem. Sens.* 27:216–224, 2001.

Shafer-Peltier, K. E., A. S. Haka, J. T. Motz, M. Fitzmaurice, R. R. Dasari, and M. S. Feld. Model-based biological Raman spectral imaging. *J. Cell. Biochem.* 39(Suppl.):125–137, 2002.

Sharonov, G. V., A. V. Feofanov, O. V. Bocharova, M. V. Astapova, V. I. Dedukhova, B. V. Chernyak, D. A. Dolgikh, A. S. Arseniev, V. P. Skulachev, and M. P. Kirpichnikov. Comparative analysis of proapoptotic activity of cytochrome c mutants in living cells. *Apoptosis* 10:797–808, 2005.

Shonat, R. D., E. S. Wachman, W. H. Niu, A. P. Koretsky, and D. L. Farkas. Near-simultaneous hemoglobin saturation and oxygen tension maps in mouse brain using an AOTF microscope. *Biophys. J.* 73:1223–1231, 1997.

Shonat, R. D., Wachman, E. S., Niu, W., Koretsky, A. P. and Farkas, D. L. Near-simultaneous hemoglobin saturation and oxygen tension maps in mouse cortex during amphetamine stimulation. *Adv. Exp. Med. Biol.* 454:149–158, 1998. Siboni, G., C. Rothmann, B. Ehrenberg, and Z. Malik. Spectral imaging of mc540 during murine and human colon carcinoma cell differentiation. *J. Histochem. Cytochem.* 49:147–153, 2001.

Sinha, A., and G. Barbastathis. Broadband volume holographic imaging. *Appl. Opt.* 43:5214–5221, 2004.

Sorg, B. S., B. J. Moeller, O. Donovan, Y. Cao, and M. W. Dewhirst. Hyperspectral imaging of hemoglobin saturation in tumor microvasculature and tumor hypoxia development. *J. Biomed. Opt.* 10:044004-1–11, 2005.

Steller, W., J. Einenkel, L.- C. Horn, U.- D. Braumann, H. Binder, R. Salzer, and C. Krafft. Delimitation of squamous cell cervical carcinoma using infrared microspectroscopic imaging. *Anal. Bioanal. Chem.* 384:145–154, 2006.

Strang, G. *Linear Algebra and Its Applications*. New York: Harcourt Brace Jovanovich, 1988.

Suhre, D. R., M. Gottlieb, L. H. Taylor, and N. T. Melamed. Spatial-resolution of imaging noncollinear acousto-optic tunable filters. *Opt. Eng.* 31:2118–2121, 1992.

Sunamura, M., A. Maruyama, T. Tsuji, and R. Kurane. Spectral imaging detection and counting of microbial cells in marine sediment. *J. Microbiol. Meth.* 53:57–65, 2003.

Süstrunk, S., R. Buckley, and S. Swen. *Standard RGB Color Spaces*. Online. http://ivrgwww.epfl.ch/publications/sbs99.pdf. 1999.

Tang, H. W., G. Q. Chen, J. S. Zhou, and Q. S. Wu. Hadamard transform fluorescence image microscopy using one-dimensional movable mask. *Anal. Chim. Acta* 468:27–34, 2002.

ThalerC., S. V. Koushik, P. S. Blank, and S. S. Vogel. Quantitative multiphoton spectral imaging and its use for measuring resonance energy transfer. *Biophys. J.* 89:2736–2749, 2005.

Timlin, J. A., D. M. Haaland, M. B. Sinclair, A. D. Aragon, M. J. Martinez, and M. Werner-Washburne. Hyperspectral microarray scanning: Impact on the accuracy and reliability of gene expression data. *BMC Genomics* 6:72, 2005.

Tomatis, S., M. Carrara, A. Bono, C. Bartoli, M. Lualdi, G. Tragni, A. Colombo, and R. Marchesini. Automated melanoma detection with a novel multispectral imaging system: Results of a prospective study. *Phys. Med. Biol.* 50:1675–1687, 2005.

Tran, S. D., J. Wang, B. C. Bandyopadhyay, R. S. Redman, A. Dutra, E. Pak, W. D. Swaim, J. A. Gerstenhaber, J. M. Bryant, C. Zheng, C. M. Goldsmith, M. R. Kok, R. B. Wellner, and B. J. Baum. Primary culture of polarized human salivary epithelial cells for use in developing an artificial salivary gland. *Tiss. Engin.* 11:172–181, 2005.

Treado, P. J., I. W. Levin, and E. N. Lewis. High-fidelity Raman imaging spectrometry: A rapid method using an acoustooptic tunable filter. *Appl. Spectrosc.* 46:1211–1216, 1992.

Treado, P. J., and M. D. Morris. Hadamard-transform Raman imaging. *Appl. Spectrosc.* 42:897–901, 1988.

Treado, P. J., and M. D. Morris. Multichannel Hadamard-transform Raman microscopy. *Appl. Spectrosc.* 44:1–4, 1990.

Ulrich, V., P. Fischer, I. Riemann, and K. König. Compact multiphoton/single photon laser scanning microscope for spectral imaging and fluorescence lifetime imaging. *Scanning* 26:217–225, 2004.

Uzunbajakava, N., A. Lenferink, Y. Kraan, E. Volokhina, G. Vrensen, J. Greve and C. Otto. Nonresonant confocal Raman imaging of DNA and protein distribution in apoptotic cells. *Biophys. J.* 84:3968–3981, 2003a.

Uzunbajakava, N., A. Lenferink, Y. Kraan, B. Willekens, G. Vrensen, J. Greve, and C. Otto. Nonresonant Raman imaging of protein distribution in single human cells. *Biopolymers* 72:1–9, 2003b.

Verstraete, M. M., B. Pinty, and R. E. Dickinson. A physical model of the bidirectional reflectance of vegetation canopies. 1. Theory. *J. Geophys. Res. Atmos.* 95:11755–11765, 1990.

Vo-Dinh, T., D. L. Stokes, M. B. Wabuyele, M. E. Martin, J. Myong Song, R. Jagannathan, M. Edward, R. J. Lee, and X. Pan. A hyperspectral imaging system for in vivo optical diagnostics hyperspectral imaging basic principles, instrumental systems, and applications of biomedical interest. *IEEE Med. Biol.* 23:40–49, 2004.

Volin, C. E., B. K. Ford, M. R. Descour, J. P. Garcia, D. W. Wilson, P. D. Maker, and G. H. Bearman. High-speed spectral imager for imaging transient fluorescence phenomena. *Appl. Opt.* 37:8112–8119, 1998.

Volin, C. E., J. P. Garcia, E. L. Dereniak, M. R. Descour, T. Hamilton, and R. McMillan. Midwave-infrared snapshot imaging spectrometer. *Appl. Opt.* 40:4501–4506, 2001.

Wachman, E. S., W. H. Niu, and D. L. Farkas. AOTF microscope for imaging with increased speed and spectral versatility. *Biophys. J.* 73:1215–1222, 1997.

Wachsmann-Hogiu, S., and D. L. Farkas. Spectroscopic non-linear microscopy. In *Handbook of Biomedical Nonlinear Microscopy* (P. So and B. Masters, eds.), pp. 461–483, New York, Oxford University Press, 2008. Waggoner, A. S., E. S. Wachman, and D. L. Farkas. Optical filtering systems for wavelength selection in light microscopy. *Current Protocols in Cytometry* 2:4, 2001.

Wang, H., Y. Fu, P. Zickmund, R. Shi, and J. X. Cheng. Coherent anti-Stokes Raman scattering imaging of axonal myelin in live spinal tissues. *Biophys. J.* 89:581–591, 2005.

Wuttig, A. Optimal transformations for optical multiplex measurements in the presence of photon noise. *Appl. Opt.* 44:2710–2719, 2005.

Yaroslavsky, A. N., V. Neel, and R. R. Anderson. Demarcation of nonmelanoma skin cancer margins in thick excisions using multispectral polarized light imaging. *J. Invest. Dermatol.* 121:259–266, 2003.

Zhang, H. B., L. B. Zeng, H. Y. Ke, H. Zheng, and Q. S. Wu. Novel multispectral imaging analysis method for white blood cell detection. *Adv. Nat. Comp., Pt 2, Proc.* 3611:210–213, 2005.

Zhang, J. Y., and Y. X. Liu. Cervical cancer detection using SVM based feature screening. *Medical Image Computing and Computer-Assisted Intervention—MICCAI 2004, Pt 2, Proc.* 3217:873–880, 2004.

Zhang, Y., C. Haskins, M. Lopez-Cruzan, J. Zhang, V. E. Centonze, and B. Herman. Detection of mitochondrial caspase activity in real time in situ in live cells. *Microsc. Microanal.* 10:442–448, 2004.

Zimmermann, T. Spectral imaging and linear unmixing in light microscopy. *Adv. Biochem. Engin. Biotechnol.* 95:245–265, 2005.

Zimmermann, T., J. Rietdorf, A. Girod, V. Georget, and R. Pepperkok. Spectral imaging and linear un-mixing enables improved FRET efficiency with a novel GFP2-YFP FRET pair. *FEBS Lett.* 531:245–249, 2002.

Zimmermann, T., J. Rietdorf, and R. Pepperkok. Spectral imaging and its applications in live cell microscopy. *FEBS Lett.* 546:87–92, 2003.

Zuurbier, C. J., C. Demirci, A. Koeman, H. Vink, and C. Ince. Short-term hyperglycemia increases endothelial glycocalyx permeability and acutely decreases lineal density of capillaries with flowing red blood cells. *J. Appl. Physiol.* 99:1471–1476, 2005.

3

Multiphoton Microscopy
in Neuroscience

FRITJOF HELMCHEN, SAMUEL S.-H. WANG,

AND WINFRIED DENK

Since the first report more than 10 years ago of two-photon excited fluorescence microscopy (Denk et al., 1990), multiphoton microscopy (MPM) has found uses in many research areas. Multiphoton microscopy has had a particularly strong impact on life sciences because several of its properties provide special advantages for studying living cells in intact tissue (forr eviews see Denk and Svoboda, 1997; Piston, 1999; König, 2000; So et al., 2000; Helmchen and Denk, 2002). In this chapterwe focus on applications in neuroscience.

The cerebral cortex in human contains about 10^{10} neurons with about 10^{14} synaptic connections among them (Braendgaard et al., 1990; Tang et al., 2001). To understand the integrated function of these structures, it is necessary to study them in their native configuration. Multiphoton microscopy has helped to reveal neural signaling in near-intact living tissue at levels ranging from individual synapses to entire neural networks (fig. 3.1 [Garaschuk et al., 2000; Wang et al., 2000b]). It has been used to investigate development, structural plasticity, modulation of excitability, and synaptic integration. Multiphoton microscopy has proved particularly powerful when used in conjunction with advances in fluorescent tissue labeling techniques, electrophysiology, and molecular biology.

The prevailing form of MPM now used in neuroscience research relies on two-photon excitation. However, because many of the properties of two-photon microscopy apply to higher orders of multiphoton excitation (MPE) as well, we will generally use the term *multi-photon microscopy*. We begin with a short description of the particularly relevant physical properties of MPE and its major modes of application. We then discuss those technical issues essential for neuroscience experiments, followed by different applications using examples from a variety of tissues. There are other notable reviews as well (Williams et al., 1994; Denk et al., 1995b; Denk, 1996; Denk and Svoboda, 1997; Mainen et al., 1999; Yuste et al., 1999a; Helmchen and Denk, 2002).

Why Use Multiphoton Excitation?

Multiphoton excitation is defined as a process in which the energy for the excitation of a single molecule comes from two or more photons that, for what is called *simultaneous MPE*, are absorbed at virtually the same time. Each photon needs to contribute only part of the total energy required for the transition, provided that the combined amount of energy is sufficient. Two-photon excitation, for example, is typically achieved with near-infrared excitation

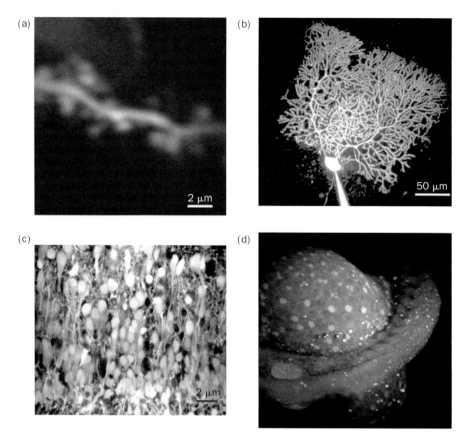

Figure 3.1 Multiphoton imaging of neural elements. (A) Synaptic level. Dendritic spines on a thin dendrite of a cortical pyramidal neuron filled with a fluorescent calcium indicator (Oregon Green BAPTA-1). Images were acquired in vivo in the somatosensory cortex of an anesthetized rat (unpublished data, F. Helmchen). (B) Cellular level. Purkinje neuron in a cerebellar brain slice filled with magnesium green via a patch pipette. (From Wang et al. [2000b].) (C) Networklevel. Population of cortical neurons labeled in a neonatal brain slice preparation using the acetoxylmethyl ester form of a calcium indicator dye. (From Garaschuket al. [2000].) (D) Organism level. A zebrafish larva is shown about 20 hours after fertilization. The egg was injected with a dextran-conjugated calcium indicator. Field of view is approximately 0.6 mm on the side (unpublished data, W. Denk).

wavelengths at 700 to 1000 nm, about twice as long as those used in single-photon excitation. Multiphoton excitation has been made practical by the advent of ultrafast lasers (*ultrafast* is loosely defined as subpicosecond; for a collection of reprints, see Gosnell and Taylor [1991]). At conventional illumination intensities, two-photon excitation is farless likely than single-photon excitation (Denk and Svoboda, 1997). This is analogous to the extremely unlikely event of two bullets colliding in midair. For multiphoton absorption, two or more photons must be absorbed within a time window of less than 10^{-15} s (Xu and Webb, 1996); however, generating significant numbers of these events with continuous illumination could destroy the specimen and consume impractical amounts of power (Booth and Hell, 1998). Ultrafast lasers generate subpicosecond pulses at megahertz rates corresponding to a duty cycle of about 10^{-5}, leading to high instantaneous beam intensity but low average beam power. In the bullet analogy, this corresponds to generating fusillades of incredible intensity separated by long, silent intervals. Indeed, on American Civil War battlefields, firing could be so intense that multiple instances of fused bullet pairs have been reported (Slade and Alexander, 1998).

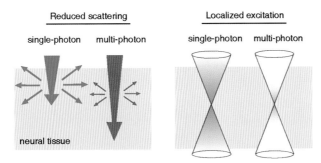

Figure 3.2 Key advantages of MPE. (Left) The long-wavelength (near-infrared) excitation light used in MPM is less scattered and penetrates deeper into tissue than ultraviolet or visible light used for single-photon excitation. (Right) Multiphoton absorption is localized because of its nonlinear dependence on illumination intensity. This confinement of excitation provides optical sectioning, reduces out-of-focus photobleaching and photodamage, and enables efficient fluorescence collection. It also permits highly localized photomanipulation in the focal volume.

One advantage of MPE is that excitation light scattering by tissue is reduced as a result of longer wavelengths (Svaasand and Ellingsen, 1983), allowing excitation to penetrate more deeply into the tissue than visible light (fig. 3.2A). Second, the multiphoton absorption probability depends strongly on the illumination intensity (with a power law exponent equal to the number of photons per molecular excitation). For example, the rate of two-photon excitation depends on the square of the excitation intensity. For this reason, the probability of excitation falls off steeply above and below the focal plane (fig. 3.2B), and excitation is essentially confined to the focal volume (Denk et al., 1995b). The optical sectioning offered by MPE is therefore inherent to the excitation process. This stands in contrast to confocal microscopy, which requires a pinhole in front of the detector to reject out-of-focus fluorescence. These two fundamental properties of MPE—longer wavelength and localized excitation—are in large part responsible for the usefulness of MPM in neuroscience.

Looking into Intact Tissue

Neural tissue, like most other living tissue, is hostile to the propagation of light. Image quality deteriorates rapidly with depth, and high-resolution optical microscopy is mostly limited to the tissue surface. For this reason, parts of the brain are often extracted and studied in vitro in an artificial environment. Although the in vitro approach enables optical studies on individual living neurons, cellular properties are affected to a varying degree by the dissection and dissociation procedures. In contrast, enhanced depth penetration and optical sectioning in MPM enable imaging of neurons with high spatiotemporal resolution in relatively thick samples of brain tissue. Multiphoton microscopy, therefore, has been applied to more and more intact preparations. In brain slices, MPM has substantially enabled the study of synaptic integration and dendritic processing, in particular on the level of individual synapses. Furthermore, MPM has made it possible to investigate dendritic excitation in the intact brain, with recent technological developments suggesting that high-resolution functional imaging is feasible even in awake and unrestrained animals.

Efficient Use of Fluorescence Light

High-resolution imaging in the intact brain would be impossible without the perifocal confinement of MPE, which allows very efficient fluorescence collection. This is because high-resolution information is contained in virtually all generated fluorescence photons, which are known to originate from the perifocal region and therefore are still useful even after being scattered (Denk et al., 1994, 1995b). With suitable detector arrangements, the collection of fluorescence light can thus be maximized without a loss of resolution. For the same reason, MPM (somewhat paradoxically, in view of the high peak intensities) is often less damaging to tissue than conventional single-photon excitation. Forthe study of living tissue, this reduction in photodamage extends the possible duration or the number of repetitions of an experiment and can even make previously impossible experiments feasible.

Application Modes

Multiphoton microscopy experiments can be divided in two groups, depending on whether fluorescence excitation or photochemical reactions are intended. In *fluorescence imaging*, MPE is used to visualize fluorophores in neural tissue. The experimental goal is to *monitor*, but not disturb, the structure or physiological state of the tissue. In *photochemical* applications, MPE is used as a tool to *manipulate* the state of the tissue. Photobleaching—usually undesired, but sometimes useful—is the simplest case of focal photochemistry. More common is the use of specifically designed caged compounds, which are molecules that have been inactivated by the covalent attachment of light-sensitive chemical groups. Before considering specific applications of either type of experiment, let us take a look at the technical aspects that need to be considered when applying MPM to brain tissue.

TECHNICAL ISSUES

Labeling

Most fluorophores traditionally used for single-photon excitation have been found to have two-photon absorption sufficient for imaging, albeit not always at exactly twice the single-photon excitation wavelength. The following sections provide a short description of fluorescence probes common in biological imaging and what they are used for.

Small Molecular Weight Indicators

Particularly useful in neuroscience are fluorescent indicators of physiological variables. Most in use today are small organic molecules that are sensitive to membrane electrical field orconcentr ations of ions (e.g., calcium, sodium, orchlor ide). With the exception of the electrical field indicators, these indicators are water soluble and are loaded into cells by three major routes: (1) injection through glass pipettes into single cells, (2) "trapping" of bath-applied precursors, and (3) direct transmembrane loading.

Loading of single cells is achieved eitherby impalement with a high-resistance glass microelectrode (Purves, 1981; Tank et al., 1988) followed by iontophoresis, or by gigaseal formation with a patch pipette, followed by whole-cell break-in and diffusion from the pipette into the cell (Pusch and Neher, 1988). Trapping requires attached groups (e.g., covalently attached acetoxylmethyl [AM] esters) (Tsien, 1989) to neutralize charged moieties on the dye molecule, allowing it to cross membranes. Intracellular enzymes then remove the protecting groups, reexposing the charge and trapping the dye inside the cells. The success of this approach depends on the presence of particular enzymes inside the target cells and access to the cell surface by the precursors. Because of incomplete uptake and dye trapping by nonneuronal structures, AM loading has been more difficult in brain slices. In a recent breakthrough, however, neocortical cell populations were labeled in living animals by direct injection of AM calcium indicatorin the brain and could be subsequently imaged in vivo using two-photon microscopy (Stosiek et al., 2003). An alternative approach is to load precursors at a location where processes (typically axons) of the cell population of interest pass. Axon tracts can be loaded in vitro (Regehr and Tank, 1991) or in vivo (Kreitzer et al., 2000). In either case, dye molecules diffuse or are actively transported along axons for up to several millimeters, yielding a region where only neuronal components are dye loaded.

Anothermeans of loading that is sometimes effective is to coax unmasked dyes into crossing the plasma membrane. This can be achieved by placing dye in close proximity to neurons followed by the application of an electrical field or a mechanical disturbance, which temporarily breaches membrane integrity. Dextran-conjugated dyes (typically 10 kD) load well this way (Gelperin and Flores, 1997; Haas et al., 2001) and are well suited for remote loading because they are transported over long distances both in anterograde and retrograde fashion (O'Donovan et al., 1993; Delaney et al., 2001). Recently, biolistic gene-gun methods originally developed for transfection have been adapted to label cells with membrane dyes (Gan et al., 2000) and intracellular indicators (Kettunen et al., 2002).

Fluorescent Proteins

The discovery and heterologous expression of fluorescent proteins have revolutionized fluorescence microscopy (Chalfie et al., 1994; for review see Sullivan and Kay, 1998; Tucker 2000). As proteins, they are genetically encodable, allowing expression in virtually any cell type, thus building a bridge between fluorescence microscopy and the tools of molecular biology and genetics. Several mutants of the original GFP with enhanced fluorescence and shifted spectra (generally referred to as *XFPs*) have been developed, permitting labeling and discrimination of multiple structures at once (Ellenberg et al., 1999).

The use of XFPs circumvents the access problem of dye loading in tissue. Genes can be transferred into cells by a variety of methods, including biolistic transfection, viral infection, and the creation of transgenic animals (Tucker, 2000). Cell type-specific promoters can be exploited, thus allowing protein to be expressed in well-defined subpopulations of cells.

A particularly significant breakthrough has been the discovery that the XFPs can be fused to other proteins without loss of fluorescence or protein function. This allows protein distribution and trafficking to be visualized directly (Shi et al., 1999). Furthermore, fusion proteins have been developed that are sensitive to various local signals such as voltage (Siegel and Isacoff, 1997; Sakai et al., 2001), chloride (Kuner and Augustine, 2000), or calcium (Miyawaki et al., 1997, 1999; Nakai et al., 2001; Truong et al., 2001).

Excitation

Excitation beam parameters for MPE follow principles similar to confocal microscopy. However, several additional considerations are relevant, which we describe in the following sections.

Beam Size at the Back Aperture

What size should the laser beam be at the back aperture of the objective? For maximum resolution, the illumination intensity should be uniform across the back aperture, making the effective NA equal to the objective's nominal NA. This, however, requires using only the central part of a Gaussian beam, resulting in substantial power loss. An alternative approach that avoids this loss would be the use of special beam-shaping optics, currently underdevelopment. Note that fordiffer ent objectives, the same relative filling of the back aperture may require different absolute beam diameters.

However, if ultimate axial resolution is not required, filling the back aperture may be neither necessary nor desirable because of photodamage considerations. The total amount of two-photon excited fluorescence (and hence the image brightness) is not altered by changes in NA if the focal volume is small compared with the dye-filled volume (Birge, 1986). The reason for this peculiar property of two-photon excitation is that, when the NA is reduced, the reduction of the two-photon excitation probability is exactly balanced by an increase in focal volume. Thus, underfilling of the back aperture results in a somewhat lower resolution (particularly along the optical axis), but does not cause a reduction of total fluorescence except for very small structures. However, reducing the effective NA may reduce the amount of photodamage, which depends on simultaneous and sequential higher order multiphoton processes.

Choosing an Excitation Wavelength

For both small-molecule and genetically encodable fluorescent probes, the relationship between single- and multiphoton spectra is not always predictable. Therefore, it is advisable to rely on specific measurements of multiphoton cross-sections (Xu and Webb, 1996). If measurements are not available, several rules of thumb can be used. First, at least some two-photon excitation usually occurs for wavelengths that, when divided by two, lie within the single-photon spectrum. Second, no excitation occurs for wavelengths longer than twice the maximum wavelength at which single-photon excitation is possible. Third, the two-photon spectrum often is broader and extends to shorter wavelengths than a rescaled version of the single-photon spectrum. Fourth, "strong" and "weak" one-photon absorbers are "very strong" and "very weak" two-photon absorbers, respectively.

Two Photons versus Three Photons

Although optical sectioning and background rejection properties of three-photon excitation are not significantly improved compared with two-photon excitation (Denk, 1996), three-photon excitation is still useful for reaching transition energies for compounds with absorption maxima in the ultraviolet range. Three-photon excitation has, therefore, been used to study ultraviolet-excitable fluorophores such as fura-2 or diamidino-2-phenylindol (DAPI) (Xu et al., 1996) as well as intrinsic cellular chromophores such as amino acids, neurotransmitters, proteins, and NADH or flavin adenine dinucleotide (FAD) (Maiti et al., 1997). A interesting recent application has been the induction of a long-lasting color change (*greening*) of DsRed by selective three-photon bleaching of the mature "red" form of DsRed, which could be useful for cell marking (Marchant et al., 2001).

Tissue Penetration

Scattering and Absorption

In most biological tissues, the optical property that limits depth penetration is not absorbance (which is very weak in the longer visible and the near-infrared wavelength range), but scattering. Neural tissue scatters light particularly strongly, leaving only a fraction of the incident light (henceforth called the *ballistic fraction*) to reach the focus unscathed. The first advantage of MPM is that the ballistic fraction, which carries the optical resolving capacity, is larger at infrared wavelengths as a result of reduced scattering compared with visible or ultraviolet light. Second, because of the strong intensity dependence of MPE, scattered light is virtually unable to excite fluorescence and hence creates a diffuse background. For the subpicosecond pulses used in MPE, it also helps that scattered light follows disperse path lengths, reducing the likelihood that multiple photons will arrive at an out-of-focus location at the same time.

Wide-angle (Rayleigh–like) scattering by small particles (for instance, synapses) attenuates the light but leaves a ballistic fraction with essentially unchanged wavefront properties.

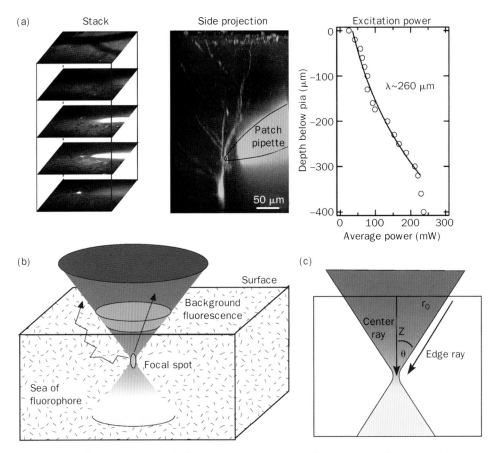

Figure 3.3 Depth limitations in multiphoton imaging. (A) Samples from a stack of fluorescence images of a layer 2/3 cortical neuron filled in vivo via a patch pipette. Large blood vessels can be seen on the surface. The soma was about 360 μm below the pial surface. About the same level of excitation was maintained throughout the stack by adjusting the illumination power manually, as shown in the right panel. An exponential fit to the data points (with the two deepest points excluded) yielded a space constant of 260 μm, which represents an estimate of the excitation scattering length (F. Helmchen, unpublished data). (B) With homogenous staining, the relative amount of fluorescence originating from near the surface increases with depth compared with fluorescence originating from the focal spot (see text). Two trajectories are shown for a ballistic and a highly scattered fluorescence photon, respectively. (C) The effective NA decreases with increasing imaging depth because edge rays are scattered more than center rays.

In this case the effective (ballistic) illumination power P diminishes with depth as

$$P(z) = P_0 e^{-z/l} \tag{3.1}$$

where z is the depth of the focal plane below the surface and l is the scattering mean free path. The two-photon excited fluorescence ($F \propto P^2$) therefore falls off with a length constant of $l/2$. Given this, the scattering mean free path can be estimated for a homogenous fluorophore distribution by determining how much laser intensity has to be used so that the measured fluorescence does not change (yielding an exponential with parameter l, fig. 3.3A). Alternatively, the fluorescence decrease with depths at constant excitation power delivered to the sample can be measured (yielding an exponential with parameter $l/2$) (Kleinfeld et al., 1998).

Because neural tissues have varying optical properties depending on the cellular content, the degree of myelination, vascularization, and so forth, the scattering (and thus depth penetration) can vary widely. A quantitative evaluation has been made in neocortical gray matter of rat, where the fluorescence fall-off of labeled vasculature has been measured (Kleinfeld et al., 1998; see correction, Kleinfeld et al., 1999). This fall-off had an exponential decay constant of about 100 μm at 830 nm, suggesting a value for l of approximately 200 μm for intact adult tissue. A recent study (Oheim et al., 2001) confirmed this and found that l increases with wavelength (1.5-fold between 760 nm and 900 nm). In brain slices, Oheim et al. (2001) found that l is substantially longer in juvenile than in adult brain, perhaps because of the incomplete myelination in juveniles. Conversely, however, in juvenile brain slices l was only 100 μm, half as long as in the intact adult brain. The reason for this is unknown.

A second factor, one that influences the ability to form a focus at all, is the presence of large optical obstacles. Cell bodies, capillaries, and fiber tracts all act as lenses to distort the optical wavefront and therefore diminish the ability to reach a diffraction-limited focus.

Limits to Depth Penetration: Focal versus Background Fluorescence

Currently, MPM routinely achieves imaging depths in intact gray matter of about 500 μm (Kleinfeld et al., 1998; Helmchen et al., 1999). Such an imaging depth can reach a number of cortical layers. For example, the most superficial cell-rich layer in neocortex (layer 2/3) is located more than 100 μm below the pial surface, beyond the reach of confocal microscopy (for a direct comparison between two-photon microscopy and confocal microscopy, see Centonze and White [1998]). Large penetration depths are reached using MPM because of the reduced scattering of infrared light, and by compensating for the remaining loss of ballistic excitation light by increasing the incident laser power. One might thus expect improvements to be limited only by the available laserpowerand by the powersustainable by the sample. However, careful consideration shows that depth penetration can actually be limited by the fact that unwanted background excitation in regions closer to the surface eventually becomes larger than the excitation within the focal volume (Ying et al., 1999; Theer et al., 2003). This background can be calculated for the case in which fluorophores are present throughout the sample (such as in XFP transgenic animals). Next let us estimate the depth at which the numbers of absorbed photons in the focal volume (N_F) and near the surface (N_B) are equal (meaning, equal contribution of signal and background to the detected fluorescence; fig 3.3B).

In the case of two-photon excitation, the numberof fluorescence photons collected per unit time (F) is given by

$$F = \phi \eta_2 N_{abs}/2 \tag{3.2}$$

where η_2 is the quantum efficiency of the dye, ϕ is the collection efficiency of the microscope system, and N_{abs} is the total numberof absorbed photons perunit time (Xu and Webb, 1996). We assume η_2 and ϕ to be constant and consideronly N_{abs}.

In a uniformly stained sample, a beam of wavelength λ will lead to a numberof photons absorbed per unit time:

$$N_F = \lambda \delta C \overline{P}^2 n\xi/(hc)^2 \tag{3.3}$$

where $\overline{P}$ is the average power of the ballistic fraction in the focal plane, δ is the two-photon cross-section, C is the fluorophore concentration, n is the refractive index, and ξ is the "two-photon advantage" as defined in Denk et al. (1995b). This numberis independent of the numerical aperture A_{NA}, as noted earlier.

A rough estimate of the total absorption near the surface can then be made as follows. If the focal plane is at a depth z from the tissue surface, the laser power at the surface needed to achieve this amount of excitation is $P_0 = \overline{P}e^{z/l}$. The excitation light is spread over a surface area of roughly $a_s = \pi(A_{NA}z/n)^2$, and significant excitation extends to a depth of at least

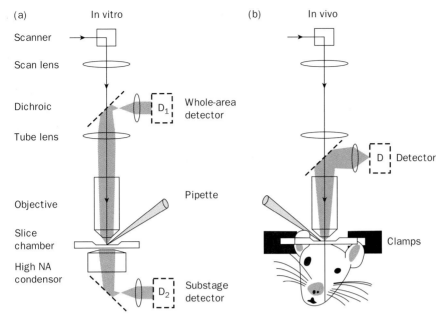

Figure 3.4 Fluorescence detection schemes. Detector positions used in neuroscience applications. (A) Dual-detector setup for multiphoton imaging in brain slices. A nondescanned, whole-area detector (D_1) is placed in the epifluorescence pathway. An additional substage detector (D_2) collects fluorescence light emerging at the far side of the brain slice through a high NA condenser. (B) For in vivo application, a single external detector is used, which can be placed close to the objective for efficient fluorescence collection.

$l/2$ (in practice, this number can be larger because low-angle scattering does destroy the ability for focal but not background MPE [Theer and Denk, in preparation]). The excitation rate for a single fluorophore at the surface is

$$r_s = \delta\xi \left[\frac{P_0}{\hbar\omega\, a_S} \right]^2 = \delta\xi \left[\frac{\lambda\overline{P}e^{z/l}}{hc\pi(A_{NA}z/n)^2} \right]^2 \tag{3.4}$$

The number of absorbed photons near the surface is then

$$N_B = Ca_s\frac{l}{2}r_S = C\pi(A_{NA}z/n)^2\frac{l}{2}\delta\xi \left[\frac{\lambda\overline{P}el}{hc\pi(A_{NA}z/n)^2} \right]^2 \tag{3.5}$$

The depth z_{max} where focal and background fluorescence are equal can then be calculated as the solution of the equation $N_F = N_B$, which simplifies to

$$2\pi(A_{NA}z_{\max})^2 = \lambda nle^{2\,z_{\max}/l} \tag{3.6}$$

or

$$z_{\max}/l = \sqrt{\frac{\lambda n}{2\pi l A_{NA}{}^2}e^{z_{\max}/l}} \tag{3.7}$$

For $l = 200$ μm, $\lambda = 900$ nm, $A_{NA} = 0.8$, and the refractive index of water ($n = 1.33$), $z_{max} \approx 1$ mm. Interestingly, z_{max} decreases as the NA is reduced (for NAs of 1.0, 0.9, 0.8, 0.7, 0.6, 0.5, 0.4, 0.3, and 0.2; values for z_{max} are 1022, 995, 965, 932, 892, 845, 786, 708, and 590 μm, respectively).

At a 1-mm depth, the ballistic power is reduced by a factor of e^5, about 150-fold. Thus, to reach even 10 mW of ballistic power at the focus, 1.5 W of average power has to reach the surface of the sample. This is at the limit of what current lasers can provide. Reaching the depth limit estimated here may require the use of lower repetition rate sources that provide higher peak power (Beaurepaire et al., 2001; Theer et al., 2003).

Reduction of the Effective Numerical Aperture

The preceding calculation assumed that rays are scattered as if they approach at a normal angle of incidence. However, rays at the edge of the illumination cone travel a longer distance to the focus than those near the optical axis (fig. 3.3C) and therefore are attenuated more by scattering. The intensity as a function of the angle θ from the optical axis can be described for an initially Gaussian beam with an NA (in the small-angle approximation) of $A_{NA} = \theta_0 n$ as

$$I(\theta, z) = e^{-\frac{z}{l \cos \theta}} e^{-\left(\frac{\theta}{\theta_0}\right)^2} \tag{3.8}$$

The cosine function can be approximated, yielding

$$I(\theta, z) = e^{-\frac{z}{l}} e^{-\left(\frac{\theta}{\theta_0}\right)^2 \left(1 + \frac{z\theta_0^2}{2l}\right)} \tag{3.9}$$

or

$$I(\theta, z) = e^{-\frac{z}{l}} e^{-\left(\frac{\theta}{\theta_{eff}}\right)^2} \tag{3.10}$$

with

$$\theta_{eff} = \theta_0 \Big/ \sqrt{1 + \frac{z}{2l} \theta_0^2} \tag{3.11}$$

Using $\theta_0 = A_{NA}/n$, these equations describe a Gaussian beam with a reduced effective NA of

$$A_{eff} = A_{NA} \Big/ \sqrt{1 + \frac{z}{2l} (A_{NA}/n)^2} \tag{3.12}$$

This analysis implies that the attenuation of peripheral illumination rays deep in tissue reduces not only resolution, but also excitation efficiency resulting from the extra power loss by a factor of $(A_{eff}/A_{NA})^2$. This effect is naturally more serious for larger apertures. For example, using $z = 500$ μm and $l = 200$ μm, for $A_{NA} = 0.9$, the NA is reduced by 20%, whereas for $A_{NA} = 0.5$, the reduction is only 8%. At $z = 1000$ μm, an NA of 0.9 is reduced to 0.61 (by 32%), whereas an NA of 0.5 is reduced to 0.43 (by only 14%). Therefore, for deep imaging, filling the back aperture may not be optimal because it only leads to a marginal increase in resolution but a large loss of excitation power.

Extending the results from the previous section allows us to calculate the corrected values for the depth penetration. At the surface, there are no significant changes, but, because of to the extra power loss, the amount of fluorescence (N_F) created near the focus is reduced by a factor of $(A_{eff}/A_{NA})^4$, leading to a modified equation 3.6:

$$2\pi (A_{NA} z_{max})^2 = \lambda n l e^{2z_{max}/l} \left(1 + \frac{z_{max}}{2l} (A_{NA}/n)^2\right)^2 \tag{3.13}$$

ora modified equation 3.7:

$$z_{max}/l = \sqrt{\frac{\lambda n}{l A_{NA}^2 2\pi}} e^{z_{max}/l} \left(1 + \frac{z_{max}}{2l} (A_{NA}/n)^2\right) \qquad (3.14)$$

ForNAs of 1.0, 0.9, 0.8, 0.7, 0.6, 0.5, 0.4, 0.3, and 0.2, we find depth limit (z_{max}) values of 824, 825, 823, 816, 803, 781, 744, 684, and 580 μm, respectively. There is now an optimum forthe NA at slightly less than 1.0. This would suggest that extreme signal resolution requires as large an NA as practical, because most dipping cone lenses have an NA of $\leq$ 1.0. However, the shallowness of the relationship indicates that, if available power is limiting or if NA-dependent aberrations or wavefront distortion reduce the focal fluorescence, illumination at lower NA might still provide better overall results.

Detection

Within the limits estimated in the previous section, most fluorescence photons emerging from the tissue are known to have originated from the focal volume even if they are scattered on theirway out. Thus, the majordesign principle of any detection system is to collect as many fluorescence photons as possible. This implies both an appropriate choice of the detector(with high quantum efficiency at the emission wavelengths and low sensitivity in the near-infrared to reject back-reflected or stray light [Tan et al., 1999]) and a careful design of the detection optics to minimize loss of fluorescence light. Maximizing the efficiency of fluorescence detection in turn reduces the excitation power needed, thereby reducing photodamage.

Ballistic versus Scattered Fluorescence Photons

Because in most biological tissue scattering is far more likely than absorption, light propagation from the focal point through a thick specimen can be thought of as a process of *photon diffusion* (Yuste et al., 1999a). As in the case of excitation light, the ballistic fraction of the fluorescence light (emerging from the focal volume) becomes smaller on its way through tissue. The ballistic fraction, however, drops much faster now because visible wavelengths are more strongly scattered. The scattering length in gray matter is about 33 to 100 μm at 630 nm (Taddeucci et al., 1996), implying that at 500 μm focal depth only a very small fraction (10^{-7} to 10^{-2}) of the emerging fluorescence would be ballistic. A confocal microscope, which uses a descanning pinhole, captures only this ballistic fraction; use of a pinhole is therefore extremely harmful to signal strength. In contrast, in MPM both scattered and ballistic photons contribute useful signal. Descanning is unnecessary and the use of whole-area (sometimes called *external*) detectors is essential (Denk et al., 1995b). To optimize further the detection optics for collection of the scattered fraction, we have to understand the shape of the light distribution emerging from the tissue surface (see also Beaurepaire and Mertz, 2002).

In the diffusion analogy, light executes a random walk in the tissue as a result of frequent scattering events (fig. 3.3B). This view is valid if a substantial number of scattering events occur, which is the case for deep ($\geq$500 μm) imaging. The distribution of light (*photon concentration*) with a point source at focal depth z_0 can then be derived using Fick's laws. For a homogenous, infinite medium, the concentration depends inversely on the radial distance from the source. In our case, the point source is embedded in scattering tissue below a plane surface. The tissue is assumed to extend to infinite depth, which is a good approximation for imaging of the intact brain. For such a semi-infinite medium and no absorption, all photons are bound to reach the tissue surface eventually, where they escape and do not return. This corresponds to a boundary condition where the photon density $D_p(z, r)$ becomes zero at the surface (z and r denote the distances along and from the optical axis, respectively). Assuming a virtual "mirror sink" outside the tissue (at $z = -z_0$ and $r = 0$), the solution forthis geometry is found to be

$$D_p(z, r) \propto \left(\frac{1}{\sqrt{(z_0 - z)^2 + r^2}} - \frac{1}{\sqrt{(z_0 + z)^2 + r^2}} \right) \tag{3.15}$$

To calculate the light intensity at the surface we have to calculate the flux, which is the spatial gradient of $D_p(z, r)$. Because the density is zero everywhere at the surface, the gradient is strictly perpendicular to the surface and we can evaluate the normal component of the gradient at the surface:

$$\left. \frac{\partial D_p(z, r)}{\partial z} \right|_{z=0} \propto \frac{2z}{(r^2 + z^2)^{3/2}} \tag{3.16}$$

This expression provides the spatial distribution of photons emerging from the sample surface. Note that the shape of this distribution is independent of the mean free path for the fluorescence photons (as long as the multiple scattering condition is fulfilled).

By spatial integration of equation 3.16 over r, we find that half of all scattered photons emerge in a circle with a radius of $\sqrt{3}$ (about 1.73) times the focal depth z_0. To capture light from such a large area, the detection system has to have a large field of view (several times the focal depth), which often is many times larger than the scanned area. Counterintuitively, the angular distribution of the emerging scattered light $[\cos(\theta)]$ is actually slightly more directed along the optical axis than the emerging ballistic light, which is isotropic, but still requires a large collection NA objective.

Large NA and large field of view are difficult to reconcile in an optical system because of off-axis aberrations. Fortunately, optical corrections are necessary only for those parts of the field of view and of the NA used forexcitation. In fact, some commercial objectives have a much largerfield of view forgather ing of light than the nominal field of view of the microscope. Forefficient detection, the detectordoes not have to be physically close to the objective lens, but detection NA and field of view do have to be properly propagated through the optical system, and projected to fit within the area and angular acceptance zone of the detector. This requirement unfortunately rules out the use of photon-counting avalanche photodiodes (APDs) and most types of spectral detectors. Consistent with the preceding reasoning is recent work demonstrating that a special high-NA (0.95), low-magnification (20×) objective gathers about 10-fold more fluorescence light from a scattering medium compared with a standard 60× (NA, 0.9) objective (Oheim et al., 2001).

In summary, the strategies for optimizing the detection system are to choose a high-NA, low-magnification (large field of view) objective and carefully design the relay optics between objective and detector. This may also include removing the aperture (intended for scattered light elimination) present at the back of most microscope objectives.

Substage Detection

For thin samples (cell culture and tissue slice preparations), a substantial part (about half) of the fluorescence emerges at the far side of the specimen. This fraction can be captured if a second fluorescence detectoris used in the condenserpath (fig. 3.4). Using this approach, the total collected signal can be increased by a factor of more than two when a high-NA oil immersion condenseris used (Koesteret al., 1999).

Data Acquisition

Most neuroscience applications involve not only stationary imaging but also require the measurement of dynamic cellular processes on various timescales. Here we describe several features of laser-scanning microscope systems that are useful or essential for neurobiological measurements.

Measuring Fast Dynamics

The scanning system should permit high-speed measurements (acquisition rates up to 1 kHz) to capture the fastest timescales in neural processing. Although a few systems achieve video rate (25–30 Hz) at full-frame resolution (e.g., Fan et al., 1999), it is often more important to be able to perform measurements along a single scan line with millisecond temporal resolution (Hernandez-Cruz et al., 1990; Yuste and Denk, 1995; Svoboda et al., 1996, 1997). The orientation and even the shape of this line should be freely selectable so that the dynamics in different cells or cellular compartments can be compared directly. In vivo, it is often more useful to use small frames containing multiple lines to avoid creating spurious intensity modulation from motion.

Volume Time-Lapse (Four-Dimensional) Imaging

Because one of the advantages of MPM is reduction of photodamage, the microscope software should allow convenient acquisition of multiple three-dimensional data stacks to allow one to follow, for example, morphological changes in the developing nervous system. If the stack extends over a large depth range, an automatic adjustment of the excitation power with imaging depth to compensate forscatter ing losses can be essential.

Multichannel Acquisition

The possibility of acquiring multiple channels simultaneously is a standard feature in most laser-scanning microscopes. Particularly helpful for colocalization studies is the fact that for different fluorophores excited by the same laser wavelength, the MPM point-spread functions are identical, with no chromatic aberration, which can be a problem for confocal microscopes. Recording a Differential interference contrast (DIC), gradient-contrast, or backscattering image in one channel is often quite useful for orientation and can be essential forpipette positioning.

Working with Living Brain Tissue

In all physiological experiments, the tissue has to be kept alive and in good condition. Optophysiological experiments complicate matters because excitation causes some amount of light-induced damage to the tissue and because specimen motion has to be kept below the optical resolution. Displacements even on a micron scale can cause artifacts in the measurements of indicator signals from small stationary structures or of subtle morphological changes in such structures.

Avoiding Motion

A common application of two-photon microscopy is the examination of living brain slices (described previously), which are prepared by rapid dissection from a wide range of animals by methods that are now standard (Dingledine, 1984). Brain slices can be kept alive for hours or, in the case of cultured (i.e., *organotypic*) slices, fordays oreven weeks. Methods are described in detail elsewhere (Gibb and Edwards, 1994; Sakmann and Neher, 1995). For an optophysiology experiment, the brain slice is transferred to an upright microscope into a recording and imaging chamber equipped with a perfusion for oxygenated artificial cerebrospinal Fluid (ACSF). Brain slices usually are held in place using small weights or by grids made of fine nylon thread. It is essential for tissue viability to maintain a sufficient rate of fluid exchange, which can be impeded by the presence of the water immersion microscope objective. Furthermore, when imaging brain slices, position drift may occur over time periods of minutes to hours as a result of tissue swelling.

With in vivo imaging (Svoboda et al., 1997), avoiding unwanted motion becomes more challenging because of animal movements. For imaging experiments through a cranial window of an anesthetized animal, first of all it is important to fix the animal's skull tightly to the microscope setup. Fixation using ear and bite bars does not, in our experience, reduce motion to a low enough level. More stable fixation is achieved by cementing a metal plate to the skull (Kleinfeld and Denk, 1999). What remains is the motion of the brain tissue relative to the skull caused mainly by heartbeat and breathing. This motion is particularly strong (up to tens of micrometers) in open cranial windows because pressure fluctuations cause the exposed surface to expand and contract. This poses a severe problem for imaging small structures such as dendrites. Pressure fluctuations often can be reduced somewhat by adjusting the posture of the animal. More helpful is the restoration of a nearly closed (isovolume) system in which the cranial window is covered with agarose and a coverglass; then pressure fluctuations do not lead to motion (Kleinfeld and Denk, 1999). If pipette access is not required, imaging through the thinned skull has been found to be possible and satisfactory (Christie et al., 2001; Grutzendler et al., 2002). Even with these measures, pulsation-induced movements can remain. Although some lateral motion can still be tolerated in line-scanning measurements or may be corrected for off-line, movements out of the focal plane are more difficult to correct for. A possible approach to stabilize morphological images against heartbeat-induced motion is to trigger the acquisition of each frame on the electrocardiogram (Helmchen, unpublished results).

Avoiding Photodamage

Multiphoton excitation increases with increased photon density in space and time—that is, when, for a given power level, either focal volume or laser pulse duration is reduced. This is limited under ideal conditions by ground-state depletion, but in living tissue, the possible amount of excitation is limited in fact by the onset of photodamage. Under high-NA illumination and prolonged scanning, power thresholds for photodamage have been determined using as criteria the degranulation of chromaffin cells (Hopt and Neher, 2001), morphological changes (Koester et al., 1999), changes in basal fluorescence (Koester et al., 1999), and intracellular calcium increase (König et al., 1999; Oehring et al., 2000; Hopt and Neher, 2001; Tirlapur et al., 2001). Photodamage depended supralinearly on laser intensity in all cases, with an exponent that sometimes appeared to be above that for the excitation. This suggests that photodamage is caused by parasitic higher order multiphoton processes. Aside from a direct photophysical effect, such as excited-state absorption, this damage may be caused by cooperativity of damage mechanisms or the saturation of protective mechanisms. In either case, the practical conclusion is that increasing the excitation rate is likely to increase the damage per excitation event.

What are concrete measures to be taken if photodamage is known or suspected to affect the experimental results? Aside from using as little light as possible, one might try increasing either focus size (reducing NA) or pulse width.

Multiphoton Photochemistry

Better Stimulating through Photochemistry

In fluorescence imaging, illumination-induced chemical change is an undesired side effect. However, in conjunction with the spatial localization of MPE, photochemical effects can be turned into a tool for controlled local *perturbation*. The simplest case is photobleaching recovery (Axelrod et al., 1976), which is applicable to virtually all fluorophores, for the measurement of transport processes. More specific is the photoconversion of purposely designed "caged" compounds, which can be used to induce controlled biochemical change (Lesterand Chang, 1977; Kaplan et al., 1978; Adams and Tsien, 1993; Wang and Augustine, 1995; Pettit et al., 1997; Furuta et al., 1999; Matsuzaki et al., 2001; Callaway and Yuste, 2002). Like fluorescence, photochemical reactions can be driven by single- or multiphoton excitation. However, only in the multiphoton case is spatial localization to subfemtoliter volumes possible.

Photobleaching

Most fluorophores bleach, i.e., they are photochemically converted to a nonfluorescent species. After bleaching by a brief increase in excitation intensity, fluorescence can recover by diffusion of intact molecules back into the bleached volume.

Uncaging

Caged compounds are photochemically labile precursors of biological signaling molecules (McCray and Trentham, 1989; Adams and Tsien, 1993). Photolysis can induce concentration jumps in less than a millisecond (Ellis-Davies, 1999) and, in the case of multiphoton uncaging, in a volume of less than 1 fL (Denk, 1994). Useful biological molecules that have been caged include neurotransmitters (Ramesh et al., 1993; Wieboldt et al., 1994; Papageorgiou et al., 1999; Matsuzaki et al., 2001), cyclic nucleotides and other phosphate-containing second messengers (Walker et al., 1989; Aarhus et al., 1995), calcium chelators (Tsien and Zucker, 1986; Kaplan and Ellis-Davies, 1988; Ellis-Davies and Kaplan, 1994; Adams et al., 1997; DelPrincipe et al., 1999a), and peptides (Sreekumar et al., 1998; Shigeri et al., 2001).

A caged compound is usable only if significant amounts can be photolyzed at illumination levels that cause little or no collateral damage (Lester and Nerbonne, 1982). Traditional caging groups typically have much lower multiphoton cross-sections (0.01 GM at 700–800 nm [Zipfel et al., 1996]) than traditional fluorophores (up to 200 GM for rhodamine 123 at 840 nm [Xu and Webb, 1996]), putting older caged compounds at the bare threshold of usability forMPE (Denk, 1994; Svoboda et al., 1996). Two-photon cross-sections have been measured for the calcium cages azid-1 (Brown et al., 1999a), DM-nitrophen (Lipp and Niggli, 1998; Brown et al., 1999a), and DMNPE-4 (DelPrincipe et al., 1999a). Recently, several promising caging groups have been synthesized that show larger two-photon cross-sections based on coumarin (0.9–1.0 GM at 740 nm [Furuta et al., 1999]) and nitroindoline (0.06 GM at 730 nm [Papageorgiou et al., 1999; Matsuzaki et al., 2001]) moieties. The improvement needed to reach routinely acceptable cross-sections may not be that great because, unlike fluorescent molecules, which are excited many times during an experiment, cage groups only need to be photolyzed once.

At intense levels of illumination, MPE can evoke action potentials directly, even in the absence of caged compound (Callaway and Yuste, 2002; Hirase et al., 2002). Such direct activation may be useful for generating action potentials in cases when glutamate receptor activation is undesirable or high illumination levels are not a concern.

Because the time course of measured responses is limited by biophysical processes such as electrotonic filtering properties of the responding cell, channel gating, and second messengerdynamics, pixel dwell times foruncaging experiments have to be much longer than for fluorescence imaging. To avoid memory effects resulting from local accumulation of liberated agonist, raster scanning can be complemented by random access scanning (Denk, 1994; Matsuzaki et al., 2001).

Dual Caging Group Chemical (Two-Photon) Uncaging

An alternate means of achieving good localization of release is not to rely on nonlinear optics but instead to rely on nonlinear chemistry by designing a molecule that generates one active signaling molecule only aftermultiple one-photon uncaging reactions (Adams and

Tsien, 1993; Pettit et al., 1997). This approach, which Pettit et al. (1997) called *chemical two-photon uncaging*, exploits the fact that some signaling molecules have multiple functional groups that can be individually protected. By this criterion, candidates for multiple caging include glutamate, γ-aminobutyric acid, caged calcium, and inositol trisphosphate. Chemical multiphoton uncaging needs ultraviolet light, with increased scattering and absorption, and hence reduced tissue penetration. Furthermore, because individual uncaging reactions are irreversible, partially photolyzed compound can accumulate and eventually compromise spatial localization. However, within these constraints, this use of multiple "logical-AND" caging groups leads to an elevation of concentration proportional to the light dose to the nth powerand hence to localization similarto that of true multiphoton absorption.

APPLICATIONS

Fluorescence Imaging

In this section we give examples of imaging applications sorted by tissue preparation procedure, with the least invasive procedures last. This roughly corresponds to the historical order of development and illustrates the trend during the past decade toward studying intact tissue. The first biological experiments using MPM were performed on isolated cells (Denk et al., 1990; Piston et al., 1994), but applications quickly moved on to tissues that were extracted but still living, including auditory epithelium, retina, and acutely cut brain slices. In all cases, MPM enabled high-resolution imaging of submicrometer structures such as stereocilia and individual central synapses.

Auditory Epithelium, the Olfactory Bulb, and Retina

Two-photon microscopy has been applied to several preparations of sensory neurons to study sensory transduction, specificity, and preprocessing. In hair cells of the vestibular system, calcium influx through transduction channels into the stereocilia was measured (Denk et al., 1995a). The precision of axonal projections from sensory neurons to the olfactory bulb has been mapped accurately using transgenic expression of GFP specific to specific olfactory receptors (Potter et al., 2001). In the antennal lobe of *Drosophila,* brain odor-evoked maps of glomerular activity could be resolved using cell-specific expression of a calcium-sensitive fluorescent protein (Wang et al., 2003).

A very rewarding specimen for MPM is the retina. Because the retina has single-photon sensitivity in the 400- to 700-nm range, conventional excitation of fluorescent calcium indicators causes unwanted activation of the phototransduction cascade to the point of completely bleaching the photopigments. However, photoreceptors are blind to near-infrared wavelengths, allowing the use of MPM to examine elements in otherfocal planes, such as ganglion or amacrine cell dendrites, in effective darkness (Denk and Detwiler, 1999; Euler et al., 2001, 2002).

Optophysiology in Brain Slices

Optical approaches to studying dendritic integration of synaptic inputs were pioneered in brain slice preparations using wide-field imaging. Neurons were loaded with a fluorescent indicator and imaged using cooled CCD cameras (Stockbridge and Ross, 1984; Connor, 1986; Tank et al., 1988) or laser-scanning confocal microscopes (Alford et al., 1993; Jaffe et al., 1994; forr eviews see Regehrand Tank, 1994; Denk et al., 1996; Yuste and Tank, 1996; Euler and Denk, 2001). However, in brain slices these approaches suffer from poor spatial resolution or are essentially restricted to the slice surface. Two-photon microscopy has made feasible high-resolution imaging of subcellular phenomena deep in brain slice preparations (Denk et al., 1995c; Yuste and Denk, 1995; for reviews see Denk and Svoboda,

1997; Mainen et al., 1999). Most important, optical studies of individual synapses in the central nervous system became possible. Synaptic contacts between central nervous system neurons are small, specialized structures about 1 μm in size. In most excitatory synapses, presynaptic terminals (*boutons*) make contact with protrusions (*spines*) from the postsynaptic dendrites (fig. 3.5A,B). Dendritic spines have long been implicated as sites of fast synaptic processing, second messenger accumulation, and long-term synaptic plasticity (Koch et al., 1992; Harris and Kater, 1994; Shepherd, 1996). In boutons and spines alike, two-photon microscopy has brought new insights to both signaling and structural dynamics. Presynaptically, calcium transients have been measured in single boutons of neocortical pyramidal neurons (Cox et al., 2000; Koester and Sakmann, 2000) and cerebellar basket cells (Tan and Llano, 1999), demonstrating that action potentials reliably invade axonal arbors.

Dendritic Spine Physiology   Current research on dendritic spines using MPM is focused on characterizing calcium signaling mechanisms (for reviews see Denk et al., 1996; Yuste et al., 2000; Sabatini et al., 2001). The major outcome of this research is the finding that dendritic spines can serve as individual biochemical compartments (fig. 3.5C) for calcium and probably othersecond messengers (Yuste and Denk, 1995; Svoboda et al., 1996; Sabatini et al., 2002). With careful experimental design, MPM permits the estimation of calcium channel numberin spines by means of optical fluctuation analysis (Sabatini and Svoboda, 2000). Single-spine calcium signals have been quantified in hippocampal, neocortical, and cerebellar neurons in response to coincident presynaptic and postsynaptic action potentials. These signals are supralinear relative to the sum of separate presynaptically and postsynaptically evoked signals (Yuste and Denk, 1995; Koesterand Sakmann, 1998; Yuste et al., 1999b; Wang et al., 2000a) (fig. 3.5C). Because calcium is a critical trigger of synaptic plasticity, these signals provide a substrate for associative coincidence detection in the central nervous system (Hebb, 1949; Linden, 1999).

Morphological Plasticity   Structural change and stability are of great interest in neurobiology. The shape of dendritic spines can be dynamic under certain circumstances (for reviews see Jontes and Smith, 2000; Matus, 2000; Yuste and Bonhoeffer, 2001), and respond to synaptic activity and neurotrophic factors (Dunaevsky et al., 1999; Engert and Bonhoeffer, 1999; Maletic-Savatic et al., 1999; Horch and Katz, 2002) (fig. 3.5D). These morphological changes may modulate the compartment properties of spines (Svoboda et al., 1996; Majewska et al., 2000), and suggest that synaptic contacts are themselves fungible and that neuronal connections may be continually rewired in response to experience and environmental change.

Embryogenesis

In developing embryos, optical monitoring can be done nondestructively for days using MPM, with such detail that individual cell fate can be followed (Mohler and Squirrell, 1999). The viability of mammalian embryos is improved dramatically by using two-photon instead of confocal microscopy, thus permitting long-term observations without compromising embryonic development (Squirrell et al., 1999).

Imaging in Living Animals

High-resolution optical microscopy is the most promising technology for revealing the role of subcellular neuronal processing in the intact brain. This was demonstrated in pioneering experiments in insects in which dendritic calcium signals evoked by sensory input were measured with CCD imaging (Borst and Egelhaaf, 1994; Sobel and Tank, 1994). The improved depth penetration of MPM allows such experiments to be done in thicker structures

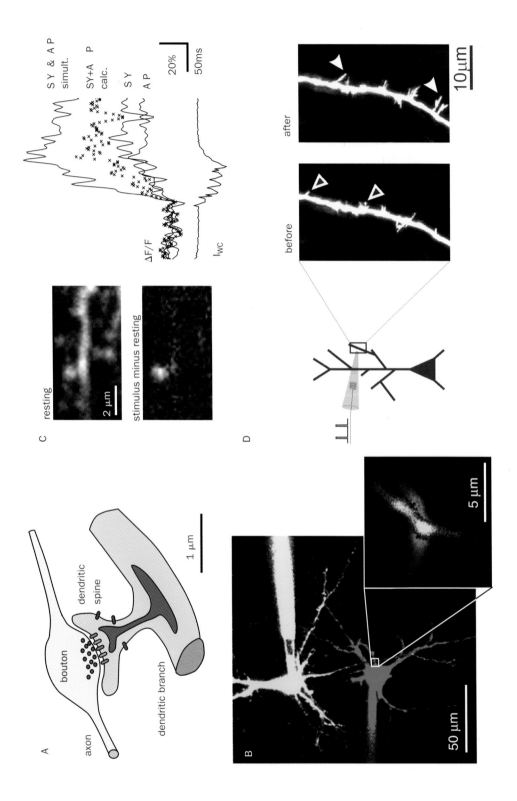

A

axon
bouton
dendritic spine
dendritic branch

1 μm

B

50 μm

5 μm

C

resting

2 μm

stimulus minus resting

ΔF/F

SY & A P
simult.

SY+A P
calc.

S Y

A P

I_WC

20%

50ms

D

before

after

10μm

and with higherr esolution, allowing questions of in vivo dynamics to be asked within the intact mammalian brain (Fig. 3.6 [Svoboda et al., 1999]).

Dendritic Excitation in Vivo   The first in vivo two-photon images of neocortical neurons in anesthetized rats were obtained using negative stain contrast obtained by soaking the cortical surface with a fluorescent dye (Denk et al., 1994). Svoboda et al. (1997) then obtained two-photon excited fluorescence images of layer 2/3 neurons individually injected with a calcium indicator. These experiments have opened the way to the measurement of sensory-evoked dendritic calcium transients in the mammalian brain in vivo (Helmchen et al., 1999; Svoboda et al., 1999; Charpak et al., 2001) (fig. 3.6). In addition, labeling of cell populations with AM-ester calcium indicators (Stosiek et al., 2003) enables in vivo measurement of neuronal network activity complementing the single-cell experiments.

Structural Dynamics in Vivo   With XFPs orotherstaining   methods that label entire cells, morphology can be tracked using four-dimensional imaging. This approach has been used to study structural dynamics of synaptic contact formation in zebrafish (Jontes et al., 2000) and of spines on virally infected pyramidal neuron dendrites in vivo (Lendvai et al., 2000). Of particular interest is the transgenic expression of XFPs under a specific promoter, which allows the labeling of all neurons sharing a common molecular marker (Ellenberg et al., 1999; Feng et al., 2000; Tucker, 2000). Structural changes can be observed even on a relatively long timescale (days to month) by repeated imaging either through a chronically implanted window or through the thinned skull. Using a mouse model of Alzheimer's disease, Christie et al. (2001) were the first to use this approach to monitor plaque size by tracking individual thioflavin S-stained senile plaques for up to 3 months. More recently, the stability of dendritic branches and spines has been investigated in the adult neocortex using two-photon imaging in transgenic mice expressing fluorescent protein in pyramidal cells (Grutzendler et al., 2002; Trachtenberg et al., 2002).

Cortical Blood Flow

Unstained red blood cells (RBCs) can be visualized by negative staining with fluorescent dye, thus appearing as moving dark shadows (Dirnagl et al., 1992). This allows the measurement of flow parameters such as RBC speed and density in individual capillaries (Kleinfeld et al., 1998) (fig. 3.6). These parameters are the underlying source of contrast in intrinsic optical and functional magnetic resonance imaging.

Figure 3.5 (facing page) Multiphoton study of dendritic spine physiology. (A) Schematic of a glutamatergic central nervous system (CNS) synapse with a presynaptic axonal bouton contacting a postsynaptic dendritic spine. (B) Dual-channel two-photon image of a CNS synapse in a neocortical slice. A dual whole-cell recording was made from a synaptically coupled pair of cortical neurons. The pyramidal neuron (pseudocolor yellow, filled with a calcium indicator) was presynaptic to the inverted pyramid (blue, filled with an Alexa dye). One of the presumed synaptic contacts is enlarged (unpublished data, kindly provided by H. Koester and B. Sakmann). (C) Dendritic spines as biochemical compartments and coincidence detectors. (Left) Subthreshold synaptic activation induces calcium accumulation restricted to an individual dendritic spine of a hippocampal CA1 pyramidal neuron filled with Calcium Green-1. (Right) Pairing of synaptic stimulation (SY) with postsynaptic action potentials (AP) produces supralinear calcium accumulations in spines. The response to the simultaneous combination (SY+AP simult.) is larger than the response to the calculated sum of the response to each individual stimulus (SY+AP calc.). (From Yuste and Denk[1995].) (D) Synaptically induced structural changes. CA1 pyramidal dendrites in an organotypic slice labeled with enhanced GFP by viral transfection. Focal electrical stimulation of a dendrite with a glass pipette produced growth of spinelike filopodia (solid white arrows) and the disappearance of other structures (open arrows). (From Maletic-Savatic et al. [1999].)

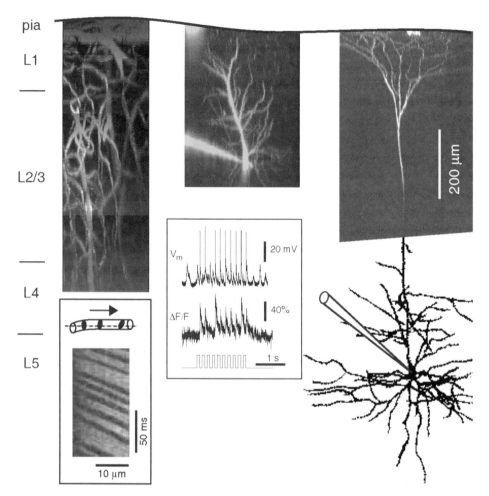

Figure 3.6 In vivo imaging in neocortex. Side projections calculated from image stacks of cortical vasculature (left), a layer 2/3 pyramidal neuron (middle), and the distal tuft of a layer 5 pyramidal neuron (right). Cortical microvasculature was stained by tail vein injection of a dextran-conjugated fluorescent dye into the blood plasma. (From Kleinfeld et al. [1998].) The left inset shows a line scan along an individual capillary with the unstained flowing blood cells appearing as darkstripes. Blood cell speed can be determined from the slopes of the stripes. Pyramidal neurons were filled with Calcium Green-1 via a somatic microelectrode. The right inset shows calcium transients ($\Delta F/F$) evoked by action potentials (membrane potential V_m) during sensory stimulation (whisker deflections). Note that imaging is limited to about 500 to 600 μm in depth. The soma and proximal part of the layer 5 neuron therefore were too deep to be imaged but are overlain from the histologic reconstruction. (Neuron images from Svoboda et al. [1999] and Helmchen et al. [1999], respectively).

Photochemistry

Intracellular Transport

Diffusional transport has been monitored using both multiphoton fluorescence recovery afterphotobleaching (FRAP) and multiphoton uncaging. Multiphoton FRAP in mammalian cytoplasm gives translational diffusion coefficients in agreement with previous estimates (Brown et al., 1999b). Applied to dendrites, multiphoton FRAP of fluorescein dextran in dendritic spines has been used to measure diffusion through spine necks (Svoboda et al., 1996) and thereby to show that spine heads are indeed biochemical compartments (Koch and

Zador, 1993). Changes in the degree of diffusional coupling between spine head and dendritic shaft have been observed directly using multiphoton FRAP (Majewska et al., 2000). Multiphoton FRAP has also been used to monitorpr otein exchange between adjoining plant plastids (Kohleret al., 1997).

Spatiotemporal Dynamics of Phototransduction

When the photoreceptors themselves are placed at the focal point, MPE can be used to generate very localized isomerization events, allowing the study of how biochemical excitation spreads within photoreceptors (Gray-Keller et al., 1999).

Receptor Mapping

The functional density of neurotransmitter receptors has been mapped with multiphoton uncaging using cholinergic (Denk, 1994) and glutamatergic (Furuta et al., 1999; Matsuzaki et al., 2001) agonists, recording as a contrast signal the whole-cell current. Glutamate receptor responsiveness has also been monitored over time on dendrites and neuronal somata using one-photon (Dodt et al., 1999) and chemical two-photon (Pettit et al., 1997; Wang et al., 2000a) uncaging. Chemical two-photon uncaging has been used to show that in cerebellar Purkinje neurons, long-term depression leads to widespread decreases in receptor sensitivity (Wang et al., 2000b). An important recent advance is the successful mapping of AMPA (alpha-amino-3-hydroxy-5-methyl-4-isoxazolepropionic acid) -type glutamate receptor sensitivity in single dendritic spines with true two-photon uncaging (Matsuzaki et al., 2001).

Caged Calcium

For the study of intracellular signaling, the principal contribution of multiphoton uncaging has been in the study of calcium release in cardiac myocytes. There, multiphoton uncaging of DM-nitrophen has been used to trigger local calcium-activated calcium release from one or a few ryanodine receptor channels (Lipp and Niggli, 1998). Calcium release triggered by multiphoton uncaging of DM-nitrophen is not followed by a refractory period, a contrast with previous studies using one-photon flash photolysis (DelPrincipe et al., 1999a).

PERSPECTIVES

Applications of MPM to neuroscience are developing quickly, driven both by the commercial availability of turnkey MPM systems and a stream of technical advances. In particular, further increases in depth penetration and novel indicators based on fluorescent proteins are likely to facilitate in vivo applications further. The following subsections describe some of the areas that we see as particularly promising.

Biochemical Variables Other Than Calcium

In addition to calcium, for which mature indicators and almost two decades of experience exist, otherions and small molecules including sodium, chloride, and cyclic adenosine monophosphate (cAMP) find increased interest as the quest for a broad understanding of biochemical signal dynamics in neurons continues. Multiphoton microscopy brings the same advantages to fluorescent indicators of other signaling molecules as it does to calcium indicators (for example, see Rose and Konnerth, 2001; Marandi et al., 2002).

Protein Trafficking

Mutant fluorescent proteins can be fused to various proteins while leaving their trafficking steps undisrupted. This approach has been used, for example, to track the delivery of glutamate receptors into spines after strong synaptic activation (Shi et al., 1999, 2001). In the future, MPM will allow such measurements to be done in intact brain under more natural stimulation conditions.

From Single Neurons to Networks

Another future target of MPM is the monitoring of activity in populations of neurons (fig. 3.1C). Optical multiunit imaging has relied on (1) intrinsic signals (Grinvald et al., 1999) orvoltage-sensitive dye signals (Wu et al., 1999), both of which show relatively small changes in signal, leading to a need forheavy averaging overspace, time, ormultiple trials; or on (2) bulk loading of calcium indicators from esterified precursors, a process that is unreliable in intact tissue. The optical sectioning provided by MPE diminishes some of the background problems with these approaches, but our hope rests on protein-based indicators, which in addition should provide neuron-specific labeling in defined neuronal populations. As this approach matures, it may—in combination with MPM—become a more perfect tool to study neural activity on almost all relevant length and timescales.

To achieve the necessary temporal resolution in sparsely labeled cell populations, new scanning modalities may need to be developed. Conceivable are high-frequency resonant galvanometerscanner s forfast image acquisition rates (Fan et al., 1999) orthe use of acousto-optical deflectors (in combination with picosecond pulses or special optics to minimize dispersive effects), which allows fast scanning of arbitrary patterns.

Freely Moving Animals

So far, in vivo subcellular imaging has been restricted to nonmoving (usually anesthetized) animals, because existing MPM systems are physically bulky. Extending such measurements to awake, behaving animals is obviously of particular interest to neuroscientists, who (after all) study the system that controls behavior. A major step in this direction is the development of a miniaturized two-photon microscope (Helmchen et al., 2001), in which the two-photon excitation light is introduced through an optical fiber and scanning is performed using resonant vibration of the fiber end. The head-mounted part of this "fiberscope" weighs about 25 g and can be easily carried by an adult rat.

Photostimulation

Improved absorption and quantum yield may eventually bring the cross-sections of caging groups close to the performance of fluorescence groups. Many isomerization reactions can be driven in both directions with different wavelengths of light. This raises the possibility, lately neglected, of reversible photoactivation (Lester and Nerbonne, 1982; Adams and Tsien, 1993).

Finally, a particularly enticing possibility is the coupling of light-sensitive proteins such as rhodopsin to effectorion channels (Deiningeret al., 1995). Such coupling has already been done for another G protein-coupled receptor, the allatostatin receptor (Lechner et al., 2002). This approach combines the advantages of genetic encoding of probes with the light-controlled activation of tissue.

Note added in proof: This chapter reflects the status of the research field in 2003. The reader is referred to several recent reviews for an account of the newest applications of multi-photon microscopy in neuroscience (Helmchen and Denk, 2005; Svoboda and Yasuda, 2006; Kerr and Denk, 2008).

REFERENCES

Aarhus, R., Gee, K., and Lee, H. C. (1995) Caged cyclic ADP-ribose: Synthesis and use. *Journal of Biological Chemistry* **270**: 7745–7749.

Adams, S. R., Lev-Ram, V., and Tsien, R. Y. (1997) A new caged Ca^{2+}, azid-1, is far more photosensitive than nitrobenzyl-based chelators. *Chemical Biology* **4**: 867–878.

Adams, S. R., and Tsien, R. Y. (1993) Controlling cell chemistry with caged compounds. *Annual Review of Physiology* **55**: 755–784.

Alford, S., Frenguelli, B. G., Schofield, J. G., and Collingridge, G. L. (1993) Characterization of Ca^{2+} signals induced in hippocampal CA1 neurones by the synaptic activation of NMDA receptors. *Journal of Physiology (London)* **469**: 693–716.

Axelrod, D., Koppel, D. E., Schlessinger, J., Elson, E., and Webb, W. W. (1976) Mobility measurement by analysis of fluorescence photobleaching recovery kinetics. *Biophysical Journal* **16**: 1055–1069.

Beaurepaire, E., and Mertz, J. (2002) Epifluorescence collection in two-photon microscopy. *Applied Optics* **41**: 5376–5382.

Beaurepaire, E., Oheim, M., and Mertz, J. (2001) Ultra-deep two-photon fluorescence excitation in turbid media. *Optics Communications* **188**: 25–29.

Birge, R. R. (1986) 2-Photon spectroscopy of protein-bound chromophores. *Accounts of Chemical Research* **19**: 138–146.

Booth, M. J., and Hell S. W. (1998) Continuous wave excitation two-photon fluorescence microscopy exemplified with the 647-nm ArKrlaserline. *Journal of Microscopy-Oxford* **190**: 298–304.

Borst, A., and Egelhaaf, M. (1994) Dendritic processing of synaptic information by sensory interneurons. *Trends in Neurosciences* **17**: 257–263.

Braendgaard, H., Evans, S. M., Howard, C. V., and Gundersen, H. J. (1990) The total number of neurons in the human neocortex unbiasedly estimated using optical dissectors. *Journal of Microscopy* **157**: 285–304.

Brown, E., Shear, J. B., Adams, S. R., Tsien, R. Y., and Webb, W. W. (1999a) Photolysis of caged calcium in femtoliter volumes using two-photon excitation. *Biophysical Journal* **76**: 489–499.

Brown, E. B., Wu, E. S., Zipfel, W., and Webb, W. W. (1999b) Measurement of molecular diffusion in solution by multiphoton fluorescence photobleaching recovery. *Biophysical Journal* **77**: 2837–2849.

Callaway, E. M., and Yuste, R. (2002) Stimulating neurons with light. *Current Opinion in Neurobiology* **12**: 587–592.

Centonze, V. E., and White, J. G. (1998) Multiphoton excitation provides optical sections from deeper within scattering specimens than confocal imaging. *Biophysical Journal* **75**: 2015–2024.

Chalfie, M., Tu, Y., Euskirchen, G., Ward, W. W., and Prasher, D. C. (1994) Green fluorescent protein as a marker for gene expression. *Science* **263**: 802–805.

Charpak, S., Mertz, J., Beaurepaire, E., Moreaux, L., and Delaney, K. (2001) Odor-evoked calcium signals in dendrites of rat mitral cells. *Proceedings of the National Academy of Sciences USA* **98**: 1230–1234.

Christie, R. H., Bacskai, B. J., Zipfel, W. R., Williams, R. M., Kajdasz, S. T., Webb, W. W., and Hyman, B. T. (2001) Growth arrest of individual senile plaques in a model of Alzheimer's disease observed by in vivo multiphoton microscopy. *Journal of Neuroscience* **21**: 858–864.

Connor, J. A. (1986) Digital imaging of free calcium changes and of spatial gradients in growing processes in single, mammalian central nervous system cells. *Proceedings of the National Academy of Sciences USA* **83**: 6179–6183.

Cox, C. L., Denk, W., Tank, D. W., and Svoboda, K. (2000) Action potentials reliably invade axonal arbors of rat neocortical neurons. *Proceedings of the National Academy of Sciences USA* **97**: 9724–9728.

Deininger, W., Kroger, P., Hegemann, U., Lottspeich, F., and Hegemann, P. (1995) Chlamyrhodopsin represents a new type of sensory photoreceptor. *EMBO Journal* **14**: 5849–5858.

Delaney, K., Davison, I., and Denk, W. (2001) Odour-evoked $[Ca^{2+}]$ transients in mitral cell dendrites of frog olfactory glomeruli. *European Journal of Neuroscience* **13**: 1658–1672.

DelPrincipe, F., Egger, M., Ellis-Davies, G. C. R., and Niggli, E. (1999a) Two-photon and UV-laser flash photolysis of the Ca^{2+} cage, dimethoxynitrophenyl-EGTA-4. *Cell Calcium* **25**: 85–91.

DelPrincipe, F., Egger, M., and Niggli, E. (1999b) Calcium signalling in cardiac muscle: Refractoriness revealed by coherent activation. *Nature Cell Biology* **1**: 323–329.

Denk, W. (1994) Two-photon scanning photochemical microscopy: Mapping ligand-gated ion channel distributions. *Proceedings of the National Academy of Sciences USA* **91**: 6629–6633.

Denk, W. (1996) Two-photon excitation in functional biological imaging. *Journal ofBiomedical Imaging* **1**: 296–304.

Denk, W., Delaney, K. R., Gelperin, A., Kleinfeld, D., Strowbridge, B. W., Tank, D. W., and Yuste, R. (1994) Anatomical and functional imaging of neurons using 2-photon laser scanning microscopy. *Journal ofNeuroscience Methods* **54**: 151–162.

Denk, W., and Detwiler, P. B. (1999) Optical recording of light-evoked calcium signals in the functionally intact retina. *Proceedings ofthe National Academy ofSciences USA* **96**: 7035–7040.

Denk, W., Holt, J. R., Shepherd, G. M., and Corey, D. P. (1995a) Calcium imaging of single stereocilia in hair cells: Localization of transduction channels at both ends of tip links. *Neuron* **15**: 1311–1321.

Denk, W., Piston, D. W., and Webb, W. W. (1995b) Two-photon molecular excitation in laser-scanning microscopy. In: *Handbook ofBiological Confocal Microscopy* (ed. Pawley, J. B.), pp. 445–458. Plenum Press, New York.

Denk, W., Strickler, J. H., and Webb, W. W. (1990) Two-photon laser scanning fluorescence microscopy. *Science* **248**: 73–76.

Denk, W., Sugimori, M., and Llinas, R. (1995c) Two types of calcium response limited to single spines in cerebellar Purkinje cells. *Proceedings ofthe National Academy ofSciences USA* **92**: 8279–8282.

Denk, W., and Svoboda, K. (1997) Photon upmanship: Why multiphoton imaging is more than a gimmick. *Neuron* **18**: 351–357.

Denk, W., Yuste, R., Svoboda, K., and Tank, D. W. (1996) Imaging calcium dynamics in dendritic spines. *Current Opinion in Neurobiology* **6**: 372–378.

Dingledine, R. (1984) *Brain Slices.* Plenum Press, New York.

Dirnagl, U., Villringer, A., and Einhaupl, K. M. (1992) In-vivo confocal scanning laser microscopy of the cerebral microcirculation. *Journal ofMicroscopy* **165**: 147–157.

Dodt, H., Eder, M., Frick, A., and Zieglgänsberger, W. (1999) Precisely localized LTD in the neocortex revealed by infrared-guided laser stimulation. *Science* **286**: 110–113.

Dunaevsky, A., Tashiro, A., Majewska, A., Mason, C., and Yuste, R. (1999) Developmental regulation of spine motility in the mammalian central nervous system. *Proceedings ofthe National Academy ofSciences USA* **96**: 13438–13443.

Ellenberg, J., Lippincott-Schwartz, J., and Presley, J. F. (1999) Dual-color imaging with GFP variants. *Trends in Cell Biology* **9**: 52–56.

Ellis-Davies, G. C. R. (1999) Basics of photoactivation. In: *Imaging Neurons* (eds. Yuste, R., Lanni, F., and Konnerth, A.), pp. 24.1–24.8. Cold Spring Harbor Press, Cold Spring Harbor, Maine.

Ellis-Davies, G. C. R., and Kaplan, J. H. (1994) Nitrophenyl-EGTA, a photolabile chelator that selectively binds Ca^{2+} with high affinity and releases it rapidly upon photolysis. *Proceedings ofthe National Academy ofSciences USA* **91**: 187–191.

Engert, F., and Bonhoeffer, T. (1999) Dendritic spine changes associated with hippocampal long-term synaptic plasticity. *Nature* **399**: 66–70.

Euler, T., and Denk, W. (2001) Dendritic processing. *Current Opinion in Neurobiology* **11**: 415–422.

Euler T., Detwiler P. B., and Denk, W. (2002) Directionally selective calcium signals in dendrites of starburst amacrine cells. *Nature* **418**: 845–852.

Euler, T., He, S., Masland, R. H. and Denk, W. (2001) Light-evoked calcium signals in on/off direction-selective ganglion cells in the rabbit retina. *Society ofNeuroscience Abstracts.* **72**: 397.4.

Fan, G. Y., Fujisaki, H., Miyakawi, A., Tsay, R.- K., Tsien, R. Y., and Ellisman, M. H. (1999) Video-rate scanning two-photon excitation fluorescence microscopy and ratio imaging with chameleons. *Biophysical Journal* **76**: 2412–2420.

Feng, G., Mellor, R. H., Bernstein, M., Keller-Peck, C., Nguyen, Q. T., Wallace, M., Nerbonne, J. M., Lichtman, J. W., and Sanes, J. R. (2000) Imaging neuronal subsets in transgenic mice expressing multiple spectral variants of GFP. *Neuron* **28**: 41–51.

Furuta, T., Wang, S. S.- H., Dantzker, J. L., Dore, T. M., Bybee, W. J., Callaway, E. M., Denk, W., and Tsien, R. Y. (1999) Brominated 7-hydroxycoumarin-4-ylmethyls: Photolabile protecting groups with biologically useful cross-sections fortwo photon photolysis. *Proceedings ofthe National Academy ofSciences USA* **96**: 1193–1200.

Gan, W. B., Grutzendler, J., Wong, W. T., Wong, R. O., and Lichtman, J. W. (2000) Multicolor "DiOlistic" labeling of the nervous system using lipophilic dye combinations. *Neuron* **27**: 219–225.

Garaschuk, O., Linn, J., Eilers, J., and Konnerth, A. (2000) Large-scale oscillatory calcium waves in the immature cortex. *Nature Neuroscience* **3**: 452–459.

Gelperin, A., and Flores, J. (1997) Vital staining from dye-coated microprobes identifies new olfactory interneurons for optical and electrical recording. *Journal ofNeuroscience Methods* **72**: 97–108.

Gibb, A. J., and Edwards, F. A. (1994) Patch clamp recording from cells in sliced tissue. In: *Microelectrode Techniques: The Plymouth Workshop Handbook* (ed., Ogden, D.), pp. 255–274. The Company of Biologists Limited, Cambridge.

Gosnell, T. R., and Taylor, A. J. (1991) *Selected Papers on Ultrafast Laser Technology.* SPIE Optical Engineering Press, Bellingham.

Gray-Keller, M., Denk, W., Shraiman, B., and Detwiler, P. B. (1999) Longitudinal spread of second messenger signals in isolated rod outer segments of lizards. *Journal of Physiology (London)* **519**: 679–692.

Grinvald, A., Shmuel, A., Vanzetta, I., Shtoyerman, E., Shoham, D., and Arieli, A. (1999) Intrinsic signal imaging in the neocortex. In: *Imaging Neurons* (eds., Yuste, R., Lanni, F., and Konnerth, A.), pp. 45.1–45.17. Cold Spring Harbor Press, Cold Spring Harbor, Maine.

Grutzendler, J., Kasthuri, N., and Gan, W.-B. (2002) Long-term dendritic spine stability in the adult cortex. *Nature* **420**: 812–816.

Haas, K., Sin, W.-C., Javaherian, A., Li, Z., and Cline, H. T. (2001) Single-cell electroporation for gene transfer in vivo. *Neuron* **29**: 583–591.

Harris, K. M., and Kater, S. B. (1994) Dendritic spines: Cellular specializations imparting both stability and flexibility to synaptic function. *Annual Reviews of Neuroscience* **17**: 341–371.

Hebb, D. O. (1949) *Organization of Behavior: A Neuropsychological Theory.* Wiley, Hoboken, New Jersey.

Helmchen, F., and Denk, W. (2005) Deep tissue two-photon microscopy. *Nature Methods* **2**: 932–940.

Helmchen, F., and Denk, W. (2002) New developments in multiphoton microscopy. *Current Opinion in Neurobiology* **12**: 593–601.

Helmchen, F., Fee, M. S., Tank, D. W., and Denk, W. (2001) A miniature head-mounted two-photon microscope: High-resolution brain imaging in freely moving animals. *Neuron* **31**: 903–912.

Helmchen, F., Svoboda, K., Denk, W., and Tank, D. W. (1999) In vivo dendritic calcium dynamics in deep-layer cortical pyramidal neurons. *Nature Neuroscience* **2**: 989–996.

Hernandez-Cruz, A., Sala, F., and Adams, P. R. (1990) Subcellular calcium transients visualized by confocal microscopy in a voltage-clamped vertebrate neuron. *Science* **247**: 858–862.

Hirase, H., Nikolenko, V., Goldberg, J. H., and Yuste, R. (2002) Multiphoton stimulation of neurons. *Journal of Neurobiology* **51**: 237–247.

Hopt, A., and Neher, E. (2001) Highly nonlinear photodamage in two-photon fluorescence microscopy. *Biophysical Journal* **80**: 2029–2036.

Horch, H. W., and Katz, L. C. (2002) BDNF release from single cells elicits local dendritic growth in nearby neurons. *Nature Neuroscience* **5**: 1177–1184.

Jaffe, D. B., Fisher, S. A., and Brown, T. H. (1994) Confocal laser scanning microscopy reveals voltage-gated calcium signals within hippocampal dendritic spines. *Journal of Neurobiology* **25**: 220–233.

Jontes, J. D., Buchanan, J., and Smith, S. J. (2000) Growth cone and dendrite dynamics in zebrafish embryos: Early events in synaptogenesis imaged in vivo. *Nature Neuroscience* **3**: 231–237.

Jontes, J. D., and Smith, S. J. (2000) Filopodia, spines, and the generation of synaptic diversity. *Neuron* **27**: 11–14.

Kaplan, J. H., and Ellis-Davies, G. C. R. (1988) Photolabile chelators for the rapid photorelease of divalent cations. *Proceedings of the National Academy of Sciences USA* **85**: 6571–6575.

Kaplan, J. H., Forbush, B., 3rd, and Hoffman, J. F. (1978) Rapid photolytic release of adenosine 5′-triphosphate from a protected analogue: Utilization by the Na:K pump of human red blood cell ghosts. *Biochemistry* **17**: 1929–1935.

Kerr, J. N., and Denk, W. (2008) Imaging in vivo: Watching the brain in action. *Nature Reviews Neuroscience* **9**: 195–205.

Kettunen, P., Dema, J., Lohmann, C., Kasthuri, N., Gong, Y., Wong, R. O., and Gan, W. B. (2002) Imaging calcium dynamics in the nervous system by means of ballistic delivery of indicators. *Journal of Neuroscience Methods* **119**: 37–43.

Kleinfeld, D., and Denk, W. (1999) Two-photon imaging of neocortical microcirculation. In: *Imaging Neurons* (eds., Yuste, R., Lanni, F., and Konnerth, A.), pp. 23.1–23.15. Cold Spring Harbor Press, Cold Spring Harbor, Maine.

Kleinfeld, D., Mitra, P. P., Helmchen, F., and Denk, W. (1998) Fluctuations and stimulus-induced changes in blood flow observed in individual capillaries in layers 2 through 4 of rat neocortex. *Proceedings of the National Academy of Sciences USA* **95**: 15741–15746.

Kleinfeld, D., Mitra, P. P., Helmchen, F., and Denk, W. (1999) Fluctuations and stimulus-induced changes in blood flow observed in individual capillaries in layers 2 through 4 of rat neocortex (correction). *Proceedings of the National Academy of Sciences USA* **96**: 8307.

Koch, C., and Zador, A. (1993) The function of dendritic spines: Devices subserving biochemical rather than electrical compartmentalization. *Journal of Neuroscience* **13**: 413–422.

Koch, C., Zador, A., and Brown, T. H. (1992) Dendritic spines: Convergence of theory and experiment. *Science* **256**: 973–974.

Koester, H. J., Baur, D., Uhl, R., and Hell, S. W. (1999) Ca^{2+} fluorescence imaging with pico- and femtosecond two-photon excitation: Signal and photodamage. *Biophysical Journal* **77**: 2226–2236.

Koester, H. J., and Sakmann, B. (1998) Calcium dynamics in single spines during coincident pre- and postsynaptic activity depend on relative timing of back-propagating action potentials and subthreshold excitatory postsynaptic potentials. *Proceedings ofthe National Academy ofSciences USA* **95**: 9596–9601.

Koester, H. J., and Sakmann, B. (2000) Calcium dynamics associated with action potentials in single nerve terminals of pyramidal cells in layer 2/3 of the young rat neocortex. *Journal ofPhysiology (London)* **529**: 625–646.

Kohler, R. H., Cao, J., Zipfel, W. R., Webb, W. W., and Hanson, M. R. (1997) Exchange of protein molecules through connections between higherplant plastids. *Science* **276**: 2039–2042.

König, K. (2000) Multiphoton microscopy in life sciences. *Journal ofMicroscopy* **200**: 83–104.

König, K., Becker, T. W., Fischer, P., Riemann, I., and Halbhuber, K.- J. (1999) Pulse-length dependence of cellular response to intense near-infrared laser pulses in multiphoton microscopes. *Optics Letters* **24**: 113–115.

Kreitzer, A. C., Gee, K. R., Archer, E. A., and Regehr, W. G. (2000) Monitoring presynaptic calcium dynamics in projection fibers by in vivo loading of a novel calcium indicator. *Neuron* **27**: 25–32.

Kuner, T., and Augustine, G. J. (2000) A genetically encoded ratiometric indicator for chloride: Capturing chloride transients in cultured hippocampal neurons. *Neuron* **27**: 447–459.

Lechner, H. A., Lein, E. S., and Callaway, E. M. (2002) A genetic method for selective and quickly reversible silencing of mammalian neurons. *Journal ofNeuroscience* **22**: 5287–5290.

Lendvai, B., Stern, E. A., Chen, B. and Svoboda, K. (2000) Experience-dependent plasticity of dendritic spines in the developing rat barrel cortex in vivo. *Nature* **404**: 876–881.

Lester, H. A., and Chang, H. W. (1977) Response of acetylcholine receptors to rapid photochemically produced increases in agonist concentration. *Nature* **266**: 373–374.

Lester, H. A., and Nerbonne, J. M. (1982) Physiological and pharmacological manipulations with light flashes. *Annual Reviews ofBiophysics and Bioengineering* **11**: 151–175.

Linden, D. J. (1999) The return of the spike: Postsynaptic action potentials and the induction of LTP and LTD. *Neuron* **22**: 661–666.

Lipp, P., and Niggli, E. (1998) Fundamental calcium release events revealed by two-photon excitation photolysis of caged calcium in Guinea-pig cardiac myocytes. *Journal ofPhysiology (London)* **508**: 801–809.

Mainen, Z. F., Maletic-Savatic, M., Shi, S. H., Hayashi, Y., Malinow, R., and Svoboda, K. (1999) Two-photon imaging in living brain slices. *Methods* **18**: 231–239.

Maiti, S., Shear, J. B., Williams, R. M., Zipfel, W. R., and Webb, W. W. (1997) Measuring serotonin distribution in live cells with three-photon excitation. *Science* **275**: 530–532.

Majewska, A., Tashiro, A., and Yuste, R. (2000) Regulation of spine calcium dynamics by rapid spine motility. *Journal ofNeuroscience* **20**: 8262–8268.

Maletic-Savatic, M., Malinow, R., and Svoboda, K. (1999) Rapid dendritic morphogenesis in CA1 hippocampal dendrites induced by synaptic activity. *Science* **283**: 1923–1927.

Marandi, N., Konnerth, A., and Garaschuk, O. (2002) Two-photon chloride imaging in neurons of brain slices. *Pflügers Archive* **445**:357–365.

Marchant, J. S., Stutzmann, G. E., Leissring, M. A., LaFerla, F. M., and Parker, I. (2001) Multiphoton-evoked colorchange of DsRed as an optical highlighterforcellularand subcellularlabeling. *Nature Biotechnology* **19**: 645–649.

Matsuzaki, M., Ellis-Davies, G. C. R., Nemoto, T., Miyashita, Y., Iino, M., and Kasai, H. (2001) Dendritic spine geometry is critical for AMPA receptor expression in hippocampal CA1 pyramidal neurons. *Nature Neuroscience* **4**: 1086–1092.

Matus, A. (2000) Actin-based plasticity in dendritic spines. *Science* **290**: 754–758.

McCray, J. A., and Trentham, D. R. (1989) Properties and uses of photoreactive caged compounds. *Annual Review in Biophysics and Biophysical Chemistry* **18**: 239–270.

Miyawaki, A., Griesbeck, O., Heim, R., and Tsien, R. Y. (1999) Dynamic and quantitative Ca^{2+} measurements using improved chameleons. *Proceedings ofthe National Academy ofSciences USA* **96**: 2135–2140.

Miyawaki, A., Llopis, J., Heim, R., McCaffery, J. M., Adams, J. A., Ikura, M., and Tsien, R. Y. (1997) Fluorescent indicators for Ca^{2+} based on green fluorescent proteins and calmodulin. *Nature* **388**: 882–887.

Mohler, W. A., and Squirrell, J. M. (1999) Multiphoton imaging of embryonic development. In: *Imaging Neurons* (eds., Yuste, R., Lanni, F., and Konnerth, A.), pp. 21.1–21.11. Cold Spring Harbor Press, Cold Spring Harbor, Maine.

Nakai, J., Ohkura, M., and Imoto, K. (2001) A high signal-to-noise Ca^{2+} probe composed of a single green fluorescent protein. *Nature Biotechnology* **19**: 137–141.

O'Donovan, M. J., Ho, S., Sholomenko, G., and Yee, W. (1993) Real-time imaging of neurons retrogradely and anterogradely labelled with calcium-sensitive dyes. *Journal ofNeuroscience Methods* **46**: 91–106.

Oehring, H., Riemann, I., Fischer, P., Halbhuber, K. J., and Konig, K. (2000) Ultrastructure and reproduction behaviour of single CHO-K1 cells exposed to near infrared femtosecond laser pulses. *Scanning* **22**: 263–270.

Oheim, M., Beaurepaire, E., Chaigneau, E., Mertz, J., and Charpak, S. (2001) Two-photon microscopy in brain tissue: Parameters influencing the imaging depth. *Journal ofNeuroscience Methods* **111**: 29–37.

Papageorgiou, G., Ogden, D. C., Barth, A., and Corrie, J. E. T. (1999) Photorelease of carboxylic acids grom 1-acyl-7-nitroindolines in aqueous solution: Rapid and efficient photorelease of L-glutamate. *Journal ofthe American Chemical Society* **121**: 6503–6504.

Pettit, D. L., Wang, S. S., Gee, K. R., and Augustine, G. J. (1997) Chemical two-photon uncaging: A novel approach to mapping glutamate receptors. *Neuron* **19**: 465–471.

Piston, D. W. (1999) Imaging living cells and tissues by two-photon excitation microscopy. *Trends in Cell Biology* **9**: 66–69.

Piston, D. W., Kirby, M. S., Cheng, H., Lederer, W. J., and Webb, W. W. (1994) Two-photon-excitation fluorescence imaging of three-dimensional calcium-ion activity. *Applied Optics* **33**: 662–669.

Potter, S. M., Zheng, C., Koos, D. S., Feinstein, P., Fraser, S. E., and Mombaerts, P. (2001) Structure and emergence of specific olfactory glomeruli in the mouse. *Journal ofNeuroscience* **21**: 9713–9723.

Purves, R. D. (1981) *Microelectrode Methods for Intracellular Recording and Ionophoresis.* Academic Press, London.

Pusch, M., and Neher, E. (1988) Rates of diffusional exchange between small cells and a measuring patch pipette. *Pflügers Archive* **411**: 204–211.

Ramesh, D., Wieboldt, R., Niu, L., Carpenter, B. L., and Hess, G. P. (1993) Photolysis of a protecting group for the carboxyl function of neurotransmitters within 3 microseconds and with product quantum yield of 0.2. *Proceedings ofthe National Academy ofSciences USA* **90**: 11074–11078.

Regehr, W. G., and Tank, D. W. (1991) Selective fura-2 loading of presynaptic terminals and nerve cell processes by local perfusion in mammalian brain slice. *Journal ofNeuroscience Methods* **37**: 111–119.

Regehr, W. G., and Tank, D. W. (1994) Dendritic calcium dynamics. *Current Opinion in Neurobiology* **4**: 373–382.

Rose, C., and Konnerth, A. (2001) NMDA receptor-mediated Na^+ signals in spines and dendrites. *Journal ofNeuroscience* **21**: 4207–4214.

Sabatini, B. L., Maravall, M., and Svoboda, K. (2001) Ca^{2+} signaling in dendritic spines. *Current Opinion in Neurobiology* **11**: 349–356.

Sabatini, B. L., Oertner, T. G., and Svoboda, K. (2002) The life cycle of Ca^{2+} ions in dendritic spines. *Neuron* **33**: 439–452.

Sabatini, B. L., and Svoboda, K. (2000) Analysis of calcium channels in single spines using optical fluctuation analysis. *Nature* **408**: 589–593.

Sakai, R., Repunte-Canonigo, V., Raj, C. D., and Knopfel, T. (2001) Design and characterization of a DNA-encoded, voltage-sensitive fluorescent protein. *European Journal ofNeuroscience* **13**: 2314–2318.

Sakmann, B., and Neher, E. (1995) *Single-channel recording.* Plenum Press, New York.

Shepherd, G. M. (1996) The dendritic spine: A multifunctional integrative unit. *Journal ofNeurophysiology* **75**: 2197–2210.

Shi, S.- H., Hayashi, Y., Esteban, J. A., and Malinow, R. (2001) Subunit-specific rules governing AMPA receptor trafficking to synapses in hippocampal pyramidal neurons. *Cell* **105**: 331–343.

Shi, S.- H., Hayashi, Y., Petralia, R. S., Zaman, S. H., Wenthold, R. J., Svoboda, K., and Malinow, R. (1999) Rapid spine delivery and redistribution of AMPA receptors after synaptic NMDA receptor activation. *Science* **284**: 1811–1816.

Shigeri, Y., Tatsu, Y., and Yumoto, N. (2001) Synthesis and application of caged peptides and proteins. *Pharmacol Ther* **91**: 85–92.

Siegel, M. S., and Isacoff, E. Y. (1997) A genetically encoded optical probe of membrane voltage. *Neuron* **19**: 735–741.

Slade, J., and Alexander, J. (1998) *Firestorm at Gettysburg: Civilian Voices.* ShifferPublishing Ltd., Atglen.

So, P. T., Dong, C. Y., Masters, B. R., and Berland, K. M. (2000) Two-photon excitation fluorescence microscopy. *Annual Reviews in Biomedical Engineering* **2**: 399–429.

Sobel, E. C., and Tank, D. W. (1994) In vivo Ca^{2+} dynamics in a cricket auditory neuron: An example of chemical computation. *Science* **263**: 823–826.

Squirrell, J. M., Wokosin, D. L., White, J. G., and Bavister, B. D. (1999) Long-term two-photon fluorescence imaging of mammalian embryos without compromising viability. *Nature Biotechnology* **17**: 763–767.

Sreekumar, R., Ikebe, M., Fay, F. S., and Walker, J. W. (1998) Biologically active peptides caged on tyrosine. *Methods in Enzymology* **291**: 78–94.

Stockbridge, N., and Ross, W. N. (1984) Localized Ca^{2+} and calcium-activated potassium conductances in terminals of a barnacle photoreceptor. *Nature* **309**: 266–268.

Stosiek, C., Garaschuk, O., Holthoff, K., and Konnerth, A. (2003) *In vivo* two-photon calcium imaging of neuronal networks. *Proceedings of the National Academy of Sciences U.S.A.* **100**: 7319–7324.

Sullivan, K. F., and Kay, S. A. (1998) *Green Fluorescent Proteins: Methods in Cell Biology.* Academic Press, San Diego, Calif.

Svaasand, L. O., and Ellingsen, R. (1983) Optical properties of human brain. *Photochemistry and Photobiology* **38**: 293–299.

Svoboda, K., Denk, W., Kleinfeld, D., and Tank, D. W. (1997) In vivo dendritic calcium dynamics in neocortical pyramidal neurons. *Nature* **385**: 161–165.

Svoboda, K., Helmchen, F., Denk, W., and Tank, D. W. (1999) Spread of dendritic excitation in layer 2/3 pyramidal neurons in rat barrel cortex in vivo. *Nature Neuroscience* **2**: 65–73.

Svoboda, K., Tank, D. W., and Denk, W. (1996) Direct measurement of coupling between dendritic spines and shafts. *Science* **272**: 716–719.

Svoboda, K., and Yasuda, R. (2006) Principles of two-photon excitation microscopy and its applications to neuroscience. *Neuron* **50**: 823–839.

Taddeucci, A., Martelli, F., Barilli, M., Ferrari, M., and Zaccanti, G. (1996) Optical properties of brain tissue. *Journal of Biomedical Optics* **1**: 117–123.

Tan, Y. P., and Llano, I. (1999) Modulation by K^+ channels of action potential-evoked intracellular Ca^{2+} concentration rises in rat cerebellar basket cell axons. *Journal of Physiology* **520**: 65–78.

Tan, Y. P., Llano, I., Hopt, A., Würriehausen, F., and Neher, E. (1999) Fast scanning and efficient photodetection in a simple two-photon microscope. *Journal of Neuroscience Methods* **92**: 123–135.

Tang, Y., Nyengaard, J. R., De Groot, D. M., and Gundersen, H. J. (2001) Total regional and global number of synapses in the human brain neocortex. *Synapse* **41**:258–273.

Tank, D. W., Sugimori, M., Connor, J. A., and Llinas, R. R. (1988) Spatially resolved calcium dynamics of mammalian Purkinje cells in cerebellar slice. *Science* **242**: 773–777.

Tirlapur, U. K., Konig, K., Peuckert, C., Krieg, R., and Halbhuber, K. J. (2001) Femtosecond near-infrared laser pulses elicit generation of reactive oxygen species in mammalian cells leading to apoptosis-like death. *Experimental Cell Research* **263**: 88–97.

Theer, P., and Denk, W. (2006) On the fundamental imaging-depth limit in two-photon microscopy. *Journal of the Optical Society of America, A: Optics Image Science and Vision* **23**: 3139–3149.

Theer, P., Hasan, M. T., and Denk, W. (2003) Two-photon imaging to a depth of 1000 μm in living brain using a $Ti:Al_2O_3$ regenerative amplifier. *Optics Letters* **28**: 1022–1024.

Trachtenberg, J. T., Chen, B. E., Knott, G. W., Feng, G., Sanes, J. R., Welker, E., and Svoboda, K. (2002) Long-term in vivo imaging of experience-dependent synaptic plasticity in adult cortex. *Nature* **420**: 788–794.

Truong, K., Sawano, A., Mizuno, H., Hama, H., Tong, K. I., Mal, T. K., Miyawaki, A., and Ikura, M. (2001) FRET-based in vivo Ca^{2+} imaging by a new calmodulin-GFP fusion molecule. *Nature Structural Biology* **8**: 1069–1073.

Tsien, R. Y. (1989) Fluorescent probes of cell signaling. *Annual Reviews in Neuroscience* **12**: 227–253.

Tsien, R. Y., and Zucker, R. S. (1986) Control of cytoplasmic calcium with photolabile tetracarboxylate 2-nitrobenzhydrol chelators. *Biophysical Journal* **50**: 843–853.

Tucker, K. L. (2000) In vivo imaging of the mammalian nervous system using fluorescent proteins. *Histochem Cell Biology* **115**: 31–39.

Walker, J. W., Feeney, J., and Trentham, D. R. (1989) Photolabile precursors of inositol phosphates: Preparation and properties of 1-(2-nitrophenyl)ethyl esters of myo-inositol 1,4,5-trisphosphate. *Biochemistry* **28**: 3272–3280.

Wang, J. W., Wong, A. M., Flores, J., Vosshall, L. B., and Axel, R. (2003) Two-photon calcium imaging reveals an odor-evoked map of activity in the fly brain. *Cell* **112**: 271–282.

Wang, S. S., and Augustine, G. J. (1995) Confocal imaging and local photolysis of caged compounds: Dual probes of synaptic function. *Neuron* **15**: 755–760.

Wang, S. S.- H., Denk, W., and Häusser, M. (2000a) Coincidence detection in single dendritic spines mediated by calcium release. *Nature Neuroscience* **3**: 1266–1273.

Wang, S. S., Khiroug, L., and Augustine, G. J. (2000b) Quantification of spread of cerebellar long-term depression with chemical two-photon uncaging of glutamate. *Proceedings of the National Academy of Sciences USA* **97**: 8635–8640.

Wieboldt, R., Gee, K. R., Niu, L., Ramesh, D., Carpenter, B. K., and Hess, G. P. (1994) Photolabile precursors of glutamate: Synthesis, photochemical properties, and activation of glutamate receptors on a microsecond time scale. *Proceedings of the National Academy of Sciences USA* **91**: 8752–8756.

Williams, R. M., Piston, D. W. and Webb, W. W. (1994) Two-photon molecular excitation provides intrinsic 3-dimensional resolution for laser-based microscopy and microphotochemistry. *FASEB Journal* **8**: 804–813.

Wu, J.- W., Cohen, L. B., Tsau, Y., Lam, Y.- W., Zochowski, M., and Falk, C. X. (1999) Imaging with voltage-sensitive dyes: Spike signals, population signals, and retrograde transport. In: *Imaging Neurons* (eds., Yuste, R., Lanni, F., and Konnerth, A.), pp. 49.1–49.12. Cold Spring Harbor Press, Cold Spring Harbor, Maine.

Xu, C., and Webb, W. W. (1996) Measurement of two-photon excitation cross sections of molecular fluorophores with data from 690 to 1050 nm. *Journal ofthe Optical Society ofAmerica B* **13**: 481–491.

Xu, C., Zipfel, W., Shear, J. B., Williams, R. M., and Webb, W. W. (1996) Multiphoton fluorescence excitation: New spectral windows forbiological nonlinearmicr oscopy. *Proceedings ofthe National Academy ofthe Sciences USA* **93**: 10762–10768.

Ying, J., Liu, F., and Alfano, R. R. (1999) Spatial distribution of two-photon-excited fluorescence in scattering media. *Applied Optics-OT* **38**: 224–229.

Yuste, R., and Bonhoeffer, T. (2001) Morphological changes in dendritic spines associated with long-term synaptic plasticity. *Annual Reviews in Neuroscience* **24**: 1071–1089.

Yuste, R., and Denk, W. (1995) Dendritic spines as basic functional units of neuronal integration. *Nature* **375**: 682–684.

Yuste, R., and Tank, D. W. (1996) Dendritic integration in mammalian neurons, a century after Cajal. *Neuron* **16**: 701–716.

Yuste, R., Lanni, F., and Konnerth, A. (eds.) (1999a) *Imaging Neurons: A Laboratory Manual.* Cold Spring Harbor Press, Cold Spring Harbor, Maine.

Yuste, R., Majewska, A., Cash, S. S., and Denk, W. (1999b) Mechanisms of calcium influx into hippocampal spines: Heterogeneity among spines, coincidence detection by NMDA receptors, and optical quantal analysis. *Journal of Neuroscience* **19**: 1976–1987.

Yuste, R., Majewska, A., and Holthoff, K. (2000) From form to function: Calcium compartmentalization in dendritic spines. *Nature Neuroscience* **3**: 653–659.

Zipfel, W. R., Williams, R. M., and Webb, W. W. (1996) Release of caged bioeffectormolecules by two photon excitation: Excitation spectra, absorption cross sections and practicalities of some common caging groups. *Biophysical Journal* **70**: MP248.

4

Messenger RNA Imaging in Living Cells for Biomedical Research

DAHLENE FUSCO, EDOUARD BERTRAND, AND ROBERT H. SINGER

RATIONALE FOR LIVE CELL IMAGING OF MESSENGER RNA

Recent advances in messengerRNA (mRNA) visualization technology have placed biomedical imaging at an interface between molecularbiology and cellularbiology. It is now increasingly possible to study in vivo gene expression at the transcript level. Analyses of mRNA expression, movement, interactions, and localization will enhance understanding of cellular responses to various conditions and will complement studies of protein expression. Live cell mRNA imaging technology can add new information about where and when transcription and translation occur, as well as provide a single cell expression profile that cannot be achieved with microchip or biochemical analyses. Simultaneous analysis of multiple transcription sites can provide a single cell profile of gene expression that can be linked precisely to cellular morphology (Levsky et al., 2002), and observation of these transcription sites in living cells will provide novel information about the time course of gene expression in normal and pathological samples.

There are multiple medical applications for mRNA imaging, extending from studies of disease mechanism to diagnostic and treatment applications. One of the most significant contributions of mRNA studies to the identification of a disease mechanism was the analysis of trinucleotide repeat transcript foci in nuclei of myotonic dystrophy cells and tissues (Taneja et al. 1995; Davis et al., 1997). This study revealed that dystrophia myotonica-protein kinase (DMPK) transcripts containing increased numbers of cytosine-thymine-guanosine (CTG) repeats accumulate in the nucleus in stable, long-lived clusters. Nuclear accumulation of mutant transcripts is consistent with a model of DM loss of function resulting from a lack of nuclearexpor t, along with possible impedance of othermRNA exports. Another disease mechanism that can be studied using mRNA visualization is the process of metastasis. MessengerRNA localization patterns appearto be related to metastatic capability because nonmetastatic tumorcells differin theirmRNA localization and motility characteristics compared with metastatic tumor cells (Shestakova et al., 1999). Messenger RNA analyses are also being applied to the study of fragile X mental retardation, in which the RNA binding protein fragile X mental retardation protein (FMRP) is disrupted, providing a connection between mRNA regulation and normal neuronal development (Ashley et al., 1993; Siomi et al., 1993). Translational dysregulation of mRNAs normally associated with

FMRP may be the proximal cause of fragile X syndrome, and one proposed model for FMRP function is that it shuttles mRNAs from the nucleus to postsynaptic sites where transcripts are held in a translationally inactive form until further stimulation alters FMRP function (Brown et al., 2001). Several other mRNA binding proteins have been shown to play a role in mRNA trafficking in developing neurons. These neuronal mRNA binding proteins include zip code binding proteins one and two (ZBP-1, ZBP-2 [Gu et al., 2002; Ross et al., 1997]), which play a role in the dendritic trafficking of β-actin mRNA. Live-cell mRNA studies can also provide important insights in virology. Subcellular localization and tracking of viral sequences, combined with labeling of relevant proteins and other structures, will elucidate viral attack mechanisms. Viral RNA studies will also help clarify the cellular life cycle of viral RNA as well as host cell defense mechanisms, and will allow comparison of cell phenotype with amount of viral RNA per cell as infection progresses.

Diagnostic applications of mRNA studies include the monitoring of gene function and activity in living cells and tissues to determine levels of endogenous gene expression as well as that of infectious and gene therapeutic agents. The diagnostic potential for mRNA studies has increased dramatically with the development of molecular beacons, which can discriminate between oligonucleotide targets differing by a single base pair mismatch (Tyagi and Kramer, 1996). High-resolution sequence recognition has the potential for use in rapid, minimally invasive identification of the strain of an infectious agent or of drug resistance in living cells. Messenger RNA imaging technology also has the potential to provide new methods for treatment of disease. Gene therapeutic regimes can be bettermonitor ed forfunction and specificity through localization and quantification of delivered reagents. Also, antisense oligonucleotides are being developed for treatment of viral infection with greater specificity than currently available antiviral reagents (Lever, 2001; Manns et al., 2001; Orr, 2001), which often affect host cell gene expression in addition to viral production. Antisense treatment for neuroprotection after hemorrhage and for chemotherapy is also in progress (Capoulade et al. 2001; Mayne et al., 2001). In summary, the increasing availability and reliability of technology for the detection of mRNA in living cells has opened an expansive potential forpr oviding specific, effective treatment; performing high-resolution medical diagnoses; and understanding disease mechanisms.

MessengerRNA detection in individual, living cells has several advantages overother techniques forthe study of gene expression. Forexample, in comparison with population analyses, such as northern blotting, single-cell studies can describe the amount of virus in individual infected cells, or the percentage of cells that are infected. By imaging mRNA in living cells instead of fixed cells, it is possible to determine the timing of mRNA expression and dynamics, as well as different patterns of mRNA movement and interactions throughout the lifetime of a transcript. Because mRNA movement appears to be a rapid, rare event, the ability to examine a series of time points becomes important for the observation of mRNA transit. In addition, live-cell imaging avoids the possible introduction of artifact during cell fixation.

Foryear s, the majorlimitation of imaging mRNA in living cells has been a dearth of available technology. Recent years have yielded significant advances in four major categories of mRNA visualization technology, including microinjection of fluorescent RNA (Ainger et al., 1993; Ferrandon et al., 1994), hybridization of fluorescent oligonucleotide probes (Politz et al., 1998, 1999), the use of cell-permeant dyes that bind RNA, and sequence-specific mRNA recognition by a GFP fusion protein (Bertrand et al., 1998). Each of these detection techniques are described next, followed by a summary of the biological information in the nucleus and the cytoplasm gained through each technique.

EXISTING TECHNOLOGIES FOR THE VISUALIZATION
OF RNA IN LIVING CELLS

Microinjection of Fluorescent RNA

RNA can be seen in living cells by using a variety of approaches based on fluorescent labelling (fig. 4.1). Fluorescent RNA can be synthesized by in vitro transcription with phage polymerases in the presence of fluorescent nucleotide analogues for microinjection into living cells. The introduction of new nucleotides coupled with the Alexa dyes has significantly increased the brightness and hence detection of the resulting RNAs (Wilkie and Davis, 2001). Likewise, transcription with a mixture of unmodified and amino-allyl nucleotides generates RNA that can be chemically coupled to activated fluorophores (Wang et al., 1991). Because amino-allyl nucleotides are incorporated at a higher frequency than fluorescent analogues, very bright RNA can be obtained. The main disadvantage of microinjecting fluorescently labeled mRNA is the introduction of nonendogenous material into the cell. If a modified nucleotide is incorporated at a position involved in an interaction with a "trans-acting" factor, the biological properties of the resulting RNA are likely to be altered. For instance, such fluorescent RNA is usually not translated or localized correctly. Nevertheless, this approach has proved successful in a number of cases (Ainger et al., 1993; Theurkauf and Hazelrigg, 1998; Wilkie and Davis, 2001).

Fluorescent Oligonucleotide Probes (Fluorescent in Vivo Hybridization)

Fluorescent oligonucleotide probes have been developed to follow unmodified, endogenous cellular RNA. An oligonucleotide complementary to the desired sequence is covalently linked to fluorochromes and can contain chemical modifications that increase cell penetration and hybrid stability. Oligonucleotides are delivered across the plasma membrane and hybridize with the target sequence to identify RNA directly in living cells (Politz et al., 1995, 1998). Multiple methods of oligonucleotide delivery have been tested, including pinocytosis, fusion with RBC ghosts, scrape loading, glass bead loading, liposomal delivery, enzymatic cell permeabilization methods, and—the most efficient method to date—microinjection using glass pipettes (Dirks et al., 2001).

Six categories of fluorescent oligonucleotide probes exist, including conventional DNA probes, caged DNA probes, linear 2' O methyl (2'OMe) RNA probes, peptide nucleic acid (PNA) probes, molecular beacons, and probes with a linked fluorophore with fluorescence properties that change upon association with the target mRNA strand. A caveat of antisense oligodeoxyribonucleotides is that they can prevent translation by blocking ribosome movement along the mRNA or by inducing RNase hydrogen cleavage of the RNA/oligo hybrid, altering expression of target mRNA. In addition, fluorescent oligonucleotides have been found to localize exclusively to the nucleus in some cases (Dirks et al., 2001).

Conventional Oligodeoxynucleotide Probes

Conventional oligodeoxynucleotide probes (ODNs) are DNA probes with fluorescence properties that do not change significantly upon hybridization. Advantages of conventional fluorescent ODNs are their ability to detect endogenous mRNA, their relative stability (as long as 18 hours [Politz et al., 1995]), and the ease with which they are introduced into the cell. Oligodeoxynucleotide probes can be used to perform FRET by simultaneously introducing two oligonucleotides labeled with different fluorophores and recording the loss of donorfluor escence and/orgain of acceptorfluor escence. A potential problem of FRET detected as loss of donorfluor escence is that othercauses may lead to loss of fluorescence within the cell (Sokol et al., 1998). The major disadvantage of ODNs is low sensitivity resulting from a low signal-to-noise ratio as a result of the presence of

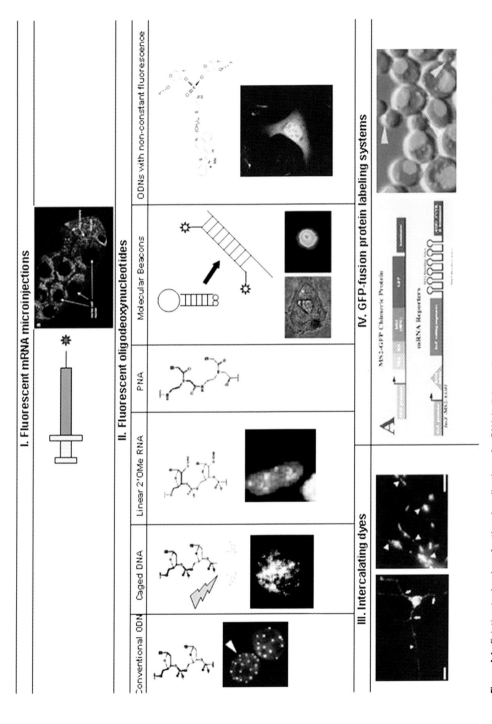

Figure 4.1 Existing technology for the visualization of mRNA in living cells, including images obtained using those technologies (Knowles et al., 1996; Knowles and Kosik, 1997; Bertrand et al., 1998; Theurkauf and Hazelrigg, 1998; Politz et al., 1999; Lorenz et al., 2000; Dirks et al., 2001; Perlette and Tan, 2001; Privat et al., 2001).

a nonhybridized probe. Other criticisms of ODNs are susceptibility to nucleases, relatively low affinity and specificity for target, slow hybridization kinetics, and low ability to bind to structured targets (Molenaar et al., 2001). A final disadvantage of conventional oligonucleotide probes is their tendency to diffuse into and remain trapped in the nucleus (Leonetti et al., 1991). One method used to maintain oligonucleotides in the cytoplasm is fusion to macromolecules such as streptavidin (Tsuji et al., 2001), which prevents passage through nuclear pores. The most promising conventional oligonucleotides to date are phosphorothioate oligodeoxynucleotides, in which a nonbridging oxygen of the phosphodiesternucleic acid is replaced by sulfur to increase stability while maintaining RNAase hydrogen sensitivity, and the ability to shuttle between the nucleus and the cytoplasm (Lorenz et al., 2000).

Caged Fluorophores

Caged fluorophores (Politz et al., 1999) are conventional ODNs that have been attached to a chemical moiety, such as two O-nitrobenzene, which locks the fluorophore into its nonfluorescent tautomer (Pederson, 2001). Caging groups can be removed by photolytic cleavage, allowing specific photoactivation of a small number of highly localized oligonucleotides. To ensure that only the hybridized probe is analyzed, caged oligo (dT) [oligonucleotide (deoxythymidine)], is first incubated with cells in culture media, and then free probe is washed away for 1 hour before the experiment is performed. The main advantage of caged fluorophores is that the signal-to-noise ratio is improved over that of conventional fluorescent in vivo hybridization because only a localized fraction of probe becomes fluorescent after spot photolysis. Those oligonucleotides that are activated but not hybridized quickly diffuse to extinction in the large cellular volume. Because hybridized probe is associated with larger RNA molecules, it adopts the movement pattern of its RNA target, which can thus be analyzed. Disadvantages of caged fluorophores are the small area that can be analyzed at a given instant as well as the need for fur thermanipulation after intr oduction of oligonucleotides.

Linear 2′ O Methyl RNA Probes

Linear 2′ OMe RNA probes are a variant upon traditional oligonucleotides (Carmo-Fonseca et al., 1991) for the detection of small nuclear RNAs. 2′ O methyl RNA probes can be synthesized using standard 2′ OMe phosphoramidite monomers covalently linked to fluorescent dyes at the 5′ end via a succinimidyl ester derivative. A study was performed to compare 2′ OMe RNA with ODNs, and it found that the former were more nuclease resistant and showed greater affinity and specificity, faster hybr idization kinetics, and enhanced binding to structured targets (Molenaar et al., 2001). Comparison of the performance of linear 2′ OMe RNA probes and that of molecular beacons found no improvements with the molecularbeacon used in the study (Molenaar et al., 2001). A disadvantage of 2′ OMe probes is their tendency to accumulate in the nucleus and their lack of cytoplasmic hybridization (Dirks et al., 2001).

Peptide Nucleic Acid Probes

Peptide nucleic acid probes are DNA analogues in which the phosphodiester backbone has been replaced by a 2-aminoethyl–glycine linkage. Peptide nucleic acid probes are uncharged, resistant to enzymatic cleavage, and form extremely stable hybrids with DNA and RNA, which should make them ideal candidates for the detection of mRNA in living cells. Unfortunately, to date, all use of PNA probes in living cells has shown the absence of specific binding and distribution throughout the nucleus (Dirks et al., 2001).

Molecular Beacons

Molecular beacons are oligodeoxynucleotides with a fluorophore and a quencher linked to the $3'$ and $5'$ ends, and are held proximal by an intermediate stem loop structure in the presence of micromolar Mg^{2+} and at appropriate temperatures (Tyagi and Kramer, 1996). In the absence of a specific target sequence, the stem loop structure holds the two fluorophores sufficiently proximal for fluorescent quenching to occur. In the presence of a specific target, the beacon unfolds to bind in a linear conformation, providing sufficient distance between fluorophores to inhibit quenching and enable detection. A major advantage of molecular beacons is the detection of target hybridization without the need to distinguish hybridized from nonhybridized probe, because molecular beacons can increase fluorescence up to 200 times upon hybridization (Tyagi and Kramer, 1996; Tyagi et al., 1998). Furthermore, molecular beacons do not require additional manipulation after they have been introduced, in contrast to reagents such as caged oligonucleotides. Last, molecular beacons are highly specific, capable of differentiating between as little as a single base pair mismatch in solution, and have a wide range of annealing temperatures, resulting in high specificity (Tyagi and Kramer, 1996). One disadvantage of molecular beacons is their instability resulting from degradation by RNAse hydrogen, within approximately 45 minutes (Sokol et al., 1998), although improved beacon stability has been cited through the use of phosphorothioate oligonucleotides and $2'$ O methylation (Gene Link, New York). A second disadvantage of molecular beacons is the observed nonspecific opening, orbr eathing, of theirstem loop by mechanisms otherthan hybridization (Molenaaret al., 2001).

Oligonucleotides with Nonconstant Fluorescent Properties

The final category of oligonucleotide probes are those with fluorescence properties that change upon association with the target mRNA strand. There are two methods by which fluorescent signal could alter with hybridization: either through a change in spectral characteristics or through an increase in fluorescent quantum yield. Dyes with the former characteristic are highly sensitive to DNA sequence; those with the latter property have proved efficient for detection of RNA. Cyanine dyes such as oxazole yellow and thiazole orange, or "light-up probes," have been found to increase fluorescent quantum yield by as much as 15 times upon hybridization. Thiazole orange can be linked via an alkyl chain to an internucleotidic phosphate group of an oligo-α-thymidylate, and labeled probes can be microinjected or internalized through diffusion (Privat et al., 2001). Alpha-isomeric oligo $2'$-DNA probes labeled with thiazole orange should be stable for up to 20 hours (Thuong et al., 1987). To date, only light-up probes hybridizing to poly(A) and ribosomal RNA have been used successfully. When thiazole orange probes to poly(A) RNA were microinjected into the cytoplasm of adherent fibroblast cells, fluorescence was detected in the cytoplasm and the nucleus with the exclusion of nucleoli. Fluorescent speckles were observed in both the nucleus and the cytoplasm. Both fluorescence levels and number of speckles were greater in the nucleus than the cytoplasm, indicating a possible lack of probe retention in the cytoplasm compared with the nucleus. However, this method has the advantage of allowing visualization of endogenous cellular RNA with improved signal-to-background noise compared with conventional oligonucleotide probes and greater resistance to degradation than molecular beacons. Disadvantages are the possibility that a probe will diffuse irreversibly into the nucleus, and that there will be lesser signal-to-noise enhancement compared with molecularbeacons.

Cell Permeant Dyes

SYTO-14 is a membrane-permeable nucleic acid stain that has been used to study RNA in living cells. SYTO-14 is advantageous because it will detect the endogenous RNA population, it is not limited to the nucleus, and its use requires no subsequent intervention after introduction into cells. A major disadvantage of cell permeant RNA binding dyes is their lack of specificity resulting from identical recognition of all RNAs.

RNA Tagged with Sequence-Specific Green Fluorescent Protein Fusion Proteins

It is possible to visualize a specific transcript by fusion of GFP with a sequence-specific RNA binding protein, such as the coat protein of phage MS2, and insertion of the RNA sequence recognized by this protein in the RNA of interest. When both the MS2-GFP fusion protein and the reporter RNA are coexpressed, MS2-GFP binds to the reporter in living cells and GFP fluorescence constitutes a reliable indicator of RNA localization (Bertrand et al., 1998). To remove the noise of unbound MS2-GFP, it is possible either to express a low level of MS2-GFP such that most of the MS2-GFP molecules are bound to the RNA, or to include a localization signal in the MS2-GFP protein that will exclude unbound molecules from the area of interest (Bertrand et al., 1998). This technique has several advantages: (1) it is possible to visualize RNA that is synthesized and processed by the cell; (2) it is possible to use the technique in organisms that cannot be microinjected, or that do not allow penetration of macromolecules; and (3) it is possible to increase the sensitivity of RNA detection by "multimerizing" the binding site in the RNA reporter. The detection limit of this technique has reached the level of single mRNA molecules through the use of MS2 repeats containing 6 to 24 binding sites (i.e., 12–48 GFP molecules) (Fusco et al., 2003). An analogous detection system has been designed to visualize genes in living nuclei, using a lac i-GFP fusion protein and an array of lac o binding sites inserted in the genome. This system first demonstrated that single molecule detection is possible, and requires as little as 60 GFP molecules (Robinett et al., 1996). The combination of these two systems, forthe simultaneous visualization of gene activation and mRNA trafficking in living cells, is in progress.

However, the MS2-GFP system does involve the visualization of a nonendogenous mRNA, introduces a nonendogenous protein into the RNA movement process, and does not allow cellular control of expression levels. Although the first two disadvantages are inherent to the MS2 labeling system, the third disadvantage can be addressed by designing inducible cell lines with specific control of MS2 reporter levels. Alternatively, a GFP-labeled RNA binding protein can be used. Theurkauf and Hazelrigg (1998) tagged *bicoid* mRNA with a GFP–exuperantia fusion. Although this system avoids introduction of an exogenous protein, other than GFP, it is difficult to determine whether a protein that is known to bind an mRNA is in fact persistently bound to that mRNA, thus presenting the question of whether all GFP fusion protein movements are indicative of mRNA movements. The MS2-GFP labeling system provides a stable link between GFP and mRNA ($K_d = 39nM$, in vitro [Lago et al., 1998]), allowing reliable identification of mRNA particles throughout trafficking behavior.

ANALYSIS OF RNA DYNAMICS IN LIVING CELLS

Aftersuccessful visualization of the RNA of interest, one must choose a method foranalyzing RNA dynamics. When single RNA molecules are detectable, direct tracking of individual RNA by time-lapse microscopy is applicable. When single RNA molecules are not detectable, two general techniques can be used: photobleaching and fluorescent correlation spectroscopy (FCS).

Photobleaching Techniques

Photobleaching techniques involve irradiation with intense light leading to the irreversible inactivation of fluorophores. If a highly focused laser beam is used, it is possible to bleach a small area of the cell specifically (width, 0.2 μm). By measuring the exchange of fluorescent and nonfluorescent molecules, it is possible to extract information such as diffusion coefficient, residency time, and whether molecules behave as a homogeneous population or separate into several distinct kinetic entities (Axelrod et al., 1976; Phair and Misteli, 2000). One method that uses photobleaching is fluorescence recovery after photobleaching, or FRAP. In FRAP, a small area of the cell is photobleached, and the recovery of fluorescent signal is followed by time-lapse microscopy. The percent of final recovery yields the fraction of immobile molecules, and the curve of recovery is directly related to the diffusion coefficient in the compartment analyzed (Axelrod et al., 1976). In some cases, heterogeneous populations of molecules can also be separated. A second method, fluorescence loss in photobleaching (orFLIP), relies on the continuous bleaching of one spot within the cell. Molecules moving into the spot are bleached, and the fluorescence in other areas of the cells decreases as unbleached molecules move out of these areas to refeed the bleached spot. Fluorescence loss in photobleaching data are complementary to FRAP data, and their analyses can be very informative for the rates at which molecules leave a given compartment. Residency times can also be estimated using simple mathematical models (Phairand Misteli, 2000).

Fluorescence Correlation Spectroscopy

Fluorescence correlation spectroscopy is a fluctuation analysis method used to measure molecular transport and chemical kinetics by counting the molecules entering and leaving a small interrogation volume over successive time intervals (Wang et al., 1991). This technique is a powerful analysis method for studying the dynamics of highly dispersed fluorescent molecules in a living cell (Politz et al., 1998). Depending on the diffusion coefficients, sampling times range from 200 ns to 3500 s. With sufficient sampling, molecules coming in or moving out of the spot provoke small fluctuations in fluorescence intensity, which are recorded. The rate at which molecules enter and exit is related to their diffusion coefficient, which can be calculated using autocorrelation curves (Politz et al., 1998). Several populations of molecules with distinct properties can be resolved using FCS. In addition, two-color FCS can generate information on the interactions between two tracked molecules because interacting molecules will move together in and out of the sampled volume.

Time-Lapse Microscopy/Particle Tracking

In certain cases it is possible to use time-lapse microscopy to follow RNA movements. This requires either single molecule sensitivity or the concentration of many RNA molecules in a multimolecule aggregate, as previously observed in some cases (Ainger et al., 1993; Ferrandon et al., 1994; Bertrand et al., 1998; Wilkie and Davis, 2001). The single-molecule dynamics measured in this method are complementary to the population dynamics obtained through FRAP or FCS. It is difficult to estimate diffusion coefficients from time-lapse data because laborious analysis of a large number of molecules is required, but the temporal sequence of movements can provide novel information. For example, if a molecule is alternatively immobile and mobile on a relatively short timescale, photobleaching techniques may average these behaviors whereas both will be apparent by time-lapse microscopy. The ability to follow the fate of a molecule can reveal an ordered sequence of events. Also, time-lapse techniques can reveal rare but functionally important behaviors that could not be detected otherwise. For instance, *Ash1* mRNA visualized in the cytoplasm of living yeast (Bertrand et al., 1998) is localized at the bud tip, after transport by a specific myosin motor on actin cables. However, because the transport is very rapid and lasts for less than a minute of the entire mRNA lifetime, the proportion of actively transported mRNA molecules is, on average, very low and does not exceed a few percent of the total molecules. Such a minor population would very likely not be detected by photobleaching and FCS techniques.

If single-molecule resolution is achieved, a detailed analysis of individual particle movements can yield information related to the mechanism of mRNA transport. Analysis can include collection of position, velocity, and acceleration values for the determination of movement type, such as active transport, brownian motion, or corralled diffusion. In the case of active transport, movement analysis can help identify candidate motors involved in transport. For example, if a particle is observed to display actin-dependent movement at a constant speed of approximately 200 to 400 nm/s, a myosin motor is likely to be involved (Cheney et al., 1993). Other interesting characteristics of mRNA particles that can be obtained from time-lapse images in living cells include particle size, location, diffusion coefficient, mean squared displacement, and relation to other cellular components, such as the cytoskeleton, all of which can be useful in determining the mechanism of mRNA movement and localization. If images are of high quality but particles still cannot be tracked, it is possible to perform two-dimensional deconvolutions on movies to improve the signal-to-noise ratio further and to facilitate particle tracking (Huygens, Bitplane, The Netherlands).

In the nucleus, the majority of live-cell mRNA studies to date have examined the general polyadenylated RNA population using fluorescent oligodT probes (Politz et al., 1995, 1998, 1999), but a variety of well-characterized nuclear RNA or DNA binding proteins have been studied as GFP fusions (Huang et al., 1998; Misteli et al., 2000; Phairand Misteli, 2000; Chen and Huang, 2001). In contrast, in the cytoplasm, an increasing number of specific transcripts are being studied in living cells. These studies are significant because they provide unique information about the dynamics of mRNA movement and help elucidate the mechanisms of mRNA particle assembly, transport, and localization, as well as crucial mRNA interactions with proteins and the cytoskeleton. This information can be obtained in single cells for endogenous, engineered, and viral transcripts, greatly widening our current understanding of mRNA behavior.

IN VIVO MESSENGER RNA ANALYSES IN THE NUCLEUS AND CYTOPLASM

Nuclear RNA Studies

Poly(A) RNA in the Nucleus

Studies of poly(A) RNA in the nucleus of living cells have contributed to the understanding of RNA trafficking in the nucleus as well as that of the nucleoplasm itself. Such studies have begun to address the debate between *solid phase* versus *free diffusion* models of nuclear organization and RNA transport (Blobel, 1985; Schroder et al., 1987; Agutter, 1994), and the question of whether RNAs diffuse freely throughout the nucleus or undergo active transport from transcription sites to nuclear pores (Boulon et al., 2002).

Fluorescence correlation spectroscopy analysis of nuclear poly(A) RNA has revealed that a fraction of poly(A) transcripts have surprisingly high mobility (Politz et al., 1998). Diffusion coefficient measurements of oligo(dT) probes in living nuclei have distinguished two poly(A) populations. The first population has a diffusion coefficient of 4×10^{-7} cm^2/s, similar to oligo(dA), and likely represents free oligo(dT) probe. This rate is very similar to that measured in solution, suggesting the presence of an aqueous phase of low viscosity in the nucleoplasm. The second population of labeled transcripts displays a heterogenous set of slower diffusion coefficients ranging from 1×10^{-7} to 1×10^{-10} cm^2/s. Of these slower diffusing transcripts, one third have a diffusion coefficient less than 6×10^{-9} cm^2/s, and the others have an average coefficient of 9×10^{-8} cm^2/s. The latter value corresponds to the value expected for an average-size heterogeneous ribonucleoprotein (hnRNP) (7.5 kb) that is randomly folded and diffusing in an aqueous environment (Pederson, 1999). Thus, this population could correspond to freely diffusing poly(A) mRNA. In contrast, the most slowly moving molecules may be attached to large, nearly immobile macromolecular complexes.

Studies using fluorescence recovery after photobleaching with the same fluorescent probes generated a less detailed view of the oligonucleotide mobility spectrum because of the difficulty of sorting the various populations of moving oligonucleotides in living nuclei (Politz et al., 1998). However, the average diffusion coefficient was in good agreement with that obtained by FCS, also suggesting that the nucleoplasm contains an aqueous phase that allows rapid mobility of large molecules.

Politz et al. (1999) used oligo(dT) labeled with caged fluorescein to observe the movements of poly(A) RNA in nuclei of living rat L6 myoblasts. Poly(A) RNA was observed to leave the uncaging spot in all directions with a mean square displacement that varied linearly with time, and the measured apparent diffusion coefficient was the same for movement at 37°C and 23°C, indicating a random diffusive process. Three-dimensional imaging in the presence of uncaged oligo(dT) and Hoescht–labeled chromosomes showed that, excluding nucleoli, poly(A) RNA could access most of the nonchromosomal space in the nucleus. A significant fraction of the probe ($\approx$30%) did not leave the initial activated area, suggesting that it was bound to poly(A) RNA engaged in large complexes, such as those involved in transcription and processing. Movement of poly(A) RNA out of the initial spot followed a diffusive model ($x^2 = 2Dt$), at a long timescale (150 s) and forlong-r ange movements (5 μm). However, the diffusion coefficient measured (D $= 6 \times 10^{-9}$ cm^2/s) reflected only the slowest moving population revealed using FCS and was slower than all measurements obtained using FRAP. A possible explanation forthis discrepancy lies in the fact that diffusion coefficients obtained by FRAP or FCS are measured in a small volume, whereas the diffusion coefficient, D, obtained by uncaging is estimated from movements over the entire nucleus. Possibly, the nucleoplasm contains microdomains that allow aqueouslike diffusion with D around 1×10^{-7} cm^2/s, such as that measured by FCS and FRAP, but diffusion out of these microdomains would be limiting. This model is consistent with our current view of the nuclear structure wherein chromatin occupies a large fraction of the nuclear volume ($\approx$50%) and limits the nucleoplasm to a reticular network of interconnected channels (Politz et al., 1999). Thus, this network could reduce long-range diffusion, especially if it has a low connectivity.

In addition to studies of poly(A) RNA in the nucleus, several specific transcripts have been observed in living cells. These studies are significant because they address the question of whetherRNA trafficking in the nucleus is sequence specific (Blobel, 1985) orwhether it occurs in an equivalent fashion for all transcripts. In future studies it will be possible to determine whether mRNA transport particles or *granules* form in the nucleus or the cytoplasm. The location of granule formation will help identify the proteins involved in this process. In addition, colocalization of transcripts within the nucleus can be examined to identify transcript groups that undergo coordinated posttranscriptional regulation.

Human β-Globin and Rat Proenkephalin Premessenger RNAs

The first microinjection of fluorescently labeled mRNA was performed (Wang et al., 1991) in normal rat kidney epithelial cells. This study used fluorescence microscopy to show the in vivo, intron-dependent localization of rhodamine-labeled human β-globin and rat proenkephalin pre-mRNAs to loci defined by monoclonal antibodies against small nuclear ribonucleoproteins (Sm) and the spliceosome component (SC-35). Later experiments showed that splicing of endogenous mRNA is cotranscriptional and that spliced introns may be specifically recruited to speckles (Zhang et al., 1994).

Small Nuclear RNA (snRNA)

The in vivo distribution of small nuclear ribonucleoproteins (snRNPs) has been analyzed by microinjecting fluorochrome-labeled antisense probes into the nuclei of live HeLa and 3T3 cells (Carmo-Fonseca et al., 1991). Probes for U2 and U5 snRNAs specifically label the same discrete nuclear foci whereas a probe for U1 snRNA shows widespread nucleoplasmic labeling, excluding nucleoli, in addition to labeling foci. A probe for U3 snRNA specifically labels nucleoli. These in vivo data confirm that mammalian cells have nuclear foci that contain spliceosomal snRNPs. Colocalization studies, both in vivo and in situ, demonstrate that the spliceosomal snRNAs are present in the same nuclear foci. These foci are also stained by antibodies that recognize snRNP proteins, m3G-cap structures, and the splicing factorU2AF, but are not stained by anti-SC-35 oranti-La antibodies. U1 snRNP and the splicing factorU2AF closely colocalize in the nucleus, both before and afteractinomycin D treatment, suggesting that they may both be part of the same complex in vivo. The specific examination of U snRNA transcripts provided a novel description of RNA localization within the nucleus, complementing the general movement description obtained through studies of poly(A) RNA.

Vesicular Stomatitis Virus N Protein Messenger RNA and Cellular Myelocytomatosis Protein Messenger RNA

A study performed in rat embryo fibroblasts showed that probes to vesicular stomatitis virus N protein mRNA and c-myc mRNA displayed diffusional rate intranuclear movements with a concentration on nuclear structures (Leonetti et al., 1991). Nuclear movements were not affected by adenosine triphosphate depletion, temperature shift, or the presence of excess unlabeled oligomer. These results are consistent with those obtained through studies of poly(A) RNA and snRNA, indicating that transcript movement throughout the nucleus is diffusional and that transcripts accumulate on structures throughout the nucleus before entering the cytoplasm.

In conclusion, all the nuclear mRNA trafficking data obtained converge toward a central role of diffusion in the nuclear movements of RNA. This is further confirmed by the fact that the movements observed did not require energy in the form of adenosine triphosphate, and were also not modified by a small decrease in temperature (Politz et al., 1998, 1999). As described next, these data are also in good agreement with the study of the movements of nuclearRNA binding proteins.

RNA Binding Proteins in the Nucleus

Green fluorescent protein fusion protein observations have dramatically expanded our view of molecular dynamics within the nucleus. Although time-lapse microscopy of fluorescent nuclear proteins has permitted study of movements of nuclear organelles (Misteli et al., 1997; Boudonck et al., 1999; Muratani et al., 2002), photobleaching techniques have revealed the highly dynamic nature of their steady-state equilibrium (Phair and Misteli, 2000). Several nuclear RNA binding proteins have now been studied, including the splicing factor SF2/ASF and the small nuleolar RNP (snoRNP) protein fibrillarin (Phair and Misteli, 2000; Chen and Huang, 2001). Each factordisplays a specific localization related to its function, with SF2/ASF being concentrated in the so-called *speckles,* and fibrillarin in nucleoli (Phair and Misteli, 2000). Both proteins were found to be highly mobile within the nucleoplasm (Phairand Misteli, 2000; Chen and Huang, 2001). Diffusion coefficients were in the range of 0.2 to 0.5×10^{-9} cm^2/s, which was similarto the value measured for poly(A) movements by uncaging techniques.

Although photobleaching techniques have revealed a surprisingly dynamic behavior of nuclear RNA binding proteins, time-lapse microscopy has, in contrast, underscored a low mobility of several nuclear compartments (Misteli et al., 1997; Boudonck et al., 1999; Platani et al., 2000). Visualization of speckles through an SF2/ASF-GFP fusion protein showed that they were mostly immobile (Misteli et al., 1997). This is not in contradiction with the high mobility of the SF2/ASF protein itself and the dynamic nature of this compartment. In fact, these time-lapse studies have also shown that the size and shape of speckles is continuously and rapidly changing, and does respond to changes in gene expression (Misteli et al., 1997; Eils et al., 2000).

Another small compartment that has been studied by time-lapse microscopy is the Cajal body (Boudonck et al., 1999; Platani et al., 2000; Muratani et al., 2002). Cajal bodies, which are present at a few copies per cell, are involved in the biogenesis of small RNP such as snRNP and snoRNP (Gall, 2000; Darzacq et al., 2002; Verheggen et al., 2002), which transit through this compartment during their maturation (Narayanan et al., 1999; Sleeman and Lamond, 1999). Green fluorescent protein tagging of coilin, a resident protein of this compartment, has allowed the study of its dynamics in both plant and animal cells (Boudonck et al., 1999; Platani et al., 2000). Cajal bodies were observed to move within the nucleoplasm, approaching velocities of 1 μm/min forthe fastest species (but close to 0.1 μm/min on average). Movements did not appear completely random, allowing possible transient contacts with nucleoli. It should be noted that such contacts occur at high frequency in some cells, and depend on physiological conditions, as originally described by Ramon y Cajal (1903). The otherinter esting aspect of these studies was that Cajal bodies could also occasionally split or fuse. Remarkably, asymmetric splitting events were observed in which some Cajal body constituents such as fibrillarin moved into only one fraction of the split Cajal body (Platani et al., 2000).

Studies of nuclearmRNA using GFP have, to date, focused on the visualization of RNA binding proteins rather than RNA itself because of the artifact introduced into nuclear transcription through transient expression of nonintegrated plasmids. The development of stable cell lines expressing integrated copies of the MS2 repeat loops will allow analysis of nuclearmRNA movements using the MS2-GFP system in the nearfutur e.

Cytoplasmic RNAs

In contrast to nuclearstudies, cytoplasmic studies of mRNA behaviorusing the MS2-GFP system have been possible. Along with the MS2 system, multiple transcript labeling techniques have begun to identify the mechanism of cytoplasmic mRNA localization in various organisms and to characterize the travel unit of cytoplasmic mRNA. Messenger RNA localization has been observed to occur in such diverse systems as *Drosophila* oocytes, *Xenopus* oocytes, yeast, neuronal axons and oligodendrocytes, and primary embryonic fibroblasts (Kuhl and Skehel, 1998; Oleynikov and Singer, 1998; Mowry and Cote, 1999; Chartrand et al., 2001; Dreyfuss et al., 2002), and is related to such varied processes as development, mating-type switching, neuronal development, and cellular motility. Although the mechanism of mRNA localization is not yet fully understood for any single system, the synchronous study of all these systems is enabling a steady rate of progress.

In Neurons

Poly(A) RNA   The movement of endogenous RNA was directly visualized in neuronal processes of living cells using the intercalating dye SYTO-14 (Knowles et al., 1996). Labeled RNA granules colocalized with poly(A+) mRNA, the 60S ribosomal subunit, and elongation factor1 α, implying the translational potential of granules. The average rate of granule movement was 0.1 μm/s and, with additional time in culture granule movements, was increasingly likely to be retrograde. Subsequently, Knowles and Kosik (1997) showed that SYTO-14-labeled mRNA granules in rat cortical neurons increased distal translocation in response to neurotrophin-3, and decreased translocation in the presence of the tyrosine kinase inhibitor K252a, demonstrating a relationship between extracellular signals and RNA translocation in living neurons (Knowles et al., 1996).

Myelin Basic Protein Messenger RNA   One of the first studies in which fluorescently labeled mRNA was microinjected into the cytoplasm of living cells was performed by Ainger et al. (1993), who microinjected myelin basic protein (MBP) mRNA into neurites and observed the formation of granules that were transported along oligodendrocytes and localized to the myelin compartment. Analysis of transport and localization patterns of various MBP deletion and mutation constructs resulted in the identification of separate sequences required for MBP mRNA transport and localization, termed the *RNA transport sequence* (RTS— now known as the *A2RE*, or *hnRNP A2 response element* [Carson et al., 2001]) and *RNA localization region* (or RLS), respectively. Time-lapse confocal imaging of A2RE granule movements along microtubules indicated sustained anterograde translocation velocities of ≈0.2 μm/s. More recent analyses of A2RE RNA granule movements, performed with rapid confocal line scanning through a single granule on a microtubule, demonstrated a rapid pattern of back-and-forth vibrations along the microtubule axis. Through mean squared displacement analysis of these vibrations, it was observed that A2RE granules undergo *corralled diffusion,* or movement of a defined distance before reversal of direction (Carson et al., 2001). Occasionally an A2RE particle will escape corralled diffusion and undergo long-range displacement before recapture.

Studies of A2RE are significant not only for their role in MBP transport but also in studies of HIV-1. The HIV-1 genome contains two A2RE-like sequences, within the major homology region of the *gag* gene and in a region of overlap between the *vpr* and *tat* genes (Mouland et al., 2001). Interestingly, a single base change in the A2RE-2 sequences of HIV (A8G) decreases binding of hnRNP A2 as well as RNA transport. The A2RE-2 A8G polymorphism was found in only 3 of 1074 HIV-1 sequences analyzed, one of which came from a long-term nonprogressor, providing the possibility that viral function may require A2RE-2-mediated RNA trafficking.

Ca2+/Calmodulin-Dependent Protein Kinase II   The movements of GFP-labeled CaMKII mRNA particles were observed in rat hippocampal neurons in the first use of the MS2-GFP system in somatic cells. Interestingly, this study showed that anterograde particle movements increased in response to neuronal depolarization, in comparison with oscillatory and retrograde movements (Rook et al., 2000). This study established a link between an external signal and neuronal mRNA movement type. In addition, studies of CaMKII trafficking may elucidate the mechanism of long-term plasticity. Recently obtained evidence implies that coassembly of CaMKII and neurogranin into the same granules may coordinate expression of these two proteins, which are both required for long-term plasticity at synapses (Pak et al., 2000). This finding implicates cotrafficking of mRNA as a mechanism for coordinated gene expression in neurons.

β-Actin Messenger RNA   Examination of GFP-labeled β-actin ZBP-1 provided insights regarding the translocation of β-actin mRNA in neurons (Zhang et al., 2001)[4]. Live-cell imaging of enhanced GFP (EGFP)-ZBP-1 revealed that GFP-labeled particles move in a rapid, bidirectional fashion that is reduced 10-fold by antisense oligonucleotides to the β-actin zip code.

In Yeast

Ash1 RNA   *Ash1* mRNA was visualized in live *Saccharomyces cerevisiae* in the first use of the MS2-GFP labeling system. The GFP-*Ash1* particle required specific *ASH1* sequences for formation and localization to the bud tip (Bertrand et al., 1998). Mutations in the *Ash1*-associated src homology 2 domain-containing transforming protein E (She) proteins were found to disrupt particle localization, and She2 and She3 mutants also inhibited particle formation. Video microscopy showed that She1p/Myo4p moved mRNA particles to the bud tip at velocities of 200 to 440 nm/s. Further studies of the *Ash1* mRNA transport process

(Beach et al., 1999) showed that in cells lacking Bud6p/Aip3p orShe5p/Bni1p, *Ash1* mRNA particles traveled to the bud but failed to remain at the bud tip, revealing a distinction between mRNA transport and localization similar to that described in oligodendrocytes (Ainger et al., 1997). These findings not only characterized the movement of the *Ash1* mRNA particle but also contributed significantly to the identification of proteins, including a motor, that are required for *Ash1* particle transport.

In Drosophila

bicoid   An alternative to fusing GFP to mRNA through the MS2 system is the use of GFP-tagged RNA binding proteins. The dynamic behavior of *bicoid* mRNA in living *Drosophila* oocytes was visualized with such a mechanism using a GFP–exuperantia fusion (Theurkauf and Hazelrigg, 1998). The results of this study elaborate the relationship between mRNA transport and the cytoskeleton, demonstrating that particle movement through nurse cells is microtubule dependent, whereas particle movement from nurse cells to the oocyte, through ring canals, is cytoskeleton independent. Lastly, upon entering the oocyte, particle movements again require microtubules for transport to the oocyte cortex.

Additional information regarding the *bicoid* localization mechanism came from microinjection of fluorescent *bicoid* transcripts into *Drosophila* oocytes (Cha et al., 2001). Interestingly, fluorescent *bicoid* mRNA injected into the oocyte displays non-polar microtubule-dependent transport to the closest cortical surface, but *bicoid* mRNA injected into the nurse cell cytoplasm, withdrawn, and injected into a second oocyte shows microtubule-dependent transport to the anterior cortex. This study identified the requirement of a nurse cell cytoplasmic component for proper trafficking of *bicoid*. It is interesting to note that a component that is present in the nurse cell cytoplasm allows reconnection of *bicoid* to the microtubule network after a microtubule-independent transit through the ring canal, indicating that transcripts bear some mark, possibly a bound protein, indicating their prior location even as they continue transit into other compartments. Thus, transcripts in a given compartment may be able to be distinguished by theirpast behaviorbased on their current set of interaction partners. This concept of transcript historesis has also been proposed with respect to the behavior of CaMKII granules (Rook et al., 2000). In this case, it was suggested that granule location at a specific point in time reflected the activation history of a synapse, allowing identification of stimuli that were in the past.

Wingless and Pair-Rule Transcripts   Injection of fluorescently labeled *wingless* and pair-rule transcripts into living *Drosophila* syncytial blastoderm embryos (Wilkie and Davis, 2001) showed that fluorescent apical RNAs specifically assemble into particles that approach the minus end of microtubules using the motor protein cytoplasmic dynein and its associated dynactin complex.

In Mammalian Cell Lines

c-Fos mRNA   The behavior of transiently expressed c-fos mRNA was observed in living Cos-7 cells through microinjection of fluorescently labeled oligonucleotides (Tsuji et al., 2001). Nuclear accumulation was prevented by fusion of oligos to streptavidin, and oligos were stable for up to 30 minutes. When two oligos labeled with different fluorescent molecules hybridized simultaneously to a c-fos transcript, FRET was detected throughout the cytoplasm, with prevalence in perinuclear regions.

An advantage of using two distinct ODNs and performing FRET analysis is the increased specificity conferred by selecting for dual hybridization events.

vav, β-Actin, and β One Adrenergic Receptor Messenger RNA   Microinjection of molecular beacons targeted to the *vav* protooncogene in K562 human leukemia cells followed by confocal microscopy within 15 minutes of microinjection demonstrated that molecular beacons could detect transcripts in the cytoplasm of living cells, with a detection limit of ≈ 10 molecules of mRNA (Sokol et al., 1998). A laterstudy analyzed molecularbeacons microinjected into the cytoplasm of living kangaroo rat kidney cells (PtKs) for hybridization to β-actin and β one adrenergic receptor mRNA (Perlette and Tan, 2001). This study showed that most beacon hybridization occurred within 10 minutes of microinjection, and maximum intensity was obtained by 15 minutes and maintained until 40 minutes aftermicr oinjection.

SV40, Human Growth Hormone, and β-Actin Messenger RNA   Recent studies (Fusco et al., submitted) have extended the analysis of β-actin mRNA to the single-molecule level using the MS2-GFP fusion system. Dynamic behaviorof β-actin 3' untranslated region (UTR) mRNA was compared with that of human growth hormone (hGH) and SV40 transcripts in Cos-7 cells. These three transcripts were chosen because β-actin mRNA, hGH mRNA, and SV40 mRNA represent transcripts of three distinct categories: endogenous sequences that have been observed to localize, endogenous sequences that have not been observed to localize, and nonendogenous sequences that have not been observed to localize, respectively. It was found that all three transcript types displayed four categories of movement: static, diffusional, corralled diffusional, and directed. Interestingly, transcripts containing the β-actin 3' UTR were observed to undergo directed movement more often than SV40 and hGH transcripts, indicating that the mechanism formRNA localization may involve a sequence-dependent increase in motorlike movements. Although Cos-7 cells do not themselves localize β-actin mRNA, they do contain the β-actin 3' UTR binding protein, ZBP-1, which may be sufficient to enable the primary steps of localization and the observed increase in sequence-dependent directed movements.

In Primary Cells

To determine the mechanism of mRNA localization better, it is necessary to study this process directly in primary cells, such as fibroblasts or neurons, where localization is known to occur. These studies are difficult because of the low transfection efficiency of primary cells and the thickness of their cytoplasm, which complicates particle tracking. Nevertheless, such studies are in progress (Fusco et al., 2003) and it is hoped that they will yield significant insights into the mechanism of mRNA localization in somatic cells, along with the role of this process in wound healing, metastasis, and neuronal development.

CONCLUSION

The field of live-cell mRNA analysis is in a period of reorientation, slowly shifting focus from the development of new visualization technology to the application of that technology to important medical questions. Significant advances have been made already, including the identification of a diffusionlike transport of poly(A) RNA in the nucleus (Politz et al., 1998, 1999), the detailed characterization of mRNA transport and localization particles in yeast (Bertrand et al., 1998) and in *Drosophila* oocytes (Theurkauf and Hazelrigg, 1998; Cha et al., 2001), signal transduction mechanisms related to mRNA transport in neurons (Rook et al., 2000) and in fibroblasts (Latham et al., 2001), and sequence-dependent movement differences in mammalian cells lines (Fusco et al., 2003). Despite these advances, majorquestions remain. For example, what is the route taken by mRNA molecules from the transcription site to the nuclear pore? Is this route identical for all transcripts? Does this route include motorized transport at any step? Is there an intermediate site of accumulation between the transcription site and the nuclear pore? Also, where are the multimolecule mRNA granules (so often observed) formed? In the nucleus or the cytoplasm? In the cytoplasm, what is the mechanism of mRNA localization, and does it include a sequence-specific process of directed movement, of packaging, of anchoring, or some combination of these processes?

Is the mechanism the same for all localized transcripts and what are the signal transduction pathways involved in stimulating localization? Last, what are the medical consequences of disrupted mRNA trafficking and the medical applications of altered mRNA trafficking? As the technology described in this chapterbecomes betterpr acticed and applied, we will doubtless find answers to many of these questions and advance our understanding of molecular and cellular biology in the process. Furthermore, the future will yield new developments in imaging cells in live animals (Farina et al., 1998) and will eventually unveil the details of the RNA life cycle.

REFERENCES

Agutter, P. S. Models for solid-state transport: Messenger RNA movement from nucleus to cytoplasm. *Cell Biol Int* **18**, 849–858 (1994).

Ainger, K., et al. Transport and localization of exogenous myelin basic protein mRNA microinjected into oligodendrocytes. *J Cell Biol* **123**, 431–441 (1993).

Ainger, K., et al. Transport and localization elements in myelin basic protein mRNA. *J Cell Biol* **138**, 1077–1087 (1997).

Ashley, C. T., Jr., Wilkinson, K. D., Reines, D., and Warren, S. T. FMR1 protein: Conserved RNP family domains and selective RNA binding. *Science* **262**, 563–566 (1993).

Axelrod, D., Koppel, D. E., Schlessinger, J., Elson, E., and Webb, W. W. Mobility measurement by analysis of fluorescence photobleaching recovery kinetics. *Biophys J* **16**, 1055–1069 (1976).

Beach, D. L., Salmon, E. D., and Bloom, K. Localization and anchoring of mRNA in budding yeast. *Curr Biol* **9**, 569–578 (1999).

Bertrand, E., et al. Localization of ASH1 mRNA particles in living yeast. *Mol Cell* **2**, 437–445 (1998).

Blobel, G. Gene gating: A hypothesis. *Proc Natl Acad Sci USA* **82**, 8527–8529 (1985).

Boudonck, K., Dolan, L., and Shaw, P. J. The movement of coiled bodies visualized in living plant cells by the green fluorescent protein. *Mol Biol Cell* **10**, 2297–2307 (1999).

Boulon, S., Basyuk, E., Blanchard, J., Bertrand, E., and Verheggen, C. Intra-nuclear RNA trafficking: Insights from live cell imaging. *Biochimie* **84**, 805–813 (2002).

Brown, V., et al. Microarray identification of FMRP-associated brain mRNAs and altered mRNA translational profiles in fragile X syndrome. *Cell* **107**, 477–487 (2001).

Capoulade, C., et al. Apoptosis of tumoral and nontumoral lymphoid cells is induced by both mdm2 and p53 antisense oligodeoxynucleotides. *Blood* **97**, 1043–1049 (2001).

Carmo-Fonseca, M., et al. In vivo detection of snRNP-rich organelles in the nuclei of mammalian cells. *EMBO J* **10**, 1863–1873 (1991).

Carson, J. H., Cui, H., and Barbarese, E. The balance of power in RNA trafficking. *Curr Opin Neurobiol* **11**, 558–563 (2001).

Cha, B. J., Koppetsch, B. S., and Theurkauf, W. E. In vivo analysis of *Drosophila bicoid* mRNA localization reveals a novel microtubule-dependent axis specification pathway. *Cell* **106**, 35–46 (2001).

Chartrand, P., Singer, R. H., and Long, R. M. RNP localization and transport in yeast. *Annu Rev Cell Dev Biol* **17**, 297–310 (2001).

Chen, D., and Huang, S. Nucleolarcomponents involved in ribosome biogenesis cycle between the nucleolus and nucleoplasm in interphase cells. *J Cell Biol* **153**, 169–176 (2001).

Cheney, R. E., et al. Brain myosin-V is a two-headed unconventional myosin with motor activity. *Cell* **75**, 13–23 (1993).

Darzacq, X., et al. Cajal body-specific small nuclear RNAs: A novel class of 2′-O-methylation and pseudouridylation guide RNAs. *EMBO J* **21**, 2746–2756 (2002).

Davis, B. M., McCurrach, M. E., Taneja, K. L., Singer, R. H., and Housman, D. E. Expansion of a CUG trinucleotide repeat in the 3′ untranslated region of myotonic dystrophy protein kinase transcripts results in nuclear retention of transcripts. *Proc Natl Acad Sci USA* **94**, 7388–7393 (1997).

Dirks, R. W., Molenaar, C., and Tanke, H. J. Methods for visualizing RNA processing and transport pathways in living cells. *Histochem Cell Biol* **115**, 3–11 (2001).

Dreyfuss, G., Kim, V. N., and Kataoka, N. Messenger-RNA-binding proteins and the messages they carry. *Nat Rev Mol Cell Biol* **3**, 195–205 (2002).

Eils, R., Gerlich, D., Tvarusko, W., Spector, D. L., and Misteli, T. Quantitative imaging of pre-mRNA splicing factors in living cells. *Mol Biol Cell* **11**, 413–418 (2000).

Farina, K. L., et al. Cell motility of tumor cells visualized in living intact primary tumors using green fluorescent protein. *Cancer Res* **58**, 2528–2532 (1998).

Ferrandon, D., Elphick, L., Nusslein-Volhard, C., and St Johnston, D. Staufen protein associates with the 3′ UTR of bicoid mRNA to form particles that move in a microtubule-dependent manner. *Cell* **79**, 1221–1232 (1994).

Fusco, D., Accornero, N., Lavoie, B., Shenoy, S. M., Blanchard, J., Singer, R. H., and Bertrand, E. Single mRNA molecules demonstrate probabilistic movement in living mammalian cells. *Curr. Biol.* 13, 161–167 (2003).

Gall, J. G. Cajal bodies: The first 100 years. *Annu Rev Cell Dev Biol* **16**, 273–300 (2000).

Gu, W., Pan, F., Zhang, H., Bassell, G. J., and Singer, R. H. A predominantly nuclear protein affecting cytoplasmic localization of beta-actin mRNA in fibroblasts and neurons. *J Cell Biol* **156**, 41–51 (2002).

Huang, S., Deerinck, T. J., Ellisman, M. H., and Spector, D. L. The perinucleolar compartment and transcription. *J Cell Biol* **143**, 35–47 (1998).

Knowles, R. B., et al. Translocation of RNA granules in living neurons. *J Neurosci* **16**, 7812–7820 (1996).

Knowles, R. B., and Kosik, K. S. Neurotrophin-3 signals redistribute RNA in neurons. *Proc Natl Acad Sci USA* **94**, 14804–14808 (1997).

Kuhl, D., and Skehel, P. Dendritic localization of mRNAs. *Curr Opin Neurobiol* **8**, 600–606 (1998).

Lago, H., Fonseca, S. A., Murray, J. B., Stonehouse, N. J., and Stockley, P. G. Dissecting the key recognition features of the MS2 bacteriophage translational repression complex. *Nucl Acids Res* **26**, 1337–1344 (1998).

Latham, V. M., Yu, E. H., Tullio, A. N., Adelstein, R. S., and Singer, R. H. A Rho-dependent signaling pathway operating through myosin localizes beta-actin mRNA in fibroblasts. *Curr Biol* **11**, 1010–1016 (2001).

Leonetti, J. P., Mechti, N., Degols, G., Gagnor, C., and Lebleu, B. Intracellular distribution of microinjected antisense oligonucleotides. *Proc Natl Acad Sci USA* **88**, 2702–2706 (1991).

Lever, A. M. Gene therapy for HIV. *Sex Transm Infect* **77**, 93–96 (2001).

Levsky, J., Shenoy, S. M., Pezo, R. C., and Singer, R. H. Single-cell gene expression profiling. *Science* **297**, 836–840 (2002).

Lorenz, P., Misteli, T., Baker, B. F., Bennett, C. F., and Spector, D. L. Nucleocytoplasmic shuttling: A novel in vivo property of antisense phosphorothioate oligodeoxynucleotides. *Nucl Acids Res* **28**, 582–592 (2000).

Manns, M. P., Cornberg, M., and Wedemeyer, H. Current and future treatment of hepatitis C. *Indian J Gastroenterol* **20** (Suppl. 1), C47–C51 (2001).

Mayne, M., et al. Antisense oligodeoxynucleotide inhibition of tumor necrosis factor-alpha expression is neuroprotective after intracerebral hemorrhage. *Stroke* **32**, 240–248 (2001).

Misteli, T., Caceres, J. F., and Spector, D. L. The dynamics of a pre-mRNA splicing factor in living cells. *Nature* **387**, 523–527 (1997).

Misteli, T., Gunjan, A., Hock, R., Bustin, M., and Brown, D. T. Dynamic binding of histone H1 to chromatin in living cells. *Nature* **408**, 877–881 (2000).

Molenaar, C., et al. Linear 2′ O-methyl RNA probes for the visualization of RNA in living cells. *Nucl Acids Res* **29**, E89 (2001).

Mouland, A. J., et al. RNA trafficking signals in human immunodeficiency virus type 1. *Mol Cell Biol* **21**, 2133–2143 (2001).

Mowry, K. L., and Cote, C. A. RNA sorting in *Xenopus* oocytes and embryos. *FASEB J* **13**, 435–445 (1999).

Muratani, M., et al. Metabolic-energy-dependent movement of PML bodies within the mammalian cell nucleus. *Nat Cell Biol* **4**, 106–110 (2002).

Narayanan, A., Speckmann, W., Terns, R., and Terns, M. P. Role of the box C/D motif in localization of small nucleolar RNAs to coiled bodies and nucleoli. *Mol Biol Cell* **10**, 2131–2147 (1999).

Oleynikov, Y., and Singer, R. H. RNA localization: Different zipcodes, same postman? *Trends Cell Biol* **8**, 381–283 (1998).

Orr, R. M. Technology evaluation: Fomivirsen, Isis Pharmaceuticals Inc/CIBA vision. *Curr Opin Mol Ther* **3**, 288–294 (2001).

Pak, J. H., et al. Involvement of neurogranin in the modulation of calcium/calmodulin-dependent protein kinase II, synaptic plasticity, and spatial learning: A study with knockout mice. *Proc Natl Acad Sci USA* **97**, 11232–11237 (2000).

Pederson, T. Movement and localization of RNA in the cell nucleus. *FASEB J* **13** (Suppl. 2), S238–S242 (1999).

Pederson, T. Fluorescent RNA cytochemistry: Tracking gene transcripts in living cells. *Nucl Acids Res* **29**, 1013–1016 (2001).

Perlette, J., and Tan, W. Real-time monitoring of intracellular mRNA hybridization inside single living cells. *Anal Chem* **73**, 5544–5550 (2001).

Phair, R. D., and Misteli, T. High mobility of proteins in the mammalian cell nucleus. *Nature* **404**, 604–609 (2000).

Platani, M., Goldberg, I., Swedlow, J. R., and Lamond, A. I. In vivo analysis of Cajal body movement, separation, and joining in live human cells. *J Cell Biol* **151**, 1561–1574 (2000).

Politz, J. C., Browne, E. S., Wolf, D. E., and Pederson, T. Intranuclear diffusion and hybridization state of oligonucleotides measured by fluorescence correlation spectroscopy in living cells. *Proc Natl Acad Sci USA* **95**, 6043–6048 (1998).

Politz, J. C., Taneja, K. L., and Singer, R. H. Characterization of hybridization between synthetic oligodeoxynucleotides and RNA in living cells. *Nucl Acids Res* **23**, 4946–4953 (1995).

Politz, J. C., Tuft, R. A., Pederson, T., and Singer, R. H. Movement of nuclear poly(A) RNA throughout the interchromatin space in living cells. *Curr Biol* **9**, 285–291 (1999).

Privat, E., Melvin, T., Asseline, U., and Vigny, P. Oligonucleotide-conjugated thiazole orange probes as "light-up" probes formessengerr ibonucleic acid molecules in living cells. *Photochem Photobiol* **74**, 532–541 (2001).

Ramon y Cajal, S. Un sencillo metodo de coloracion selectiva del reticulo protoplasmico y sus efectos en los diversos organos nerviosos de vertebrados e invertebrados. *Trab Lab Invest Biol* **2**, 129–221 (1903).

Robinett, C. C., et al. In vivo localization of DNA sequences and visualization of large-scale chromatin organization using lac operator/repressor recognition. *J Cell Biol* **135**, 1685–1700 (1996).

Rook, M. S., Lu, M., and Kosik, K. S. CaMKIIalpha 3′ untranslated region-directed mRNA translocation in living neurons: Visualization by GFP linkage. *J Neurosci* **20**, 6385–6393 (2000).

Ross, A. F., Oleynikov, Y., Kislauskis, E. H., Taneja, K. L., and Singer, R. H. Characterization of a beta-actin mRNA zipcode-binding protein. *Mol Cell Biol* **17**, 2158–2165 (1997).

Schroder, H. C., Bachmann, M., Diehl-Seifert, B., and Muller, W. E. Transport of mRNA from nucleus to cytoplasm. *Prog Nucl Acid Res Mol Biol* **34**, 89–142 (1987).

Shestakova, E. A., Wyckoff, J., Jones, J., Singer, R. H., and Condeelis, J. Correlation of beta-actin messenger RNA localization with metastatic potential in rat adenocarcinoma cell lines. *Cancer Res* **59**, 1202–1205 (1999).

Siomi, H., Siomi, M. C., Nussbaum, R. L., and Dreyfuss, G. The protein product of the fragile X gene, FMR1, has characteristics of an RNA-binding protein. *Cell* **74**, 291–298 (1993).

Sleeman, J. E., and Lamond, A. I. Newly assembled snRNPs associate with coiled bodies before speckles, suggesting a nuclearsnRNP maturation pathway. *Curr Biol* **9**, 1065–1074 (1999).

Sokol, D. L., Zhang, X., Lu, P., and Gewirtz, A. M. Real time detection of DNA · RNA hybridization in living cells. *Proc Natl Acad Sci USA* **95**, 11538–11543 (1998).

Taneja, K. L., McCurrach, M., Schalling, M., Housman, D., and Singer, R. H. Foci of trinucleotide repeat transcripts in nuclei of myotonic dystrophy cells and tissues. *J Cell Biol* **128**, 995–1002 (1995).

Theurkauf, W. E., and Hazelrigg, T. I. In vivo analyses of cytoplasmic transport and cytoskeletal organization during *Drosophila* oogenesis: Characterization of a multi-step anterior localization pathway. *Development* **125**, 3655–3666 (1998).

Thuong, N. T., Asseline, U., Roig, V., Takasugi, M., and Helene, C. Oligo(alpha-deoxynucleotides) covalently linked to intercalating agents: Differential binding to ribo- and deoxyribopolynucleotides and stability towards nuclease digestion. *Proc Natl Acad Sci USA* **84**, 5129–5133 (1987).

Tsuji, A., et al. Development of a time-resolved fluorometric method for observing hybridization in living cells using fluorescence resonance energy transfer. *Biophys J* **81**, 501–515 (2001).

Tyagi, S., Bratu, D. P., and Kramer, F. R. Multicolor molecular beacons for allele discrimination. *Nat Biotechnol* **16**, 49–53 (1998).

Tyagi, S., and Kramer, F. R. Molecular beacons: Probes that fluoresce upon hybridization. *Nat Biotechnol* **14**, 303–308 (1996).

Verheggen, C., et al. Mammalian and yeast U3 snoRNPs are matured in specific and related nuclear compartments. *EMBO J* **21**, 2736–2745 (2002).

Wang, J., Cao, L. G., Wang, Y. L., and Pederson, T. Localization of pre-messenger RNA at discrete nuclear sites. *Proc Natl Acad Sci USA* **88**, 7391–7395 (1991).

Wilkie, G. S., and Davis, I. *Drosophila* wingless and pair-rule transcripts localize apically by dynein-mediated transport of RNA particles. *Cell* **105**, 209–219 (2001).

Zhang, G., Taneja, K. L., Singer, R. H., and Green, M. R. Localization of pre-mRNA splicing in mammalian nuclei. *Nature* **372**, 809–812 (1994).

Zhang, H. L., et al. Neurotrophin-induced transport of a beta-actin mRNP complex increases beta-actin levels and stimulates growth cone motility. *Neuron* **31**, 261–275 (2001).

5

Building New Fluorescent Probes

ALAN S. WAGGONER, LAUREN A. ERNST, AND BYRON BALLOU

CURRENT STATE-OF-THE-ART

Fluorescent probe-based cellular imaging is a powerful, quantitative, widely accepted technology for cell biologists. Fluorescent probes and digital imaging—interdependent technologies—grew rapidly during the past two decades. Commercially available instrumentation and fluorescent probe reagents of many types are widely used by cell biologists to dissect the workings of living cells. The technology has matured, and the best descriptions of methods are often found in manufacturers' literature (e.g., Molecular Probes, Inc., Clontech, and microscope and flow cytometry companies). The large number of specific probes and the wide choice of chemical properties and emission wavelengths mark a major increase in ourability to monitorbiological events, mostly gained during the past two decades.

During the past few years, GFP and its variants (CFP, YFP, DsRed, and other fluorescent proteins) have made a huge impact on the studies of intracellular processes in living cells (Griffin et al., 2000; Zhang et al., 2002; Lippincott-Schwartz and Patterson, 2003; Miyawaki et al., 2003). Research is now underway in a number of laboratories to extend the capabilities of fluorescent proteins, for example, as selective binding probes (Peelle et al., 2001), as intracellular biosensors (Violin et al., 2003; Yu et al, 2003; Dooley et al., 2004, Hanson et al., 2004), and as probes for movement of macromolecules by in situ activation of fluorescence (Patterson and Lippincott-Schwartz, 2002).

Anothernew fluorescence technology is quantum dots; semiconductornanocr ystals with diameters in the 5- to 10-nm range. These materials combine high absorptivity and high quantum yield, giving unprecedented brightness. The absorptivity of quantum dots increases as the wavelength of exciting light decreases; this property gives efficient excitation, widely separated spectral windows for excitation and emission, and the ability to excite many different emission colors with a single wavelength. When manufactured to have uniform size and composition, quantum dots have narrow emission bands, allowing many fluorophores to be used in a given experiment with minimal spectral overlap (Alivisatos, 1996; Watson et al., 2003). By varying composition and size, quantum dots with emission maxima from 460 to 860 nm have been prepared and used for biological labeling. Longer wavelengths are likely soon to be available. Last, quantum dots have favorable properties for two-photon excitation (Larson et al., 2003) and for Förster energy transfer (Medintz et al., 2003; Clapp et al., 2004). Their relatively long fluorescence lifetime can be used to reduce background by time gating (Dahan et al., 2001). Their chief advantages are their high brightness and resistance to photobleaching. Quantum dots are sufficiently bright to allow single-molecule imaging without specialized equipment (Dahan et al., 2003; Lidke et al., 2004), and are sufficiently stable to allow long-term observation. The propensity of quantum dots to blink can distinguish quantum dots from other bright objects (Dahan et al., 2003).

These almost magical properties ensure that quantum dots will displace familiar fluorescent probes for many types of labeling in vivo and in vitro. On the other hand, although uncoated quantum dots are small, they are massive, with high molecular weights ($\approx 5 \times 10^5$) and high density (specific gravity, ≈ 6). Moreover, quantum dots must be coated to protect them from an aqueous environment. The use of quantum dots in biology depends mostly on the development of surface coatings that protect the surface, minimize aggregation, and allow conjugation. These surface materials substantially increase the size and molecular weight of quantum dots, which may restrict movement of labeled macromolecules and interfere with the ability of quantum dots to cross cell membranes or to penetrate dense tissues.

Last, the semiconductor materials used to date are composed of heavy metals, and it is possible that breakdown of quantum dots could lead to toxicity (Derfus et al., 2004; Kim et al., 2004; Lidke et al., 2004). In our experience, these dangers are minimal; we have observed no signs of localized toxicity in labeled tissues after many months, even though the tissues remained fluorescent (Ballou et al., 2004).

Work is also underway for alternative methods to create intracellular fluorescent labels and biosensors. Creating genetically encoded sequences that bind fluorochromes selectively is one such approach. Using FlAsH technology (fluorescein with As(III) substituents on the $4'$- and $5'$- positions), a nonfluorescent dye is diffused into cells and becomes fluorescent when bound to a specific tetracysteine motif engineered into a target protein (Griffin et al., 2000; Adams et al., 2002; Sosinsky et al., 2003). Yet another recent approach, also from the Tsien laboratory, uses nucleic acid aptamers that bind a nonfluorescent dye to make a fluorescent species (Babendure et al., 2003).

Other new approaches to generating fluorescent signals in situ include fluorettes, short peptide sequences that bind to nonfluorescent fluorogens and render them fluorescent (Rozinov and Nolan, 1998), antibody binding sites that act similarly (Simeonov et al., 2000), or other proteins that may be engineered using fluorochromes that act as sensors (Richieri et al., 1999), which might be whole antibodies, antigen-binding fragments, or scFv's (single-chain variable fragment antibodies).

NEED FOR NEW PROBES

One might ask, then, why synthesis of new fluorescent organic chemical probes is worthwhile. The answers are straightforward: to make low-molecular weight probes that are less likely to affect biological activity, to allow access to specific sites on biomolecules that larger probes will not fit (there is probably no nook or cranny that a specific probe cannot be designed to fit, although the process of making the probe may be extremely laborious), and to improve small organic probes to increase sensitivity, allow a wider spectral range, and provide increased stability, especially for near-infrared and infrared probes.

For all types of probes, the goals are to have sufficient fluorescence at the tagged site while minimizing background, and to have the fluorescent signal generated immediately (this is typically a problem with GFPs, which may be slow to mature).

It is not the purpose of this review to discuss the existing classes of probes and their relative utility. Much of this information can be obtained from the Molecular Probes catalog or from review articles (e.g., Tsien and Waggoner, 1995). Instead, we discuss the prospects for developing core fluorophores that fluoresce in the near-infrared region of the spectrum and that can be used to create fluorescent labels and indicators for deep imaging in tissues.

Advances in imaging microscopy methods and fluorescent probes have greatly enhanced our understanding of the structure and function of cells in culture. Now these methods are being called upon to analyze fluorescence signals from deeper regions (>100 μm) in tissues. Confocal techniques have grown particularly fast because of the capability to perform optical sectioning and visualize at greater depths in specimens by means of two-photon excitation (Tsien and Waggoner, 1995; Williams et al., 1995). However, there remain limitations to the sensitivity and resolution that can be obtained at depths greater than 100 μm in tissues. Light scattering and sample autofluorescence contribute to the limitations. In animal models and humans, the presence of hemoglobin makes deep imaging even more difficult. To circumvent these limitations in mammalian tissues, advances are needed in the fluorescent reagents that act as probes to report information such as antigen location, physiological events, and the presence of labeled immune cells. A realistic goal would be to have the capability of imaging as few as 1 to 10 labeled cells at a depth of 1000 μm.

INFRARED PROBES FOR DEEP IMAGING: WHY THE INFRARED?

Why might infrared-fluorescing probes be useful? There are several important reasons. First, cell autofluorescence predominates at the blue region of the spectrum but also extends into longer wavelengths, depending on cell type. Most reports are anecdotal, but factors of a 10 to 100 increase in the ratio of signal to background fluorescence as the probe fluorescence is shifted from 500 to near 700 nm are not unusual. There may still be significant gains by moving further into the infrared because very few endogenous cell and tissue compounds fluoresce in the near infrared (except for photosynthetic systems of plants and bacteria). Synthesis of near-infrared probes is therefore an area of active research (Strekowski et al., 1996a; Flanagan et al., 1997; Pham et al., 2003).

Second, for deep tissue imaging, long-wavelength light travels farther in tissues. There is a spectral window (fig. 5.1) between the end of the major components of hemoglobin absorption (about 650 nm) and the limit imposed by increased water absorbance (fig. 5.2). Probes that can be excited and emit in this window are more likely to be useful for imaging tumors, the vascular system, and other cells and organs in living animals.

Optical imaging in thick specimens and living tissues is limited by high light scattering, absorption by hemoglobin in the visible to far-red wavelengths, and absorption by water and other tissue components in the infrared. Myoglobin and other heme proteins, bilirubin, and ferritin also absorb in some animal tissues; chlorophyll and other photosynthetic pigments are major absorbers in plants. Background fluorescence resulting from pyridine nucleotides, flavins, and lipofuscins limits sensitivity in animals, peaking in the blue but extending into the red. Fluorescence of photosynthetic pigments in plants extends into the infrared.

For animal tissues, shifting to red and near-infrared wavelengths markedly increases the tissue thickness through which reasonable resolution can be obtained. The important early work of Wan et al. (1981) showed that light transmission through body wall tissue improved by several powers of 10 from blue to near infrared, increasing to 0.5% to 5% percentimeterat 800 nm.

Unfortunately, water absorption increases steeply in the near infrared and becomes prohibitive for wavelengths much above 1.3 μm (Curcio and Petty, 1951; Hale and Querry, 1973; Irvine and Pollack, 1968) (fig. 5.2). This high absorption not only attenuates the signal, but limits the power that can be input as a result of local heating. The most favorable spectral windows in water are less than 920 nm, and in the small window of reduced absorption between 1000 and 1100 nm.

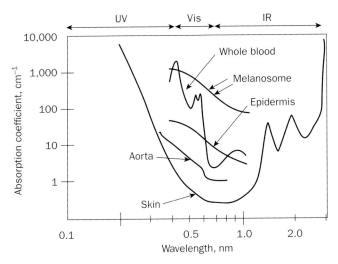

Figure 5.1 Absorption coefficients of whole blood, melanosome, and skin. The minima of light absorption by biological samples are between 600 and 1000 nm. IR, infrared; UV, ultraviolet; Vis, visible. (Redrawn from Terpetschnig and Wolfbeis [1998].)

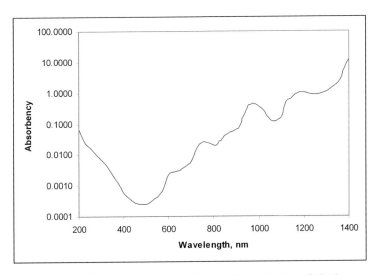

Figure 5.2 Water absorption in the near infrared. (From Irving and Pollack [1968].)

However, light passage through tissue is governed by a combination of scatter and absorption; depending on wavelength, signal attenuation may vary by four orders of magnitude from one tissue to another (some examples are shown in fig. 5.1 [Terpetschnig and Wolfbeis, 1998]). The data summarized in figures 5.1 and 5.2 refer only to light attenuation and not to image degradation. It is possible to imagine extreme cases in which light is highly absorbed, but not scattered, in which case an image could be recovered by increasing illumination or exposure time; or a case of pure scatter, when much of the light can be recovered, but the image is irreversibly degraded (intralipid at visible wavelengths is a good example of high scatter with little absorption). Rayleigh and compound scattering are generally reduced by using longer wavelength light, although over some ranges of particle size and refractive index, scatter may be enhanced by increasing wavelength.

Considerable effort has gone into recovering information from highly scattering media using several techniques, including time-resolved and tomographic methods (Ballou et al., 1997; Chance, 1998; Mahmood et al., 1999; Zhang et al., 2000; Ntziachristos and Chance, 2001; Ntziachristos et al., 2002; Sevick-Muraca et al., 2002; Bremer et al., 2003; Eppstein et al., 2003; Houston et al., 2003; Ke et al., 2003; Lim et al., 2003; Ntziachristos et al., 2003; Thompson et al., 2003; Weisslederand Ntziachristos, 2003). On the otherhand, where depth is limited (submillimeterdistances), light attenuation by waterabsor ption is relatively unimportant and longer wavelengths may be used to minimize scatter. Thirteen hundred nanometers of light has been used for optical coherence tomography (OCT) (Huang et al., 1991; Pan and Farkas, 1997, 1998; Fujimoto et al., 1998; Fujimoto et al., 2000; Pan et al., 2001; Drexler, 2004) and confocal scanning reflectance microscopy (Rajadhyaksha et al., 1995a,b; Gonzalez et al., 1998; Koenig et al., 1999).

Integrating the effects of scattering and absorption on both excitation and emission, Lim et al. (2003) investigated several wavelength bands for detection in tissues of varying scatterand hemoglobin content, at a depth of ≈ 0.5 cm, using both modeling and experiment. Despite waterabsor bency, fortissues with a high hemoglobin content, considerable advantage was obtained by spectral windows centered on 1320 nm compared with ≈ 840 nm, ≈ 1110 nm, or1680 nm; forlow hemoglobin content, 840- and 1110-nm-centered windows were better. Which wavelength is better depends on the tissue and on whether resolution or detection is the more important goal.

A number of optical techniques produce interpretable images at depths in tissue ranging from micrometers to centimeters. Confocal scanning reflectance microscopy, as well as confocal and two- or multiphoton scanning fluorescence microscopy are useful up to a limit set by scattering and the working distance of moderately high-aperture objectives, roughly 200 to 400 μm. Despite the substantial improvement over confocal microscopy conferred by two- or multiphoton techniques, several problems still remain. Despite the exquisite resolution offered by multiphoton methods, high-aperture lenses are required to focus sufficient energy in a small volume to induce fluorescence, which limits improvement in depth. Lesser problems include autofluorescence in the excitation volume (although not in the rest of the field) and phototoxicity (Williams et al., 1994). Attempts to use still longer wavelengths will encounter the barrier of water absorption and local heating, although there is still ample room for improvement. Finally, except for quantum dots, the ability to use multiple probes is limited.

There is another reason for considering near-infrared fluorescent probes. A number of laboratories are attempting to correlate combinations of biochemical events, locations of proteins and organelles, and overall cell behavior. This requires multiple probes having different emission wavelengths. Imaging systems must contain corresponding fluorescence filters (or the equivalent) to discriminate between the different fluorescent signals. This method has been called *multiparameter imaging, multispectral imaging, hyperspectral imaging,* and *multicolor imaging.* Unfortunately, most absorption and emission spectra are relatively broad (≈ 50 nm half-maximum) and have tails extending well beyond their peak maxima, so that it is difficult to discriminate signals quantitatively from the probes when their spectra overlap too much. This limits the number of probes that can be imaged in the visible region of the spectrum. Having probes in the near-infrared spectrum may allow another two to three probes to be used for correlation experiments.

LIMITATIONS OF EXISTING NEAR-INFRARED FLUOROPHORES

Why are there so few fluorescent infrared probes available now? Consider first the class of organic infrared dyes. Many polymethine fluorophores in the infrared show broadened

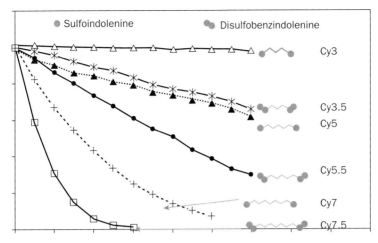

Figure 5.3 Decreasing photostability of cyanine dyes as the length of the polymethine chain increases. Dyes were exposed separately to a constant white light flux for the indicated time, then absorbance was measured. (Data and drawing by R. Mujumdar.)

absorption and emission spectra with reduced extinction coefficients. Part of the apparent broadening is an artifact of using a wavelength scale instead of an energy linear scale. However, much of the effect is probably the result of the ability of the polymethines to exist in many cis-isomer states (Brooker and Vittum, 1957; Brooker, 1962). The cis-isomers appear to reduce the fluorophore quantum yield significantly. Rigid polycyclic chromophores would not exhibit these phenomena, but they have other problems. Polycyclic aromatics are planar and very nonpolar, so that they strongly tend to aggregate and drop out of solution. Moreover, if the polycyclic chromophore does not have a sufficient number of optimally placed electron-rich atoms, it is unlikely to have a near-infrared absorption or emission component. Size is another factor. Generally, polycyclic chromophores require more atoms and double bonds in a structure to shift the emission to long wavelengths, and the large size always stands as a possible perturbant in a fluorescent probe experiment. In what follows, we discuss mainly the cyanine fluorochromes (Flanagane et al., 1997; Mishra et al., 2000), but the principles apply to other fluorochromes.

Last, one of the major factors limiting existing infrared chromophores is their low photostability (fig. 5.3). In imaging microscopy, lack of photostability is often a major limitation. The infrared fluorophores also are often chemically less stable. Little research has been conducted that illuminates the chemical–physical basis of infrared instability. Various approaches are underway to investigate mechanisms and to create more stable organic fluorophores.

DESIGNING NEW PROBES THAT EMIT IN THE NEAR INFRARED

All fluorescent probes contain two fundamental components that have to be properly engineered by organic synthesis to create the final useful fluorescent unit: the core fluorophore structure (fig. 5.4) and the very important substituents, or R-groups, attached to the dye to make it reactive, water soluble, lipid soluble, a good DNA intercalator, and so forth. The R-groups might include reactive groups, such as isothiocyanate, maleimide, or succinimidyl ester; or they may be used to adjust solubility (e.g., sulfonic acid groups for

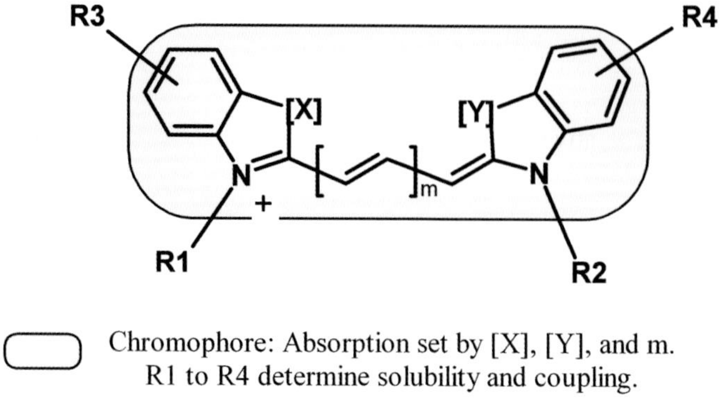

Chromophore: Absorption set by [X], [Y], and m.
R1 to R4 determine solubility and coupling.

Figure 5.4 Basic structure of cyanine fluorochromes.

aqueous solubility) and intracellular location (e.g., long alkyl chains, in the case of fluorescent membrane labels, such as DiI (1,1′-dioctadecyl-3,3,3′,3′-tetramethylindocarbocyanine perchlorate).

The core fluorophore, on the other hand, determines the spectral properties of the fluorophore in different microenvironments where the probe might be located. Spectral properties include absorbance as a function of wavelength, fluorescence as a function of wavelength, extinction, quantum yield, and excited state lifetime. Forfluor escent labels, it is usually important that these properties be independent of the microenvironment of the probe so that the label signal remains constant in all regions of the cell. Fluorescent indicator probes, on the other hand, are designed so that one or more of the spectral properties changes with the microenvironment. For example, a calcium indicator must change its absorbance oremission properties depending on whethercalcium is bound. Forcalcium probes, a key function of the R-groups is to provide calcium selectivity.

It is the goal of the probe chemist to manipulate the core fluorophore and the R-groups to create the probes' spectral properties and biological specificity. The following properties are always in the back of the chemist's mind as this engineering takes place (Waggoner, 1986):

- Detection sensitivity: detection of as few copies of probe per cell as possible
- Reporting the cell activity of interest (and only that specific activity)
- Photostability forlong imaging periods
- Low toxicity and phototoxicity
- Easy delivery to the right location in the right cell
- Different emission wavelengths so that multiple probes can be used to correlate different activities in cells
- Expandable technology; the ideal probe is applicable to many functions
- Ease of use: robust and available to the global research community

DESIGNING THE CORE CHROMOPHORE

Good near-infrared dyes are more difficult to design and synthesize than dyes that fluoresce in the visible region. The extended conjugation systems that are required decrease the chemical stability of the chromophore; syntheses designed for dyes in the ultraviolet–

visible region do not necessarily work for near-infrared dyes. The more complex structures required to obtain good near-infrared dyes are new challenges for dye chemists.

To obtain high brightness, absorbency and quantum yield must be maximized, whereas extended polyene systems are needed. These requirements are antithetical. Internal conversion is a major nonradiative relaxation pathway in polymethine dyes resulting primarily from the nonplanar structure of the molecule that arises from a conformationally "loose" polymethine chain linking the heteroaromatic fragments. Long extended chains allow for increased out-of-plane motion; increased flexibility and cis-trans isomerism reduce both quantum yield and absorbancy as conjugated chains grow longer. The longer the chain, the greater the possibility of nonradiative decay. As a result, the quantum yields of fluorochromes with longer emission wavelengths are substantially lower than those of the best dyes in the visible and near-ultraviolet region of the spectrum.

The remedy for this difficulty may be "rigidization" of the core fluorophore. Restricting the molecule to a planar conformation through covalent bridging of polyene chains or aromatic fragments can dramatically reduce the rate of this internal conversion process (Tredwell and Keary, 1979).

Rigidization may be pursued by covalently locking the molecule in a desired configuration (Waggoner et al., 1996; Karandikar et al., 1998; Mujumdar et al., 1999; Waggoner and Mujumdar, 1999, 2000), or by designing the molecule to assume the desired configuration in an appropriate surround. Examples include the large increase of fluorescence of some cyanine dyes on binding to nucleic acids (Lee et al., 1986) and the increase in fluorescence by immersion in a nonpolar environment.

PROTECTING THE CORE FLUOROPHORE

Chemical instability and photodegradation, both reversible and irreversible, deplete the dye population under prolonged irradiation. The principal mechanism for the photochemical degradation of cyanine dyes in aqueous oxygenated media has been established. Light absorption by the ground-state dye yields a singlet excited state that undergoes intersystem crossing to generate excited triplet-state dye, which in turn undergoes energy transfer to molecular oxygen to produce singlet molecular oxygen and the ground-state dye (Byers et al., 1976). Singlet molecular oxygen attacks the first exocyclic double bond of the polymethine chain adjacent to the aromatic moieties at the ends. Addition across this bond breaks the conjugation of the dye, which results in fission of the chain. Thus, cyanine dye photostability is related to the structure of the dye. Increasing the polymethine chain length reduces photostability. For many cyanine dyes, the effect of the aromatic residues and heteroatom substituents within those residues on photostability has been determined. For Cy3 dyes, the photostability increases along the aromatic cores benzoxazole, benzothiazole and 3,3′-dimethylindolenine (Jun et al., 1996). Fluorination of the heteroaromatic end groups yields improved photostability (Renikuntla et al., 2004).

For optical recording applications, several approaches have been used to improve the stability of the dyes (Emmelius et al., 1991). These approaches involve the inhibition of singlet oxygen activity by the addition of singlet oxygen quenchers and covalent attachment of the quencher to the dye; similar approaches can also be used with dyes intended for use in biological systems. Also, cyanine dyes are stabilized by the introduction of cyclic units into the polymethine chain (Strekowski et al., 1996b). Last, steric prevention of a singlet oxygen attack on the chromophore increases the dye stability, as exemplified by inclusion complexes of cyanine dyes with cyclodextrins in water (Guether and Reddington, 1997; Buston et al., 2000). Shielding of the polymethine chain by cyclodextrins inhibits access to oxygen and solvent species, protecting the dye (Matsuzawa et al., 1994).

DESIGNING THE R-GROUPS TO OPTIMIZE THE UTILITY OF THE INFRARED PROBE

Solubility

Foraqueous use, adding polarsubstituents on the end groups is the most usual method of increasing solubility. With the cyanines, the most generally effective substituents are the sulfonic acids; however, hydroxyl, amino, and quaternized amino groups have all been used (Mishra et al., 2000), as have sugar residues (Reddington and Waggoner, 1998). Dyes intended for use in hydrophobic environments (e.g., as polarity sensors) do not require such substituents. Last, dyes that are to bind to membranes, but remain on the surface, require both hydrophobic and hydrophilic substituents (e.g., the charged diI dyes.)

Inhibiting Dye Aggregation

Many polymethine dyes when placed in aqueous solvents form ground-state aggregates and show solvent-dependent photophysics, in which both effects result in reduced fluorescence quantum efficiencies and shortened fluorescence lifetimes (West and Pearce, 1965; Makio et al., 1980). The aggregated dyes are nonfluorescent. Aggregation in aqueous medium is suppressed by groups that provide the dye with a better sphere of solvation, such as the negatively charged sulfonic acid groups; out-of-plane substituents also reduce dye aggregation. Last, cyanines having perfluorinated end groups show reduced aggregation in aqueous solutions (Renikuntla et al., 2004). Other groups that convert the dye to make a reactive labeling reagent or to form a physiological indicator can be added as necessary.

These reflections are the result of many years of experience in dye synthesis for numerous applications. We hope that our thoughts prove useful to others. Engineering new fluorochromes is still an exciting enterprise, and we hope that many unanticipated methods and techniques will emerge to surprise us.

REFERENCES

Adams, S. R., Campbell, R. E., Gross, L. A., Martin, B. R., Walkup, G. K., Yao, Y., Llopis, J., and Tsien, R. Y. (2002) New biarsenical ligands and tetracysteine motifs for protein labeling in vitro and in vivo: Synthesis and biological applications. *J Am Chem Soc* 124:6063–6076.

Alivisatos, A. P. (1996) Semiconductor clusters, nanocrystals, and quantum dots. *Science (Washington, D.C.)* 271: 933–937.

Babendure, J. R., Adams, S. R., and Tsien, R. Y. (2003) Aptamers switch on fluorescence of triphenylmethane dyes. *J Am Chem Soc* 125:14716–14717.

Ballou, B., Fisher, G. W., Hakala, T. R., and Farkas, D. L. (1997) Tumor detection and visualization using cyanine fluorochrome-labeled antibodies. *Biotechnol Prog* 13:649–658.

Ballou, B., Lagerholm, B. C., Ernst, L. A., Bruchez, M. P., and Waggoner, A. S. (2004) Noninvasive imaging of quantum dots in mice. *Bioconjug Chem* 15:79–86.

Bremer, C., Ntziachristos, V., and Weissleder, R. (2003) Optical-based molecular imaging: Contrast agents and potential medical applications. *Eur Radiol* 13:231–243.

Brooker, L. G. S. (1962) The search for longer conjugated chains in cyanine dyes. In *Recent Progress in the Chemistry ofNaural and Synthetic Colouring Matters and Related Fields*. T. S. Gore, B. S. Joshi, S. V. Sunthankar, and B. D. Tilak, eds. New York: Academic Press, 573–587.

Brooker, L. G. S., and Vittum, P. W. (1957) A century of progress in the synthesis of dyes for photography. *J Phot Sci* 5:71–88.

Buston, J. E. H., Young, J. R., and Anderson, H. L. (2000) Rotaxane-encapsulated cyanine dyes: Enhanced fluorescence efficiency and photostability. *Chem Commun (Cambridge)* 11:905–906.

Byers, G. W., Gross, S., and Henrichs, P. M. (1976) Direct and sensitized photooxidation of cyanine dyes. *Photochem Photobiol* 23:37–43.

Chance, B. (1998) Near-infrared images using continuous, phase-modulated, and pulsed light with quantitation of blood and blood oxygenation. *Ann N Y Acad Sci* 838:29–45.

Clapp, A. R., Medintz, I. L., Mauro, J. M., Fisher, B. R., Bawendi, M. G., and Mattoussi, H. (2004) Fluorescence resonance energy transfer between quantum dot donors and dye-labeled protein acceptors. *J Am Chem Soc* 126:301–310.

Curcio, J. A., and Petty, C. C. (1951) The near infrared absorption spectrum of liquid water. *J Opt Soc Am* 41:302–304.

Dahan, M., Laurence, T., Pinaud, F., Chemla, D. S., Alivisatos, A. P., Sauer, M., and Weiss, S. (2001) Time-gated biological imaging by use of colloidal quantum dots. *Opt Lett* 26:825–827.

Dahan, M., Levi, S., Luccardini, C., Rostaing, P., Riveau, B., and Triller, A. (2003) Diffusion dynamics of glycine receptors revealed by single-quantum dot tracking. *Science* 302:442–445.

Derfus, A. M., Chan, W. C. W., and Bhatia, S. N. (2004) Probing the cytotoxicity of semiconductor quantum dots. *Nano Lett* 4:11–18.

Dooley, C. M., Dore, T. M., Hanson, G. T., Jackson, W. C., Remington, S. J., and Tsien, R. Y. (2004) Imaging dynamic redox changes in mammalian cells with green fluorescent protein indicators. *J Biol Chem* 279:22284–22293.

Drexler, W. (2004) Ultrahigh-resolution optical coherence tomography. *J Biomed Opt* 9:47–74.

Emmelius, M., Pawlowski, G., and Vollmann, H. W. (1991) Materials for optical data storage. *R Soc Chem* 88:183–200.

Eppstein, M. J., Fedele, F., Laible, J., Zhang, C., Godavarty, A., and Sevick-Muraca, E. M. (2003) A comparison of exact and approximate adjoint sensitivities in fluorescence tomography. *IEEE Trans Med Imag* 22:1215–1223.

Flanagan, J. H., Jr., Khan, S. H., Menchen, S., Soper, S. A., and Hammer, R. P. (1997) Functionalized tricarbocyanine dyes as near-infrared fluorescent probes for biomolecules. *Bioconjug Chem* 8:751–756.

Fujimoto, J. G., Bouma, B., Tearney, G. J., Boppart, S. A., Pitris, C., Southern, J. F., and Brezinski, M. E. (1998) New technology for high-speed and high-resolution optical coherence tomography. *Ann N Y Acad Sci* 838:95–107.

Fujimoto, J. G., Pitris, C., Boppart, S. A., and Brezinski, M. E. (2000) Optical coherence tomography: An emerging technology forbiomedical imaging and optical biopsy. *Neoplasia* 2:9–25.

Gonzalez, S., Rajadhyaksha, M., and Anderson, R. R. (1998) Non-invasive (real-time) imaging of histologic margin of a proliferative skin lesion in vivo. *J Invest Dermatol* 111:538–539.

Griffin, B. A., Adams, S. R., Jones, J., and Tsien, R. Y. (2000) Fluorescent labeling of recombinant proteins in living cells with FlAsH. *Methods Enzymol* 327:565–578.

Guether, R., and Reddington, M. V. (1997) Photostable cyanine dye b-cyclodextrin conjugates. *Tetrahedron Lett* 38:6167–6170.

Hale, G. M., and Querry, M. R. (1973) Optical constants of water in the 200-nm to 200-μm wavelength region. *Appl Opt* 12:555–563.

Hanson, G. T., Aggeler, R., Oglesbee, D., Cannon, M., Capaldi, R. A., Tsien, R. Y., and Remington, S. J. (2004) Investigating mitochondrial redox potential with redox-sensitive green fluorescent protein indicators. *J Biol Chem* 279:13044–13053.

Houston, J. P., Thompson, A. B., Gurfinkel, M., and Sevick-Muraca, E. M. (2003) Sensitivity and depth penetration of continuous wave versus frequency-domain photon migration near-infrared fluorescence contrast-enhanced imaging. *Photochem Photobiol* 77:420–430.

Huang, D., Swanson, E. A., Lin, C. P., Schuman, J. S., Stinson, W. G., Chang, W., Hee, M. R., Flotte, T., Gregory, K., Puliafito, C. A., and Fujimoto, J. G. (1991) Optical coherence tomography. *Science* 254:1178–1181.

Irvine, W. M., and Pollack, J. B. (1968) Infrared optical properties of water and ice spheres. *Icarus* 8:324.

Jun, L., Ping, C., and Deshui, Z. (1996) Study on the photooxidation of near-infrared absorbing cyanine dye. *Proc SPIE Int Opt Eng* 2931:62–66.

Karandikar, B. M., Waggoner, A. S., and Mujumdar, R. B. (1998) Rigidized monomethine cyanines. U.S. patent no. 5852191.

Ke, S., Wen, X., Gurfinkel, M., Charnsangavej, C., Wallace, S., Sevick-Muraca, E. M., and Li, C. (2003) Near-infrared optical imaging of epidermal growth factor receptor in breast cancer xenografts. *Cancer Res* 63:7870–7875.

Kim, S., Lim, Y. T., Soltesz, E. G., De Grand, A. M., Lee, J., Nakayama, A., Parker, J. A., Mihaljevic, T., Laurence, R. G., Dor, D. M., Cohn, L. H., Bawendi, M. G., and Frangioni, J. V. (2004) Near-infrared fluorescent type II quantum dots forsentinel lymph node mapping. *Nat Biotechnol* 22:93–97.

Koenig, F., Gonzalez, S., White, W. M., Lein, M., and Rajadhyaksha, M. (1999) Near-infrared confocal laser scanning microscopy of bladder tissue in vivo. *Urology* 53:853–857.

Larson, D. R. , Zipfel, W. R., Williams, R. M., Clark, S. W., Bruchez, M. P., Wise, F. W., and Webb, W. W. (2003) Water-soluble quantum dots for multiphoton fluorescence imaging in vivo. *Science* 300:1434–1436.

Lee, L. G., Chen, C. H., and Chiu, L. A. (1986) Thiazole orange: A new dye for reticulocyte analysis. *Cytometry* 7:508–517.

Lidke, D. S., Nagy, P., Heintzmann, R., Arndt-Jovin, D. J., Post, J. N., Grecco, H. E., Jares-Erijman, E. A., and Jovin, T. M. (2004) Quantum dot ligands provide new insights into erbB/HER receptor-mediated signal transduction. *Nat Biotechnol* 22:198–203.

Lim, Y. T., Kim, S., Nakayama, A., Stott, N. E., Bawendi, M. G., and Frangioni, J. V. (2003) Selection of quantum dot wavelengths forbiomedical assays and imaging. *Mol Imaging* 2:50–64.

Lippincott-Schwartz, J., and Patterson, G. H. (2003) Development and use of fluorescent protein markers in living cells. *Science* 300:87–91.

Mahmood, U., Tung, C. H., Bogdanov, A., Jr., and Weissleder, R. (1999) Near-infrared optical imaging of protease activity fortumordetection. *Radiology* 213:866–870.

Makio, S., Kanamaru, N., and Tanaka, J. (1980) The J-aggregate 5,5′,6,6′-tetrachloro-1,1′-diethyl-3,3′-bis(4)-sulfobutylbenzimidazolocarbocyanine sodium-salt in aqueous-solution. *Bull Chem Soc Jap* 53:3120–3124.

Matsuzawa, Y., Tamura, S., Matsuzawa, N., and Ata, M. (1994) Light-stability of a beta-cyclodextrin (cyclomaltoheptaose) inclusion complex of a cyanine dye. *J Chem Soc Faraday Trans* 90:3517–3520.

Medintz, I. L., Clapp, A. R., Mattoussi, H., Goldman, E. R., Fisher, B., and Mauro, J. M. (2003) Self-assembled nanoscale biosensors based on quantum dot FRET donors. *Nat Mater* 2:630–638.

Mishra, A., Behera, R. K., Behera, P. K., Mishra, B. K., and Behera, G. B. (2000) Cyanines during the 1990s: A review . *Chem Rev* 100:1973–2011.

Miyawaki, A., Sawano, A., and Kogure, T. (2003) Lighting up cells: Labelling proteins with fluorophores. *Nat Cell Biol* September(Suppl.):S1–S7.

Mujumdar, R. B., Waggoner, A. S., and Karandikar, B. M. (1999) Monomethine cyanines rigidized by a two-carbon chain. U.S. patent no. 5986093.

Ntziachristos, V., Bremer, C., Graves, E. E., Ripoll, J., and Weissleder, R. (2002) In vivo tomographic imaging of near-infrared fluorescent probes. *Mol Imaging* 1:82–88.

Ntziachristos, V., Bremer, C., and Weissleder, R. (2003) Fluorescence imaging with near-infrared light: New technological advances that enable in vivo molecularimaging. *Eur Radiol* 13:195–208.

Ntziachristos, V., and Chance, B. (2001) Probing physiology and molecular function using optical imaging: Applications to breast cancer. *Breast Cancer Res* 3:41–46.

Pan, Y., and Farkas, D. L. (1997) *In vivo* imaging of biological tissues using 1300 nm optical coherence tomography. *Proc. SPIE* 2983:93–100.

Pan, Y.- T., and Farkas, D. L. (1998) Dual-color, 3-D imaging of biological tissues using optical coherence tomography. *J Biomed Optics* 3:446–455.

Pan, Y., Lavelle, J. P., Bastacky, S. I., Meyers, S., Pirtskhalaishvili, G., Zeidel, M. L., and Farkas, D. L. (2001) Detection of tumorigenesis in rat bladders with optical coherence tomography. *Med Phys* 28:2432–2440.

Patterson, G. H., and Lippincott-Schwartz, J. (2002) A photoactivatable GFP for selective photolabeling of proteins and cells. *Science* 297:1873–1877.

Peelle, B., Gururaja, T. L., Payan, D. G., and Anderson, D. C. (2001) Characterization and use of green fluorescent proteins from *Renilla mulleri* and *Ptilosarcus guernyi* forthe human cell display of functional peptides. *J Protein Chem* 20:507–519.

Pham, W., Lai, W. F., Weissleder, R., and Tung, C. H. (2003) High efficiency synthesis of a bioconjugatable near-infrared fluorochrome. *Bioconjug Chem* 14:1048–1051.

Rajadhyaksha, M., Grossman, M., Esterowitz, D., and Webb, R. H. (1995a) In-vivo confocal scanning laser microscopy of human skin: Melanin provides strong contrast. *J Invest Dermatol* 104:946–952.

Rajadhyaksha, M., Grossman, M., Webb, R. H., and Anderson, R. R. (1995b) Video-rate confocal scanning laser microscopy of live human skin. *J Invest Dermatol* 104:618.

Reddington, M. V., and Waggoner, A. S. (1998) Preparation of aminodeoxy glycoconjugated fluorescent labeling reagents. Wyoming patent no. 9849176.

Renikuntla, B. R., Rose, H. C., Eldo, J., Waggoner, A. S., and Armitage, B. A. (2004) Improved photostability and fluorescence properties through polyfluorination of a cyanine dye. *Organic Lett* 6:909–912.

Richieri, G. V., Ogata, R. T., and Kleinfeld, A. M. (1999) The measurement of free fatty acid concentration with the fluorescent probe ADIFAB: A practical guide for the use of the ADIFAB probe. *Mol Cell Biochem* 192:87–94.

Rozinov, M. N., and Nolan, G. P. (1998) Evolution of peptides that modulate the spectral qualities of bound, small-molecule fluorophores. *Chem Biol* 5:713–728.

Sevick-Muraca, E. M., Houston, J. P., and Gurfinkel, M. (2002) Fluorescence-enhanced, near infrared diagnostic imaging with contrast agents. *Curr Opin Chem Biol* 6:642–650.

Simeonov, A., Matsushita, M., Juban, E. A., Thompson, E. H., Hoffman, T. Z., Beuscher, A. E., 4th, Taylor, M. J., Wirsching, P., Rettig, W., McCusker, J. K., Stevens, R. C., Millar, D. P., Schultz, P. G., Lerner, R. A., and Janda, K. D. (2000) Blue-fluorescent antibodies. *Science* 290:307–313.

Sosinsky, G. E., Gaietta, G. M., Hand, G., Deerinck, T. J., Han, A., Mackey, M., Adams, S. R., Bouwer, J., Tsien, R. Y., and Ellisman, M. H. (2003) Tetracysteine genetic tags complexed with biarsenical ligands as a tool for investigating gap junction structure and dynamics. *Cell Commun Adhes* 10:181–186.

Strekowski, L., Lipowska, M., Gorecki, T., Mason, J. C., and Patonay, G. (1996a) Functionalization of near-infrared cyanine dyes. *J Heterocyclic Chem* 33:1685–1688.

Strekowski, L., Lipowska, M., Gorecki, T., Mason, J. C., and Patonay, G. (1996b) Functionalization of near-infrared cyanine dyes. *J Heterocyclic Chem* 33:1685–1688.

Terpetschnig, E., and Wolfbeis, O. S. (1998) Luminescent probes for NIR sensing applications. *NATO ASI Ser* 52:161–182.

Thompson, A. B., Hawrysz, D. J., and Sevick-Muraca, E. M. (2003) Near-infrared fluorescence contrast-enhanced imaging with area illumination and area detection: The forward imaging problem. *Appl Opt* 42:4125–4136.

Tredwell, C. J., and Keary, C. M. (1979) Picosecond time-resolved fluorescence lifetimes of the polymethine and related dyes. *Chem Phys* 43:307–316.

Tsien, R. J., and A. S. Waggoner. (1995) Fluorophores for confocal microscopy. In *Handbook ofBiological Confocal Microscopy.* J. B. Pawley, ed. New York: Plenum Press, 269–279

Violin, J. D., Zhang, J., Tsien, R. Y., and Newton, A. C. (2003) A genetically encoded fluorescent reporter reveals oscillatory phosphorylation by protein kinase C. *J Cell Biol* 161:899–909.

Waggoner, A. S. (1986) Fluorescent probes for analysis of cell structure, function, and health by flow and imaging cytometry. In *Applications ofFluorescence in the Biomedical Sciences.* D. L. Taylor, A. S. Waggoner, R. F. Murphy, F. Lanni, and R. R. Birge, eds.. New York: Alan R. Liss, 2–28.

Waggoner, A. , Karandikar, B., and Mujumdar, R. (1996) Rigidized monomethine cyanine dyes, their preparation and use. European patent no. 747448.

Waggoner, A. S., and Mujumdar, R. (1999) Preparation of rigidized trimethine cyanine dyes and their use as fluorescent markers. Wyoming patent no. 9931181.

Waggoner, A. S., and Mujumdar, R. (2000) Rigidized trimethine cyanine dyes. U.S. patent no. 6133445.

Wan, S., Anderson, R. R., and Parrish, J. A. (1981) Analytical modeling for the optical properties of the skin with in vitro and in vivo applications. *Photochem Photobiol* 34:493–499.

Watson, A., Wu, X., and Bruchez, M. (2003) Lighting up cells with quantum dots. *Biotechniques* 34:296–300, 302–303.

Weissleder, R., and Ntziachristos, V. (2003) Shedding light onto live molecular targets. *Nat Med* 9:123–128.

West, W., and Pearce, S. (1965) Dimeric state of cyanine dyes. *J Phys Chem* 69:1894–1903.

Williams, R. M., Piston, D. W., and Webb, W. W. (1994) Two-photon molecular excitation provides intrinsic 3-dimensional resolution for laser-based microscopy and microphotochemistry. *FASEB J* 8:804–813.

Yu, D., Baird, G. S., Tsien, R. Y., and Davis, R. L. (2003) Detection of calcium transients in *Drosophila* mushroom body neurons with camgaroo reporters. *J Neurosci* 23:64–72.

Zhang, J., Campbell, R. E., Ting, A. Y., and Tsien, R. Y. (2002) Creating new fluorescent probes for cell biology. *Nat Rev Mol Cell Biol* 3:906–918.

Zhang, Q., Ma, H., Nioka, S., and Chance, B. (2000) Study of near infrared technology for intracranial hematoma detection. *J Biomed Opt* 5:206–213.

6
Imaging Membrane Potential with Voltage-Sensitive Dyes

DEJAN VUCINIC, EFSTRATIOS KOSMIDIS, CHUN X. FALK,
LAWRENCE B. COHEN, LESLIE M. LOEW,
MAJA DJURISIC, AND DEJAN ZECEVIC

An optical measurement of membrane potential using a molecular probe can be beneficial in a variety of circumstances. One advantage is the possibility of simultaneous measurements from many locations. This is especially important in the study of the nervous system and the heart in which many parts of an individual cell, many cells, or many regions are simultaneously active. In addition, optical recording offers the possibility of recording from processes that are too small or fragile for electrode recording.

Several different optical properties of membrane-bound dyes are sensitive to membrane potential, including fluorescence, absorption, dichroism, birefringence, FRET, nonlinear second harmonic generation, and resonance Raman absorption. However, because the vast majority of applications have involved fluorescence or absorption, these will be the only subjects of this review. All the optical signals described in this chapter are "fast" signals (Cohen and Salzberg, 1978) that are presumed to arise from membrane-bound dye; they follow changes in membrane potential with time courses that are rapid compared with the rise time of an action potential. Figure 6.1 illustrates the kind of result that is used to define a voltage-sensitive dye. In a model preparation, the giant axon from a squid, these optical signals are fast, following membrane potential with a time constant of less than 10 μsec (Loew et al., 1985) and their size is linearly related to the size of the change in potential (e.g., Gupta et al., 1981). Thus, these dyes provide a direct, fast, and linear measure of the change in membrane potential of the stained membranes.

Several voltage-sensitive dyes (fig. 6.2) have been used to monitor changes in membrane potential in a variety of preparations.

Figure 6.2 illustrates four different chromophores (the merocyanine dye, XVII, was used for the measurement illustrated in fig. 6.1). For each chromophore, approximately 100 analogues have been synthesized in an attempt to optimize the signal-to-noise ratio that can be obtained in a variety of preparations. (This screening was made possible by synthetic efforts of three laboratories: Jeff Wang, Ravender Gupta, and Alan Waggoner then at Amherst College; Rina Hildesheim and Amiram Grinvald at the Weizmann Institute; and Joe Wuskell and Leslie Loew at the University of Connecticut Health Center.) For each of the four chromophores illustrated in figure 6.2, there were 10 or 20 dyes that gave approximately the same signal size on squid axons (Gupta et al., 1981). However, dyes that had nearly identical signal size on squid axons could have very different responses on other preparations, and thus tens of dyes usually have to be tested to obtain the largest possible

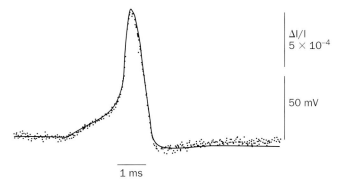

Figure 6.1 Changes in absorption (dots) of a giant axon stained with a merocyanine dye, XVII (fig. 6.2), during a membrane action potential (smooth trace) recorded simultaneously. The change in absorption and the action potential had the same time course. In this and subsequent figures, the size of the vertical line represents the stated value of the fractional change in intensity of transmitted ($\Delta I/I$) or fluorescent ($\Delta F/F$) light. The response time constant of the light measuring system was 35 μsec; 32 sweeps were averaged. The dye is available from Nippon Kankoh-Shikiso Kenkyusho (Okayama, Japan).

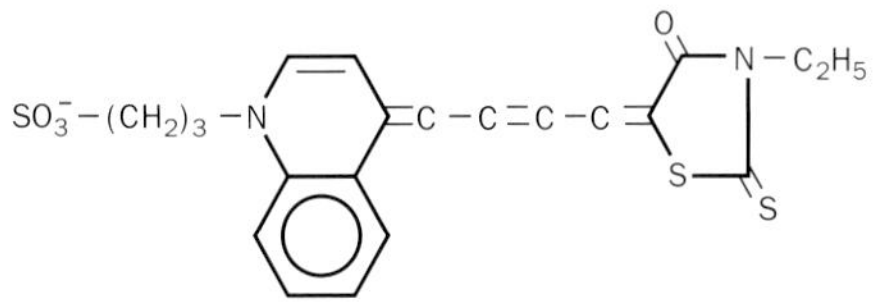

XVII, Merocyanine, Absorption, Birefringence

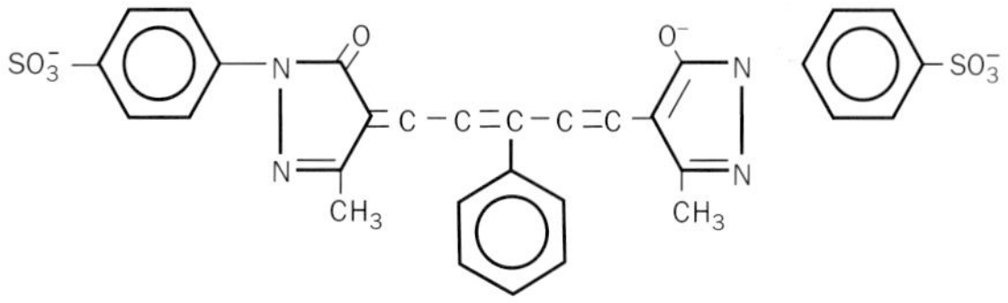

RH155, Oxonol, Absorption

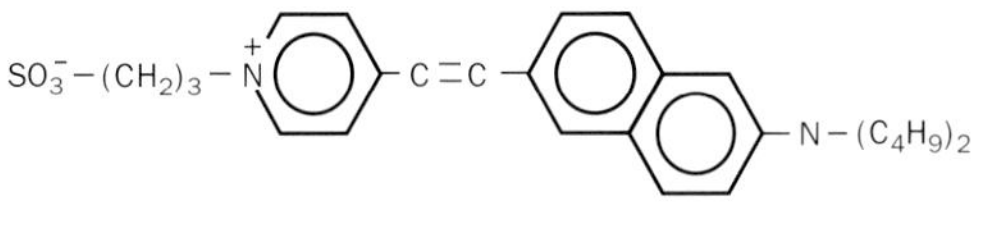

Di–4–ANEPPS, Styryl, Fluorescence

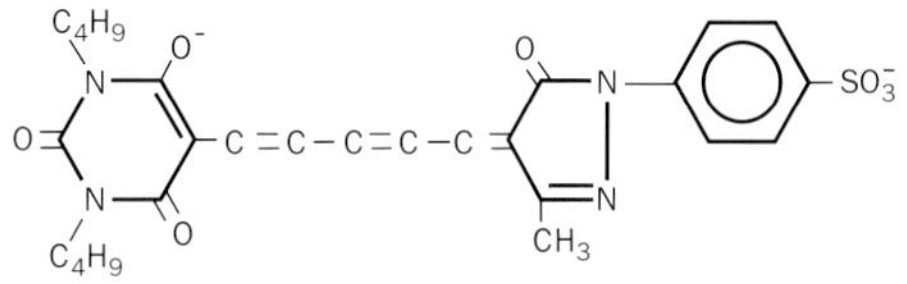

XXV, Oxonol, Fluorescence, Absorption

Figure 6.2 Examples of four different chromophores that have been used to monitor membrane potential. The merocyanine dye, XVII (WW375), and the oxonol dye, RH155, are commercially available as NK2495 and NK3041 from Nippon Kankoh-Shikiso Kenkyusho Co. Ltd. (Okayama, Japan). The oxonol, XXV (WW781), and styryl, di-4-ANEPPS, are available commercially as dye R-1114 and D-1199 from Molecular Probes (Junction City, Oregon).

signal. A common failing was that the dye did not penetrate through connective tissue or along intercellular spaces to the membrane of interest.

The following rules of thumb seem to be useful. First, each of the chromophores is available with a fixed charge that is either a quaternary nitrogen (positive) or a sulfonate (negative). Generally the positive dyes have given larger signals when dye is applied extracellularly in vertebrate preparations. Second, each chromophore is available with carbon chains of several lengths. The more hydrophilic dyes (methyl or ethyl) work best if the dye has to penetrate through a compact tissue (vertebrate brain).

A number of studies have been conducted to determine the molecular mechanisms that result in potential-dependent optical properties. Evidence supporting three different mechanisms (for different dyes) has been presented. These are dipole rotation, electrochromism,

PARTS OF A NEURON

Many detectors – One neuron

· Potential changes in dendrites
· Microinject dye: stain one neuron

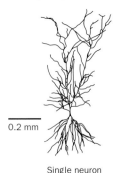

0.2 mm

Single neuron

INDIVIDUAL NEURONS

One detectors – One neuron

· Follow spike activity of many
  individual neurons
· Bath-applied dye

0.5 mm

Invertebrate ganglion

POPULATION SIGNALS

One detector – Many neurons

· Signals are population average
· Vertebrate brain
· Bath-applied or transported dye

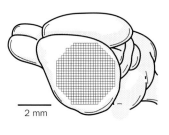

2 mm

Figure 6.3 Schematic drawings of the three kinds of measurements described as examples. On the left is an individual cortical hippocampal CA1 pyramidal cell. Each pixel of the 464-element photodiode array or 6400-pixel CCD camera receives light from a small part of the dendrite, axon, or cell body of the neuron. An optical measurement of membrane potential provides important information about how the neuron converts its synaptic input into its spike output. In the middle is a drawing of a slice through an invertebrate ganglion with its cell bodies in a cortex around the outside and a neuropil in the middle. Here, each detector receives light from one or a small number of cell bodies. A voltage-sensitive dye measurement of spike activity while the ganglion is generating a behavior provides important information about how the behavior is generated. On the right is an illustration of a vertebrate brain with the superimposed 464-element photodiode array (used in all three examples). In this circumstance, each pixel of the array receives light from thousands of cells and processes. The signal is the population average of the change in membrane potential in those cells and processes. (The image of the hippocampal neuron was taken from Mainen et al. [(1996].)

and a potential sensitive monomer–dimer equilibrium. Dye mechanisms are discussed in Waggoner and Grinvald (1977), Loew et al. (1985), and Fromherz et al. (1991).

We begin with examples of results obtained from measurements addressing three quite different neurobiological problems. In one example, when the light level was low, the camera was a fast, back-illuminated CCD camera. In the remaining two instances, the camera was a photodiode array with only 464 elements. Although this seems like very low spatial resolution, surprisingly, the camera resolution was not limiting in these examples. On the positive side, the photodiode array has an outstanding dynamic range. With an incident intensity of more than 10^{10} photons/pixel/frame, it measures signals that were a fractional change ($\Delta I/I$) of one part in 10^5. Both cameras have frames rates faster then 1 kfps, fast enough to measure most biologically important signals. The optical signals in the three example measurements are not large; they represent fractional changes in light intensity ($\Delta I/I$) of from 10^{-4} to 3×10^{-2}. Nonetheless, they can be measured with an acceptable signal-to-noise ratio after attention to details of the measurement, which are described later in the chapter.

Because our own research interests are in the area of neurobiology, the three examples presented here come from that area. However, the information presented about optimizing the signal-to-noise ratio of the measurements is applicable to other areas of biology. Figure 6.3 (Mainen et al., 1996) illustrates three qualitatively different applications in neurobiology in which imaging membrane potential has been useful.

First, let's view the left panel in figure 6.3. To see how a neuron integrates its synaptic input into its action potential output, one needs to be able to measure membrane potential everywhere synaptic input occurs and at the places where spikes are initiated. Second, in the middle panel of figure 6.3, to understand how a nervous system generates a behavior, it is important to measure simultaneously the action potential activity of many (all) of the participating neurons. And third, in the right panel of figure 6.3, responses to sensory stimuli and generation of motor output in the vertebrate brain are often accompanied by synchronous activation of many neurons in widespread brain areas; voltage-sensitive dye recordings allow simultaneous measurement of population signals from many areas. In these three instances, optical recordings have provided kinds of information about the function of the nervous system that were previously unobtainable.

The next part of the chapter describes the experimental details that are important in obtaining the signal-to-noise ratios achieved in the experiments just described. We discuss signal type, dyes, light sources, photodetectors, and optics.

THREE EXAMPLES

Processes of an Individual Neuron

Understanding the biophysical properties of single neurons and how they process information is fundamental to understanding how the brain works (fig. 6.3, left panel). With the development of new measuring techniques that allow more direct investigation of individual nerve cells, it has become widely recognized, especially during the past 20 years, that dendritic membranes of many vertebrate central nervous system neurons contain active conductances such as voltage-activated Na^+, Ca^{2+}, and K^+ channels (e.g., Stuart and Sakmann, 1994; Magee and Johnston, 1995; Magee et al., 1995; Spruston et al., 1995). An important consequence of active dendrites is that regional electrical properties of branching neuronal processes will be extraordinarily complex, dynamic, and, in the general case, impossible to predict in the absence of detailed measurements.

To obtain such a measurement one would, ideally, like to be able to monitor, at multiple sites, subthreshold events as they travel from the sites of origin on neuronal processes and summate at particular locations to influence action potential initiation. This goal has not been achieved in any neuron, vertebrate or invertebrate, because of the technical limitations of experimental measurements that use electrodes. To achieve better spatial resolution, it is necessary to turn from direct electrical recording to indirect optical measurements using voltage-sensitive dyes. Recently, the sensitivity of intracellular voltage-sensitive dye techniques for monitoring neuronal processes in situ has been dramatically improved (by a factor of about 100), allowing direct recording of subthreshold and action potential signals from the neurites of individual neurons (Antic and Zecevic, 1995; Zecevic, 1996; Antic et al., 1999; 2000). The improvement in the signal-to-noise ratio is based on previous experience from other laboratories (e.g., Davila et al., 1974; Grinvald et al., 1987) and on (1) finding an intracellular dye that provides a relatively large fractional change in fluorescence and (2) improvements in the apparatus to increase the incident light intensity, to lower the noise, and to filtermor e efficiently.

An Invertebrate Neuron

A typical result of a multisite voltage-sensitive dye recording is shown in figure 6.4. The image of the cell was projected onto the array of photodiodes as indicated in figure 6.4A. This panel represents multisite recording of action potential signals from axonal branches Br2, Br3, and Br4, evoked by a transmembrane current step (fig. 6.4B). Optical signals associated with action potentials, expressed as fractional changes in fluorescent light intensity ($\Delta F/F$), were between 3×10^{-3} and 3×10^{-2} in recordings from the processes. With these measurements it is straightforward to determine the direction and velocity of action potential propagation in neuronal processes. In figure 6.4C, recordings from different locations, scaled to the same height, are compared to determine the site of origin of the action potential. The earliest action potential, in response to soma stimulation, was generated near

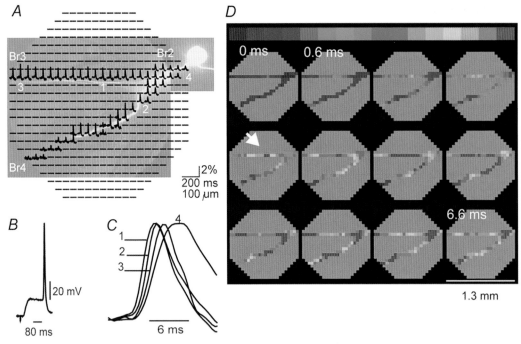

Figure 6.4 (A) Giant metacerebral neuron from the left cerebral ganglion 12 hours after injection with the fluorescent voltage-sensitive dye JPW1114. The cell body and main processes are clearly visible in the unfixed preparation. Excitation wavelength, 540 ± 30 nm; dichroic mirror, 570 nm; long-pass barrier filter, 610 nm. Voltage-sensitive dye recording of action potential signals from elements of the photodiode array positioned over the image of axonal arborizations of a metacerebral cell in the left cerebral ganglion. Axonal branches are marked Br 1 through 4. (B) Spikes were evoked by transmembrane current steps delivered through the recording microelectrode in the soma. Each optical trace in A represents 70 ms of recording centered around the peak of the spike as indicated by the time bar in B. Each diode received light from a $50 \times 50\text{-}\mu m^2$ area in the object plane. (C) Recordings from four different locations indicated in A, scaled to the same height, are compared to determine the site of the origin of the action potential and the direction of propagation. (D) Color-coded representation of the data shown in A indicating the size and location of the primary spike trigger zone and the pattern of spike propagation. Consecutive frames represent data points that are 1.6 ms apart. The color scale is relative with the peak of the action potential for each detector shown in red. (Modified from Zecevic [1996].)

location 2, in the axonal branch Br4, situated in the cerebral–buccal connective outside the ganglion. The spike propagated orthodromically from the site of initiation toward the periphery in branch Br4 and antidromically toward the soma and into branch Br3 in the external lip nerve. The direction of propagation is clear from the color-coded representation of the data (fig. 6.4D), which shows the spatial distribution of membrane potential in the neuronal processes at nine different times separated by 1.6 ms. Red color corresponds to the peak of the action potential. The views show the position of the action potential trigger zone at location 2 and ortho- and antidromic spread of the nerve impulse from the site of initiation. The earliest spike was evoked approximately 1 mm away from the soma. The spike initiation segment in the axon is roughly 300 μm in length and remote from the soma. It appears that under normal conditions, slow depolarizing voltage pulses applied to the soma are spread electrotonically into the processes with little attenuation. These depolarizing pulses initiate action potentials at remote sites in the processes that are characterized by higher excitability than that of neighboring segments.

Light scattering in the ganglion limits the maximum useful spatial resolution in this kind of measurement. Thus the 24 × 24 pixel resolution of NeuroPlex (RedShirtImaging, LLC, Fairfield, Connecticut) appears to be adequate in this circumstance.

On the basis of similar measurements, we recently determined that this nerve cell has multiple trigger zones that can be independently activated. The precise pattern of action potential initiation and propagation within the whole branching structure of a neuron can be analyzed by multisite recording. The information about spatial and temporal dynamics of neuronal signals can be used to constrain the choice of channel densities and geometric factors in biophysical models that are used to describe functional properties of neurons.

A Vertebrate Neuron

It is of considerable interest to apply the same technique to dendrites of vertebrate central nervous system neurons in brain slices. The initial experiments were carried out on pyramidal neurons in neocortical brain slices from P14–18 rats using a 464-element photodiode array. The experiments provided several important methodological results. First, it was established that it is possible to deposit the dye into vertebrate neurons without staining the surrounding tissue. Second, pharmacological effects of the dye were completely reversible if the staining pipette was pulled out and the cell was allowed to recover for 1 or 2 hours. Third, the level of photodynamic damage already allows meaningful measurements and could be reduced further. Last, the sensitivity of the dye was comparable with that achieved in the experiments on invertebrate neurons (Zecevic, 1996). In these preliminary experiments, the dye spread roughly for 500 μm into dendritic processes within 2 hours.

An example of voltage-sensitive dye recording from a pyramidal neuron is shown in figure 6.5I. The fluorescent image of the cell was projected onto the octagonal photodiode array as illustrated in figure 6.5IA. The neuron was stimulated, by depolarizing the cell body to produce a burst of two action potentials. Each trace in figure 6.5IB represents the output of one photodiode for 44 ms. As evident from the figure, the optical signals were found in the regions of the array that correspond closely to the geometry of the cell. Optical signals associated with action potentials, expressed as fractional changes in fluorescent light intensity ($\Delta F/F$), were between 1% and 2% in recordings from neuronal processes. Figure 6.5IC shows a comparison of the electrical recordings from the soma (smooth line) with the optical signals filtered to eliminate high-frequency noise (dashed line). There is a good agreement between time courses of electrical and optical recordings. In figure 6.5ID, recordings from different locations, scaled to the same height, are compared on an expanded timescale to establish the pattern of action potential propagation. Each trace is a spatial average from two adjacent detectors. Both spikes in the burst originated in the proximity of the soma and back-propagated along the apical and basolateral dendrites. The pattern of action potential propagation is shown in a color-coded display for another pyramidal neuron in figures 6.5IE and IF.

An improvement in the sensitivity of voltage imaging by a factor of about 10 was achieved by using newly developed, fast, low-noise CCD camera (NeuroCCD, RedShirt-Imaging) in place of the photodiode array used in experiments shown in figure 6.5I. This allowed experiments in which subthreshold synaptic potentials were monitored at the site of origin in the dendritic tuft of mitral cells. The thin dendrites in the tuft are not accessible to microelectrode recording. Voltage imaging is the only available method for studying signal integration in terminal dendritic branches. An example of voltage imaging with the fast CCD camera from a mitral cell in a slice of the olfactory bulb of the rat is shown in figure 6.5II. In this series of experiments we tested whether the integration of synaptic signals at the site of origin (tuft) could be monitored in mitral cells in the absence of action potentials. The image of the distal part of the primary dendrite of a mitral cell stained with voltage-sensitive dye was projected onto the CCD camera. The light intensity changes were monitored at a frame rate of 3.7 KHz. The olfactory nerve was stimulated with an intensity that evoked subthreshold responses. In the measurement shown in figure 6.5IIB, a pair of stimuli delivered to the olfactory nerve resulted in two excitatory post-synaptic potentials (EPSPs) that summated but remained subthreshold. The temporal characteristics of action potentials and synaptic potentials recorded from the distal dendritic regions are directly obtained from optical data. The amplitude calibration in terms of membrane potential (measured in millivolts) can be obtained from optical data in mitral cell experiments using an action potential as a calibrating signal. It has been established by direct electrical measurements that action potential in the primary dendrite has constant amplitude that can be determined from patch electrode recording from the soma or proximal dendrite

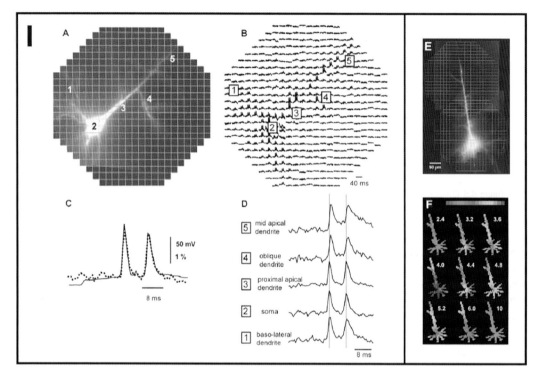

Figure 6.5 (IA) Outline of the 464-element photodiode array superimposed over the fluorescent image of a pyramidal neuron. (IB) Single trial recording of action potential-related optical signals. Each trace represents the output of one diode. Traces are arranged according to the disposition of the detectors in the array. Each trace represents 44 ms of recording. Each diode received light from a 14×14-μm^2 area in the object plane. Spikes were evoked by a somatic current pulse. (IC) Comparison of electrical and optical recordings. Spatial average of optical signals from eight individual diodes from the somatic region (dotted line) are superimposed on the electrical recording from the soma (solid line). (ID) Action potentials from individual detectors at different locations along the basal, oblique, and apical dendrites. Traces from different locations are scaled to the same height. The increasing delay between the signal from the somatic region and most proximal dendritic segments reflects the propagation velocity (0.22 m/s). At a distance of 230 μm from the soma the half width increased from 1.7 to 2.3 ms for the first spike and from 2.2 to 4.6 ms for the second spike in the burst. (Modified from Antic et al. [1999].) (IE, IF) Fluorescent image of another pyramidal neuron (IE) and a color-coded representation of the back propagation of an action potential evoked in the soma region (IF). (II) Optical recording of the evoked potentials from the site of origin (the terminal tuft of the primary dendrite). (IIA) Fluorescence CCD image of the distal part of a primary dendrite of an individual mitral cell stained with voltage-sensitive dye. High-resolution image obtained with conventional CCD camera; low-resolution frame obtained by CCD used for voltage imaging. (IIB) Optical recording from two different regions (red and blue rectangles) on the primary dendrite showing signals corresponding to an action potential evoked by direct stimulation of an axon using an extracellular electrode in the external plexiform layer (EPL) and two summating evoked potentials induced by a pair of closely spaced volleys to the olfactory nerve layer (ON). Traces on the left were filtered by two passes of a 2-1-2 smoothing routine. (IIC) The red, unfiltered trace on the left was used to determine the amplitude of the optical signal corresponding to a 100-mV spike, as measured by the patch electrode in the soma. The action potential signal was used as a calibration for determining the size of synaptic potentials in terms of membrane potential.

II

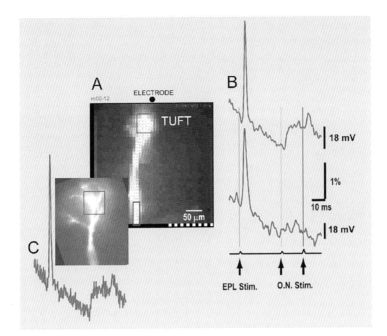

Figure 6.5 (*Continued*)

(Chen et al., 1997). In figure 6.5II, the stimulus to the olfactory nerve was preceded by a spike in the mitral cell evoked by another stimulating electrode positioned in the external plexiform layer. The red, unfiltered trace on the left in figure 6.5IIC was used to determine the amplitude of the optical signal corresponding to a 100-mV spike, as measured by the patch electrode in the soma. The action potential signal was used as a calibration for determining the size of synaptic potentials in terms of membrane potential.

Action Potentials from Individual Neurons in an *Aplysia* Ganglion

Nervous systems are made up of large numbers of neurons, and many of these are active during the generation of behaviors (fig. 6.3, middle panel). The original motivation for developing optical methods formonitor ing activity was the hope that they could be used to record all the action potential activity of all the neurons in simpler invertebrate ganglia during behaviors (Davila et al., 1973). Techniques that use microelectrodes to monitor activity are limited in that they can observe single cell activity in only as many cells as one can simultaneously place electrodes (typically fourorfewerneur ons). Obtaining information about the activity of many cells is essential for understanding the roles of the individual neurons in generating behavior and for understanding how nervous systems are organized.

In the first attempt to use voltage-sensitive dyes in ganglia (Salzberg et al., 1973), we were fortunate to be able to monitor activity in a *single* neuron because the photodynamic damage was severe and the signal-to-noise ratio small. Now, however, with better dyes and methods, the spike activity of hundreds of individual neurons can be recorded simultaneously (Zecevic et al., 1989; Nakashima, et al., 1992). In the experiments described later, we measured the spike activity of up to 50% of the approximately 1000 cells in the *Aplysia* abdominal ganglion. Opisthobranch molluscs have been a preparation of choice for this kind of measurement because their central nervous systems have relatively few, relatively large neurons and almost all the cell bodies are fully invaded by the action potential. In addition, opisthobranch ganglia are organized with cell bodies on the outside and neuropil in the center. These characteristics are important because the dyes we used stained only the outer, cell body, layer, and the signal-to-noise ratio for action potential recording would be reduced if the cell bodies did not have a full-size action potential.

The 464-element silicon photodiode array was placed in the image plane formed by a microscope objective of 25×0.4 NA. A single-pole high-pass and a four-pole low-pass Bessel filter in the amplifiers limited the frequency response to 1.5 to 200 Hz. We used the isolated siphon preparation developed by Kupfermann et al. (1971). Considerable effort was made to find conditions that maximized dye staining while causing minimal pharmacological effects on the gill withdrawal behavior. Intact ganglia were incubated in a 0.15 mg/mL solution of the oxonol dye RH155 (fig. 6.2; orits diethyl analogue) using a protocol developed by Nakashima et al. (1992). A light mechanical stimulation (1–2 g) was delivered to the siphon.

The signal-to-noise ratio in the measurements from individual cell bodies is optimal when the light from one cell body falls on one photodetector. More photodetectors would mean that the light from an individual cell would be divided onto more than one detector (with a concomitant reduction in signal-to-noise ratio). Fewer photodetectors than cells would mean that the light from more than one neuron would fall on one detector with a reduction in signal-to-noise ratio for the individual neurons. Thus, an array of only 464 detectors is approximately optimal for a ganglion of 1000 neurons.

Because the image of a ganglion is formed on a rectilinear diode array, there is no simple correspondence between images of cells and photodetectors. The light from larger cell bodies will fall on several detectors, and its activity will be recorded as simultaneous events on neighboring detectors. In addition, because these preparations are multilayer, most detectors will receive light from several cells. Thus, a sorting step is required to determine the activity in neurons from the spike signals on individual photodiodes. In the top right of figure 6.6, the raw data from seven photodiodes from an array are shown.

The activity of four neurons (shown in the raster diagram at the bottom) can account for the spike signals in the top section. Two problems are illustrated in this figure. Both arise from the signal-to-noise ratio. First, there may be an additional spike on detector 116 just before the stimulus (at the arrow), but the signal-to-noise ratio is not large enough to be certain. Second, after the stimulus, there is a great deal of activity that obscures small signals. Both problems suggest that the optical recordings are not complete; attempts to determine the completeness suggested that we were able to detect the activity of about 50% of the neurons in the *Aplysia* abdominal ganglion (Wu et al., 1994a). The fractional change in transmitted light ($\Delta I/I$) in these three signals ranges from about 10^{-4} to 1.5×10^{-3}; the noise is substantially smaller.

The result of a complete analysis is shown in the raster diagram of figure 6.7. The 0.5-s mechanical stimulus started at the time indicated by the line labeled "Stim." There are 135 neurons with activity that was detected optically. Similar results have been obtained by Nakashima et al. (1992). Because this recording is only 50% complete, the actual number of activated neurons during the gill withdrawal reflex was estimated to be about 300.

We were surprised at the large number of neurons that were activated by light touch. Furthermore, a substantial number of the remaining neurons were likely to be either inhibited by the stimulus or to receive a large subthreshold excitatory input. It is almost as if the *Aplysia* nervous system is designed such that every cell in the abdominal ganglion cares about this (and perhaps every) sensory stimulus. In addition, more than 1000 neurons in other ganglia are activated by this touch (Tsau et al., 1994). Clearly, information about this very mild and localized stimulus is propagated widely in the *Aplysia* nervous system. These results force a more pessimistic view of the current understanding of the neuronal basis of apparently simple behaviors in relatively simple nervous systems. Elsewhere we present arguments suggesting that the abdominal ganglion may function as a distributed system (Tsau et al., 1994; Wu et al., 1994b).

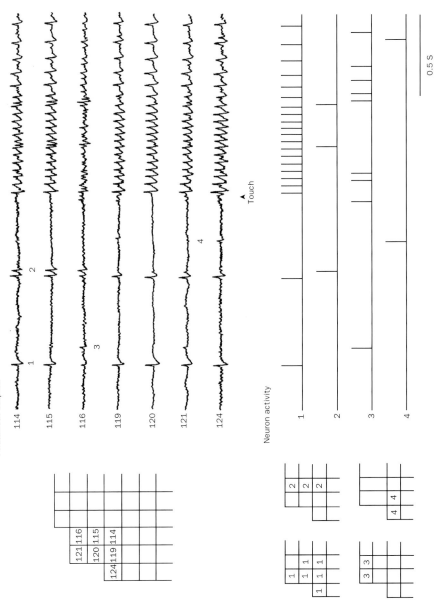

Figure 6.6 Optical recordings from a portion of a photodiode array from an Aplysia abdominal ganglion. The drawing to the left represents the relative position of the detectors and their displayed activity. In the top right, the original data from seven detectors are illustrated. The number to the right of each trace identifies the detector from which the trace was taken. The bottom section of the figure shows the raster diagram illustrating the results of our sorting this data into the spike activity of four neurons. At the number 1 in the top section, there are synchronously occurring spike signals on all seven detectors. A synchronous event of this kind occurs more than 20 times; we presume that each event represents an action potential in one relatively large neuron. The activity of this cell is represented by the vertical lines on trace 1 of the bottom section. The activity of a second cell is indicated by small signals at the number 4 on 119 and its neighbor, 124. The activity of this cell is represented by the vertical lines on trace 4 of the bottom section. The activity of neurons 2 and 3 was similarly identified. (Modified from Zecevic et al. [1989].)

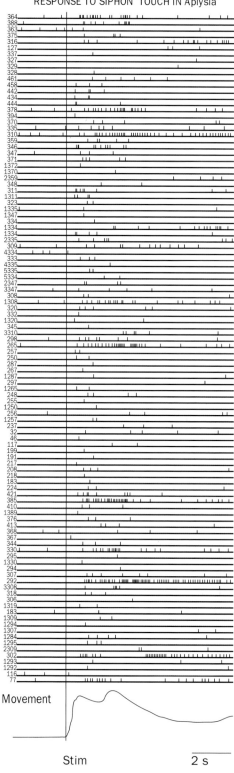

Figure 6.7 Raster diagram of the action potential activity recorded optically from an *Aplysia* abdominal ganglion during a gill withdrawal reflex. The 0.5-s touch to the siphon began at the time of the line labeled "Stim." In this recording, activity in 135 neurons was measured. We think this recording is incomplete and that the actual number of active neurons was between 250 and 300. Most neurons were activated by touch, but one, number 4334 (about a third down from the top), was inhibited. This inhibition was seen in repeated trials in this preparation. (Modified from Wu et al. [1994b].)

Recently, similar measurements have been made from a vertebrate system: the ganglia of the submucous plexus of the guinea pig small intestine (Obaid et al., 1999). It will be interesting to see whether this vertebrate preparation also functions in a distributed manner.

Population Signals from the Turtle Olfactory Bulb

In the experiments on *Aplysia* ganglia just described, each photodetector received light from one or a small number of neurons. In contrast, in measurements from the turtle olfactory bulb (fig. 6.3, right panel), each detector received light from a volume of the bulb that includes thousands of neurons and processes. The resulting signal will be a population average of the change in membrane potential of all these cells. Populations signals have been recorded from many preparations (e.g., Grinvald et al., 1982b; Orbach and Cohen, 1983; Sakai et al., 1985; Kauer, 1988; Cinelli and Salzberg, 1992; Albowitz and Kuhnt, 1993); the results from turtle are described because of our familiarity with them.

Odor stimuli have long been known to induce stereotyped local field potential responses consisting of sinusoidal oscillations of 10 to 80 Hz riding on top of a slow "DC" signal. Since its first discovery in the hedgehog (Adrian, 1942), odor-induced oscillations have been seen in phylogenetically distant species including locust, (Laurent and Naraghi, 1994), turtle (Beuerman, 1975), and monkey (Hughes and Mazurowski, 1962). We measured the voltage-sensitive dye signal that accompanies these oscillations in the box turtle; our measurements allowed a more detailed visualization of the spatiotemporal characteristics of the oscillations.

The turtles were first anesthetized by placing them in ice for 2 hours. A craniotomy was performed over the olfactory bulb. The dura and arachnoid mater were then carefully removed to facilitate staining. The exposed olfactory bulb was stained by covering it with dye solution (RH414, 0.02–0.2 mg/mL) for60 minutes. Excess dye was then washed away with turtle saline. The odor output from the olfactometer was approximately square shaped and had a latency of about 100 ms from the onset of the command pulse to the solenoid controlling odor delivery.

We optimized the optics for measurements of epifluorescence at low magnification. In this circumstance, the intensity reaching the objective image plane is proportional to the fourth power of the objective NA (Inoue, 1986). Because conventional microscope optics have small NAs at low magnifications, we assembled a microscope (macroscope) based on a 25-mm focal length, 0.95 f, C-mount camera lens (used with the C-mount end facing the preparation) (Salama, 1988; Ratzlaff and Grinvald, 1991; Kleinfeld and Delaney, 1996) (Macroscope, RedShirtImaging). With a magnification of 4×, the intensity reaching the photodetectorwas 100 times largerwith this lens than with a conventional 4×, 0.16-NA microscope lens. Additional details of the apparatus are given in Wu and Cohen (1993), Wu et al. (1998), and Lam et al. (2000).

With this 4× magnification, each detector will receive light from an area of the object plane that is $170 \times 170\,\mu m^2$. Because of light scattering and out-of-focus light, it would not be possible to achieve much better spatial resolution even if the camera had more pixels.

Multiple Components of the Odor-Induced Response

In figure 6.8, the recordings from seven selected diodes in a single trial are shown. The location of these diodes is indicated on the image by the numbered squares on the left.

In rostral locations (detectors 1 and 2), we found a single oscillation with a relatively high frequency. On a diode from a middle location (detector 4) there was a relatively brief, short-latency oscillation, and on a diode from the caudal bulb (detector 7) the oscillation was of a lower frequency and long latency. In areas between two regions, the recorded oscillations were combinations of two signals: rostral/middle in detector 3 and middle/caudal in detectors 5 and 6. We named the three oscillations rostral, middle, and caudal. The fractional

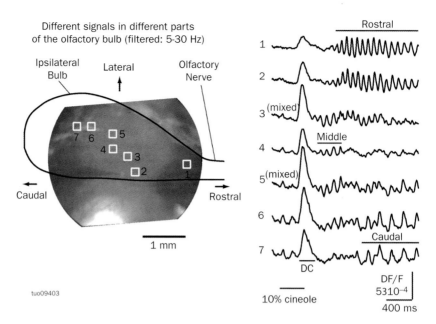

Figure 6.8 Simultaneous optical recordings from seven different areas of an olfactory bulb. An image of the olfactory bulb is shown on the left. Signals from seven selected pixels are shown on the right. The positions of these pixels are labeled with squares and numbers on the image of the bulb. All seven signals have a filtered version of the DC signal at the time indicated by the bar-labeled DC. The oscillation in the rostral region has a high frequency and relatively long latency and duration (detectors 1 and 2). The oscillation from the middle region has a high frequency and short latency and duration (detector 4). The oscillation from the caudal region has a lower frequency and the longest latency (detector 7). The signal from detectors between these regions (detectors 3, 5, and 6) appear to contain a mixture of two components. The horizontal line labeled "10% cineole" indicates the time of the command pulse to the odor solenoid. The data are filtered by high-pass digital RC (5 Hz) and low-pass Gaussian (30 Hz) filters. (Modified from Lam et al. [2000].)

change in fluorescence ($\Delta F/F$) in these three signals ranged from about 2×10^{-4} to 10^{-3}; the noise is substantially smaller. In addition, a DC signal, which appears as a single peak after high-pass filtering in figure 6.8, was observed over most of the ipsilateral olfactory bulb. The rostral oscillation had a long latency and a high frequency. The middle oscillation had a short latency and a frequency that was similar to the rostral oscillation. The caudal oscillation had a lower frequency and the longest latency. In additional to differences in frequency and latency, the three oscillations also had different shapes: The rostral and caudal oscillations had relatively sharp peaks whereas the middle oscillation was more sinusoidal (Lam et al., 2000).

The rostral oscillation is a propagating wave that propagates in the rostral-to-caudal direction. Similarly, the caudal oscillation was also shown to propagate in some instances (Lam et al., 2000). Prechtl et al. (1997) also found that propagation was a characteristic of oscillations in turtle visual cortex.

Surprisingly, the oscillations in turtle appear to be independent of the odorant. Figure 6.9 shows three comparisons of the rostral oscillation for two different odorants and two comparisons each for the middle and the caudal oscillations. Even though the odorants are qualitatively different in terms of their chemical structure and perception by humans, their frequency, amplitude, latency, and envelopes are quite similar (Lam et al., 2003).

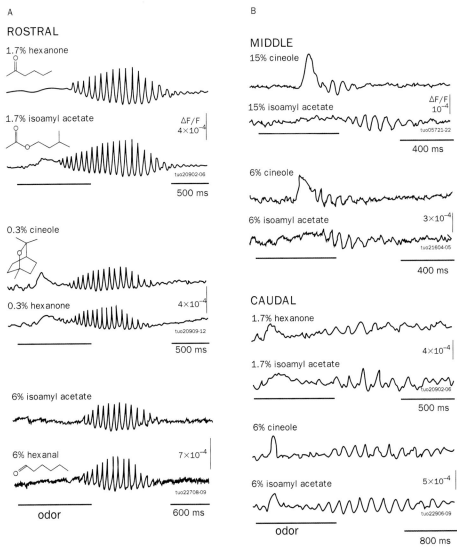

Figure 6.9 (A) Comparisons of the envelopes of the rostral oscillations; three pairs including four different odorants from two preparations are illustrated. The envelopes are not obviously different even though the chemical structures and the human perception of the odorants are quite distinct. The data were bandwidth filtered from 5 to 30 Hz to reduce the size of the DC component and to reduce high-frequency noise. (B) Comparisons of the envelopes of the middle and caudal oscillations; two pairs including two different odorants from two preparations are illustrated for the middle oscillation, two pairs including three different odorants from two preparations are illustrated for the caudal oscillation. The envelopes of the middle and caudal oscillation are not obviously different even though the chemical structures and human perception of the odorants are quite distinct. (Modified from Lam et al. [2003].)

Oscillations are not restricted to the olfactory system. Despite their ubiquity, the roles and functions of oscillation are not well understood. However, in view of their ubiquity and the fact that they involve a very large number of neurons, it is reasonable to speculate that oscillations may have an important role in perception. Our data show that the odor-induced oscillations in the olfactory bulb are substantially more complicated than anticipated and that multiple functional population domains are processing olfactory input in parallel. In addition, we find that these oscillations do not appear to carry information about odor quality; examples can be found where qualitatively different odorants elicit nearly identical oscillations.

Measuring voltage-sensitive population signals from in vivo mammalian preparations is sometimes more difficult than in the turtle because the noise from the heartbeat and respiration are larger, and because mammalian preparations are not as easily stained as those of lower vertebrates. Two methods for reducing the movement artifacts from heartbeat are, together, quite effective. First, a subtraction procedure is used when two recordings are made but only one of the trials has a stimulus (Orbach et al., 1985). Both recordings are triggered from the upstroke of the electrocardiogram, so both should have similar heartbeat noise. When the trial without the stimulus is subtracted from the trial with the stimulus, the heartbeat artifact is reduced. Second, an air-tight chamber is fixed onto the skull surrounding the craniotomy (Blasdel and Salama, 1986). When this chamber is filled with silicon oil and closed, movements resulting from heartbeat and respiration are substantially reduced. Using both methods reduces the noise from these movement artifacts enough so that it is no longer the main source of noise in the measurements.

METHODS

The three examples given earlier involved fractional intensity changes that ranged from 10^{-4} to 3×10^{-2}. To measure these signals, the noise in the measurements had to be an even smaller fraction of the resting intensity. In the sections that follow, some of the considerations necessary to achieve such a low noise are outlined. One topic that was discussed in earlier reviews is not covered here: evidence that pharmacological effects and photodynamic damage resulting from the voltage-sensitive dyes are manageable (see, for example, Cohen and Salzberg, 1978; Waggoner, 1979; Salzberg, 1983; Cohen and Lesher, 1986; Grinvald et al., 1988).

Signal Type

Sometimes it is possible to decide in advance which kind of optical signal will give the best signal-to-noise ratio, but in other situations an experimental comparison is necessary. The choice of signal type often depends on the optical characteristics of the preparation. Birefringence signals are relatively large in preparations that, like axons, have a cylindrical shape and radial optic axis. However, in preparations with spherical symmetry (e.g., cell soma), the birefringence signals in adjacent quadrants will cancel (Boyle and Cohen, 1980).

Thick preparations (e.g., mammalian cortex) also dictate the choice of signal. In this circumstance, transmitted light measurements are not easy (a subcortical implantation of a light guide would be necessary), and the small size of the absorption signals that are detected in reflected light (Ross et al., 1977; Orbach and Cohen, 1983) mean that fluorescence would be optimal (Orbach et al., 1985). Anotherfactorthat affects the choice of absorption or fluorescence is that the signal-to-noise ratio in fluorescence is more strongly degraded by dye bound to extraneous material.

Fluorescence signals have most often been used in monitoring activity from tissue-cultured neurons. Both kinds of signals have been used in measurements from ganglia and brain slices. Fluorescence has always been used in measurements from intact brains.

AMPLITUDE OF THE VOLTAGE CHANGE

The voltage-sensitive dye signals discussed in this chapter are presented as the fractional intensity change ($\Delta I/I$). These signals give information about the time course of the potential change but no direct information about its magnitude. In some instances, indirect information about the magnitude of the voltage change can be obtained (e.g., Orbach et al., 1985; Delaney et al., 1994; Antic and Zecevic, 1995). Another approach is the use of ratiometric measurements at two independent wavelengths (Gross et al., 1994). However, to determine the amplitude of the voltage change from a ratio measurement one must know the fraction of the fluorescence that results from dye not bound to active membrane—a requirement that is only approximately met in special circumstances (e.g., tissue culture).

MEASURING TECHNOLOGY

Noise

Shot Noise

The limit of accuracy with which light can be measured is set by the shot noise arising from the statistical nature of photon emission and detection. Fluctuations in the number of photons emitted per unit time occur, and if an ideal light source (e.g., a tungsten filament) emits an average of 10^{16} photons/ms, the root mean square deviation in the number emitted is the square root of this number or 10^8 photons/ms. Because of shot noise, the signal-to-noise ratio is directly proportional to the square root of the number of measured photons and inversely proportional to the square root of the bandwidth of the photodetection system (Braddick, 1960; Malmstadt et al., 1974). The basis for the square root dependence on intensity is illustrated in figure 6.10.

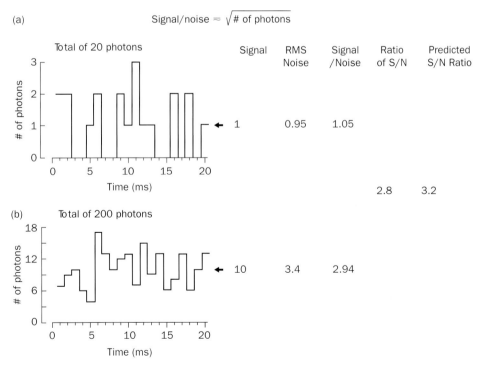

Figure 6.10 Plots of the results of using a table of random numbers to distribute 20 photons (A) or 200 photons (B) into 20 time bins. The result illustrates the fact that when more photons are measured, the signal-to-noise ratio (S/N) is improved. On the right, the signal-to-noise ratio is measured for the two results. The ratio of the two signal-to-noise ratios was 0.43. This is close to the ratio predicted by the relationship that the signal-to-noise ratio is proportional to the square root of the measured intensity. (Redrawn from Wu and Cohen [1993].)

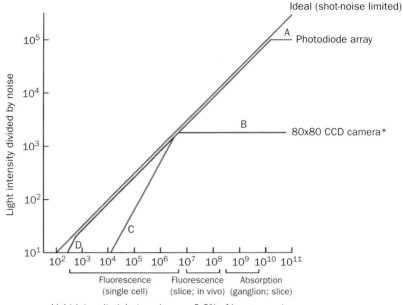

Figure 6.11 The ratio of light intensity divided by the noise in the measurement as a function of light intensity in photons per millisecond per 0.2% of the object plane. The theoretical optimal signal-to-noise ratio (green line) is the shot noise limit. Two camera systems are shown: a photodiode array with 464 pixels (blue lines) and a cooled, back-illuminated, 2-kHz frame rate, 74×74-pixel CCD camera (red lines). The photodiode array provides an optimal signal-to-noise ratio at higher intensities whereas the CCD camera is better at lower intensities. The approximate light intensity per detector in fluorescence measurements from a single neuron, fluorescence measurements from a slice or in vivo preparation, and in absorption measurements from a ganglion or a slice are indicated along the x-axis. The signal-to-noise ratio for the photodiode array falls away from the ideal at high intensities (A) because of extraneous noise, and at low intensities (C) because of darknoise. The lower darknoise of the cooled CCD allows it to function at the shot noise limit at lower intensities until read noise dominates (D). The CCD camera saturates (*) at intensities more than 5×10^6 photons/ms/0.2% of the object plane.

In figure 6.10A, the result of using a random number table to distribute 20 photons into 20 time windows is shown. In figure 6.10B, the same procedure was used to distribute 200 photons into the same 20 bins. Relative to the average light level there is more noise in the top trace (20 photons) than in the bottom trace (200 photons). On the right side of figure 6.10, the measured signal-to-noise ratios are listed; the improvement from A to B is similar to that expected from the square root relationship. This square root relationship is indicated by the dotted line in figure 6.11, which plots the light intensity divided by the noise in the measurement versus the light intensity. In a shot noise-limited measurement, improvement in the signal-to-noise ratio can only be obtained by (1) increasing the illumination intensity, (2) improving the light-gathering efficiency of the measuring system, or (3) reducing the bandwidth.

Because only a small fraction of the 10^{16} photons/ms emitted by a tungsten filament source is measured, a signal-to-noise ratio of 10^8 (discussed earlier) cannot be achieved. A partial listing of the light losses follows. A 0.9-NA lamp collector lens would collect 0.1% of the light emitted by the source. Only 0.2 of that light is in the visible wavelength range; the remainder is infrared (heat). Limiting the incident wavelengths to those that have the signal means that only 0.1 of the visible light is used. Thus, the light reaching the preparation might typically be reduced to 10^{13} photons/ms. If the light-collecting system that forms the image has high efficiency (e.g., in an absorption measurement), about 10^{13} photons/ms will reach the image plane. In a fluorescence measurement there will be much less light measured because (1) only a fraction of the incident photons are absorbed by the fluorophores, (2) only a fraction of the absorbed photons appear as emitted photons, and (3) only a fraction of the emitted photons are collected by the objective. If the camera has a quantum efficiency of 1.0, then, in absorption, a total of 10^{13} photoelectrons/ms will be measured. With a camera of 1000 pixels, there will be 10^{10} photoelectrons/ms/pixel. The shot noise will be 10^5 photoelectrons/ms/pixel; thus, the very best that can be expected is a noise that is 10^{-5} of the resting light (a signal-to-noise ratio of 100 dB). The extra light losses in a fluorescence measurement will further reduce the maximum obtainable signal-to-noise ratio.

One way to compare the performance of different camera systems and to understand their deviations from the optimum (shot noise limited) is to determine the light intensity divided by the noise in the measurement and plot that versus the number of photons reaching the each pixel of the camera. The green line in figure 6.11 is the plot for an ideal camera. At high light intensities, this ratio is large and thus small changes in intensity can be detected. Forexample, at 10^{10} photons/ms, a fractional intensity change of 0.1% can be measured with a signal-to-noise ratio of 100. On the other hand, at low intensities the ratio of intensity divided by noise is small and only large signals can be detected. For example, at 10^4 photons/ms, the same fractional change of 0.1% can be measured with a signal-to-noise ratio of one only after averaging 100 trials.

In addition, figure 6.11 compares the performance of two particular camera systems— a photodiode array (blue line) and a cooled CCD camera (red line) with the shot noise ideal (green line). The photodiode array approaches the shot noise limitation over the range of intensities from 3×10^6 to 10^{10} photons/ms. This is the range of intensities obtained in absorption measurements and fluorescence measurements on in vitro slices and intact brains. On the other hand, the cooled CCD camera approaches the shot noise limit over the range of intensities from 5×10^3 to 5×10^6 photons/ms. This is the range of intensities obtained from fluorescence experiments on individual cells and neurons. In the discussion that follows we indicate the aspects of the measurements and the characteristics of the two camera systems that cause them to deviate from the shot noise ideal. The two camera systems we chose to illustrate in figure 6.11 have relatively good dark noise and saturation characteristics; other cameras would be dark noise limited at higher light intensities and would saturate at lower intensities.

Extraneous Noise

A second type of noise, termed *extraneous* or *technical noise,* is more apparent at high light intensities at which the sensitivity of the measurement is high because the fractional shot noise and dark noise are low. One type of extraneous noise is caused by fluctuations in the output of the light source (discussed later). Two other sources of extraneous noise are vibrations and movement of the preparation. A number of precautions for reducing vibrational noise have been described (Salzberg et al., 1977; London et al., 1987). The pneumatic isolation mounts on many vibration isolation tables are more efficient in reducing vertical vibrations than in reducing horizontal movements. One possibility is air-filled soft rubber tubes (Newport Corporation, Irvine, California). For more severe vibration problems, Minus K Technology (Inglewood, California) sells vibration isolation tables with very low resonant frequencies. Nevertheless, it has been difficult to reduce vibrational noise to less than 10^{-5} of the total light. With this amount of vibrational noise, increases in measured intensity beyond 10^{10} photons/ms would not improve the signal-to-noise ratio. For this reason, the performance of the photodiode array system is shown reaching a ceiling in figure 6.11 (segment A, blue line).

Dark Noise

Dark noise degrades the signal-to-noise ratio at low light levels. Because the CCD camera is cooled, its dark noise is substantially lower than that of the photodiode array system. The larger dark noise in the photodiode array accounts for the fact that segment C in figure 6.11 is substantially to the right of segment D.

Light Sources

Three kinds of sources have been used. Tungsten filament lamps are a stable source, but their intensity is relatively low, particularly at wavelengths less than 480 nm. Arc lamps are somewhat less stable, but can provide more intense illumination. Until recently, measurements made with laserillumination have been substantially noisier.

Tungsten Filament Lamps

It is not difficult to provide a power supply stable enough so that the output of the bulb fluctuates by less than one part in 10^5. In absorption measurements, in which the fractional changes in intensity are relatively small, only tungsten filament sources have been used. On the other hand, fluorescence measurements often have larger fractional changes that will better tolerate light sources with systematic noise, and the measured intensities are lower, making improvements in signal-to-noise ratio from brighter sources attractive.

Arc Lamps

Opti-Quip, Inc. (Highland Mills, New York) provides 150- and 250-W xenon power supplies, lamp housings, and arc lamps with noise that is in the range of one part in 10^4. The 150-W lamp yielded two to three times more light at 520 ± 45 nm than a tungsten filament bulb and, in turn, the 250-W bulb was two to three times brighter than the 150-W bulb. The extra intensity is especially useful for fluorescence measurements from single neurons in which the light intensity is low and the dark noise is dominant. In that situation, the signal-to-noise ratio improves linearly with intensity.

Lasers

It has been possible to take advantage of one useful characteristic of laser sources. In preparations with minimal light scattering, the laser output can be focused onto a small spot, allowing measurement of membrane potential from small processes in tissue-cultured neurons (Bullen et al., 1997). However, there may be excess noise resulting from laser speckle (Dainty, 1984).

Optics

Numerical Aperture

The need to maximize the numberof measured photons has been a dominant factorin the choice of optical components. In epifluorescence, both the excitation light and the emitted light pass through the objective, and the intensity reaching the photodetector is proportional to the fourth power of NA (Inoue, 1986). Clearly, NA is an important consideration in the choice of lenses. Direct comparison of the intensity reaching the image plane has shown that the light-collecting efficiency of an objective is not completely determined by the stated magnification and NA. Differences of a factor of *five* between lenses of the same specification have been seen. We presume that these differences depend on the number of lenses, the coatings, and absorbances of glasses and cements. We recommend empirical tests of several lenses for efficiency.

Depth of Focus

Salzberg et al. (1977) determined the effective depth of focus for a 0.4-NA objective lens by recording an optical signal from a neuron when it was in focus and then moving the neuron out of focus by various distances. They found that the neuron had to be moved 300 μm out of focus to cause a 50% reduction in signal size. Using 0.5-NA optics, 100 μm out of focus led to a reduction of 50% (Kleinfeld and Delaney, 1996).

Light Scattering and Out-of-Focus Light

Light scattering can limit the spatial resolution of an optical measurement. London et al. (1987) measured the scattering of 705-nm light in *Navanax* buccal ganglia. They found that inserting a ganglion in the light path caused light from a 30-μm spot to spread so that the diameter of the circle of light that included intensities more than 50% of the total was roughly 50 μm. The spread was greater, to about 100 μm, with light of 510 nm. Figure 6.12 illustrates the results of similar experiments carried out on the salamander olfactory bulb. Figure 6.12A indicates that when no tissue is present, essentially all the light (750 nm) from a small spot falls on one detector. Figure 6.12C illustrates the result when a 500-μm-thick slice of olfactory bulb is present. The light from the small spot is spread to about 200 μm. Mammalian cortex appears to scatter more than the olfactory bulb. Thus, light scattering will cause considerable blurring of signals in adult vertebrate preparations.

A second source of blurring is signal from regions that are out of focus. For example, if the active region is a cylinder (a column) perpendicular to the plane of focus, and the objective is focused at the middle of the cylinder, then the light from the in-focus plane will have the correct diameter at the image plane. However, the light from the regions above and below are out of focus and have a diameter that is too large. Figure 6.12B illustrates the effect of moving the small spot of light 500 μm out of focus. The light from the small spot is spread to about 200 μm. Thus, in preparations with considerable scattering or with out-of-focus signals, the actual spatial resolution may be limited by the preparation and not by the numberof pixels in the imaging device.

Confocal and Two-Photon Confocal Microscopes

The confocal microscope (Petran and Hadravsky, 1966) substantially reduces both the scattered and out-of-focus light that contributes to the image. A recent modification using two-photon excitation of the fluorophore further reduces out-of-focus fluorescence and photobleaching (Denk et al., 1995). With both types of microscope one can obtain images from intact vertebrate preparations with much better spatial resolution than was achieved with ordinary microscopy. Although these microscopes have been used successfully to monitor changes in calcium concentration inside small processes of neurons (Eilers et al., 1995; Yuste and Denk, 1995), their sensitivity is relatively poor. There are no reports of their use to measure the small signals obtained with voltage-sensitive dyes of the type discussed in this chapter. On the otherhand, slowervoltage-sensitive dye signals have been measured confocally (Loew, 1993).

The generation of fluorescent photons in the two-photon confocal microscope is not efficient. We compared the signals from Calcium Green-1 in the mouse olfactory bulb using two-photon and ordinary microscopy. In this comparison (fig. 6.13 [Friedrich and Korsching, 1997; Wachowiak and Cohen, 2001]), the number of photons contributing to the intensity measurement in the two-photon confocal microscope was about 1000 times smaller than the number measured with the conventional microscope and a NeuroCCD-SM camera. As a result, the signal-to-noise ratio in the NeuroCCD-SM recording is much larger. Moreover, four trials were averaged in the two-photon measurement shown in figure 6.13, whereas the result from the NeuroCCD-SM camera was from a single trial. In addition, the NA of the lens used for the two-photon measurement was 0.8 whereas that used in the ordinary microscope was only 0.5. If a correction for these two factors is applied, the two-photon measurement would have a signal-to-noise ratio six times smaller than that shown.

The factors that contribute to the relatively small number of photons in the two-photon measurement are (1) the incident light interacts with many fewer dye molecules because only a thin section receives high intensity illumination and (2) Calcium Green-1 has a low two-photon cross-section, which results in a low optical efficiency. This low efficiency cannot be overcome by increasing the incident intensity because higher intensity will heat the preparation. On the other hand, the advantages of two-photon microscopy are clear; rejection of scattered light and very shallow depth of focus results in much better x-/y- and z-axis resolution. The two kinds of imaging systems are optimal for different niches in the parameter space of imaging.

Random Access Fluorescence Microscopy

Bullen et al. (1997) have used acousto-optic deflectors to construct a random scanning microscope and were able to measure signals from parts of cultured hippocampal neurons. To reduce the effects of fluctuations in the laser output, the fluorescence signals were divided by the output of a photodetectorsampling the incident light. Relatively large signal-to-noise

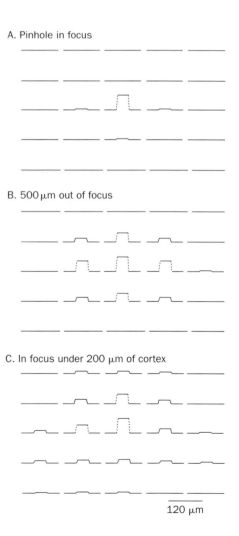

Figure 6.12 Effects of focus and scattering on the distribution of light from a point source onto the array. (A) A 40-μm pinhole in aluminum foil covered with saline was illuminated with light at 750 nm. The pinhole was in focus. More than 90% of the light fell on one detector. (B) The stage was moved downward by 500 μm. Light from the out-of-focus pinhole was now seen on several detectors. (C) The pinhole was in focus but was covered by a 500-μm slice of salamander cortex. Again, the light from the pinhole was spread over several detectors. A 10 × 0.4-NA objective was used. Kohler illumination was used before the pinhole was placed in the object plane. The recording gains were adjusted so the largest signal in each of the three trials would be approximately the same size in the figure. (Redrawn from Orbach and Cohen [1983].)

(a)

Calcium Green-1 signals measured with a 2-photon microscope and an ordinary fluorescence microscpe with a NeuroCCD-SM camera

2-photo microscope
(4 trials)

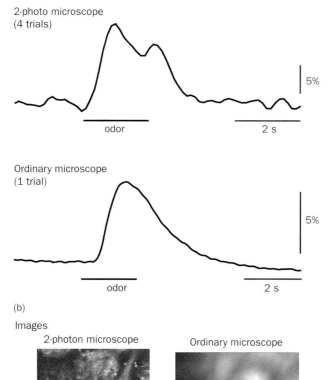

5%

odor   2 s

Ordinary microscope
(1 trial)

5%

odor   2 s

(b)

Images

2-photon microscope   Ordinary microscope

Figure 6.13 Comparison of signals from a two-photon microscope and an ordinary fluorescence microscope. The signal-to-noise ratios were substantially larger using an ordinary microscope and a NeuroCCD-SM camera. On the other hand, the rejection of scattered light and very shallow depth of focus results in much better x-/y- and z-axis resolution using the two-photon microscope. (A) Odorant elicited signals in in vivo mouse preparations using the two imaging systems. In both cases, the signals were the spatial average of the light from one glomerulus. (B) The images formed by the two kinds of microscope. The image made with the ordinary microscope covers a two times larger area of the bulb. The two-photon image is the total intensity; the ordinary microscope image is the image of the signal. The olfactory receptor neuron nerve terminals in the mouse olfactory bulb were stained with Calcium Green-1 (Friedrich and Korsching, 1997; Wachowiakand Cohen, 2001; Friedrich, Wachowiak, and Cohen, Max Planck Institute for Medical Research, Heidelberg, and Yale University, unpublished results.)

ratios were obtained using voltage-sensitive dyes. This method can have the advantage that only a small proportion of the preparation is illuminated, thereby reducing the photodynamic damage from the very bright laser light source. However, this method will probably be restricted to preparations such as cultured neurons, in which there is relatively little light scattering.

Photodetectors

Because the signal-to-noise ratio in a shot noise-limited measurement is proportional to the square root of the number of photons converted into photoelectrons (discussed earlier), quantum efficiency is important. Silicon photodiodes have quantum efficiencies approaching the ideal (1.0) at wavelengths where most dyes absorb or emit light (500–900 nm). In contrast, only specially chosen vacuum photocathode devices (phototubes, photomultipliers, or image intensifiers) have a quantum efficiency as high as 0.5. Thus, in shot noise-limited situations, a silicon diode will have a larger signal-to-noise ratio. Photographic film has an even smaller quantum efficiency (0.01) (Shaw, 1979) and thus has not been used for the kinds of measurements discussed in this chapter.

Imaging Devices

Many factors must be considered in choosing an imaging system. Perhaps the most important considerations are the requirements for spatial and temporal resolution. Because the signal-to-noise ratio in a shot noise-limited measurement is proportional to the number of measured photons, increases in either temporal or spatial resolution reduce the signal-to-noise ratio. Our discussion considers systems that have frame rates near 1 kHz. In most of these systems, the camera has been placed in the objective image plane of a microscope. However, Tank and Ahmed (1985) suggested a scheme by which a hexagonal close-packed array of optical fibers is positioned in the image plane, and individual photodiodes are connected to the other end of the optical fibers. NeuroPlex-II, a 464-pixel photodiode array camera (RedShirtImaging) is based on this scheme.

Silicon Diode Arrays

Parallel Readout Arrays

Diode arrays with 256 to 1020 elements are now in use in several laboratories (e.g., Iijima et al., 1989; Zecevic et al., 1989; Nakashima, et al., 1992; Hirota et al., 1995). These arrays are designed for parallel readout: each detector followed by its own amplifier with an output that can be digitized at frame rates of more than 1 kHz. Although the need to provide a separate amplifierforeach diode element limits the numberof pixels in parallel readout systems, it contributes to the very large (10^5) dynamic range that can be achieved. Two parallel readout array systems are commercially available: Argus-50 (256 pixels; Hamamatsu Photonics K.K., www.hpk.co.jp) and NeuroPlex-II (464 pixels; RedShirtImaging, www.redshirtimaging.com).

Serial Readout Arrays

By using a serial readout, the number of amplifiers is greatly reduced. In addition, it is simpler to cool CCD chips to reduce the dark noise. However, because of saturation, currently available CCD cameras are not optimal for the higher intensities available in some neurobiological experiments (fig. 6.11). The high-intensity limit of the CCD camera is set by the light intensity that fills the electron wells on the CCD chip. The well depth of commercially available CCD chips is less than 10^6 e–. This accounts forthe bending overof the CCD camera performance at segment B in figure 6.11. A dynamic range of even 10^3 is not easily

Table 6-1. Characteristics of fast CCD and CMOS camera systems (as reported by the manufacturer)

	Frame rate (Hz), full frame	Well size ($\times 1000$ e)	Read noise, electrons	Back illumination	Bits a to d	Pixels
SciMedia MiCAM01[a]	1600	–	–	No	12	92×64
Dalsa CA-D1-128[b]	756	–	350	No	12	128×128
RedShirtImaging NeuroCCD-SM, CardioCCD[c]	2000	200	20	Yes	14	80×80
RedShirtImaging NeuroCCD-SM256[c]	100	750	30	Yes	14	256×256
TILL Photonics, IMAGO[d]	140	35	14	No	12	160×120
SciMedia MiCAM Ultima[a]CMOS	10,000	10,000	1000	No	14	100×100

[a]http://www.scimedia.co.jp
[b]http://www.dalsa.com.
[c]http://www.redshirtimaging.com.
[d]http://www.TILL-photonics.com.

achieved with a CCD camera. A CCD camera will not be optimal for measurements of absorption or for fluorescence measurements on in vitro slices and intact brains (fig. 6.11). The light intensity would have to be reduced with a consequent decrease in signal-to-noise ratio. On the other hand, CCD cameras are close to ideal for measurements from individual neurons stained with internally injected dyes.

SciMedia (Taito-ku, Tokyo) now markets a complementary metal-oxide semiconductor (CMOS) camera with a well depth of 10^7 e–. Table 6.1 compares several fast CCD cameras and this CMOS camera. The CMOS camera will be optimal over a range of intensities that is intermediate between the CCD and photodiode array systems illustrated in figure 6.11.

Vacuum Photocathode Cameras

Although the lowerquantum efficiencies of vacuum photocathodes is a disadvantage, these devices may have lowerdar k noise. One of these, a Radechon (Kazan and Knoll, 1968), has an output proportional to the changes in the input. Vacuum photocathode cameras have been used in relatively low time resolution recordings from mammalian cortex (Blasdel and Salama, 1986) and salamander olfactory bulb (Kauer, 1988; Cinelli et al., 1995) and are used in the Imager 2001 (Optical Imaging, Inc., www.opt-imaging.com).

FUTURE DIRECTIONS

Because the light-measuring apparatus is already reasonably optimized, any improvement in the sensitivity of the voltage-sensitive dye measurements will need to come from the development of better dyes and/or investigating signals from additional optical properties of the dyes. The dyes in figure 6.2 and the vast majority of those synthesized are of the general class named *polyenes* (Hamer, 1964), a group that was first used to extend the wavelength response of photographic film. It is possible that improvements in signal size can be obtained with new polyene dyes (see Waggoner and Grinvald [1977] and Fromherz et al. [1991] for a discussion of maximum possible fractional changes in absorption and fluorescence). On the other hand, the fractional change on squid axons has not increased in recent years (Gupta et al., 1981; Cohen, Grinvald, Kamino, and Salzberg, unpublished results), and most improvements (e.g., Grinvald et al., 1982a; Antic and Zecevic, 1995; Momose-Sato et al., 1995; Tsau et al., 1996) have involved synthesizing analogues that work well on new preparations.

The best of the styryl and oxonol polyene dyes have fluorescence changes of 10% to 20%/100 mV in situations in which the staining is specific to the membrane with the potential that is changing (Grinvald et al., 1982a; Loew et al., 1992; Rohr and Salzberg, 1994). Recently, Gonzalez and Tsien (1995) introduced a new scheme for generating voltage-sensitive signals using two chromophores and energy transfer. Although these fractional changes were also in the range of 10%/100 mV, more recent results are about 30% (Gonzalez and Tsien, personal communication). However, to achieve fast response times (>100 Hz), one of the chromophores must be very hydrophobic and thus does not penetrate into brain tissue. As a result, it has not been possible to measure fast signals with this pair of dyes in intact tissues (Gonzalez and Tsien, personal communication; Obaid and Salzberg, personal communication).

Bouevitch et al. (1993) and Ben-Oren et al. (1996) found that membrane potential changed the nonlinear second harmonic generation from styryl dyes in cholesterol bilayers and in fly eyes. Large (50%) fractional changes were measured. It is hoped that improvements in the illumination intensity will result in larger signal-to-noise ratios.

Ehrenberg and Berezin (1984) have used resonance Raman to study surface potential; these methods might also be applicable for measuring transmembrane potential.

Neuron-type Specific Staining

An important new direction is the development of methods for neuron-type specific staining. Three quite different approaches have been tried. First, the use of retrograde staining procedures has recently been investigated in the embryonic chick and lamprey spinal cords (Tsau et al., 1996). An identified neuron class (motoneurons) was selectively stained. In lamprey experiments, spike signals from individual neurons were sometimes measured (Hickie et al., 1996). Further efforts at optimizing this staining procedure is needed. Second is the use of cell-type specific staining developed for fluorescein by Nirenberg and Cepko (1993). It might be possible to use similartechniques to stain cells selectively with voltage-sensitive or ion-sensitive dyes. Third, Siegel and Isacoff (1997) constructed a genetically encoded combination of a potassium channel and GFP. When introduced into a frog oocyte, this molecule had a (relatively slow) voltage-dependent signal with a fractional fluorescence change of 5%. More recently, Sakai et al. (2001) and Ataka and Pieribone (2002) have developed similar constructs with very rapid kinetics. Neuron-type specific staining would make it possible to determine the role of specific neurons types in generating the input/output function of a brain region.

Optical recordings already provide unique insights into brain activity and organization. Improvements in sensitivity or selectivity would make these methods more powerful.

ACKNOWLEDGMENTS

We are indebted to our collaborators Vicencio Davila, Amiram Grinvald, Kohtaro Kamino, Ying-wan Lam, Bill Ross, Brian Salzberg, Matt Wachowiak, Alan Waggoner, Jian-young Wu, Joe Wuskell, and Michal Zochowski fornumer ous discussions about optical methods. The experiments carried out in our laboratories were supported by National Institutes of Health grants DC05259, NS42739, and GM35063.

REFERENCES

Adrian, E. D. (1942). Olfactory reactions in the brain of the hedgehog. J. Physiol., 100, 459–473.

Albowitz, B., and U. Kuhnt. (1993). Spread of epileptiform potentials in the neocortical slice: Recordings with voltage-sensitive dyes. Brain Res., 631, 329–333.

Antic, S., G. Major, and D. Zecevic. (1999). Fast optical recording of membrane potential changes from dendrites of pyramidal neurons. J. Neurophysiol., 82, 1615–1621.

Antic, S., and D. Zecevic. (1995). Optical signals from neurons with internally applied voltage-sensitive dyes. J. Neurosci., 15, 1392–1405.

Antic, S., J. P. Wuskell, L. Loew, and D. Zecevic. (2000). Functional profile of the giant metacerebral neuron of helix aspersa: Temporal and spatial dynamics of electrical activity in situ. J. Physiol. (Lond.), 527, 55–69.

Ataka, K., and V. A. Pieribone. (2002). A genetically-targetable fluorescent probe of channel gating with rapid kinetics. Biophys. J., 82, 509–516.

Ben-Oren, I., G. Peleg, A. Lewis, B. Minke, and L. Loew. (1996). Infrared nonlinear optical measurements of membrane potential in photoreceptor cells. Biophys. J., 71, 1616–1620.

Beuerman, R. W. (1975). Slow potentials of the turtle olfactory bulb in response to odor stimulation of the nose. Brain Res., 97(1), 61–78.

Blasdel, G. G., and G. Salama. (1986). Voltage-sensitive dyes reveal a modular organization in monkey striate cortex. Nature, 321, 579–585.

Bouevitch, O., A. Lewis, I. Pinevsky, J. Wuskell, and L. Loew. (1993). Probing membrane potential with nonlinear optics. Biophys. J., 65, 672–679.

Boyle, M. B., and L. B. Cohen. (1980). Birefringence signals that monitor membrane potential in cell bodies of molluscan neurons. Fed. Proc., 39, 2130.

Braddick, H. J. J. (1960). Photoelectric photometry. Rep. Prog. Physics, 23, 154–175.

Bullen, A., S. S. Patel, and P. Saggau. (1997). High-speed, random-access fluorescence microscopy: I. High resolution optical recording with voltage-sensitive dyes and ion indicators. Biophys. J., 73, 477–491.

Chen, W. R., J. Midtgaard, and G. M. Shepherd. (1997). Forward and backward propagation of dendritic impulses and their synaptic control in mitral cells. Science, 278, 463–467.

Cinelli, A. R., S. R. Neff, and J. S. Kauer. (1995). Salamander olfactory bulb neuronal activity observed by video rate, voltage-sensitive dye imaging. I. Characterization of the recording system. J. Neurophysiol., 73, 2017–2032.

Cinelli, A. R., and B. M. Salzberg. (1992). Dendritic origin of late events in optical recordings from salamander olfactory bulb. J. Neurophysiol., 68, 786–806.

Cohen, L. B., and S. Lesher. (1986). Optical monitoring of membrane potential: Methods of multisite optical measurement. Soc. Gen. Physiol. Ser., 40, 71–99.

Cohen, L. B., and B. M. Salzberg. (1978). Optical measurement of membrane potential. Rev. Physiol. Biochem. Pharmacol., 83, 35–88.

Dainty, J. C. (1984). *Laser Speckle and Related Phenomena.* Springer-Verlag, New York.

Davila, H. V., L. B. Cohen, B. M. Salzberg, and B. B. Shrivastav. (1974). Changes in ANS and TNS fluorescence in giant axons from *Loligo*. J. Memb. Biol., 15, 29–46.

Davila, H. V., B. M. Salzberg, L. B. Cohen, and A. S. Waggoner. (1973). A large change in axon fluorescence that provides a promising method for measuring membrane potential. Nat. New Biol., 241, 159–160.

Delaney, K. R., A. Gelperin, M. S. Fee, J. A. Flores, R. Gervais, D. W. Tank, and D. Kleinfeld. (1994). Waves and stimulus-modulated dynamics in an oscillating olfactory network. Proc. Natl. Acad. Sci. USA, 91, 669–673.

Denk, W., D. W. Piston, and W. Webb. (1995). Two-photon molecular excitation in laser-scanning microscopy. In *Handbook ofBiological Confocal Microscopy,* J. W. Pawley, ed. Plenum Press, New York, 445–458.

Ehrenberg, B., and Y. Berezin. (1984). Surface potential on purple membranes and its sidedness studied by resonance Raman dye probe. Biophys. J., 45, 663–670.

Eilers, J., G. Callawaert, C. Armstrong, and A. Konnerth. (1995). Calcium signaling in a narrow somatic sub-membrane shell during synaptic activity in cerebellar Purkinje neurons. Proc. Natl. Acad. Sci. USA, 92, 10272–10276.

Friedrich, R. W., and S. I. Korsching. (1997). Combinatorial, and chemotropic odorant coding in the zebrafish olfactory bulb visualized by optical imaging. Neuron, 18, 737–752.

Fromherz, P., K. H. Dambacher, H. Ephardt, A. Lambacher, C. O. Muller, R. Neigl, H. Schaden, O. Schenk, and T. Vetter. (1991). Fluorescent dyes as probes of voltage transients in neuron membranes: Progress report. Ber. Bunsenges. Phys. Chem., 95, 1333–1345.

Gonzalez, J. E., and R. Y. Tsien. (1995). Voltage sensing by fluorescence energy transfer in single cells. Biophys. J., 69, 1272–1280.

Grinvald, A., R. D. Frostig, E. Lieke, and R. Hildesheim. (1988). Optical imaging of neuronal activity. Physiol. Rev., 68, 1285–1366.

Grinvald, A., R. Hildesheim, I. C. Farber, and L. Anglister. (1982a). Improved fluorescent probes for the measurement of rapid changes in membrane potential. Biophys. J., 39, 301–308.

Grinvald, A., A. Manker, and M. Segal. (1982b). Visualization of the spread of electrical activity in rat hippocampal slices by voltage-sensitive optical probes. J. Physiol. (Lond.), 333, 269–291.

Grinvald, A., B. M. Salzberg, V. Lev-Ram, and R. Hildesheim. (1987). Optical recording of synaptic potentials from processes of single neurons using intracellular potentiometric dyes. Biophys. J., 51, 643–651.

Gross, E., R. S. Bedlack, and L. M. Loew. (1994). Dual-wavelength ratiometric fluorescence measurements of the membrane dipole potential. Biophys. J., 67, 208–216.

Gupta, R. K., B. M. Salzberg, A. Grinvald, L. B. Cohen, K. Kamino, S. Lesher, M. B. Boyle, A. S. Waggoner, and C. H. Wang. (1981). Improvements in optical methods for measuring rapid changes in membrane potential. J. Memb. Biol., 58, 123–137.

Hamer, F. M. (1964). *The Cyanine Dyes and Related Compounds.* Wiley, New York.

Hickie, C., P. Wenner, M. O'Donovan, Y. Tsau, J. Fang, and L. B. Cohen. (1996). Optical monitoring of activity from individual and identified populations of neurons retrogradely labeled with voltage-sensitive dyes. Abs. Soc. Neurosci., 22, 321.

Hirota, A., K. Sato, Y. Momose-Sato, T. Sakai, and K. Kamino. (1995). A new simultaneous 1020-site optical recording system for monitoring neural activity using voltage-sensitive dyes. J. Neurosci. Methods, 56, 187–194.

Hughes, J. R., and J. A. Mazurowski. (1962). Studies on the supracallosal mesial cortex of unanesthetized, conscious mammals. II Monkey. B. Responses from the olfactory bulb. Electroencephalogr. Clin. Neurophysiol., 14, 635–645.

Iijima, T., M. Ichikawa, and G. Matsumoto. (1989). Optical monitoring of LTP and related phenomena. Abs. Soc. Neurosci., 15, 398.

Inoue, S. (1986). *Video Microscopy.* Plenum Press, New York, 128.

Kauer, J. S. (1988). Real-time imaging of evoked activity in local circuits of the salamander olfactory bulb. Nature, 331, 166–168.

Kazan, B., and M. Knoll. (1968). *Electronic Image Storage.* Academic Press, New York.

Kleinfeld, D., and K. R. Delaney. (1996). Distributed representation of vibrissa movement in the upper layers of somatosensory cortex revealed with voltage-sensitive dyes. J. Comp. Neurol., 375, 89–108.

Kupfermann, I., Pinsker, H., Castelucci, V., and Kandel, E. R. (1971). Central and peripheral control of gill movements in Aplysia. Science, 174, 1252–1256.

Lam, Y.- w., L. B. Cohen, M. Wachowiak, and M. R. Zochowski. (2000). Odors elicit three different oscillations in the turtle olfactory bulb. J. Neurosci., 20, 749–762.

Lam, Y.- w., L. B. Cohen, and M. R. Zochowski. (2003). Effect of odorant quality on the three oscillations and the DC signal in the turtle olfactory bulb. Eur. J. Neurosci., 17, 436–446.

Laurent, G., and M. Naraghi. (1994). Odorant-induced oscillations in the mushroom bodies of the locust. J. Neurosci., 14(5 Pt 2), 2993–3004.

Loew, L. M. (1993). Confocal microscopy of potentiometric fluorescent dyes. Methods Cell Biol., 38, 195–209.

Loew, L. M., L. B. Cohen, J. Dix, E. N. Fluhler, V. Montana, G. Salama, and J. Y. Wu. (1992). A napthyl analog of the aminostyryl pyridinium class of potentiometric membrane dyes shows consistent sensitivity in a variety of tissue, cell, and model membrane preparations. J. Memb. Biol., 130, 1–10.

Loew, L. M., L. B. Cohen, B. M. Salzberg, A. L. Obaid, and F. Bezanilla. (1985). Charge-shift probes of membrane potential: Characterization of aminostyrylpyridinium dyes on the squid giant axon. Biophys. J., 47, 71–77.

London, J. A., D. Zecevic, and L. B. Cohen. (1987). Simultaneous optical recording of activity from many neurons during feeding in *Navanax.* J. Neurosci., 7, 649–661.

Magee, J. C., G. Christofi, H. Miyakawa, B, Christie, N. Lasser-Ross, and D. Johnston (1995). Subthreshold synaptic activation of voltage-gated calcium channels mediate a localized calcium influx into dendrites of hippocampal pyramidal neurons. J. Neurophysiol., 74, 1335–1342.

Magee, J. C., and D. Johnston. (1995). Synaptic activation of voltage-gated channels in the dendrites of hippocampal pyramidal neurons. Science, 268, 301–304.

Mainen, Z. F., N. T. Carnevalle, A. M. Zador, B. J. Claiborne, and T. H. Brown. (1996). Electronic architecture of hippocampal CA1 pyramidal neurons based on three-dimensional reconstructions. J. Neurophysiol., 76, 1904–1923.

Malmstadt, H. V., C. G. Enke, S. R. Crouch, and G. Harlick. (1974). *Electronic Measurements for Scientists.* Benjamin, Menlo Park, Calif.

Momose-Sato, Y., K. Sato, T. Sakai, A. Hirota, K. Matsutani, and K. Kamino. (1995). Evaluation of optimal voltage-sensitive dyes for optical measurement of embryonic neural activity. J. Memb. Biol., 144, 167–176.

Nakashima, M., S. Yamada, S. Shiono, M. Maeda, and F. Sato. (1992). 448-Detector optical recording system: Development and application to *Aplysia* gill-withdrawal reflex. IEEE Trans. Biomed. Eng., 39, 26–36.

Nirenberg, S., and C. Cepko. (1993). Targeted ablation of diverse cell classes in the nervous system in vivo. J. Neurosci., 13, 3238–3251.

Obaid, A. L., T. Koyano, J. Lindstrom, T. Sakai, and B. M. Salzberg. (1999). Spatiotemporal patterns of activity in an intact mammalian network with single-cell resolution: Optical studies of nicotinic activity in an enteric plexus. J. Neurosci., 19, 3073–3093.

Orbach, H. S., and L. B. Cohen. (1983). Optical monitoring of activity from many areas of the in vitro and in vivo salamander olfactory bulb: A new method for studying functional organization in the vertebrate central nervous system. J. Neurosci., 3, 2251–2262.

Orbach, H. S., L. B. Cohen, and A. Grinvald. (1985). Optical mapping of electrical activity in rat somatosensory and visual cortex. J. Neurosci., 5, 1886–1895.

Petran, M., and M. Hadravsky. (1966). Czechoslovakian patent 7720.

Prechtl, J. C., L. B. Cohen, B. Peseran, P. P. Mitra, and D. Kleinfeld. (1997). Visual stimuli induce waves of electrical activity in turtle cortex. Proc. Natl. Acad. Sci. USA, 94, 7621–7626.

Ratzlaff, E. H., and A. Grinvald. (1991). A tandem-lens epifluorescence microscope: Hundred-fold brightness advantage forwide-field imaging. J. Neurosci. Methods, 36, 127–137.

Rohr, S., and B. M. Salzberg. (1994). Multiple site optical recording of transmembrane voltage in patterned growth heart cell cultures: Assessing electrical behavior, with microsecond resolution, on a cellular and subcellular scale. Biophys. J., 67, 1301–1315.

Ross, W. N., B. M. Salzberg, L. B. Cohen, A. Grinvald, H. V. Davila, A. S. Waggoner, and C. H. Wang. (1977). Changes in absorption, fluorescence, dichroism, and birefringence in stained giant axons: Optical measurement of membrane potential. J. Memb. Biol., 33, 141–183.

Sakai, T., A. Hirota, H. Komuro, S. Fujii, and K. Kamino. (1985). Optical recording of membrane potential responses from early embryonic chick ganglia using voltage-sensitive dyes. Brain Res., 349, 39–51.

Sakai, R., V. Repunte-Canonigo, C. D. Raj, and T. Knopfel. (2001). Design and characterization of a DNA-encoded, voltage-sensitive fluorescent protein. Eur. J. Neurosci., 13, 2314–2318.

Salama, G. (1988). Voltage-sensitive dyes and imaging techniques reveal new patterns of electrical activity in heart and cortex. SPIE Proc., 94, 75–86.

Salzberg, B. M. (1983). Optical recording of electrical activity in neurons using molecular probes. In *Current Methods in Cellular Neurobiology*. J. L. Barker and J. F. McKelvy, eds. Wiley, New York, 139–187.

Salzberg, B. M., H. V. Davila, and L. B. Cohen. (1973). Optical recording of impulses in individual neurones of an invertebrate central nervous system. Nature, 246, 508–509.

Salzberg, B. M., A. Grinvald, L. B. Cohen, H. V. Davila, and W. N. Ross. (1977). Optical recording of neuronal activity in an invertebrate central nervous system: Simultaneous monitoring of several neurons. J. Neurophysiol., 40, 1281–1291.

Shaw, R. (1979). Photographic detectors. Appl. Opt. Optical Eng., 7, 121–154.

Siegel, M. S., and E. Y. Isacoff. (1997). A genetically encoded optical probe of membrane voltage. Neuron, 19, 735–741.

Spruston, N., Y. Schiller, G. Stuart, and B. Sakmann. (1995). Activity-dependent action potential invasion and calcium influx into hippocampal CA1 dendrites. Science, 268, 297–300.

Stuart, G. J., and B. Sakmann (1994). Active propagation of somatic action potentials into neocortical pyramidal cell dendrites. Nature, 367, 69–72.

Tank, D., and Z. Ahmed. (1985). Multiple-site monitoring of activity in cultured neurons. Biophys. J., 47, 476A.

Tsau, Y., P. Wenner, M. J. O'Donovan, L. B. Cohen, L. M. Loew, and J. P. Wuskell. (1996). Dye screening and signal-to-noise ratio for retrogradely transported voltage-sensitive dyes. J. Neurosci. Methods, 70, 121–129.

Tsau, Y., J. Y. Wu, H. P. Hopp, L. B. Cohen, D. Schiminovich, and C. X. Falk. (1994). Distributed aspects of the response to siphon touch in *Aplysia:* Spread of stimulus information and cross-correlation analysis. J. Neurosci., 14, 4167–4184.

Wachowiak, M., and L. B. Cohen. (2001). Representation of odorants by receptor neuron input to the mouse olfactory bulb. Neuron, 32, 725–737.

Waggoner, A. S. (1979). Dye indicators of membrane potential. Annu. Rev. Biophys. Bioeng., 8, 47–68.

Waggoner, A. S., and A. Grinvald. (1977). Mechanisms of rapid optical changes of potential sensitive dyes. Annu. N Y Acad. Sci., 303, 217–241.

Wu, J. Y., and L. B. Cohen. (1993). Fast multisite optical measurements of membrane potential. In *Fluorescent and Luminescent Probes for Biological Activity*. W. T. Mason, ed. Academic Press, London, 389–404.

Wu, J. Y., L. B. Cohen, and C. X. Falk. (1994a). Neuronal activity during different behaviors suggests a distributed neuronal organization in the *Aplysia* abdominal ganglion. Science, 263, 820–823.

Wu, J. Y., Y. W. Lam, C. X. Falk, L. B. Cohen, J. Fang, L. Loew, J. C. Prechtl, D. Kleinfeld, and Y. Tsau. (1998). Voltage-sensitive dyes for monitoring multi-neuronal activity in the intact CNS. J. Histochem., 30, 169–187.

Wu, J. Y., Y. Tsau, H. P. Hopp, L. B. Cohen, A. C. Tang, and C. X. Falk. (1994b). Consistency in nervous systems: Trial-to-trial and animal-to-animal variations in the response to repeated application of a sensory stimulus in *Aplysia*. J. Neurosci., 14, 1366–1384.

Yuste, R., and W. Denk. (1995). Dendritic spines as basic functional units of neuronal integration. Nature, 375, 682–684.

Zecevic, D. (1996). Multiple spike-initiation zones in single neurons revealed by voltage-sensitive dyes. Nature, 381, 322–325.

Zecevic, D., J. Y. Wu, L. B. Cohen, J. A. London, H. P. Hopp, and C. X. Falk. (1989). Hundreds of neurons in the *Aplysia* abdominal ganglion are active during the gill-withdrawal reflex. J. Neurosci., 9, 3681–3689.

7

Biomedical Imaging Using
Optical Coherence Tomography

JAMES G. FUJIMOTO, YU CHEN, AND AARON AGUIRRE

Optical coherence tomography (OCT) is an emerging optical imaging modality for biomedical research and clinical medicine. Optical coherence tomography can perform high-resolution, cross-sectional tomographic imaging in materials and biological tissues by measuring the echo time delay and magnitude of back-reflected or backscattered light (Huang et al., 1991a; Fujimoto, 2003). In medical applications, OCT has the advantage that it can perform *optical biopsy,* imaging tissue structure in situ and in real time, without the need to remove and process specimens as in conventional excisional biopsy and histopathology. Optical coherence tomography can achieve axial image resolutions of 1 to 15µm; one to two orders of magnitude higher than standard ultrasound imaging. This chapter provides an overview of OCT technology and applications.

Optical coherence tomographic imaging is analogous to ultrasound imaging, except that it uses light instead of sound. Figure 7.1 shows how OCT works. Optical coherence tomography performs cross-sectional imaging by measuring the time delay and magnitude of optical echoes at different transverse positions. The dimensions of internal structures can be determined by measuring the "echo" time for light to be back-reflected or backscattered from structures at different depths or axial distances (axial scans). A cross-sectional image is generated by scanning the optical beam in the transverse direction and performing successive axial scan measurements. This generates a two-dimensional data set that is a measurement of the back-reflection or backscattering in a cross-sectional plane through the material or biological tissue.

Optical coherence tomography was first demonstrated more than decade ago, in 1991 (Huang et al., 1991a). Imaging was performed ex vivo in the human retina and in atherosclerotic plaque as examples of imaging in transparent, weakly scattering media and in highly scattering media. Figures 7.2 and 7.3 show examples of early OCT images. These early OCT images had an axial image resolution of ≈ 15 µm, which is almost one order of magnitude finer than standard ultrasound. Imaging was performed with infrared light at an ≈ 800-nm wavelength. The OCT images are displayed using a false color scale, in which varying amounts of backscattered or back-reflected light are displayed as different colors on a rainbow color scale. The vertical direction in the images corresponds to the direction of the incident optical beam, the axial depth direction. The light signals detected in OCT imaging are extremely small, typically between 10^{-10} and 10^{-5} of the incident light. Because the signal varies over four to five orders of magnitude, it is convenient to use a log scale to

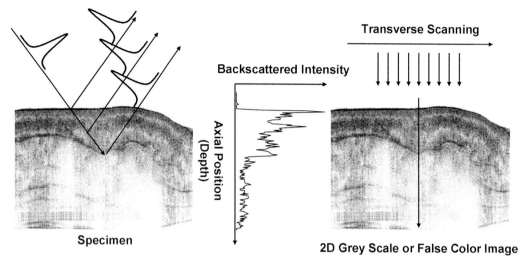

Figure 7.1 Optical coherence tomography generates cross-sectional images by measuring the magnitude and echo time delay backscattered or back-reflected light from at different transverse positions. This generates a two-dimensional or three-dimensional data set that can be displayed as a grayscale or false color image.

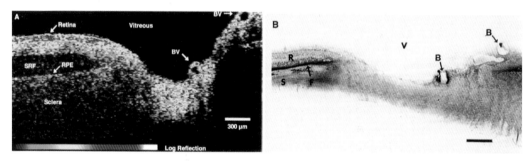

Figure 7.2 Optical coherence tomographic image of the human retina ex vivo. Imaging was performed at a 830-nm wavelength with 17 μm axial resolution in air, corresponding to 15 μm in tissue. The OCT image (A) is displayed using a log false color scale with a signal level ranging between 4×10^{-10} to 10^{-6}, or –94 dB to –60 dB of the incident light intensity. The OCT image of the retina shows the contour of the optic diskas well as retinal vasculature near the diskregion. The retinal pigment epithelium (RPE) is highly scattering. The retinal nerve fiber layer can also be seen as a highly scattering layer. Postmortem retinal detachment with subretinal fluid (SRF) accumulation can be seen in this ex vivo image. Histology is shown in (B). V, vitreal space; R, retina; S, sclera; B, blood vessels. (Huang et al., 1991a).

display the image. The log display expands the dynamic range, but compresses relative variations in signal.

Figure 7.2 shows an ex vivo OCT image of the human retina in the region of the optic disk with corresponding histology (Huang et al., 1991a). The OCT image shows the contour of the optic nerve head as well as the retinal vasculature. The retinal nerve fiber layer can also be visualized, which suggested the application of OCT forglaucoma detection. This image was performed ex vivo, so there is postmortem retinal detachment with subretinal fluid accumulation. Optical coherence tomography in ophthalmology has emerged as the largest and most successful clinical application to date.

Figure 7.3 shows an ex vivo OCT image of the human coronary artery with corresponding histology (Huang et al., 1991a). This result demonstrated that OCT imaging could be performed in tissues that strongly scatter light. The OCT image of the coronary artery

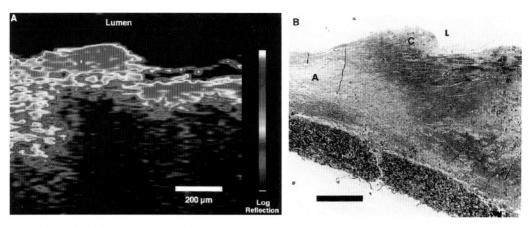

Figure 7.3 Optical coherence tomographic image of human artery ex vivo. The OCT image (A) of the coronary artery shows fibrocalcific plaque on the right three quarters and fibroatheromatous plaque on the left side of the specimen. The fatty-calcified plaque scatters light and therefore attenuates the OCT beam, limiting the image penetration depth. Histology is shown in (B). L, lumen; C, fatty-calcified plaque; A fibroatheromatous plaque; M media. (Huang et al., 1991a).

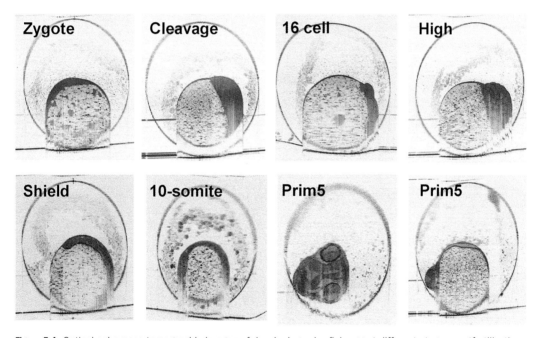

Figure 7.4 Optical coherence tomographic images of developing zebrafish egg at different stages postfertilization. Optical coherence tomography performs optical biopsy, enabling repeated imaging of the same specimen over time without the need to sacrifice and process the specimen, as in histology.

shows fibrocalcific plaque on the right three quarters of the specimen and fibroatheromatous plaque on the left side. The plaque scatters light and therefore attenuates the OCT beam, limiting the image penetration depth. Intravascular OCT imaging using fiber optic imaging catheters for the detection of vulnerable plaque and the guidance of interventional procedures such as stenting is currently an active area of medical research.

Figure 7.4 shows a more recent example of ultrahigh-resolution OCT images of a developing zebra fish egg in vivo. Imaging was performed with 5-μm axial resolution at a 1300-nm wavelength.

In this example, the OCT images are displayed using a grayscale. These images were obtained from the same specimen at different times. This demonstrates the ability of OCT to visualize structures noninvasively without the need to sacrifice and process specimens, as in conventional biopsy and histopathology. Optical coherence tomographic imaging in developmental biology is an example of an application in fundamental biomedical research.

OPTICAL COHERENCE TOMOGRAPHY COMPARED WITH OTHER IMAGING TECHNOLOGIES

To understand the possible applications of OCT imaging in biomedicine, it is helpful to compare OCT with other imaging techniques such as ultrasound and microscopy, as shown in figure 7.5. The resolution of ultrasound imaging depends on the frequency or wavelength of the sound waves (Erbel et al., 1998; Kremkau, 1998; Szabo, 2004; Hedrick et al., 2005). For standard clinical ultrasound, sound wave frequencies cover a wide range (3–40 MHz) and yield spatial resolutions of ≈0.1 to 1 mm. Ultrasound has the advantage that sound waves at this frequency are easily transmitted into biological tissues and therefore it is possible to image deep within the body. High-resolution ultrasound that uses high-frequency ultrasound has been developed and used in the laboratory as well as in some clinical applications such as intravascular imaging. Resolutions of 15 to 20 μm and finerhave been achieved with ultrasound frequencies of ≈100 MHz. However, high-frequency ultrasound is strongly attenuated in biological tissues, and attenuation coefficients increase approximately proportional to the frequency. Therefore, high-frequency ultrasound has limited imaging depths of only a few millimeters.

Microscopy and confocal microscopy are examples of imaging techniques that have extremely high resolutions, approaching 1 μm. Imaging is typically performed in an *en face* plane and resolutions are determined by the diffraction limit of light. Imaging depth in biological tissue is limited because image signal and contrast are significantly degraded by

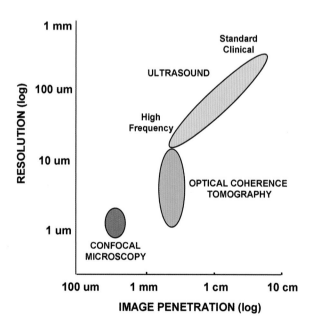

Figure 7.5 Comparison of resolution and imaging depth for ultrasound, OCT, and confocal microscopy. Ultrasound can achieve high resolution by using higher frequency sound waves; however, ultrasonic attenuation also increases, thus limiting image penetration depth. Optical coherence tomographic image resolution is determined by the coherence length or bandwidth of the light source. Resolutions ranging from 15 μm to 1 μm are possible. Image penetration depth is determined by attenuation from scattering and is limited to 2 to 3 mm. Confocal microscopy can have micrometer and submicrometer resolution, but imaging depths are limited to a few hundred micrometers.

optical scattering. In most biological tissues, imaging can be performed to depths of only a few hundred micrometers.

Optical coherence tomography fills the gap between microscopy and ultrasound. The axial resolution in OCT is determined by the bandwidth of the light source used for imaging. Optical coherence tomography has axial image resolutions ranging from 1 to 15 μm, approximately 10 to 100 times finer resolution than standard ultrasound imaging. The inherently high resolution of OCT imaging enables the visualization of tissue structure on the level of architectural morphology. Optical coherence tomography is ideally suited for ophthalmology, because the eye is easily accessible optically. The principal disadvantage of OCT imaging is that light is highly scattered by most biological tissues and attenuation from scattering limits image penetration depths to ≈2 mm. Tissue light scattering limits image penetration, irrespective of axial image resolution, as shown in figure 7.5. On the positive side, because OCT is an optical technology, it can be integrated with a wide range of instruments such as endoscopes, catheters, laparoscopes, or needles that enable imaging inside the body.

Imaging Using Light versus Sound

Optical coherence tomographic imaging is analogous to ultrasound imaging except that it uses light instead of sound. There are several different embodiments of OCT, but essentially OCT performs imaging by measuring the magnitude and echo time delay of back-reflected or backscattered light from internal microstructures in materials or tissues. Optical coherence tomography images are two-dimensional or three-dimensional data sets that represent optical back-reflection or backscattering in a cross-sectional plane or volume. When a beam of sound or light is directed onto tissue, it is back-reflected or backscattered differently from structures that have different acoustic or optical properties, as well as from boundaries between structures. The dimensions of these structures can be determined by measuring the "echo" time it takes for sound or light to return from different axial distances.

In ultrasound, the axial measurement of distance or range is called *A-mode scanning,* whereas cross-sectional imaging is called *B-mode scanning.* The principal difference between ultrasound and optical imaging is that the speed of light is much faster than sound. The speed of sound is approximately 1500 m/s, whereas the speed of light is approximately 3×10^8 m/s. The measurement of distances with a 100-μm resolution, a typical resolution in ultrasound, requires a time resolution of ≈100 ns, well within the limits of electronic detection. Ultrasound technology has dramatically advanced in recent years with the availability of high-performance and low-cost analogue-to-digital converters and digital signal processing technology. Unlike sound, the echo time delays of light are extremely fast. The measurement of distances with a 10-μm resolution, a typical resolution in OCT imaging, requires a time resolution of ≈30 fs (30×10^{-15} s). Direct electronic detection is impossible on this timescale, and indirect measurement methods such as optical correlation or interferometry must be used.

MEASURING LIGHT ECHOES USING INTERFEROMETRY

Interferometry enables measurement of the echo time delay of back-reflected or backscattered light with high sensitivity and high dynamic range. These techniques are analogous to coherent optical detection in optical communications. Classic OCT systems are based on low-coherence interferometry or white-light interferometry, an optical measurement technique first described by Sir Isaac Newton. Low-coherence interferometry has been applied to measure optical echoes and backscattering in optical fibers and waveguide devices

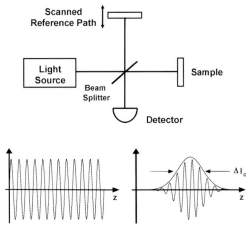

Figure 7.6 The echo time delay of light can be measured by using low-coherence interferometry. A Michelson–type interferometer is used to perform a correlation measurement. Back-reflected or backscattered light from the tissue being imaged is correlated with light that travels a known reference path delay. Interferometric detection is sensitive to field rather than intensity and is analogous to optical heterodyne detection.

(Takada et al., 1987; Youngquist et al., 1987; Gilgen et al., 1989). The first studies using low-coherence interferometry in biological systems, to measure eye length, were performed in 1988 (Fercher et al., 1988). Measurements of corneal thickness were also demonstrated using low-coherence interferometry (Huang et al., 1991b).

Recently there haven been important advances in OCT using techniques known as *Fourier domain detection*. These techniques are described in more detail at the end of this chapter. Here we describe the classic method of detection using a scanning interferometer, known as *time domain detection*. Figure 7.6 shows a schematic diagram of a Michelson–type interferometer. The measurement or signal beam is reflected from the biological specimen or tissue being imaged and a reference beam is reflected from a reference mirror that is being scanned over a calibrated path delay. The beams are interfered and a detector measures the intensity, or the square, of the electromagnetic field. The time delay that the beam travels in the reference arm can be controlled by varying the position of the reference mirror. Interference effects will be observed in the intensity output of the interferometer when the relative path length is changed by scanning the reference mirror. If the light source is coherent (narrow line width) with a long coherence length, then interference fringes will be observed for a wide range of relative path lengths of the reference and measurement arms. In optical ranging or OCT, it is necessary to measure precisely the absolute distance and dimensions of structures within the material or biological tissue. In this case, low-coherence light (broad bandwidth) is used. Low-coherence light can be characterized as having statistical discontinuities in phase overa distance known as the *coherence length*. The coherence length is a measure of the coherence and is inversely proportional to the frequency bandwidth.

When low-coherence light is used in the interferometer, interference is observed only when the path lengths of the reference and measurement arms are matched to within the coherence length. For path length mismatches greater than the coherence length, the electromagnetic fields from the two beams are uncorrelated, and there is no interference. The interferometer measures the field autocorrelation of the light. In OCT imaging, the coherence length determines the axial or depth resolution. The magnitude and echo time delay of the reflected light can be measured by scanning the reference mirror delay and demodulating the interference signal from the interferometer. Because the interference signal is measured as a function of time and echoes are measured sequentially, this detection technique is also known as *time domain detection*.

IMAGE RESOLUTION, DETECTION SENSITIVITY, AND IMAGE GENERATION

The axial image resolution in OCT is determined by the coherence length of the light source. In contrast to standard microscopy, OCT can achieve fine axial resolution independent of the beam focusing and spot size. The coherence length is proportional to the width of the field autocorrelation measured by the interferometer, and the envelope of the field autocorrelation is related to the Fourier transform of the power spectrum. Thus, the width of the autocorrelation function, or the axial resolution, is inversely proportional to the width of the power spectrum. For a Gaussian-shaped spectrum, the axial resolution Δz is $\Delta z = (2\ln2/\pi)(\lambda^2/\Delta\lambda)$, where Δz and $\Delta\lambda$ are the full-widths-at-half-maximum (FWHM) of the autocorrelation function and power spectrum, respectively, and λ is the source center wavelength (Swanson et al., 1992). Because axial resolution is inversely proportional to the bandwidth of the light source, broad bandwidth optical sources are required to achieve high axial resolution. Figure 7.7 shows a plot of axial resolution versus bandwidth for light sources at different wavelengths.

The transverse resolution in OCT imaging is the same as in optical microscopy and is determined by the diffraction-limited spot size of the focused optical beam. The diffraction-limited minimum spot size is inversely proportional to the numerical aperture (NA) or the focusing angle of the beam. The transverse resolution is $\Delta x = (4\lambda/\pi)(f/d)$, where d is the spot size on the objective lens and f is its focal length. Fine transverse resolution can be obtained by using a large NA that focuses the beam to a small spot size. At the same time, the transverse resolution is also related to the depth of focus or the confocal parameter b, which is $2z_R$, or two times the Rayleigh range: $2z_R = \pi\Delta x^2/2\lambda$. Thus, increasing the transverse resolution produces a decrease in the depth of focus, which is similar to conventional microscopy.

Figure 7.8 shows the relationship between focused spot size and depth of focus for low and high NA focusing. Typically, OCT imaging is performed with low NA focusing to have a large depth of field. In this case, the confocal parameter is larger than the coherence length, $b > \Delta z$, and low-coherence interferometry is used to achieve axial resolution. The image resolution is determined by the coherence length in the axial direction and the spot size in the transverse direction. In contrast to microscopy, OCT can achieve fine axial resolution independent of the NA. This feature is particularly powerful for applications such as ophthalmic imaging or catheter/endoscope imaging, in which high NAs are not available. However, operation with low NA also limits the transverse resolution because focused spot sizes are large.

It is also possible to perform OCT with high NA focusing and to achieve fine transverse resolutions. This results in a decreased depth of focus. This is the typical operating regime

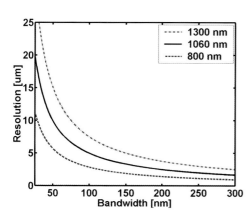

Figure 7.7 Axial resolution versus bandwidth of light sources for center wavelengths of 800 nm, 1000 nm, and 1300 nm. Micron-scale axial resolution requires extremely broad optical bandwidths. Bandwidth requirements increase dramatically for longer wavelengths.

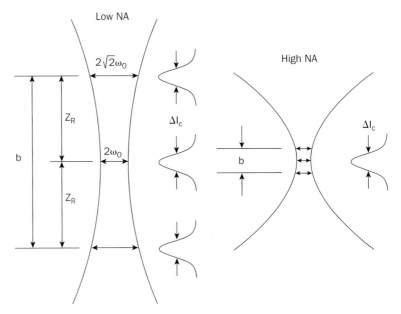

Figure 7.8 Transverse resolution in OCT. This schematic shows low and high NA focusing limits and a trade-off between transverse resolution versus depth of focus. Optical coherence tomographic imaging is usually performed with low NA focusing, with the confocal parameter much longer than the coherence length. The high NA focusing limit achieves fine transverse resolution, but has reduced depth of field.

for microscopy or confocal microscopy. In this case, the depth of focus can be shorter than the coherence length, $b < \Delta z$, and the depth of focus can be used to differentiate backscattered or back-reflected signals from different depths. This mode of operation is known as *optical coherence microscopy* (Izatt et al., 1994, 1996; Aguirre et al., 2003). Optical coherence microscopy has the advantage of achieving extremely fine transverse image resolution and is useful for imaging scattering systems because the coherence gating rejects scattered light in front and behind the focal plane more effectively than confocal gating alone.

The detection sensitivity of the OCT system determines the imaging performance for different applications. When OCT imaging is performed in very weakly scattering media, the imaging depth is not strongly limited, because there is very little attenuation of the incident beam. In this case, the sensitivity of the OCT detection determines a limit on the smallest signals that can be detected. Forexample, in ophthalmic imaging, high sensitivity is essential to image retinal structures that are nominally transparent and have very low backscattering. Image contrast is produced because of differences in the backscattering properties of different tissues. For retinal imaging, the ANSI standards govern the maximum permissible light exposure and set limits for both the sensitivity of OCT imaging as well as image acquisition speeds (Swanson et al., 1993; Hee et al., 1995a).

Optical coherence tomographic imaging in highly scattering media requires high sensitivity and dynamic range. In this case, the detection sensitivity determines the maximum depth to which imaging can be performed. Most biological tissues are highly scattering, and OCT imaging in tissues otherthan the eye is possible using longeroptical wavelengths, which reduce scattering and increase image penetration depths (Parsa et al., 1989; Schmitt et al., 1994; Fujimoto et al., 1995; Brezinski et al., 1996). Figure 7.9 shows an example of OCT imaging in a human epiglottis ex vivo comparing 800-nm with 1300-nm imaging wavelengths (Fujimoto et al., 1995). The dominant absorbers in most tissues are melanin and hemoglobin, which have absorption in the visible and near infrared wavelength

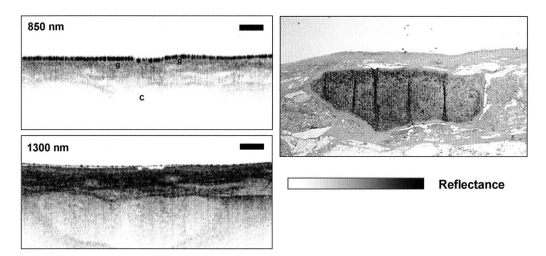

Figure 7.9 Optical coherence tomographic image penetration depth. Optical coherence tomographic images of human epiglottis ex vivo performed at 850-nm and 1300-nm wavelengths. Scattering is reduced at longer wavelengths, thereby increasing the image penetration depth. Superficial glandular structures (g) can be seen in images performed at 850-nm and 1300-nm wavelengths, but the underlying cartilage (c) is better visualized with the 1300-nm wavelength. With a detection sensitivity of ≈90 to 100 dB, image penetration depths of up to 2 to 3 mm is possible in most scattering tissues. The bar equals 500 μm (Brezinski et al., 1996).

range. Water absorption becomes appreciable for wavelengths approaching 1.9 to 2 μm. In most tissues, scattering at near-infrared wavelengths is one to two orders of magnitude higher than absorption. Scattering decreases for longer wavelengths, so that OCT image penetration increases (Parsa et al., 1989). For example, if a tissue has a scattering coefficient in the range of ≈40/cm at 1300 nm, then the round-trip attenuation from scattering alone from a depth of 3 mm is $e^{-24} \approx 4 \times 10^{-11}$. Thus, if the detection sensitivity is -100 dB, backscattered or back-reflected signals are attenuated to the sensitivity limit when the image depth is 2 to 3 mm. Because the attenuation is an exponential function of depth, increasing the detection sensitivity would not appreciably increase the imaging depth.

Optical coherence tomographic data are typically displayed as a two-dimensional grayscale orfalse colorimage. Figure 7.2 showed an example of an OCT image of the retina. The image displayed consists of 200 pixels (horizontal) by 500 pixels (vertical). The vertical direction corresponds to the direction of the incident optical beam and the axial data sets. The backscattered signal ranges from approximately –50 dB, the maximum signal, to –100 dB, the sensitivity limit. Because the signal varies over orders of magnitude, it is convenient to use the log of the signal to display the image. The retina is an example of a tissue that has weak signals. Figure 7.9 showed an OCT image of epiglottis ex vivo as an example of highly scattering tissue. In a highly scattering material or tissue, light is rapidly attenuated with propagation depth, resulting in a degradation of signal in the image. In addition, there can be significant speckle or other noise arising from the microstructural features of the material or specimen. For these reasons, it is more common to display OCT images in highly scattering materials or tissues using a grayscale. The use of grayscale display avoids artifacts that can occur with false color.

OPTICAL COHERENCE TOMOGRAPHY TECHNOLOGY AND SYSTEMS

Optical coherence tomography has the advantage that it can be implemented using compact fiber optic components and integrated with a wide range of medical instruments. Optical

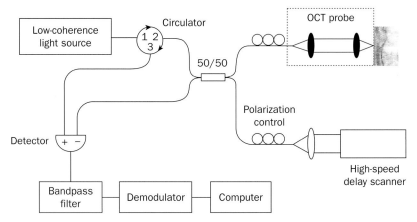

Figure 7.10 Fiber optic Michelson interferometer. Optical coherence tomography has the advantage that it uses many components and techniques from photonics. This schematic shows a Michelson interferometer with a circulator for dual-balanced detection. This configuration adds the heterodyne signal from the interference of the sample and reference arms and subtracts excess noise from the light source.

coherence tomographic systems can be considered from a modular viewpoint in terms of integrated hardware and software modules with various functionalities. The OCT system can be divided into an imaging engine (consisting of an interferometer, light source, and detector) as well as imaging devices and probes.

In general, the imaging engine can be any device that performs high-resolution and high-sensitivity detection of back-reflected or backscattered optical echoes. Most OCT systems to date have used a scanning reference delay interferometer using a low-coherence light source. Figure 7.10 shows a schematic of an OCT system using a fiber optic Michelson–type interferometer with time domain detection. A low-coherence light source is coupled into the interferometer. One arm of the interferometer emits a beam that is directed and scanned on the sample being imaged, whereas the other arm of the interferometer is a reference arm with a scanning delay. The interferometer uses a circulator to collect two interference signals that occur out of phase. When these two signals are detected and subtracted, the desired interference signals add, and excess noise from the light source subtract. This configuration is known as dual balanced detection and is used in optical communications systems.

There are many different embodiments of the interferometer and imaging engine for specific applications such as polarization diversity (insensitive) imaging, polarization-sensitive imaging, and Dopplerflow imaging. Dopplerflow imaging has been performed using imaging engines that directly detect the interferometric output rather than demodulating the interference fringes (Chen et al., 1997a,b; Izatt et al., 1997; Yazdanfar et al., 1997). Polarization-sensitive detection techniques have been demonstrated using a dual-channel interferometer (De Boer et al., 1997; Everett et al., 1998). These techniques permit quantitative imaging of the birefringence properties of structures. Conversely, polarization diversity or polarization-insensitive interferometers can also be built using detection techniques from coherent heterodyne optical detection.

Figure 7.11 shows an example of an imaging engine for endoscopic OCT imaging (Herz et al., 2004). This system uses a modified commercial OCT instrument (LightLab Imaging, Westford, Massachusetts) and a circulator-based, polarization diversity interferometer configuration. The light source is broadband laser and is coupled to an optical circulator and a 90/10 fiber coupler, which transmits 90% of the light to the sample arm of the interferometer. The reference arm of the interferometer uses a mechanical delay line with a rapid scanning rotary mirror to achieve an axial scan repetition rate of 3125 scans/s.

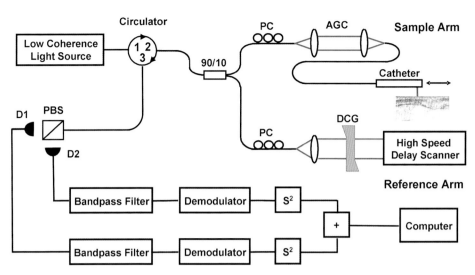

Figure 7.11 Schematic of a polarization diversity OCT system for endoscopic imaging. The system uses a circulator-based interferometer for efficient power delivery and detection. A dual-channel design obtains a polarization-independent signal (the square root of the sum of squares of the two polarization channels) that is not affected by stress on the fiber optic imaging catheter/endoscope. PS, polarization controller; AGC, air-gap coupling; DCG, dispersion compensating glasses; PBS, polarizing beam splitter.

To balance the dispersion between the sample and reference arms of the interferometer, dispersion-compensating glasses are used in the reference arm and an air-gap coupling is used in the sample arm. The interference signal is divided into two orthogonal polarization channels by a polarizing beam splitter, and the two detector outputs are digitally demodulated. A polarization diversity OCT signal can be obtained from the square root of the sum of the squares of the two polarization channels.

Recently, otherimaging engines using Fourierdomain detection methods have been demonstrated that are based on spectral analysis of broadband light sources as well as frequency-swept, narrow-line width sources. These detection techniques have the advantage that they can achieve very high sensitivities and imaging speeds because they detect all the light echo signals simultaneously and do not require mechanical scanning of an optical delay. These detection methods are described in more detail at the end of this chapter.

Light Sources and Ultrahigh-Resolution Imaging

The bandwidth of the light source determines the axial resolution of the OCT system. For clinical applications, superluminescent diodes (SLDs) are most commonly used because they are compact and relatively inexpensive. Ophthalmic OCT instruments use commercially available, compact GaAs SLDs that operate near 800 nm with bandwidths of ≈30 nm and achieve axial resolutions of ≈10 μm in tissue with output powers of a few milliwatts. Imaging applications in scattering tissues require the use of longer wavelengths. Commercially available SLDs operating at 1.3 μm can achieve axial resolutions of ≈10 to 15 μm in tissue with output powers of several tens of milliwatts.

For research applications, femtosecond lasers are powerful light sources for ultrahigh-resolution OCT imaging because they can generate extremely broad bandwidths across a range of wavelengths in the near infrared. Figure 7.7 showed the axial resolution in air for bandwidths at centerwavelengths of 800 nm, 1000 nm, and 1300 nm. These wavelengths can be generated using solid-state femtosecond lasers, such as the $Ti:Al_2O_3$, Nd:Glass or Yb fiber, and Cr:Forsterite lasers.

The Kerr lens mode-locked Ti:Al$_2$O$_3$ laseris the most commonly used laserin femtosecond optics and ultrafast phenomena. Early OCT imaging studies using Ti:Al$_2$O$_3$ lasers demonstrated axial image resolutions of $\approx$4 μm (Bouma et al., 1995). In recent years, high-performance Ti:Al$_2$O$_3$ lasers have been made possible through the development of double-chirped mirror technology (Kartner et al., 1997). Double-chirped mirrors can compensate high-order dispersion and have extremely broadband reflectivity, enabling the generation of few cycle optical pulses. With these recent advances, Ti:Al$_2$O$_3$ lasers achieve pulse durations of $\approx$5 fs, corresponding to only two optical cycles and octave bandwidths at 800 nm (Morgneret al., 1999; Sutteret al., 1999; Ellet al., 2001). Using state-of-the-art Ti:Al$_2$O$_3$ lasers, OCT axial image resolutions of $\approx$1 μm have been demonstrated (Drexler et al., 1999).

Figure 7.12 shows a comparison of the optical bandwidth and interferometer output traces showing the axial resolution of an SLD and a Ti:Al$_2$O$_3$ laser(Dr exleret al., 1999). Optical bandwidths of $\approx$260 nm are transmitted through the OCT instrument, corresponding to axial image resolution of $\approx$1 μm. Figure 7.13 shows an example of an ultrahigh-resolution OCT image of the human retina (Drexler et al., 2001). The figure shows a comparison between standard-resolution OCT with 10-μm axial resolution performed using an SLD light source and ultrahigh-resolution OCT with $\approx$3-μm axial resolution performed using a femtosecond laser light source. Measurements were performed with exposures of less than 750 μW (within ANSI standards for safe retinal exposure and consistent with the exposure levels used in commercial OCT clinical instruments). These images show that dramatic improvements in image quality are possible with ultrahigh-resolution OCT using state-of-the-art light sources.

Femtosecond Ti:Al$_2$O$_3$ lasers, in combination with high nonlinearity, air–silica microstructure fibers or tapered fibers can generate a broadband continuum that spans the visible to the near-infrared wavelength range. These fibers have enhanced nonlinearity because of their dispersion characteristics, which shift the zero dispersion to shorter wavelengths, and small core diameters, which provide tight mode confinement. Microstructure and tapered fibers have been used with femtosecond Ti:Al$_2$O$_3$ lasers to achieve bandwidths spanning the visible and near infrared (Birks et al., 2000; Ranka et al., 2000). Continuum generation from a femtosecond Ti:Al$_2$O$_3$ laser with air–silica microstructure photonic crystal fibers was demonstrated to achieve OCT axial image resolutions of 2.5 μm in the spectral region 1.2 to 1.5 μm (Hartl et al., 2001), resolutions of 1.3 μm in the spectral region 800 to 1400 nm (Wang et al., 2003b), and to record resolutions of less than 1 μm in the spectral region of 550 to 950 nm (Povazay et al., 2002). Although they have outstanding performance, femtosecond laser light sources are relatively costly and complex. Recent advances have demonstrated that Ti:Al$_2$O$_3$ lasers can operate with much lower pump powers than previously thought possible, thereby greatly reducing cost (Kowalevicz et al., 2002b; Unterhuber et al., 2003). At the same time, there have been advances in multiplexed SLD technology, and bandwidths approaching 150 nm can now be achieved, but with limited output power(Kowalevicz et al., 2002a).

The Ti:Al$_2$O$_3$ laser has the advantage that it generates very broad bandwidths, but because most biological tissues are optically scattering, longer wavelengths of 1300 nm are used formost OCT imaging applications because they enable deeperimaging than shorter wavelengths. Most commercial OCT systems at 1300-nm wavelengths use SLD light sources that have bandwidths of $\approx$50 to 80 nm, yielding axial image resolutions of $\approx$10 to 15 μm in tissue. Kerr lens mode-locked femtosecond Cr^{4+}:Forsterite lasers operate at wavelengths near1300 nm. The Cr^{4+}:Forsterite laser material has lower gain than Ti:Al$_2$O$_3$, but has the advantage that it can be directly pumped at 1-μm wavelengths using compact Yb fiber lasers. The first OCT imaging studies using femtosecond Cr^{4+}:Forsterite lasers performed many years ago demonstrated OCT axial image resolutions of 5 to 10 μm by coupling the femtosecond laseroutput into a nonlinearfiberand broadening the spectrum

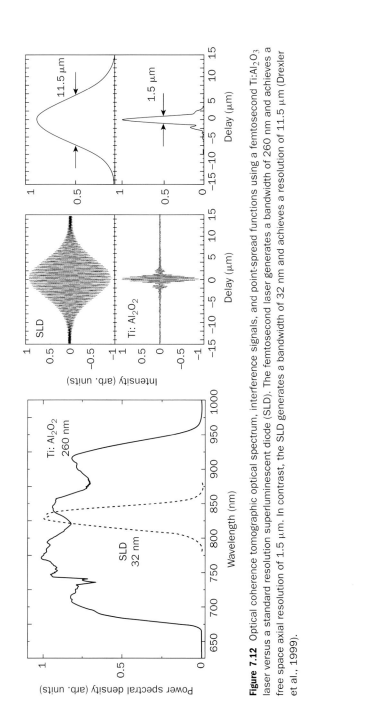

Figure 7.12 Optical coherence tomographic optical spectrum, interference signals, and point-spread functions using a femtosecond Ti:Al$_2$O$_3$ laser versus a standard resolution superluminescent diode (SLD). The femtosecond laser generates a bandwidth of 260 nm and achieves a free space axial resolution of 1.5 μm. In contrast, the SLD generates a bandwidth of 32 nm and achieves a resolution of 11.5 μm (Drexler et al., 1999).

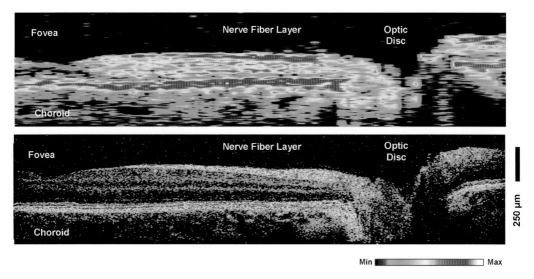

Figure 7.13 Comparison of standard and ultrahigh-resolution OCT images of the normal human retina. The axial resolution is 10 μm using an SLD light source and 3 μm using a femtosecond laser light source. The ultrahigh-resolution OCT image enables visualization of all major retinal layers (Drexler et al., 2001).

by self-phase modulation (Bouma et al., 1996). Early Cr^{4+}:Forsterite laser technology was challenging to use because the laser required intracavity prisms for dispersion compensation. However, with the development of double-chirped mirror technology, Kerr lens mode-locked Cr^{4+}:Forsterite performance improved significantly and it became possible to generate pulse durations as short as 14 fs and bandwidths of up to 250 nm directly from the laser(Chudoba et al., 2001). The Cr^{4+}:Forsterite laser can also be operated with longer duration pulses and broad bandwidths obtained using nonlinear self-phase modulation in optical fibers. Bandwidths of ≈180 nm, corresponding to OCT axial image resolutions as fine as ≈3.7 μm in tissue, have been achieved (Herz et al., 2004).

Finally, imaging at 1000-nm wavelengths is an attractive compromise between the fine axial resolution but limited image penetration available at 800-nm wavelengths versus the reduced resolution but increased image penetration at 1300-nm wavelengths (Wang et al., 2003a). A wide range of commercially available femtosecond laser sources, including mode-locked Nd:Glass lasers and Yb fiber lasers, are available at 1000 nm. Commercial femtosecond lasers in combination with high nonlinearity optical fibers are an attractive and robust approach for achieving bandwidths necessary for ultrahigh-resolution OCT imaging. Recent results demonstrated the use of a commercially available femtosecond Nd:Glass laser (HighQ LaserProductions, Hohenems, Austria) and a nonlinear fiber to generate bandwidths of 200 nm centered around 1050 nm, corresponding to axial image resolutions of ≈3.5 μm in tissue (Bourquin et al., 2003). This system is well suited for in vivo ultrahigh-resolution OCT imaging studies that would be performed outside the research laboratory.

Imaging Instruments and Probes

Because OCT technology is fiberoptic based, it can be easily integrated with many optical diagnostic instruments. Optical coherence tomography has been integrated with slit lamps and fundus cameras for ophthalmic imaging of the retina (Hee et al., 1995a; Puliafito et al., 1995). Figure 7.14 shows a schematic diagram of an instrument design for OCT retinal imaging. An objective lens relay images the retina onto an intermediate image plane. This retinal image can be relay imaged onto a video camera to enable real-time operator viewing of the retina. The OCT beam is coupled to the optical path of the instrument using a beam

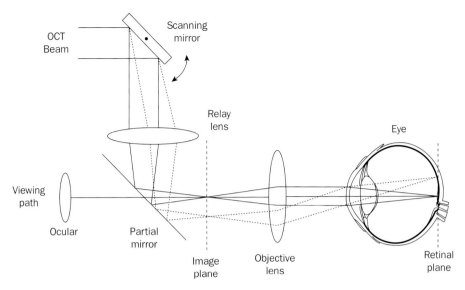

Figure 7.14 Schematic of an OCT instrument for imaging of the retina. An objective lens relays images of the retina to a plane inside the OCT instrument. Operator viewing of the fundus is performed by imaging with a video camera. A computer-controlled, two-axis scanning mirror positions and scans the OCT beam. A relay lens focuses the OCT beam onto the image plane, and the objective lens directs the OCT beam through the pupil onto the retina. The OCT beam is focused on the retina by adjusting the objective lens. During imaging, when the OCT beam is scanned, the beam pivots about the pupil of the eye to minimize vignetting.

splitter and is focused onto the intermediate image plane using a relay lens. The focused OCT beam is then imaged onto the retina by the objective lens and the patient's eye. The transverse spot size of the OCT beam on the retina is typically $\approx 20\,\mu\text{m}$ and is limited by the pupil and aberrations in the patient's eye. Prior to OCT image acquisition, the instrument must be focused by adjusting the objective lens position so that the OCT beam is focused on the retina. This also focuses the video image of the retina. The instrument also emits an illumination beam that illuminates the retina to enable video viewing of the retina. The transverse position of the OCT beam is scanned by two perpendicularly oriented, x/y scanning mirrors. The optical system is designed so that when the mirror angles are scanned, the beam pivots about the pupil of the eye, as shown in figure 7.14. This prevents the OCT beam from being vignetted by the pupil and it enables access to a wide field of view on the retina. To ensure that the OCT pivots about the pupil of the eye, the patient's eye must be located at a given distance from the objective lens.

Optical coherence tomographic imaging can be performed at different locations on the retina by controlling the scanning of the OCT beam. The operator can view the video retinal image in real time, which is displayed in a window on the computer monitor. When the OCT beam is scanned, it produces a scan pattern on the retina that is visible to the operator, enabling precise aiming. The OCT beam is also visible to the patient as a thin red line, with a position in the patient's visual field that corresponds to the points on the retina that are being scanned. Because the scanning pattern of the OCT beam on the retina can be controlled by computer, different scan patterns can be designed and adopted as part of the diagnostic protocol for specific retinal diseases.

The ability to register OCT images with *en face* features and pathology is an important consideration formany applications. Similardesign principles apply forOCT imaging using low-NA microscopes. Low-NA microscopes have been used for imaging developmental biology specimens in vivo as well as for surgical imaging applications (Boppart et al., 1996; Brezinski et al., 1997; Boppart et al., 1998).

Other OCT imaging instruments include forward imaging devices that perform one- or two-dimensional beam scanning. Rigid laparoscopes use relay imaging with Hopkins–type relay lenses or graded index rod lenses. Optical coherence tomography can be integrated with laparoscopes to permit internal body OCT imaging with a simultaneous *en face* view of the region being imaged (Boppart et al., 1997; Sergeev et al., 1997). Handheld imaging probes have also been demonstrated (Boppart et al., 1997; Feidchtein et al., 1998). These devices resemble light pens and use piezoelectric or galvanometric beam scanning. Handheld probes can be used in open-field surgical situations to enable the clinician to view subsurface tissue structure by aiming the probe at the desired location. These devices can also be integrated with conventional scalpels or laser surgical devices to permit simultaneous, real-time viewing as tissue is being resected. Recently, there has been increased interest in the use of Micro-Electro-Mechanical Systems (MEMS) scanning devices for OCT imaging probes. These devices enable one- or two-dimensional beam scanning and are a promising technology for developing miniature OCT imaging devices (Zara and Smith, 2002; Xie et al., 2003a,b; George, 2004; Jain et al., 2004; Tran et al., 2004).

Anotherclass of OCT imaging probes is flexible endoscope orcatheterdevices for internal body imaging. Small-diameter scanning endoscopes and catheters can be developed using fiber optics (Tearney et al., 1996). The first clinical studies of internal body imaging were performed almost 10 years ago using a novel, forward-scanning flexible probe that could be used in conjunction with a standard endoscope, bronchoscope, or trocar (Sergeev et al., 1997; Feidchtein et al., 1998). This device used a miniature magnetic scanner to perform one-dimensional, forward beam scanning with a 1.5- to 2-mm-diameter probe. Transverse scanning OCT catheters/endoscopes have also been developed and demonstrated (Tearney et al., 1996, 1997b; Fujimoto et al., 1999). Figure 7.15 shows an example of an OCT catheter/endoscope. The catheter/endoscope has a single-mode optical fiber encased in a hollow rotating torque cable, coupled to a distal lens and a microprism that directs the OCT beam radially outward. The cable and distal optics are enclosed in a transparent waterproof housing. The OCT beam is scanned by proximally rotating the cable to perform imaging in a radarlike pattern, cross-sectionally through vessels or hollow organs. Imaging may also be performed in a longitudinal plane using a push–pull movement of the fiber optic assembly.

Figure 7.16 shows an early example of an in vivo catheter/endoscope rotary scan OCT image of the gastrointestinal tract in the New Zealand White rabbit. Early results demonstrated endoscopic OCT imaging of the pulmonary, gastrointestinal, and urinary tracts, and

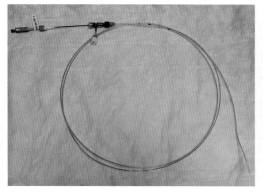 

Figure 7.15 Photograph of prototype OCT catheter/endoscope for imaging inside the body. A single-mode fiber lies within a moving speedometer cable enclosed in a protective plastic sheath. The distal end focuses the beam at 90 deg from the axis of the catheter.

arterial imaging (Tearney et al., 1997b; Fujimoto et al., 1999). Imaging was performed at a 1300-nm wavelength with an axial resolution of $\approx$10 μm. To achieve high-speed imaging at several frames per second, a novel high-speed, phase-scanning optical delay line was used (Tearney et al., 1997a). The two-dimensional image data were displayed using a polar coordinate transformation and inverse grayscale. The structure of the esophagus is clearly delineated, with well-organized layers of the squamous epithelium (ep), lamina propria (lp), muscularis mucosa (mm), submucosa (sm), and inner(im) and outermuscular(om) layers.

Figure 7.17 shows a recent example of an ultrahigh-resolution in vivo catheter/ endoscope longitudinal scan OCT image of the gastrointestinal tract of a New Zealand White

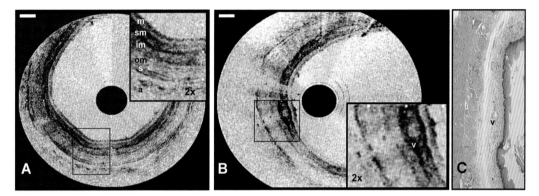

Figure 7.16 (A–C) Optical coherence tomographic catheter/endoscope image of the esophagus of a New Zealand White rabbit in vivo. The image clearly differentiates the layers of the esophagus, including the mucosa (m), submucosa (sm), inner muscularis (im), and outer muscularis (om) (Tearney et al., 1997b).

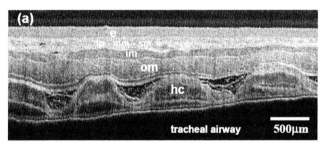

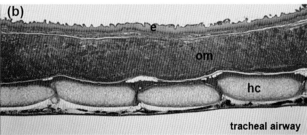

Figure 7.17 Optical coherence tomographic catheter/endoscope image of the esophagus and trachea of a New Zealand White rabbit in vivo and compartive histology. The image was acquired from the esophagus and shows the squamous epithalial structure of the esophagus including the epithelium (e), muscularis mucosa (mm), inner muscularis (im) and outer muscularis (om) as well as the hyaline cartilage (hc) rings of the trachea (Herz et al., 2004).

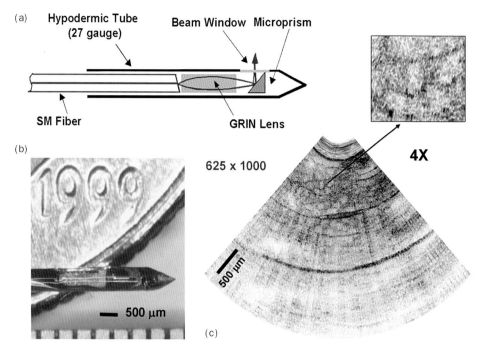

Figure 7.18 (A) Optical coherence tomographic imaging needle. The needle consists of a single-mode optical fiber with a graded index lens and a microprism to focus and deflect the beam. A thin-wall, 27-gauge stainless steel hypodermic tube houses the optical fiber and distal optics. (B) Photo of a prototype OCT needle on a U.S. dime shows the size of the needle. (C) Optical coherence tomographic image of a hamster leg muscle in vivo acquired with the needle. Muscle fascicles are clearly discerned. The OCT needle enables imaging of a 4- to 5-mm-diameter cylindrical region in solid tissues. The scale bar is 500 μm (Li et al., 2000b).

rabbit performed with 5-μm axial resolution (Herz et al., 2004). The structure of the esophagus is clearly delineated and the longitudinal scan enables the generation of a large image region. Image penetration is sufficient to see through the esophagus into the trachea, including cartilage rings. These types of catheter/endoscope OCT imaging probes are small enough to fit in the accessory port of a standard endoscope or bronchoscope and to permit an *en face* view of the area being simultaneously imaged with OCT. Optical coherence tomography catheters/endoscopes can be also used independently as in intravascular imaging.

Even smaller needle imaging devices can be developed (Li et al., 2000b). Recently, a 27-gauge OCT imaging needle was developed for performing OCT imaging in solid tissues or tumor s. A small-diameter needle can be inserted into solid tissues with minimal trauma. Because imaging can be performed to depths of 2 to 3 mm in most tissues, image needles enable the imaging of a cylindrical volume of tissue with a radius of 2 to 3 mm or a diameter of 4 to 6 mm and with arbitrary length. Figure 7.18 shows an OCT imaging needle constructed using an optical fiber, graded index lens and microprism, and in vivo OCT imaging of hamster leg muscle, an example of solid tissue. Ultimately, OCT imaging devices could be integrated with excisional biopsy devices such as needle or core biopsy devices to enable first-look imaging of the tissue to be excised, thereby reducing sampling errors.

OPTICAL COHERENCE TOMOGRAPHY AND OPTICAL BIOPSY

Optical coherence tomography has limited image penetration depths of 2 to 3 mm in most tissues. This imaging depth is shallow compared with ultrasound, but is comparable with the depth over which many biopsies are performed. In addition, many diagnostically important changes of tissue morphology occur at the epithelial surfaces of organ lumens. The capability to perform *optical biopsy*—the in situ, real-time imaging of tissue morphology—could be important in a variety of clinical scenarios, such as assessing tissue pathology in situations when conventional excisional biopsy is hazardous or impossible, guiding conventional excisional biopsy to reduce false-negative rates from sampling errors, guiding surgical or microsurgical intervention, and monitoring changes in pathology, which are markers of disease progression. The next sections discuss representative examples of these clinical applications.

Optical Coherence Tomographic Imaging in Ophthalmology

Optical coherence tomography was first applied for imaging in the eye. To date, OCT has had the largest clinical impact in ophthalmology. The first in vivo OCT images of the human eye were demonstrated in 1993 (Fercher et al., 1993; Swanson et al., 1993). Optical coherence tomography enables the noncontact, noninvasive imaging of the anterior eye as well as the retina (Hee et al., 1995a; Puliafito et al., 1995; Schuman et al., 2004). Numerous clinical studies have been performed and a journal search reveals hundreds of peer-reviewed publications on OCT in ophthalmology.

Figure 7.19 shows an example of an OCT image of the normal retina of a human subject. Optical coherence tomography provides a cross-sectional view of the retina with unprecedented resolution. Although the retina is almost transparent and has extremely low optical backscattering, the high sensitivity of OCT allows extremely small backscattering from features such as interretinal layers and the vitreal–retinal interface to be imaged. The retinal pigment epithelium, which contains melanin, and the choroid, which is highly vascular, are visible as highly scattering structures in the OCT image.

Because OCT retinal images have extremely high resolutions, imaging is sensitive to the patient's eye motion during the measurement. Image processing algorithms are be used to measure axial eye motion and to correct for axial motion artifacts (Swanson et al., 1993). Motion correction algorithms have played in important role in enabling clinically useful OCT imaging. Axial motion can be well corrected, but transverse eye motion caused by changes in fixation will cause uncertainty in the registration of OCT images to features on the retina and remains difficult to correct.

Optical coherence tomography has been demonstrated for the diagnosis and monitoring of a variety of macular diseases (Puliafito et al., 1995), including macular edema

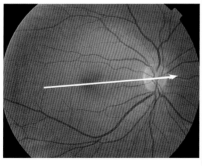 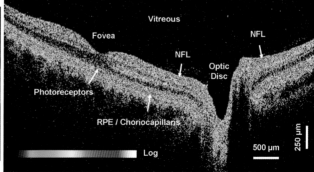

Figure 7.19 Optical coherence tomographic image of a normal retina showing the fovea and optic disc. The OCT scan location is shown on a corresponding photograph of the retina. The fovea and optic disc are identifiable by their characteristic morphology, and the layered structure of the retina is apparent. The retinal nerve fiber layer (NFL) is highly reflective and increases in thickness toward the disc. RPE, retinal pigment epithelium.

(Hee et al., 1995b, 1998), macular holes (Hee et al., 1995c), central serous chorioretinopathy (Hee et al., 1995d), and age-related macular degeneration and choroidal neovascularization (Hee et al., 1996). The retinal nerve fiber layer thickness, an indicator of glaucoma, can be quantified in normal and glaucomatous eyes, and correlated with conventional measurements of the optic nerve structure and function (Schuman et al., 1995a,b).

Optical coherence tomography is especially promising for the diagnosis and monitoring of diseases such as glaucoma ormacularedema associated with diabetic retinopathy because it can provide quantitative information on retinal pathology, which measures disease progression. Images can be analyzed quantitatively and processed using intelligent algorithms to extract features such as retinal or retinal nerve fiber layer thickness (Hee et al., 1998). Mapping and display techniques have been developed to display OCT data in alternate forms, such as thickness maps, to aid interpretation.

As an example, figure 7.20 shows the measurement and mapping of retinal thickness. Measurement of retinal thickness in the central retina or macula is important in patients with diabetes who are prone to macular edema and diabetic retinopathy. The utility of OCT in clinical practice depends on the ability of the physician to interpret accurately and quickly the OCT results in the context of conventional clinical examination. Standard ophthalmic examination techniques give an *en face* view of the retina. Therefore, OCT topographic methods of displaying retinal and nerve fiber layer thickness, which can be directly compared with a standard view of the retina during ophthalmic examination, were developed (Hee et al., 1998). Figure 7.20 shows an example of an OCT topographic map of retinal thickness. The thickness map is constructed by performing six standard OCT scans at varying angular orientations through the fovea. The images are then segmented to detect the retinal thickness along the OCT tomograms. The retinal thickness is then linearly interpolated over the macular region and represented in false color. The diameter of the topographic map is approximately 6 mm. Retinal thickness values between 0 and 500 μm are displayed by colors ranging from blue to red. This yields a color-coded topographic map of retinal thickness that displays the retinal thickness in an *en face* view.

For quantitative interpretation, the macula was divided into nine regions including a central disc with a diameter of 500 μm (denoted as the foveal region), and an inner and outer ring, each divided into four quadrants, with outer diameters of one and two disc diameters, respectively. In an early study, average retinal thickness was measured from OCT imaging on 96 healthy individuals with statistics including mean values and standard deviations (Hee et al., 1998). Studies of a cross-section of patients are required because there is normal variation in structures such as retinal thickness across the population. The ability to reduce image information to numerical information is important because it allows a normative database to be developed and statistics to be calculated.

To perform screening or diagnosis, a normative database must first be developed and extensive cross-sectional studies performed. A diagnostic model must be developed that allows the OCT image or numerical information to be analyzed to determine the presence of or the stage of a given disease. These diagnostic criteria must perform with sufficient sensitivity and specificity. In many cases, screening or diagnosis is complicated by natural variations in the healthy population. For example, healthy subjects lose retinal nerve fibers with age, so that retinal nerve fiber layer thickness measurements for the diagnosis of glaucoma must be age adjusted. Ongoing clinical studies are addressing the application of OCT forscr eening and diagnosis of both diabetic macularedema and glaucoma, two of the leading causes of blindness.

Optical coherence tomography is a powerful technique in ophthalmology because it has the potential to diagnose early stages of disease before physical symptoms and irreversible vision loss occurs. Furthermore, repeated imaging can be easily performed to track disease progression or monitor the effectiveness of therapy. The technology was transferred

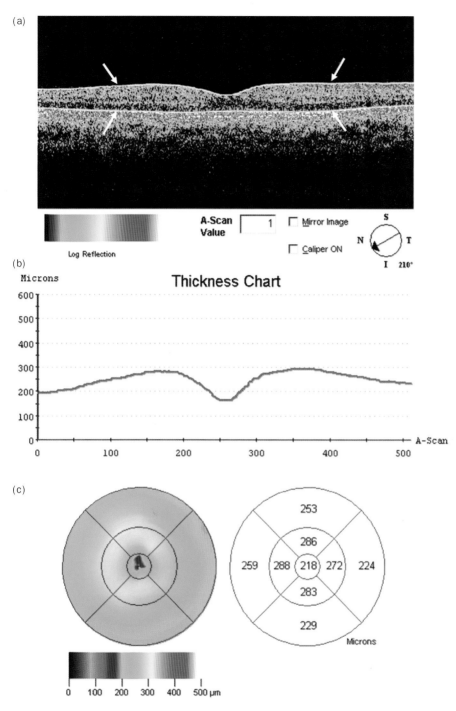

(a)

A-Scan Value [1]   ☐ Mirror Image

☐ Caliper ON

Log Reflection

S
N   T
I   210°

(b)

Microns

Thickness Chart

600
500
400
300
200
100
0

0   100   200   300   400   500

A-Scan

(c)

253
286
259 288 218 272 224
283
229

Microns

0   100   200   300   400   500 µm

Figure 7.20 Computer image processing for measuring retinal thickness. (A) Optical coherence tomographic image of the macula that has been computer processed to perform boundary detection or segmentation to measure retinal thickness. The anterior and posterior retinal surfaces are automatically identified. (B) Quantitative measurement of retinal thickness based on the segmented OCT image. (C) Topographic map of retinal thickness in the macula. The topographic map is constructed by processing multiple OCT scans in the macula, measuring the retinal thickness profile of these scans, and estimating or interpolating the retinal thickness in the regions between the scans. The retinal thickness is represented by a color table. This display has the advantage that it can be directly compared with the fundus image.

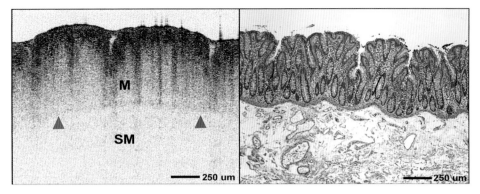

Figure 7.21 Ultrahigh-resolution OCT image of normal colon ex vivo (left) and comparison histology (right). Mucosa (M) is clearly delineated from underlying submucosa (SM) by a scattering band corresponding to the thin muscularis mucosa (arrowheads). The submucosa is visible as a less optically scattering layer. Images were acquired with 4.5-μm axial and 11-μm transverse resolution at a 1.26-μm wavelength (Hsiung et al., 2005).

to industry and introduced commercially for ophthalmic diagnostics in 1996 (Carl Zeiss Meditec, Dublin, California). A third-generation ophthalmic instrument, the StratusOCT, was introduced in 2001. Optical coherence tomography is emerging as a standard of ophthalmic care, and it is considered essential for the diagnosis and monitoring of many retinal diseases as well as glaucoma.

Optical Coherence Tomographic Imaging in Cancer Diagnosis

Optical coherence tomographic imaging also has potential applications for the detection of cancerous changes. Although excisional biopsy and histopathology are the gold standard for the diagnosis of many cancers, conventional excisional biopsy is a sampling procedure and can have unacceptably high false-negative rates (i.e., the biopsy results are normal, but the patient has cancer). Although it is unlikely that a new imaging modality such as OCT can compete with a clinically accepted standard such as excisional biopsy and histopathology, it should be possible to use imaging to guide biopsy and thereby reduce sampling error and improve sensitivity.

Optical coherence tomography can resolve changes in architectural morphology that are associated with many neoplastic changes. Early studies were performed ex vivo to correlate OCT images with histology for pathologies of the gastrointestinal, biliary, female reproductive, pulmonary, and urinary tracts (Izatt et al., 1996; Tearney et al., 1997c,d; Kobayashi et al., 1998; Pitris et al., 1998; Tearney et al., 1998; Boppart et al., 1999; Jesser et al., 1999; Pitris et al., 1999, 2000). Many early neoplastic changes are manifested by disruption of the normal glandular organization or architectural morphology of tissues. Figure 7.21 shows a recent example of an ultrahigh-resolution OCT image of the normal colon ex vivo and corresponding histology (Hsiung et al., 2005). Imaging was performed at a 1.26-μm wavelength with 4.5-μm axial resolution and an 11-μm transverse resolution. Optical coherence tomography clearly visualized the full thickness of the colon mucosa. The submucosa appeared as a lighter and less optically scattering layer. The muscularis mucosa appeared as a scattering band in the OCT image separating the mucosa and submucosa. Although the epithelial layer and single crypts were visible, they were not consistently visible in all images.

Figure 7.22 shows an ultrahigh-resolution OCT image of adenocarcinoma ex vivo and corresponding histology (Hsiung et al., 2005). Optical coherence tomographic images of carcinoma reveal loss of normal mucosal architecture and invasion of the submucosa. Highly

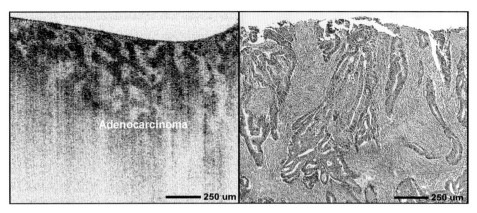

Figure 7.22 Ultrahigh-resolution OCT image of well-differentiated adenocarcinoma ex vivo (left) and comparison histology (right). Highly irregular invasive glands are visible with no clear boundary between mucosa and submucosa. Images were acquired with a 4.5-μm axial and an 11-μm transverse resolution at a 1.26-um wavelength (Hsiung et al., 2005).

scattering and irregular invasive glands were visible in OCT images of adenocarcinoma. Optical coherence tomography images also show areas with cysts and ulceration with overlying fibrinopurulent exudate. One of the key challenges for OCT imaging in the detection of neoplastic changes is the ability to achieve sufficient resolution and image contrast to differentiate clinically relevant pathologies. These examples of OCT images show distinct differences between normal and neoplastic tissues; however, these differences would not yield sufficient sensitivity and specificity to perform diagnosis based on OCT images alone.

The development of miniature, flexible catheter/endoscope imaging probes enables OCT imaging of internal organ systems. Epithelial cancers of the gastrointestinal tract, reproductive tract, and the respiratory tract represent the majority of cancers encountered in internal medicine. Many of these epithelial cancers are preceded by premalignant changes, such as dysplasia.

Figure 7.23 shows a recent example of ultrahigh-resolution endoscopic OCT imaging of the human esophagus. Imaging is performed at 1.3-μm wavelengths with $\approx$5-μm axial and $\approx$15-μm transverse resolutions, using a 1.8-mm-diameter linear-scanning catheter that is introduced into the accessory channel of a standard endoscope. Optical coherence tomographic imaging can be performed under endoscopic guidance, during which the endoscope is used to visualize the gastrointestinal tract and guide the placement of the OCT imaging catheter. The OCT imaging catheter is used to obtain a cross-sectional image of the tissue morphology. Figure 7.23 shows example images of the normal esophagus and Barrett's esophagus—a metaplastic condition associated with chronic gastroesophageal reflux. The normal esophagus is characterized by a well-organized layered structure, whereas Barrett's esophagus has loss of normal organization, with replacement by glandular structure.

Endoscopic imaging is an example of an application in which OCT imaging promises to improve diagnostic sensitivity. Gastrointestinal endoscopy has received increased attention as a result of the prevalence of esophageal, stomach, and colon cancers. In contrast to conventional endoscopy, which can only visualize surfaces, OCT can image tissue morphology beneath the tissue surface (Sergeev et al., 1997; Bouma and Tearney, 1999; Rollins and Izatt, 1999; Bouma et al., 2000; Jäckle et al., 2000a,b; Li et al., 2000a; Sivak et al., 2000; Poneros et al., 2001). Previous studies of endoscopic OCT imaging have demonstrated the ability of OCT to differentiate between abnormal gastrointestinal pathologies such as Barrett's esophagus, adenomatous polyps, and adenocarcinoma from normal tissues.

Excellent results have been obtained using OCT to differentiate Barrett's esophagus (Poneros et al., 2001). To be useful clinically, OCT would have to be able to detect more

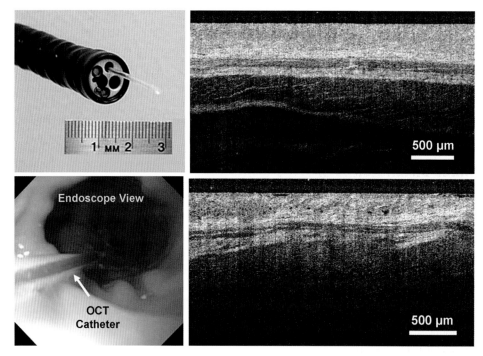

Figure 7.23 Ultrahigh-resolution endoscopic OCT imaging studies in humans. Images were acquired with less than 5 μm axial resolution. Optical coherence tomographic images in the human esophagus in vivo show normal squamous epithelium (top right) versus Barrett's esophagus (bottom right). Barrett's esophagus is characterized by glandular structure.

subtle changes, such as the presence of high-grade dysplasia, the next step in cancer progression before the development of adenocarcinoma (Pfau et al., 2003). Optical coherence tomography for detecting dysplasia in Barrett's esophagus was recently demonstrated. One study reported a sensitivity of 83% and a specificity of 75% for detection of intramucosal carcinoma and high-grade dysplasia with blinded scoring of OCT images from 55 patients (Evans et al., 2006). Another study reported a sensitivity of 68% and a specificity of 82%, with an accuracy of 78% for the detection of dysplasia in biopsies from 33 patients with Barrett's esophagus (Isenberg et al., 2005).

The development of OCT imaging forcancerdetection is extremely challenging. Conventional histopathology is an extremely powerful diagnostic technique because it enables the use of selective stains to enhance contrast between different cellular or tissue structures and provides extremely fine image resolutions, enabling the visualization of not only larger scale tissue architectural morphology, but also subcellular structure. Optical coherence tomographic imaging uses intrinsic contrast produced by differences in scattering properties of different tissue structures. Even the highest resolution OCT images with 5-μm axial resolution are still considerably lower resolution than histology and optical microscopy. On the positive side, OCT offers the advantage of enabling real-time imaging of tissue pathology in situ without the need forexcision and processing as in conventional biopsy and histopathology. When used to guide biopsy, OCT does not have to perform at the level required for independent, standalone diagnosis, but must have sufficient sensitivity and specificity so that it can improve the sensitivity of excisional biopsy.

The development of OCT forcancerdiagnosis will require detailed studies that investigate its ability to visualize and identify clinically relevant pathology. These types of studies are challenging because OCT images should be evaluated in a blinded fashion (independently, without histology), and the sensitivity and specificity of OCT imaging evaluated with a gold standard for diagnosis, usually biopsy and histopathology. Because there is considerable variation in pathology depending upon location, precise registration of OCT imaging and biopsy is required. Sufficient numbers of patients having pathology must be investigated to ensure that the sample size is large enough to have statistical significance. Because many types of dysplasia orcancerhave a low incidence, large numbers of patients must be investigated. Forthese reasons, the investigation and development of OCT for cancer diagnosis remains a challenging and ongoing area of research.

THREE-DIMENSIONAL OPTICAL COHERENCE TOMOGRAPHIC IMAGING

Perhaps one of the most exciting recent developments in OCT imaging is the extension of two-dimensional imaging to three-dimensional imaging. In addition to cross-sectional OCT imaging, it is also possible to scan a raster-type pattern and acquire multiple cross-sectional images to generate a three-dimensional OCT data set. Three-dimensional OCT data can be used to generate cross-sectional images with arbitrary orientations as well as virtual perspective renderings similar to those in magnetic resonance imaging. Figure 7.24 shows an example of three-dimensional OCT volume renderings of a normal colon and a polypoid adenoma ex vivo (Hsiung et al., 2005). Rendered three-dimensional OCT data can be viewed from a virtual surface perspective, yielding images similar to that of microscopy or magnification endoscopy. In this example, columnar epithelial morphology and crypt structures were difficult to visualize in single cross-sectional OCT images, whereas three-dimensional OCT enables morphology to be visualized using their three-dimensional structure as well as *en face* appearance. The three-dimensional OCT renderings enable normal colon versus polypoid adenoma to be differentiated in both surface views as well as cutaway views. Normal colon has an organized distribution of crypts that are uniform in size and spacing in the *en face* plane. In contrast, polypoid adenoma exhibits irregular glandular structure with disruption of normal organization.

Figure 7.25 shows another example of three-dimensional OCT imaging of a developmental biology specimen, the *Xenopus laevis* tadpole. Combined with ultrahigh-resolution

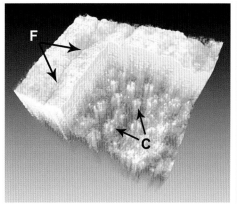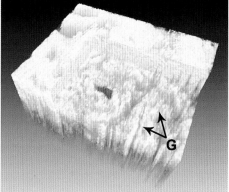

Figure 7.24 Three-dimensional OCT images of normal human colon and polypoid adenoma ex vivo. These images were constructed by raster scanning the OCT beam to acquire 275 cross-sectional images in a three-dimensional data set. Images were acquired with 3.5-μm axial and 6-μm transverse resolutions at a 1.09-μm wavelength. A fold (F) in the surface as well as crypts (C) are visible in the normal colon. The adenoma exhibits disruption of glandular structure (G), which can be visualized in the cut-away, rendered view. (Hsiung et al., 2005).

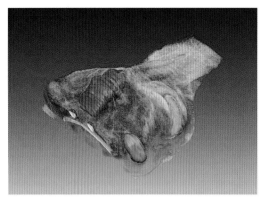

Figure 7.25 Three-dimensional OCT imaging of the *Xenopus laevis* tadpole. Three-dimensional imaging will play an increasingly important role in OCT research.

techniques, three-dimensional OCT promises to provide comprehensive information on microstructure. These results show that three-dimensional OCT can provide significantly more information than single cross-sectional images. However, endoscopic-based three-dimensional OCT imaging is challenging because it requires the development of miniature beam-scanning technologies that enable two-dimensional raster scanning in endoscopes. In addition, three-dimensional OCT requires extremely rapid acquisition speeds for real-time applications, because data sets are significantly larger than single cross-sectional images.

CONCLUSION

Optical coherence tomography is a powerful imaging technology in biomedicine because it enables real-time, in situ visualization of tissue structure and pathology without the need to excise and process specimens, as in conventional excisional biopsy and histopathology. Nonexcisional *optical biopsy* and the ability to visualize tissue morphology in real time underoper atorguidance can be used fora wide variety of applications. Optical coherence tomography has had the most clinical success in ophthalmology, where it provides structural and quantitative information that cannot be obtained by any other modality. In tissues other than the eye, optical scattering limits the image penetration depths between 2 to 3 mm. However, OCT can be interfaced to a wide range of instruments such as endoscopes, catheters, or laparoscopes, which enable the imaging of internal organ systems. Optical coherence tomography promises to have a powerful impact on many medical applications, ranging from the screening and diagnosis of neoplasia to enabling new minimally invasive surgical procedures.

APPENDIX: INTERFEROMETRIC METHODS FOR MEASURING THE ECHO TIME DELAY OF LIGHT

Optical coherence tomography uses interferometry to perform high-resolution measurements of light echoes. This appendix provides a more detailed overview of different measurement techniques. Three different types of interferometric detection techniques can be used in OCT instruments. Standard OCT instruments use an interferometer with a low-coherence light source and a scanning reference delay. This technique is known as *time domain detection* and was described earlier. It is also possible to perform detection in the Fourier domain by measuring the interference spectrum (Fercher et al., 1995).

Echo signals or axial scans are measured by "Fourier transforming" the spectrum. Fourier domain detection techniques have a powerful sensitivity and speed advantage compared with time domain detection because they essentially measure all the echoes of light simultaneously (Choma et al., 2003; De Boer et al., 2003; Leitgeb et al., 2003). There are two types of Fourier domain detection. The first type uses an interferometer with a low-coherence light source and measures the interference spectrum using a high-speed, line scan camera. This detection technique is known as *spectral/Fourier domain OCT* or *spectral OCT* (Fercher et al., 1995; Wojtkowski et al., 2002; Cense et al., 2004; Nassif et al., 2004; Wojtkowski et al., 2004, 2005). The second type uses an interferometer with a narrow-bandwidth, frequency-swept light source and detectors that measure the interference output as a function of time. This detection technique is known as *swept source/Fourier domain OCT* or *optical frequency domain imaging* (Fercher et al., 1995; Chinn et al., 1997; Yun et al., 2003; Huber et al., 2005).

Spectral/Fourier Domain Detection

Spectral/Fourier domain detection uses a broad-bandwidth light source and detects the interference spectrum from the interferometer using a line scan camera (Fercher et al., 1995; Wojtkowski et al., 2002; Cense et al., 2004; Nassif et al., 2004; Wojtkowski et al., 2004, 2005). The interferometer reference arm is stationary. Spectral/Fourier domain detection has also been referred to as *spectral OCT*. Figure 7.26 shows a schematic of how spectral/Fourier domain detection works.

The operation can be understood by recalling that a Michelson interferometer functions as a frequency filter, and the filter function depends on the path difference between the sample and reference arms. The output from a broad-bandwidth light source is split into two beams. One beam is directed onto the tissue to be imaged and is back-reflected or backscattered from internal structures at different depths. The second beam is reflected from a reference mirror with a fixed position. The measurement beam and the reference beam have a time offset determined by the path length difference, which is related to the depth of the structure in the tissue that back-reflects or backscatters the light. The interference of the

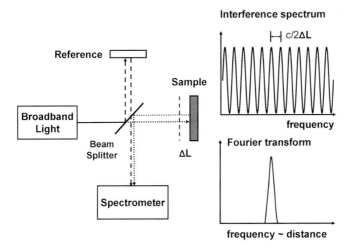

Figure 7.26 Spectral/Fourier domain OCT uses a broadband light source and a spectrometer that is read with a high-speed line scan camera. The unbalanced Michelson interferometer functions as a spectral filter that has a periodic output spectrum depending on path length mismatch ΔL. "Fourier transforming" the output interference spectrum yields a measurement of the echo magnitude and time delay.

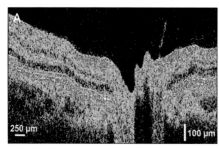

Figure 7.27 Comparison of standard OCT and high-speed, ultrahigh-resolution OCT using spectral/Fourier domain detection. (A) Standard OCT of the optic nerve head with ≈10-μm axial image resolution and 512 axial scans, acquired in ≈1.3 s. (B) High-speed, ultrahigh-resolution image with ≈2-μm axial resolution and 2048 axial scans, acquired in 0.13 s (Wojtkowski et al, 2005).

two beams will have a spectral modulation (oscillation) as a function of frequency that can be measured using a spectrometer. The periodicity of this modulation will be inversely related to the echo time delay. Therefore, different echo delays will produce different frequency modulations. The echo delays can be measured by rescaling the spectrometer output from wavelength to frequency and then Fourier transforming the interference signal. This results in an axial scan measurement of the magnitude and echo delay of light from the tissue.

Spectral/Fourier domain detection enables OCT imaging with an ≈50 times increase in imaging speed compared with standard OCT systems. In general, for a given acquisition time, high-speed imaging can increase the number of axial scans or transverse pixels perimage to yield high-definition images as well as the numberof cross-sectional images acquired in a sequence to yield three-dimensional information. Figure 7.27 shows a comparison of standard OCT and high-speed, ultrahigh-resolution OCT images of the optic nerve head of the human retina. Figure 7.27A shows a standard OCT image with a 10-μm axial resolution with 512 axial scans, acquired in ≈1.3 s. Figure 7.27B shows a high-speed, ultrahigh-resolution OCT image with ≈2-μm axial resolution and 2048 axial scans, acquired in 0.13 s. The higher resolution and greater number of transverse pixels in the high-speed, ultrahigh-resolution OCT image improve the visualization of the internal retinal structure.

In addition, high-speed OCT imaging enables the acquisition of complete three-dimensional data sets in a time comparable with that of current OCT protocols that acquire several individual images. Figure 7.28 shows an example of rendered images of the optic disc using a three-dimensional OCT data set acquired with a high-speed, ultrahigh-resolution OCT system using spectral/Fourier domain detection. Figure 7.28A shows virtual images in representative orthogonal cross-sectional planes that are generated from a three-dimensional OCT data set. Figure 7.28B shows a volume rendering of macula generated from three-dimensional OCT data. The nerve fiber layer complex is visible as the top structure with the photoreceptor outer segments and retinal pigment epithelium as the bottom structure. Visualization and rendering methods similar to those used in magnetic resonance imaging can be applied with micron-scale image resolutions.

Swept-Source/Fourier Domain Detection

Swept-source/Fourier domain detection uses a narrow-bandwidth light source that is frequency swept in time (Fercher et al., 1995; Chinn et al., 1997; Choma et al., 2003; Yun et al., 2003; Huber et al., 2005). The interferometer reference arm is stationary. Fourier domain detection has been referred to as *frequency domain interferometry* and, like low-coherence interferometry, was also developed and applied more than 10 years ago as a technique for performing high-resolution optical measurements in fiber optics and optoelectronic

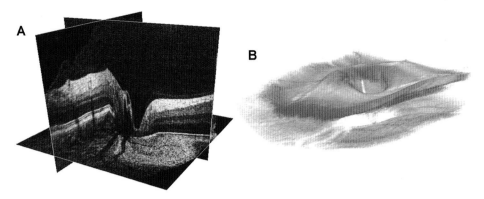

Figure 7.28 Examples of three-dimensional OCT imaging. High-speed OCT enables the acquisition of three-dimensional OCT data that can be displayed and rendered using techniques analogous to magnetic resonance imaging. (A) Examples of cross-sectional images in different orthogonal planes through the optic nerve head. (B) Example of a three-dimensional rendering of the foveal region. These images have been processed from a three-dimensional OCT data set consisting of 180 cross-sectional images of 512 axial scans each (Wojtkowski et al., 2005).

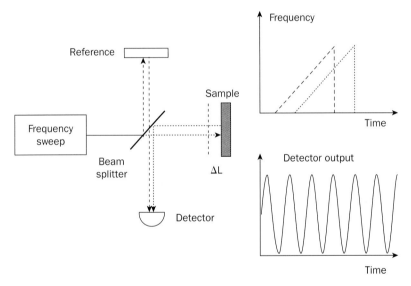

Figure 7.29 Swept-source/Fourier domain OCT uses a narrow-band, frequency-swept laser and individual detectors. The unbalanced Michelson interferometer interferes two frequency sweeps that are time delayed with respect to each other and generates a beat frequency that is proportional to the path length mismatch ΔL. Fourier transforming the beat signal from the detector yields axial scan information (echo magnitude and time delay).

components (Eickhoff and Ulrich, 1981; Barfuss and Brinkmeyer, 1989; Glombitza and Brinkmeyer, 1993). Figure 7.29 shows a schematic of how swept-source/Fourier domain detection works. The optical beam from a narrow-bandwidth, frequency-swept light source is split into two beams by an optical beam splitter. One beam is directed onto the tissue to be imaged and is back-reflected or backscattered from internal structures at different depths. The second beam is reflected from a reference mirror at a fixed delay.

The measurement beam and the reference beam have a time offset determined by the path length difference, which is related to the depth of the structure in the tissue. Because the frequency of the light is swept as a function of time, the light echoes in the measurement beam will have a frequency offset from the reference beam. When these beams interfere, a modulation or beat in intensity is produced at a frequency that is equal to the frequency offset. Therefore, different echo delays will produce different frequency modulations. The echo delays can be measured by digitizing the photodetectorsignal over a single frequency sweep of the light source, correcting any nonlinearity in the frequency sweep as a function of time and then Fourier transforming this beat frequency signal. This results in an axial scan measurement of the magnitude and echo delay of light from the tissue.

Similar to spectral/Fourier domain detection, swept-source/Fourier domain detection measures all the optical echoes at the same time, rather than sequentially as in time domain detection. This enables a dramatic improvement in detection sensitivity. Because mechanical scanning of the reference path length is not required and because detection sensitivities are increased, a significant increase in imaging speed and axial scan rate can be achieved. Although swept-source/Fourier domain is still in the research phase and many engineering issues remain to be addressed, it promises to improve dramatically the performance of OCT imaging.

ACKNOWLEDGMENTS

We acknowledge scientific contributions from Stephane Bourquin, James Connolly, Wolfgang Drexler, Jay Duker, Ingmar Hartl, Paul Herz, Pei-Lin Hsiung, Erich Ippen, Daniel Kopf, Max Lederer, Hiroshi Mashimo, Liron Pantanowitz, Joseph Schmitt, Karl Schneider, Joel Schuman, Wolfgang Seitz, Vivek Srinivasan, Kenji Taira, and Maciej Wojtkowski. This research was sponsored in part by National Science Foundation grants BES-0522845 and ECS-0501478, National Institutes of Health grants R01-CA75289–09 and R01-EY11289–20, and the Air Force Office of Scientific Research grants FA9550–040–1–0011 and FA9550–040–1–0046.

REFERENCES

Aguirre, A. D., P. Hsiung, T. H. Ko, I. Hartl, and J. G. Fujimoto. High-resolution optical coherence microscopy for high-speed, in vivo cellularimaging. *Opt Lett,* 28, 2064–2066, 2003.

Barfuss, H., and E. Brinkmeyer.. Modified optical frequency-domain reflectometry with high spatial-resolution for components of integrated optic systems. *J Lightwave Technol,* 7, 3–10, 1989.

Birks, T. A., W. J. Wadsworth, and P. S. J. Russel. Supercontinuum generation in tapered fibers. *Opt Lett,* 25, 1415–1417, 2000.

Boppart, S. A., B. E. Bouma, C. Pitris, G. J. Tearney, J. G. Fujimoto, and M. E. Brezinski. Forward-imaging instruments for optical coherence tomography. *Opt Lett,* 22, 1618–1620, 1997.

Boppart, S. A., B. E. Bouma, C. Pitris, G. J. Tearney, J. F. Southern, M. E. Brezinski, and J. G. Fujimoto. Intraoperative assessment of microsurgery with three-dimensional optical coherence tomography. *Radiology,* 208, 81–86, 1998.

Boppart, S. A., M. E. Brezinski, B. E. Bouma, G. J. Tearney, and J. G. Fujimoto. Investigation of developing embryonic morphology using optical coherence tomography. *Dev Biol,* 177, 54–63, 1996.

Boppart, S. A., A. Goodman, J. Libus, C. Pitris, C. A. Jesser, M. E. Brezinski, and J. G. Fujimoto. High resolution imaging of endometriosis and ovarian carcinoma with optical coherence tomography: Feasibility for laparoscopic-based imaging. *Br J Obstet Gynaecol,* 106, 1071–1077, 1999.

Bouma, B. E., and G. J. Tearney. Power-efficient nonreciprocal interferometer and linear-scanning fiber-optic catheter for optical coherence tomography. *Opt Lett,* 24, 531–533, 1999.

Bouma, B. E., G. J. Tearney, I. P. Bilinsky, B. Golubovic, and J. G. Fujimoto. Self-phase-modulated Kerr-lens mode-locked Cr:Forsterite laser source for optical coherence tomography. *Opt Lett,* 21, 1839–1841, 1996.

Bouma, B., G. J. Tearney, S. A. Boppart, M. R. Hee, M. E. Brezinski, and J. G. Fujimoto. High-resolution optical coherence tomographic imaging using a mode-locked Ti:Al2O3 laser source. *Opt Lett,* 20, 1486–1488, 1995.

Bouma, B. E., G. J. Tearney, C. C. Compton, and N. S. Nishioka.. High-resolution imaging of the human esophagus and stomach in vivo using optical coherence tomography. *Gastrointest Endosc,* 51(4), 467–474, 2000.

Bourquin, S., A. D. Aguirre, I. Hartl, P. Hsiung, T. H. Ko, J. G. Fujimoto, T. A. Birks, W. J. Wadsworth, U. Bunting, and D. Kopf. Ultrahigh resolution real time OCT imaging using a compact femtosecond Nd:Glass laser and nonlinear fiber. *Opt Express,* 11, 3290–3297, 2003.

Brezinski, M. E., G. J. Tearney, S. A. Boppart, E. A. Swanson, J. F. Southern, and J. G. Fujimoto. Optical biopsy with optical coherence tomography: Feasibility for surgical diagnostics. *J Surg Res,* 71, 32–40, 1997.

Brezinski, M. E., G. J. Tearney, B. E. Bouma, J. A. Izatt, M. R. Hee, E. A. Swanson, J. F. Southern, and J. G. Fujimoto. Optical coherence tomography for optical biopsy: Properties and demonstration of vascular pathology. *Circulation,* 93, 1206–1213, 1996.

Cense, B., N. Nassif, T. C. Chen, M. C. Pierce, S. Yun, B. H. Park, B. Bouma, G. Tearney, and J. F. De Boer. Ultrahigh-resolution high-speed retinal imaging using spectral-domain optical coherence tomography. *Opt Express,* 12, 2435–2447, 2004.

Chen, Z., T. E. Milner, D. Dave, and J. S. Nelson. Optical Doppler tomographic imaging of fluid flow velocity in highly scattering media. *Opt Lett,* 22, 64–66, 1997a.

Chen, Z., T. E. Milner, S. Srinivas, X. Wang, A. Malekafzali, M. J. C. van Gemert, and J. S. Nelson. Noninvasive imaging of in vivo blood flow velocity using optical Dopplertomogr aphy. *Opt Lett,* 22, 1119–1121, 1997b.

Chinn, S. R., E. A. Swanson, and J. G. Fujimoto. Optical coherence tomography using a frequency-tunable optical source. *Opt Lett,* 22, 340–342, 1997.

Choma, M. A., M. V. Sarunic, C. H. Yang, and J. A. Izatt. Sensitivity advantage of swept source and Fourier domain optical coherence tomography. *Opt Express,* 11, 2183–2189, 2003.

Chudoba, C., J. G. Fujimoto, E. P. Ippen, H. A. Haus, U. Morgner, F. X. Kärtner, V. Scheuer, G. Angelow, and T. Tschudi.. All-solid-state Cr:Forsterite laser generating 14 fs pulses at 1.3 mm. *Opt Lett,* 26, 292–294, 2001.

De Boer, J. F., B. Cense, B. H. Park, M. C. Pierce, G. J. Tearney, and B. E. Bouma. Improved signal-to-noise ratio in spectral-domain compared with time-domain optical coherence tomography. *Opt Lett,* 28, 2067–2069, 2003.

De Boer, J. F., T. E. Milner, M. J. C. van Gemert, and J. S. Nelson. Two-dimensional birefringence imaging in biological tissue by polarization-sensitive optical coherence tomography. *Opt Lett,* 22, 934–936, 1997.

Drexler, W., U. Morgner, R. K. Ghanta, F. X. Kärtner, J. S. Schuman, and J. G. Fujimoto. Ultrahigh-resolution ophthalmic optical coherence tomography. *Nat Med,* 7, 502–507, 2001.

Drexler, W., U. Morgner, F. X. Kartner, C. Pitris, S. A. Boppart, X. D. Li, E. P. Ippen, and J. G. Fujimoto. In vivo ultrahigh-resolution optical coherence tomography. *Opt Lett,* 24, 1221–1223, 1999.

Eickhoff, W., and R. Ulrich. Optical frequency-domain reflectometry in single-mode fiber. *Appl Phys Lett,* 39, 693–695, 1981.

Ell, R., U. Morgner, F. X. Kärtner, J. G. Fujimoto, E. P. Ippen, V. Scheuer, G. Angelow, T. Tschudi, M. J. Lederer, A. Boiko, and B. Luther-Davies. Generation of 5-fs pulses and octave-spanning spectra directly from a Ti:sapphire laser. *Opt Lett,* 26, 373–375, 2001.

Erbel, R., J. R. T. C. Roelandt, J. Ge, and G. Gorge. *Intravascular Ultrasound.* London: Martin Dunitz, 1998.

Evans, J. A., J. M. Poneros, B. E. Bouma, J. Bressner, E. F. Halpern, M. Shishkov, G. Y. Lauwers, M. Mino-Kenudson, N. S. Nishioka, and G. J. Tearney. Optical coherence tomography to identify intramucosal carcinoma and high-grade dysplasia in Barrett's esophagus. *Clin Gastroenterol Hepatol,* 4, 38–43, 2006.

Everett, M. J., K. Schoenenberger, B. W. Colston, Jr., and L. B. Da Silva. Birefringence characterization of biological tissue by use of optical coherence tomography. *Opt Lett,* 23, 228–230, 1998.

Feldchtein, F. I., G. V. Gelikonov, V. M. Gelikonov, R. V. Kuranov, A. M. Sergeev, N. D. Gladkova, A. V. Shakhov, N. M. Shakhova, L. B. Snopova, A. B. Terent'eva, E. V. Zagainova, Y. P. Chumakov, and I. A. Kuznetzova. Endoscopic applications of optical coherence tomography. *Opt Express,* 3, 257–270, 1998.

Fercher, A. F., C. K. Hitzenberger, W. Drexler, G. Kamp, and H. Sattmann. In vivo optical coherence tomography. *Am J Ophthalmol,* 116, 113–114, 1993.

Fercher, A. F., C. K. Hitzenberger, G. Kamp, and S. Y. Elzaiat. Measurement of intraocular distances by backscattering spectral interferometry. *Opt Commun,* 117, 43–48, 1995.

Fercher, A. F., K. Mengedoht, and W. Werner. Eye-length measurement by interferometry with partially coherent light. *Opt Lett,* 13, 1867–1869, 1988.

Fujimoto, J. G. Optical coherence tomography for ultrahigh resolution in vivo imaging. *Nat Biotechnol,* 21, 1361–1367, 2003.

Fujimoto, J. G., S. A. Boppart, G. J. Tearney, B. E. Bouma, C. Pitris, and M. E. Brezinski. High resolution in vivo intra-arterial imaging with optical coherence tomography. *Heart,* 82, 128–133, 1999.

Fujimoto, J. G., M. E. Brezinski, G. J. Tearney, S. A. Boppart, B. Bouma, M. R. Hee, J. F. Southern, and E. A. Swanson. Optical biopsy and imaging using optical coherence tomography. *Nat Med*, 1, 970–972, 1995.

George, M. Optical methods and sensors for in situ histology in surgery and endoscopy. *Minim Invas Ther Allied Technol*, 13, 95–104, 2004.

Gilgen, H. H., R. P. Novak, R. P. Salathe, W. Hodel, and P. Beaud. Submillimeter optical reflectometry. *IEEE J Lightwave Technol*, 7, 1225–1233, 1989.

Glombitza, U., and E. Brinkmeyer. Coherent frequency-domain reflectometry for characterization of single-mode integrated-optical wave-guides. *J Lightwave Technol*, 11, 1377–1384, 1993.

Hartl, I., X. D. Li, C. Chudoba, R. K. Hganta, T. H. Ko, J. G. Fujimoto, J. K. Ranka, and R. S. Windeler. Ultrahigh-resolution optical coherence tomography using continuum generation in an air–silica microstructure optical fiber. *Opt Lett*, 26, 608–610, 2001.

Hedrick, W. R., D. L. Hykes, and D. E. Starchman. *Ultrasound Physics and Instrumentation*. 4th ed. St. Louis, Mo.: ElsevierMosby, 2005.

Hee, M. R., C. R. Baumal, C. A. Puliafito, J. S. Duker, E. Reichel, J. R. Wilkins, J. G. Coker, J. S. Schuman, E. A. Swanson, and J. G. Fujimoto. Optical coherence tomography of age-related macular degeneration and choroidal neovascularization. *Ophthalmology*, 103, 1260–1270, 1996.

Hee, M. R., J. A. Izatt, E. A. Swanson, D. Huang, J. S. Schuman, C. P. Lin, C. A. Puliafito, and J. G. Fujimoto. Optical coherence tomography of the human retina. *Arch Ophthalmol*, 113, 325–332, 1995a.

Hee, M. R., C. A. Puliafito, J. S. Duker, E. Reichel, J. G. Coker, J. R. Wilkins, J. S. Schuman, E. A. Swanson, and J. G. Fujimoto. Topography of diabetic macular edema with optical coherence tomography. *Ophthalmology*, 105, 360–370, 1998.

Hee, M. R., C. A. Puliafito, C. Wong, J. S. Duker, E. Reichel, B. Rutledge, J. S. Schuman, E. A. Swanson, and J. G. Fujimoto. Quantitative assessment of macular edema with optical coherence tomography. *Arch Ophthalmol*, 113, 1019–1029, 1995b.

Hee, M. R., C. A. Puliafito, C. Wong, J. S. Duker, E. Reichel, J. S. Schuman, E. A. Swanson, and J. G. Fujimoto. Optical coherence tomography of macular holes. *Ophthalmology*, 102, 748–756, 1995c.

Hee, M. R., C. A. Puliafito, C. Wong, E. Reichel, J. S. Duker, J. S. Schuman, E. A. Swanson, and J. G. Fujimoto. Optical coherence tomography of central serous chorioretinopathy. *Am J Ophthalmol*, 120, 65–74, 1995d.

Herz, P. R., Y. Chen, A. D. Aguirre, J. G. Fujimoto, H. Mashimo, J. Schmitt, A. Koski, J. Goodnow, and C. Petersen. Ultrahigh resolution optical biopsy with endoscopic optical coherence tomography. *Opt Express*, 12, 3532–3542, 2004.

Hsiung, P. L., L. Pantanowitz, A. D. Aguirre, Y. Chen, D. Phatak, T. H. Ko, S. Bourquin, S. J. Schnitt, S. Raza, J. L. Connolly, H. Mashimo, and J. G. Fujimoto. Ultrahigh-resolution and 3-dimensional optical coherence tomography ex vivo imaging of the large and small intestines. *Gastrointest Endosc*, 62, 561–574, 2005.

Huang, D., E. A. Swanson, C. P. Lin, J. S. Schuman, W. G. Stinson, W. Chang, M. R. Hee, T. Flotte, K. Gregory, C. A. Puliafito, and J. G. Fujimoto. Optical coherence tomography. *Science*, 254, 1178–1181, 1991a.

Huang, D., J. Wang, C. P. Lin, C. A. Puliafito, and J. G. Fujimoto. Micron-resolution ranging of cornea and anterior chamber by optical reflectometry. *Lasers Surg Med*, 11, 419–425, 1991b.

Huber, R., M. Wojtkowski, K. Taira, J. G. Fujimoto, and K. Hsu. Amplified, frequency swept lasers for frequency domain reflectometry and OCT imaging: Design and scaling principles. *Opt Express*, 13, 3513–3528, 2005.

Isenberg, G., M. V. Sivak, Jr., A. Chak, R. C. Wong, J. E. Willis, B. Wolf, D. Y. Rowland, A. Das, and A. Rollins. Accuracy of endoscopic optical coherence tomography in the detection of dysplasia in Barrett's esophagus: A prospective, double-blinded study. *Gastrointest Endosc*, 62, 825–831, 2005.

Izatt, J. A., M. R. Hee, G. M. Owen, E. A. Swanson, and J. G. Fujimoto. Optical coherence microscopy in scattering media. *Opt Lett*, 19, 590–592, 1994.

Izatt, J. A., M. D. Kulkarni, H.- W. Wang, K. Kobayashi, and M. V. Sivak, Jr. Optical coherence tomography and microscopy in gastrointestinal tissues. *IEEE J Sel Top Quant Electron*, 2, 1017–1028, 1996.

Izatt, J. A., M. D. Kulkami, S. Yazdanfar, J. K. Barton, and A. J. Welch. In vivo bidirectional color Doppler flow imaging of picoliter blood volumes using optical coherence tomography. *Opt Lett*, 22, 1439–1441, 1997.

Jäckle, S., N. Gladkova, F. Feldchtein, A. Terentieva, B. Brand, G. Gelikonov, V. Gelikonov, A. Sergeev, A. Fritscher-Ravens, J. Freund, U. Seitz, S. Schröder, and N. Soehendra. In vivo endoscopic optical coherence tomography of esophagitis, Barrett's esophagus, and adenocarcinoma of the esophagus. *Endoscopy*, 32, 750–755, 2000a.

Jäckle, S., N. Gladkova, F. Feldchtein, A. Terentieva, B. Brand, G. Gelikonov, V. Gelikonov, A. Sergeev, A. Fritscher-Ravens, J. Freund, U. Seitz, S. Soehendra, and N. Schrödern. In vivo endoscopic optical coherence tomography of the human gastrointestinal tract: Toward optical biopsy. *Endoscopy,* 32, 743–749, 2000b.

Jain, A., A. Kopa, Y. T. Pan, G. K. Fedder, and H. K. Xie. A two-axis electrothermal micromirror for endoscopic optical coherence tomography. *IEEE J Sel Top Quant Electron,* 10, 636–642, 2004.

Jesser, C. A., S. A. Boppart, C. Pitris, D. L. Stamper, G. P. Nielsen, M. E. Brezinski, and J. G. Fujimoto. High resolution imaging of transitional cell carcinoma with optical coherence tomography: Feasibility for the evaluation of bladder pathology. *Br J Radiol,* 72, 1170–1176, 1999.

Kartner, F. X., N. Matuschek, T. Schibli, U. Keller, H. A. Haus, C. Heine, R. Morf, V. Scheuer, M. Tilsch, and T. Tschudi. Design and fabrication of double-chirped mirrors. *Opt Lett,* 22, 831–833, 1997.

Kobayashi, K., J. A. Izatt, M. D. Kulkarni, J. Willis, and M. V. Sivak, Jr. High-resolution cross-sectional imaging of the gastrointestinal tract using optical coherence tomography: Preliminary results. *Gastrointest Endosc,* 47, 515–523, 1998.

Kowalevicz, A. M., T. Ko, I. Hartl, J. G. Fujimoto, M. Pollnau, and R. P. Salathe. Ultrahigh resolution optical coherence tomography using a superluminescent light source. *Opt Express,* 10, 349–353, 2002a.

Kowalevicz, A. M., T. R. Schibli, F. X. Kartner, and J. G. Fujimoto. Ultralow-threshold Kerr-lens mode-locked Ti:Al$_2$O$_3$ laser. *Opt Lett,* 27, 2037–2039, 2002b.

Kremkau, F. W., *Diagnostic Ultrasound: Principles and Instruments.* 5th ed. Philadelphia, Pa.: WB Saunders, 1998.

Leitgeb, R., C. K. Hitzenberger, and A. F. Fercher. Performance of Fourier domain vs. time domain optical coherence tomography. *Opt Express,* 11, 889–894, 2003.

Li, X. D., S. A. Boppart, J. Van Dam, H. Mashimo, M. Mutinga, W. Drexler, M. Klein, C. Pitris, M. L. Krinsky, M. E. Brezinski, and J. G. Fujimoto. Optical coherence tomography: Advanced technology for the endoscopic imaging of Barrett's esophagus. *Endoscopy,* 32, 921–930, 2000a.

Li, X., C. Chudoba, T. Ko, C. Pitris, and J. G. Fujimoto. Imaging needle for optical coherence tomography. *Opt Lett,* 25, 1520–1522, 2000b.

Morgner, U., F. X. Kartner, S. H. Cho, Y. Chen, H. A. Haus, J. G. Fujimoto, E. P. Ippen, V. Scheuer, G. Angelow, and T. Tschudi. Sub-two-cycle pulses from a Kerr-lens mode-locked Ti:sapphire laser. *Opt Lett,* 24, 411–413, 1999.

Nassif, N., B. Cense, B. H. Park, S. H. Yun, T. C. Chen, B. E. Bouma, G. J. Tearney, and J. F. De Boer. In vivo human retinal imaging by ultrahigh-speed spectral domain optical coherence tomography. *Opt Lett,* 29, 480–482, 2004.

Parsa, P., S. L. Jacques, and N. S. Nishioka. Optical properties of rat liver between 350 and 2200 nm. *Appl Opt,* 28, 2325–2330, 1989.

Pfau, P. R., M. V. Sivak, Jr., A. Chak, M. Kinnard, R. C. Wong, G. A. Isenberg, J. A. Izatt, A. Rollins, and V. Westphal. Criteria for the diagnosis of dysplasia by endoscopic optical coherence tomography. *Gastrointest Endosc,* 58, 196–202, 2003.

Pitris, C., M. E. Brezinski, B. E. Bouma, G. J. Tearney, J. F. Southern, and J. G. Fujimoto. High resolution imaging of the upper respiratory tract with optical coherence tomography: A feasibility study. *Am J Respir Crit Care Med,* 157(5), 1640–1644, 1998.

Pitris, C., A. Goodman, S. A. Boppart, J. J. Libus, J. G. Fujimoto, and M. E. Brezinski. High-resolution imaging of gynecologic neoplasms using optical coherence tomography. *Obstet Gynecol,* 93, 135–139, 1999.

Pitris, C., C. Jesser, S. A. Boppart, D. Stamper, M. E. Brezinski, and J. G. Fujimoto. Feasibility of optical coherence tomography for high-resolution imaging of human gastrointestinal tract malignancies. *J Gastroenterol,* 35, 87–92, 2000.

Poneros, J. M., S. Brand, B. E. Bouma, G. J. Tearney, C. C. Compton, and N. S. Nishioka. Diagnosis of specialized intestinal metaplasia by optical coherence tomography. *Gastroenterology,* 120, 7–12, 2001.

Povazay, B., K. Bizheva, A. Unterhuber, B. Hermann, H. Sattmann, A. F. Fercher, W. Drexler, A. Apolonski, W. J. Wadsworth, J. C. Knight, P. S. J. Russell, M. Vetterlein, and E. Scherzer. Submicrometer axial resolution optical coherence tomography. *Opt Lett,* 27, 1800–1802, 2002.

Puliafito, C. A., M. R. Hee, C. P. Lin, E. Reichel, J. S. Schuman, J. S. Duker, J. A. Izatt, E. A. Swanson, and J. G. Fujimoto. Imaging of macular diseases with optical coherence tomography. *Ophthalmology,* 102, 217–229, 1995.

Ranka, J. K., R. S. Windeler, and A. J. Stentz. Visible continuum generation in air silica microstructure optical fibers with anomalous dispersion at 800 nm. *Opt Lett,* 25, 25–27, 2000.

Rollins, A. M., and J. A. Izatt. Optimal interferometer designs for optical coherence tomography. *Opt Lett,* 24, 1484–1486, 1999.

Schmitt, J. M., A. Knuttel, M. Yadlowsky, and M. A. Eckhaus. Optical-coherence tomography of a dense tissue: Statistics of attenuation and backscattering. *Phys Med Biol,* 39, 1705–1720, 1994.

Schuman, J. S., M. R. Hee, A. V. Arya, T. Pedut-Kloizman, C. A. Puliafito, J. G. Fujimoto, and E. A. Swanson. Optical coherence tomography: A new tool for glaucoma diagnosis. *Curr Opin Ophthalmol,* 6, 89–95, 1995a.

Schuman, J. S., M. R. Hee, C. A. Puliafito, C. Wong, T. Pedut-Kloizman, C. P. Lin, E. Hertzmark, J. A. Izatt, E. A. Swanson, and J. G. Fujimoto. Quantification of nerve fiberlayerthickness in normal and glaucomatous eyes using optical coherence tomography. *Arch Ophthalmol,* 113, 586–596, 1995b.

Schuman, J. S., C. A. Puliafito, and J. G. Fujimoto. *Optical coherence tomography ofocular diseases.* 2nd ed. Thorofare, N.J.: Slack, 2004.

Sergeev, A. M., V. M. Gelikonov, G. V. Gelikonov, F. I. Feldchtein, R. V. Kuranov, N. D. Gladkova, N. M. Shakhova, L. B. Suopova, A. V. Shakhov, I. A. Kuznetzova, A. N. Denisenko, V. V. Pochinko, Y. P. Chumakov, and O. S. Streltzova. In vivo endoscopic OCT imaging of precancerand cancerstates of human mucosa. *Opt Express,* 1, 432–440, 1997.

Sivak, M. V., Jr., K. Kobayashi, J. A. Izatt, A. M. Rollins, R. Ung-Runyawee, A. Chak, R. C. Wong, G. A. Isenberg, and J. Willis. High-resolution endoscopic imaging of the GI tract using optical coherence tomography. *Gastrointest Endosc,* 51(4), 474–479, 2000.

Sutter, D. H., G. Steinmeyer, L. Gallmann, N. Matuschek, F. Morier-Genoud, U. Keller, V. Scheuer, G. Angelow, and T. Tschudi. Semiconductor saturable-absorber mirror assisted Kerr-lens mode-locked Ti:sapphire laser producing pulses in the two-cycle regime. *Opt Lett,* 24, 631–633, 1999.

Swanson, E. A., D. Huang, M. R. Hee, J. G. Fujimoto, C. P. Lin, and C. A. Puliafito. High-speed optical coherence domain reflectometry. *Opt Lett,* 17, 151–153, 1992.

Swanson, E. A., J. A. Izatt, M. R. Hee, D. Huang, C. P. Lin, J. S. Schuman, C. A. Puliafito, and J. G. Fujimoto. In vivo retinal imaging by optical coherence tomography. *Opt Lett,* 18, 1864–1866, 1993.

Szabo, T. L. *Diagnostic Ultrasound Imaging: Inside Out.* Burlington, Mass.: Elsevier Academic Press, 2004.

Takada, K., I. Yokohama, K. Chida, and J. Noda. New measurement system for fault location in optical waveguide devices based on an interferometric technique. *Appl Opt,* 26, 1603–1608, 1987.

Tearney, G. J., S. A. Boppart, B. E. Bouma, M. E. Brezinski, N. J. Weissman, J. F. Southern, and J. G. Fujimoto. Scanning single-mode fiber optic catheter–endoscope for optical coherence tomography. *Opt Lett,* 21, 543–545, 1996.

Tearney, G. J., B. E. Bouma, and J. G. Fujimoto. High-speed phase- and group-delay scanning with a grating-based phase control delay line. *Opt Lett,* 22, 1811–1813, 1997a.

Tearney, G. J., M. E. Brezinski, B. E. Bouma, S. A. Boppart, C. Pitvis, J. F. Southern, and J. G. Fujimoto. In vivo endoscopic optical biopsy with optical coherence tomography. *Science,* 276, 2037–2039, 1997b.

Tearney, G. J., M. E. Brezinski, J. F. Southern, B. E. Bouma, S. A. Boppart, and J. G. Fujimoto. Optical biopsy in human gastrointestinal tissue using optical coherence tomography. *Am J Gastroenterol,* 92, 1800–1804, 1997c.

Tearney, G. J., M. E. Brezinski, J. F. Southern, B. E. Bouma, S. A. Boppart, and J. G. Fujimoto. Optical biopsy in human urologic tissue using optical coherence tomography. *J Urol,* 157, 1915–1919, 1997d.

Tearney, G. J., M. E. Brezinski, J. F. Southern, B. E. Bouma, S. A. Boppart, and J. G. Fujimoto. Optical biopsy in human pancreatobiliary tissue using optical coherence tomography. *Dig Dis Sci,* 43, 1193–1199, 1998.

Tran, P. H., D. S. Mukai, M. Brenner, and Z. P. Chen. In vivo endoscopic optical coherence tomography by use of a rotational microelectromechanical system probe. *Opt Lett,* 29, 1236–1238, 2004.

Unterhuber, A., B. Povazay, B. Hermann, H. Sattmann, W. Drexler, V. Yakovlev, G. Tempea, C. Schubert, E. M. Anger, P. K. Ahnelt, M. Stur, J. E. Morgan, A. Cowey, G. Jung, T. Le, and A. Stingl. Compact, low-cost Ti:Al$_2$O$_3$ laser for in vivo ultrahigh-resolution optical coherence tomography. *Opt Lett,* 28, 905–907, 2003.

Wang, Y., J. S. Nelson, Z. Chen, B. J. Reiser, R. S. Chuck, and R. S. Windeler. Optimal wavelength for ultrahigh-resolution optical coherence tomography. *Opt Express,* 11, 1411–1417, 2003a.

Wang, Y., Y. Zhao, J. S. Nelson, Z. Chen, and R. S. Windeler. Ultrahigh-resolution optical coherence tomography by broadband continuum generation from a photonic crystal fiber. *Opt Lett,* 28, 182–184, 2003b.

Wojtkowski, M., T. Bajraszewski, I. Gorczynska, P. Targowski, A. Kowalczyk, W. Wasilewski, and C. Radzewicz. Ophthalmic imaging by spectral optical coherence tomography. *Am J Ophthalmol,* 138, 412–419, 2004.

Wojtkowski, M., R. Leitgeb, A. Kowalczyk, T. Bajraszewski, and A. F. Fercher. In vivo human retinal imaging by Fourier domain optical coherence tomography. *J Biomed Opt,* 7, 457–463, 2002.

Wojtkowski, M., V. Srinivasan, J. G. Fujimoto, T. Ko, J. S. Schuman, A. Kowalczyk, and J. S. Duker. Three-dimensional retinal imaging with high-speed ultrahigh-resolution optical coherence tomography. *Ophthalmology,* 112, 1734–1746, 2005.

Xie, H., Y. Pan, and G. K. Fedder. Endoscopic optical coherence tomographic imaging with a CMOS-MEMS micromirror. *Sensors Actuators A (Physical),* A103, 237–241, 2003a.

Xie, T., H. Xie, G. K. Fedder, and Y. Pan. Endoscopic optical coherence tomography with new MEMS mirror. *Electronics Lett,* 39, 1535–1536, 2003b.

Yazdanfar, S., M. D. Kulkarni, and J. A. Izatt. High resolution imaging of in vivo cardiac dynamics using color Doppler optical coherence tomography. *Opt Express,* 1, 424–431, 1997.

Youngquist, R., S. Carr, and D. Davies. Optical coherence-domain reflectometry: A new optical evaluation technique. *Opt Lett,* 12, 158–160, 1987.

Yun, S. H., G. J. Tearney, B. E. Bouma, B. H. Park, and J. F. De Boer. High-speed spectral-domain optical coherence tomography at 1.3 mu m wavelength. *Opt Express,* 11, 3598–3604, 2003.

Zara, J. M., and S. W. Smith. Optical scanner using a MEMS actuator. *Sensors Actuators A (Physical),* A102, 176–184, 2002.

8

Two-Photon Fluorescence Correlation Spectroscopy

PETRA SCHWILLE, KATRIN HEINZE,
PETRA DITTRICH, AND ELKE HAUSTEIN

INTRODUCTION TO FLUORESCENCE CORRELATION SPECTROSCOPY

Motivation

Single-molecule-based fluorescence correlation spectroscopy (or FCS) is currently considered one of the most powerful complementary methods in the context of modern fluorescence microscopy. Although the information it provides is primarily of dynamic rather than spatial nature, recent FCS applications particularly in the cellular environment have raised tremendous hopes that this technique, in conjunction with confocal or two-photon imaging, can open fully new ways to investigate and understand complex biological processes on a single-molecule scale.

The difference between FCS and standard microscopic techniques might be illustrated by drawing an—admittedly, rather far-fetched—parallel to wildlife observation. To learn something about, forexample, animals in theirnatur al habitat, one might eitherobser ve them from an airplane far above and take some high-resolution images or movies, or see them move around to different locations at different times to get a "big picture" about their distributions and colocalizations at certain interesting sites. One might, on the other hand, prefer to wait behind a bush or camouflage at the spots where they can be most frequently observed and take a closerlook at theirspecific action: how fast and often they access or leave a place, how long they stay, how they interact, and how they behave as individuals. In principle, this latter approach is exactly what we want to achieve on a molecular level by applying confocal FCS analysis. In contrast to confocal laser scanning microscopy performed in a largely similar setup, the small laser-induced focal measurement volume is not rapidly scanned through the sample to acquire a detailed image, but rather is parked at a certain position of interest (fig. 8.1). If we now replace our standard photomultipliers with very sensitive APDs and just register the photon count signal with time, all temporal changes in this focal spot volume element (e.g., the entering and leaving of individual fluorescent particles resulting from diffusion or transport processes) can be accurately recorded. In concentration ranges in which the small observation volume captures only a single or very few particles at a time, FCS even enables us to investigate intramolecular dynamics, if they only induce slight brightness changes of the reporter dye. By using different labels for different molecularspecies, we can easily distinguish whetherthey move in a coordinated

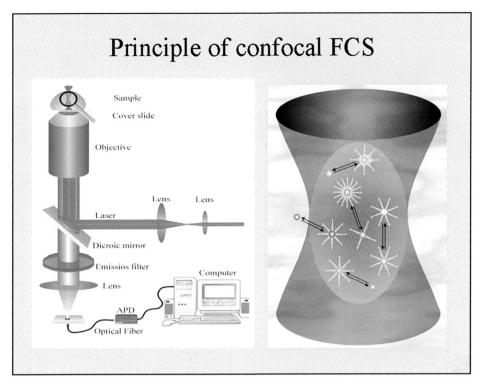

Figure 8.1 Schematic setup for a confocal FCS experiment (left). After passing an optional beam expander and being reflected by a dichroic mirror, the illuminating laser is focused into the sample solution by a microscope objective. The fluorescence is collected by the same objective and passes the dichroic mirror. Residual excitation light is removed from the beam by special bandpass filters. Last, the light is focused onto an optical fiber acting as a pinhole and directed onto the detector. To the right, molecular mechanisms are depicted that give rise to fluorescence fluctuations within the focal volume. Among them are particle movement, conformational changes, and chemical or photophysical reactions.

fashion—if they are tightly bound to each other—or whether they move independently because they do not interact. Because the measurement volume is parked, no temporal resolution is lost by the positioning process, and the only limitations of such a "single-molecule" measurement are imposed by the dead time of the detector (somewhere in the 10-ns range) and the residence time of the mobile molecules in our spot (characteristically between several 100 ns and 10 s).

Single-protein and nucleic acid molecules are believed to be the smallest functional units in biological systems, theirdynamic analysis in vitro, but also in living cells and organisms, remains one of the major goals of current biophysical research. Many successful approaches to access the level of single biomolecules have been reported during the past few years, involving their mere observation by fluorescence microscopy, but also their manipulation by use of optical tweezers or the atomic force microscope. With respect to intracellular analysis, the optical methods are certainly the most promising ones because they guarantee minimum interference with the biological system under observation. Molecules can even be probed in thermodynamic equilibrium. The scope of this chapter is to give an overview of FCS in the perspective of ultrasensitive molecular analysis in life cells. Although the ultimate goal of observing a biologically active individual molecule during its whole lifetime—and following it through the cellular environment—is still far from being reached, the wealth of information that can be gained by FCS, just recording temporal changes of the fluorescence signal resulting from spontaneous molecular processes at a certain spot is, as we will see, enormous.

Because intracellular observations are of primary interest here, this chapter is mainly devoted to a specific mode of FCS: using two-photon excitation of the fluorophores in the near infrared. Since its introduction to the biomedical sciences by Denk et al. in 1990, two-photon excitation of fluorophores at wavelengths approximately twice as large as their respective absorption maxima has revealed a number of indisputable advantages for fluorescence applications in cells and tissues. The illumination of the sample by infrared rather than visible light causes less problems with low penetration or scattering. The crucial advantage is, however, the dramatically reduced excitation volume in the sample resulting from the square dependence of the excitation process on illumination intensity. By illuminating only the spatial regime that actually contributes to the detected signal, a much better preservation of both cell and dye supply within it can be obtained. An additional advantage, as we will see later, is provided by the large overlap of two-photon excitation spectra for largely different emitting dyes.

History of Fluorescence Correlation Spectroscopy and Two-Photon Fluorescence Correlation Spectroscopy

Fluorescence correlation spectroscopy was developed in the early 1970s (Magde et al., 1972; Ehrenberg and Rigler, 1974; Elson and Magde, 1974; Magde et al., 1974) as a special case of relaxation analysis, with its underlying idea to draw information about kinetic coefficients from the way a molecular system relaxes back to equilibrium after (external) perturbations. The conceptual difference of FCS with respect to classic relaxation techniques is that the perturbations of the system are not induced by the experimentalist, but advantage is taken of minute spontaneous fluctuations of any measurable parameter around their average values, which always occur on a microscopic scale at ambient temperatures, and which are generally represented as noise patterns. A method to quantify concentration fluctuations was provided by dynamic light scattering (Berne and Pecora, 1974), which measures the dynamics of dielectric constants in liquids. However, the detection sensitivity can be much enhanced if fluorescence is used as a measurement parameter, and the resolution can be improved from small particles to single molecules. The fluctuations are quantified by temporally autocorrelating the recorded intensity signal, a mathematical procedure after which the technique is named. Autocorrelation analysis provides a measure for the self-similarity of a signal in time, and thereby describes the persistence of information carried by it. The temporal patterns by which fluctuations arise and decay yield a wealth of information about the underlying processes governing inter- and intramolecular dynamics.

At its introduction, FCS was first applied to measure diffusion and chemical kinetics of DNA–drug intercalation (Magde et al., 1972) as well as rotational diffusion (Ehrenberg and Rigler, 1974). A multitude of subsequent studies have been devoted to the determination of particle concentration and mobility, such as three- and two-dimensional diffusion (Aragón and Pecora, 1976; Fahey et al., 1977) or laminar flow (Magde et al., 1978). Although the number of publications discussing possible in vitro and also in vivo applications of FCS was rather large, early measurements suffered from poor signal quality mainly because the lack of detection sensitivity and background suppression, and large ensemble numbers. It is obvious that fluctuations can be better esolved using a smaller system, ideally on the level of single molecules. Using efficient light sources such as lasers, and ultrasensitive detection devices such as APDs, fluorescence-based detection and analysis of single molecules has lately been accomplished in various ways. The confocal detection scheme is a simple and elegant setup to restrict the system under investigation to very small molecular numbers, as introduced by Rigler et al. (Rigler and Widengren, 1990; Rigler et al., 1993) to FCS applications. An extremely small measurement volume is achieved by focusing a laser beam down to the optical resolution limit by an objective with a high NA (>0.9). Because fluorescence is only excited within the illuminated region, only molecules dwelling in the focal spot contribute to the measured signal. Photons emanating from there are imaged onto the photodiode via a field aperture (pinhole) in the image plane by which axial resolution is achieved.

To date, most FCS measurements are performed on fluorescently labeled biomolecules diffusing in aqueous buffersolution. In confocal geometries limiting the detection volume to less than 1 femtoliter(fL) (i.e., approximately the volume of an *Escherichia coli* bacterial cell), concentrations in the nanomolar range are optimal for FCS measurements. Under these circumstances, the signal fluctuations induced by molecules diffusing into or out of the focal volume are large enough to yield good signal-to-noise ratios. During the time a particle spends in the focus, chemical orphotophysical reactions orconfor mational changes may alter the emission characteristics of the fluorophore and give rise to additional fluctuations in the detected signal.

Only recently, an increasing number of intracellular measurements has been reported. The mobility of proteins and DNA orRNA fragments within the cytosol orothercell organelles belong to the most prominent measurement parameters. An increasing number of studies is now devoted to the analysis of molecular (e.g., protein–protein) interactions or translocation processes. One reason this has not been achieved earlier is certainly the poor signal quality resulting from autofluorescence and light absorption or scattering. Another problem consists of the limited supply of dye molecules within the cell, which may lead to irreversible photobleaching.

In view of these problems, the use of two-photon excitation has been proposed (fig 8.2). Two-photon excitation requires the absorption of two photons of theoretically double the wavelength usually required for the excitation, within the tiny time interval of about 1 fs (10^{-15} s). To get a reasonable probability of such three-particle events, the photon flux must be extremely high. This means that not only a high output power is required, but pulsed excitation is also usually used to get an even higherphoton density perpulse relative to the average output power. The joint probability of absorbing two photons per excitation process is proportional to the mean square of the intensity. This results in inherent depth discrimination such that only the immediate vicinity of the objective's focal spot receives sufficient intensity forsignificant fluorescence excitation.

Intracellular measurements primarily benefit from this inherent axial resolution, because under two-photon excitation, bleaching really occurs only in the focal region (Denk et al., 1990). In contrast to this, under one-photon excitation, all fluorophores residing in the double cone above and below the focal spot are excited and potentially bleached, and, as such, the increase in depth discrimination provided by the pinhole applies only to the fluorescence emission and not the excitation. Because cells and tissue also tend to be more tolerant to near-infrared radiation and there is less autofluorescence and scattering, MPE is becoming more and more popularforbiological applications, in particularforconfocal scanning microscopy.

The first measurements demonstrating the principle of combining FCS and two-photon excitation for intracellular measurements were reported as early as 1995 (Berland et al., 1995). However, the signal-to-noise levels have still been rather low in these early exper-iments, because fluorescence beads had to be used rather than single molecules, and the biological relevance of this study has thus been limited. In 1999, the first applications of intracellular two-photon FCS on a single-molecule scale were shown (Schwille et al., 1999a,b). Cytosolic as well as membrane diffusion of different fluorophores, including the clonal tag GFP, were observed, and favorable intensity and wavelength conditions to avoid photodamage and to suppress cellular autofluorescence were discussed in detail. It could be shown that at comparable signal levels, two-photon excitation clearly diminishes the long-term signal loss resulting from photobleaching in cells with limited dye resources.

Two-photon FCS

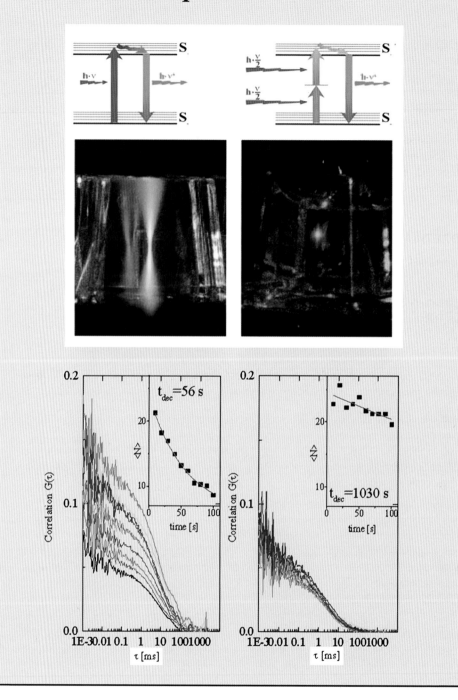

Moreover, it was shown that in plant cells or other optically problematic cell types, as well as in deepercell layers in tissue, two-photon excitation is basically the only way to obtain reasonable signals for ultrasensitive single-molecule analysis.

Recently, one of the most crucial advantages of two-photon excitation for FCS was demonstrated (Heinze et al., 2000). The two-photon-induced transition to the excited state, which is formally forbidden by the rules of symmetry, exhibits different selection rules and "vibronic" coupling. As a consequence, the two-photon excitation spectra of many common fluorophores differ considerably from their one-photon counterparts without any change in emission, which makes it possible to accomplish simultaneous excitation of spectrally distinct dyes. This possibility of dual- oreven triple-colorexcitation of single molecules by using a single infrared laser beam significantly simplifies the FCS setup if more than one molecular species is to be simultaneously observed, as with the relevant cross-correlation mode discussed later in detail. The selection of dye systems that support this rather elegant mode of two-photon FCS is a crucial task, but many possible combinations, particularly in the green and red spectral range, have so far been revealed. This rather practical aspect for possible future applications is also briefly discussed later.

THEORETICAL CONCEPTS

Autocorrelation Analysis

*Auto*correlation analysis of fluorescence fluctuations is usually performed if only a single fluorescent species is to be observed. The most prominent fluctuations are, in this case, number fluctuations in the measurement volume, which are mostly governed by Poissonian statistics. The root mean square fluctuation of the particle number N is given by

$$\frac{\sqrt{\langle (\delta N)^2 \rangle}}{\langle N \rangle} = \frac{\sqrt{\langle (N - \langle N \rangle)^2 \rangle}}{\langle N \rangle} = \frac{1}{\sqrt{\langle N \rangle}} \tag{8.1}$$

Because the relative fluctuations decrease with increasing numbers of simultaneously observed particles, it is important for decent fluctuation analysis to minimize the number of molecules in the focal volume. However, the fluorescence signal must still be higher than the residual background signal. As a rule of thumb, the temporal average of the particle number should be between 0.1 and 1000. The focal volume being about 1 fL, this corresponds to between subnanomolar($<10^{-9}$ M) and micromolar (10^{-6} M) concentrations.

Figure 8.2 (facing page) Difference between one- and two-photon excitation. For one-photon excitation, spatial resolution in the axial direction must be enforced by inserting a pinhole into the image plane. However, this can only be achieved for the detected light. The photograph on the left shows that actually all molecules within a double cone are excited, although the excitation probability is decreasing with increasing distance from the focal region. When measuring within spatially confined regions with limited chromophore supply, such as in living cells or even cellular organelles, cumulative photobleaching may lead to substantial loss of fluorescent molecules. The number of detectable chromophores can be dramatically reduced during a measurement, as can be seen from 10 consecutive autocorrelation curves recorded in 10-s time steps. After 100 s, the correlation amplitude is increased by more than 50%, meaning that the average number of dye molecules within the measurement volume is reduced accordingly. In contrast to this, two-photon excitation is achieved by quasi-simultaneous absorption of two photons with (in theory) double the wavelength required for the corresponding one-photon process. The higher photon densities necessary for a finite excitation probability can only be achieved within the close proximity of the focal spot. The photograph on the right shows this inherent sectioning. As can be seen from the autocorrelation curves, photobleaching is substantially reduced. Therefore, two-photon excitation is ideally suited for intracellular measurements.

Assuming constant excitation power, the fluctuations of the fluorescence signal are defined as the deviations from the temporal average of the signal $\delta F(t) = F(t) - \langle F(t) \rangle$, and are given for general one-photon excitation by

$$\delta F(t) = \kappa \int_V I_{ex}(\underline{r}) \cdot \Omega(\underline{r}) \cdot \delta(\sigma \cdot q \cdot C(\underline{r}, t)) \cdot dV \qquad (8.2)$$

where κ is the overall detection efficiency of the optical system, $I_{ex}(r)$ is the spatial distribution of the excitation energy with the maximum amplitude I_0, $\Omega(\underline{r})$ is the optical transfer function of the objective–pinhole combination (this parameter determines the spatial collection efficiency of the setup and is dimensionless), $\delta(\sigma \cdot q \cdot C(\underline{r}, t))$ is the dynamics of the fluorophore on the single-particle level, $\delta\sigma$ is the fluctuation in the molecularabsor ption cross-section, δq is the fluctuation in the quantum yield, and $\delta C(\underline{r}, t)$ is the fluctuation in the local particle concentration at time t (e.g., because of Brownian motion).

Determining all these parameters is extremely difficult or even impossible. To simplify equation 8.2, the convolution factorof the two dimensionless spatial optical transfer functions $I_{ex}(\underline{r})/I_0 \times \Omega(\underline{r})$ can be combined into a single function $W(\underline{r})$, which describes the spatial distribution of the emitted light. This is generally approximated by a three-dimensional Gaussian function, with $1/e^2$ widths r_w in the lateral direction and z_w in the axial direction:

$$W(\underline{r}) = \exp\left(-2\frac{x^2 + y^2}{r_w^2}\right) \exp\left(-2\frac{z^2}{z_w^2}\right) \qquad (8.3)$$

The remaining parameters κ, σ, and q can be combined with the excitation intensity amplitude I_0 into a new parameter that determines the photon count rate per detected molecule persecond: $\eta = I_0 \cdot \kappa \cdot \sigma \cdot q$. This parameter primarily determines the signal-to-noise ratio of the measurement (Koppel, 1974) and is therefore of particular practical relevance, as shown next.

Equation 8.2 can now be rewritten as

$$\delta F(t) = \int_V W(\underline{r})\delta(\eta C(\underline{r}, t)) \cdot dV \qquad (8.4)$$

Under two-photon excitation, the square dependence of the fluorescence on excitation intensity has to be taken into account. In addition to that, one has to take care of the specific mode of excitation—in ourapplications, $\approx$100-fs pulses with repetition rates in the 80-MHz regime. Under these circumstances, $\delta F(t)$ must be modified to

$$\delta F_2(t) = \frac{1}{2}\kappa g_2 \langle I_0(t) \rangle^2 \int_V S_2^2(\underline{r}) \cdot \Omega(\underline{r}) \cdot \delta\left(q \cdot \sigma_2 \cdot C(\underline{r}, t)\right) \cdot dV \qquad (8.5)$$

The factor1 /2 accounts for the requirement of two photons; σ_2 is the specific two-photon excitation cross-section. $\langle I_0(t) \rangle$ is the average intensity at the geometric focal point and $S_2(\underline{r})$ is the dimensionless spatial distribution function of the focused light in the sample space. Because the efficiency of pulsed excitation depends on $\langle I_0^2(t) \rangle$, the second-order temporal coherence $g_2 = \langle I_0^2(t) \rangle / \langle I_0(t) \rangle^2$ proportional to $(\tau_p f)^{-1}$, with pulse duration τ_p and repetition rate f, needs to be introduced. As in the one-photon mode, $\Omega(\underline{r})$ is determined by objective and pinhole transfer properties. With its inherent depth discrimination property, two-photon excitation does not, in principle, require a pinhole, such that $\Omega(\underline{r}) = 1$ can be approximated.

Similarto the one-photon case, equation 8.5 can now be simplified as follows:

$$\delta F(t) = \int_V S_2^2(\underline{r})\delta\left(\eta_2 C(\underline{r}, t)\right) \cdot dV \qquad (8.6)$$

with the count rate permolecule and second undertwo-photon excitation η_2 and the squared Gaussian distribution $S_2^2(\underline{r})$:

$$S_2^2(\underline{r}) = \exp\left(-4\frac{x^2+y^2}{r_s^2}\right)\exp\left(-4\frac{z^2}{z_s^2}\right) \tag{8.7}$$

The normalized fluctuation autocorrelation function is now defined as

$$G(\tau) = \frac{\langle \delta F(t) \cdot \delta F(t+\tau)\rangle}{\langle F(t)\rangle^2} \tag{8.8}$$

thereby analyzing the signal with respect to its self-similarity after the lag time τ (see fig. 8.3A for exemplification of the correlation process). The autocorrelation amplitude $G(0)$ is just the normalized variance of the fluctuating fluorescence signal $\delta F(t)$.

Substituting equation 8.4 into equation 8.8 forthe one-photon excitation mode yields

$$G(\tau) = \frac{\int\int W(\underline{r})W(\underline{r}')\langle \delta(\eta \cdot C(\underline{r},t))\delta\left(\eta \cdot C(\underline{r}',t+\tau)\right)\rangle dVdV'}{\left(\int W(\underline{r})\langle \delta\left(\eta \cdot C(\underline{r},t)\right)\rangle dV\right)^2} \tag{8.9}$$

We can now separate the fluctuation term $\delta\left(\eta \cdot C(\underline{r},t)\right) = C\delta\eta + \eta\delta C$. Obviously, equation 8.9 will be simplified, to a large extent, if either the concentration or the parameter η are constant for a given system. Assuming first that the chromophore's fluorescence properties are not changing within the observation time (i.e., $\delta\eta = 0$), equation 8.9 can be rewritten as

$$G(\tau) = \frac{\int\int W(\underline{r})W(\underline{r}')\langle \delta C(\underline{r},0)\delta C(\underline{r}',\tau)\rangle dVdV'}{\left(\langle C\rangle \int W(\underline{r})dV\right)^2} \tag{8.10}$$

Three-Dimensional Diffusion of a Single Species

Considering only particles of a certain fluorescence species freely diffusing in three dimensions with the diffusion coefficient D, the so-called number density autocorrelation term $\langle \delta C(\underline{r},0)\delta C(\underline{r}',\tau)\rangle$ can be calculated as

$$\langle \delta C(\underline{r},0)\delta C(\underline{r}',\tau)\rangle = \langle C\rangle(4\pi D\tau)^{-\frac{3}{2}}\exp\left(-\frac{(\underline{r}-\underline{r}')^2}{4D\tau}\right)$$

which leads to

$$G(\tau) = \frac{1}{\langle C\rangle(4\pi D\tau)^{\frac{3}{2}}} \frac{\int\int W(\underline{r})W(\underline{r}') \cdot \exp\left(-\frac{(\underline{r}-\underline{r}')^2}{4D\tau}\right)dVdV'}{\left(\int W(\underline{r})dV\right)^2} \tag{8.11}$$

Accordingly, $W(\underline{r})$ will simply be replaced by $S_2^2(\underline{r})$, and η by η_2 fortwo-photon excitation in equations 8.9 through 8.11. One can finally derive an analytical expression for the autocorrelation functions for a single freely diffusing species of molecules:

$$G(\tau) = \frac{1}{V_{eff}\langle C\rangle} \cdot \frac{1}{\left(1+\frac{\tau}{\tau_d}\right)} \cdot \frac{1}{\sqrt{1+\left(\frac{r_{w,s}}{z_{w,s}}\right)^2 \cdot \frac{\tau}{\tau_d}}} \tag{8.12}$$

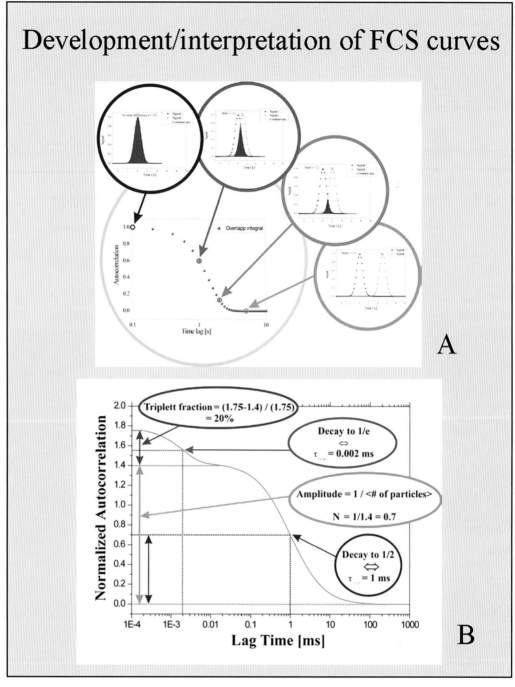

Figure 8.3 (A) The autocorrelation function of some signal F(t) basically gives a measure for its self-similarity after some lag time τ. For simplicity, let us consider only one pronounced "burst." At time $\tau = 0$ (i.e., no lag time), the normalized overlap integral is equal to one and the autocorrelation function has maximum amplitude. With increasing lag time, this overlap integral decreases gradually, until the signals no longer have any common features and the correlation function has dropped to zero. Because one molecule only correlates with itself, the correlation function for many passing molecules is simply the sum over all individual "bursts." (B) If fluorescence fluctuations are not only the result of diffusion, but also are induced by other (e.g., photophysical) processes, the autocorrelation curve reflects the individual characteristic timescales. The inverse amplitude of the diffusion term (the slowest discernible process) gives the average number of particles within the focal region. After the typical diffusion time, the diffusion process drops to half its initial amplitude.

The relationship between the characteristic lateral diffusion time τ_d of a molecule through the focal volume and the diffusion coefficient D forone-photon excitation is given by

$$\tau_d = \frac{r_w^2}{4 \cdot D} \tag{8.13}$$

and fortwo-photon excitation is given by

$$\tau_d = \frac{r_s^2}{8 \cdot D} \tag{8.13a}$$

Knowing the volume dimensions r_w and z_w forone-photon excitation, orr $_s$ and z_s fortwo-photon excitation, the diffusion coefficient can be easily derived from the characteristic decay time of the correlation function τ_d as mentioned earlier.

Under the premise of the system obeying Poissonian statistics, as outlined earlier, the correlation amplitude G(0), being the normalized variance of the number fluctuations, is thus the inverse number of particles simultaneously observed:

$$G(0) = \frac{1}{\langle N \rangle} = \frac{1}{V_{eff} \cdot \langle C \rangle} \iff \langle C \rangle = \frac{1}{V_{eff} \cdot G(0)} \tag{8.14}$$

From equations 8.11 and 8.14, the effective volume element V_{eff} can thus be determined to be

$$V_{eff} = \frac{\left(\int W(\underline{r}) dV \right)^2}{\left(\int W^2(r) dV \right)} \underset{eq.\ 8.3}{=} \frac{\left(\int \exp\left(-2\frac{x^2 + y^2}{r_w^2} \right) \exp\left(-2\frac{z^2}{z_w^2} \right) dV \right)^2}{\int \exp\left(-4\frac{x^2 + y^2}{r_w^2} \right) \exp\left(-4\frac{z^2}{z_w^2} \right) dV}$$

$$\underset{\substack{integeration \\ over\ space}}{=} \pi\sqrt{\pi} \cdot r_w^2 \cdot z_w \tag{8.15}$$

In the case of two-photon excitation, it is

$$V_{eff} = 2\sqrt{2} \cdot \pi\sqrt{\pi} \cdot r_S^2 \cdot z_S \tag{8.15a}$$

Intensity Modulations (Flickering)

Hitherto, it was assumed that the chromophore's fluorescence properties are not changing during the whole observation time (i.e., $\delta\eta = 0$). This assumption rarely holds for real dyes, especially not at higherexcitation powers. The most common cause forso-called *flickering* of the fluorescence intensity is the transition of the dye to the first excited triplet state. Because this transition is quantum mechanically forbidden, the chromophore needs a comparably long time to relax back to the ground state. During these intervals, the dye cannot emit any fluorescence photons and appears dark. Indeed, one can imagine the intersystem crossing as a series of dark intervals interrupting the otherwise (on the micro- to millisecond FCS timescale) continuous fluorescence emission of the molecule on its path through the illuminated region. Other examples for processes that induce such fast flickering phenomena are inter- or intramolecular reactions such as electron or proton transfer, altering the absorption and/or emission spectra or quenching the fluorescence.

Instead of rigorously recalculating the correct autocorrelation function for these conditions, a simplification is generally introduced. If intra- or intermolecular reactions give rise to fluorescence fluctuations on timescales much faster than those caused by particle movement, the two dynamics may just be treated as multiplicative terms:

$$G_{total}(\tau) = X_{flickering}(\tau) \cdot G_{motion}(\tau) \tag{8.16}$$

Of course, this assumption holds only for situations in which the diffusion coefficient is unaltered by the fast dynamics (Palmer and Thompson, 1987; Widengren and Rigler, 1998).

The fast dynamic process $X(\tau)$ is generally represented as an additional shoulder in the measured correlation curves at short timescales (Widengren et al., 1995), decaying as an exponential function. The overall autocorrelation function for a freely diffusing dye can then be written as

$$G_{total}(\tau) = \frac{\left(1 - F + F \cdot \exp\left(-\tau/\tau_F\right)\right)}{(1 - F)} \cdot \frac{1}{V_{eff}\langle C \rangle} \cdot \frac{1}{\left(1 + \frac{\tau}{\tau_d}\right)} \cdot \frac{1}{\sqrt{1 + \left(\frac{r_{w,s}}{z_{w,s}}\right)^2 \cdot \frac{\tau}{\tau_d}}}$$

$$(8.17)$$

The flickering phenomenon is assumed to be the result of reversible transitions between a bright (fluorescent) state B and a dark state D in which no photons are emitted:

$$B \; \underset{k_B}{\overset{k_D}{\rightleftharpoons}} \; D$$

such that τ_F and F in equation 8.17 are determined by

$$\tau_F = \frac{1}{k_D + k_B} : \text{relaxation time}$$

$$F = \frac{k_D}{k_D + k_B} : \text{average fraction of dark molecules}$$

$$\hat{=} : \text{average fraction of time a molecule spends in the dark state}$$

If the dark state D is not completely dark, the molecular emission yield η_i of the two states has to be taken into account to get the correct expression for F:

$$F = \frac{k_D k_B (\eta_B - \eta_D)^2}{(k_D + k_B)\left(k_B \eta_B^2 + k_D \eta_D^2\right)}$$

$$(8.18)$$

In figure 8.3B, an overview is given how concentrations and fluctuation timescales related to mobility or internal fluctuations can simply be estimated from the autocorrelation curves.

Different Kinds of Mobility, Multiple Species

In the cellular environment, free three-dimensional diffusion of a single fluorescent species is only rarely observed. Rather, many different other modes of particle movement, such as free two-dimensional diffusion (in membranes), restricted diffusion, or active transport, have to be taken into consideration. There also may be reactions taking place that influence the mobility of the particle (e.g., by reversible or irreversible association with larger molecules or with cellular structures). In this case, more than one fluorescent species with a certain distinct mobility pattern will be observed at the same time. Equation 8.12 must thus be generalized to take into account all different kinds of possible motion, weighted by the relative emission rates (brightness values) of the respective molecules:

$$G_{motion}(\tau) = \frac{1}{V_{eff}} \frac{\sum\limits_{\substack{all\ different \\ species\ i}} \eta_i^2 \langle C_i \rangle M_i(\tau)}{\left(\sum\limits_i \eta_i \langle C_i \rangle\right)^2}$$

$$(8.19)$$

It can be easily evidenced that the molecular brightness, quadratically entering the numerator of equation 8.19, has a large influence on the weighting of certain molecular species. Double brightness would result in fourfold higher representation, which is of large practical relevance. The motility term $M_i(\tau)$, on the otherhand, must be adapted to the particular situation. Besides the free three-dimensional diffusion discussed earlier with

$$M_i(\tau) = \frac{1}{\left(1 + \frac{\tau}{\tau_{d,i}}\right) \cdot \sqrt{1 + \left(\frac{r_{w,s}}{z_{w,s}}\right)^2 \cdot \frac{\tau}{\tau_{d,i}}}} \tag{8.19a}$$

the following cases are typically distinguished: free two-dimensional diffusion as observed on membranes

$$M_i(\tau) = \frac{1}{\left(1 + \frac{\tau}{\tau_{d,i}}\right)} \tag{8.19b}$$

and active transport with velocity v_i

$$M_i(\tau) = \exp\left(-\left(\frac{\tau \cdot v_i}{r_{w,s}}\right)^2\right) \tag{8.19c}$$

In natural membranes and inside living cells, the ideal case of Brownian diffusion often does not apply, because the movement of the particles is restricted or the mobility shows strong local changes (Schwille et al., 1999b; Wachsmuth et al., 2000). These deviations may be the result of the confinement of the particles within cellular compartments or lipid domains in membranes, but also the result of nonspecific interaction of the diffusing molecules with othermolecules orcellularstructures. Regarding diffusion within membranes, the complex composition might also lead to an altered topology and thus to a changed mobility. So far, the mathematical concept of anomalous subdiffusion has shown a good practical applicability to fit the FCS curves obtained in such nonideal environments, although the underlying phenomena are still far from being understood. Assuming anomalous diffusion, the mean square displacement is no longer directly proportional to time, but rather depends on t^α, such that $\tau/\tau_{d,i}$ in the previous equations has to be replaced by $(\tau/\tau_{anom,i})^\alpha$ (Federet al., 1996):

$$\left\langle r^2 \right\rangle \propto t^\alpha \iff \frac{\tau}{\tau_{d,i}} \longrightarrow \left(\frac{\tau}{\tau_{anom,i}}\right)^\alpha \quad with \quad \alpha < 1$$

No conventional diffusion constant can be defined in this case, and $\tau^\alpha_{anom,i} = r^2_{w,s}/\Gamma_i$, with Γ_i being a transport coefficient of the fractional time dimension.

Local confinement of the diffusion to organelles of a size comparable with that of the focal volume element requires even more sophisticated models. For example, trying to describe the particle motion within structures smaller than the focal volume element (e.g., neuronal dendrites) by the models mentioned earlier will fail completely. In such cases, complex nonanalytical solutions of the correlation function have to be considered to model the biological situation appropriately (Gennerich and Schild, 2000).

It is, to some extent, possible to distinguish the different molecular processes by the characteristic shape of the autocorrelation function. As a rule of thumb, the curves for anomalous subdiffusion decay more shallowly than those for free diffusion, whereas active transport leads to a steeperdecay. The lattermay also apply forattr active orr epulsive interactions between the particles under study.

Cross-Correlation Analysis

When performing an autocorrelation analysis, one effectively compares a measured signal with itself at some later time and looks for recurring patterns. In signal processing, it has been common practice for decades to correlate two different signals and thus, for example, get a measure for their mutual dependence. In fact, cross-correlation analysis is just a straightforward way to generalize the method described earlier. Looking out for common features of two independently measured signals, one not only removes unwanted artifacts introduced by the detector (e.g., the so-called *afterpulsing* of an APD orintensity fluctuations of the illumination source), but also provides much higher detection specificity. Moreover, the quantities to be correlated need not be both fluorescence traces. Rather, they can be any physical quantity that can be measured—or even calculated—sufficiently fast to reveal single-particle fluctuations. Indeed, it is also possible to record only the arrival times of individual photons, as done by the related method of time-correlated single photon counting, and to perform a software cross-correlation after the actual measurement has been finished to uncoverany dependencies.

Despite all the different possibilities to cross-correlate nearly any signal and parameter, two applications have proved to be especially effective (fig. 8.4). First, there is the spatial cross-correlation between fluorescence fluctuations measured in two separate volume elements and, thus, detection channels. Because independent molecules only correlate with themselves, this kind of correlation curve will reach its maximum not for small time lags, but rather for the average time a molecule needs to travel from one detection volume to the other. Thus, the flow or transport velocity of the fluorescent particles can be most comfortably determined (Brinkmeier et al., 1999).

The other even more prominent example of cross-correlation is the dual-color mode. As described briefly earlier, two spectrally different dyes are excited within the same detection element using two overlapping laser beams and separate detection pathways (Schwille et al., 1997a, Schwille, 2001). Dual-color cross-correlation is an extremely powerful tool to probe interactions between different molecular species, and a number of experiments have been carried out applying this technique to different kinds of reactions (for comparison between auto- and cross-correlation analysis to analyze molecular reactions, see fig. 8.5). As can be shown, the instrumental requirements for this mode can be dramatically simplified by two-photon excitation, where, as a result of the much broader two-photon excitation spectra, a single laserline can be used to access the spectrally distinct dyes.

Spatial Cross-Correlation

Using two instead of one volume element (fig. 8.4A), spatial cross-correlation relates signal fluctuations at two distinct spatial positions to each other. In this case, the spatiotemporal relationship between the fluorescence signals F_1 and F_2 from two volume elements—V_1 at position $\underline{r}$, and V_2 at position $\underline{r'}$—is considered by calculating the cross-correlation function

$$G_{CC}(\tau) = \frac{\langle \delta F_1(t, \underline{r}) \, \delta F_2(t + \tau, \underline{r'}) \rangle}{\langle F_1 \rangle \langle F_2 \rangle} \tag{8.20}$$

As previously described by Brinkmeier et al. (1999) for one-photon FCS, the following expression for G_x can be derived from equation 8.20:

$$G_{cc}(\tau) = \frac{\exp\left[-\frac{R^2}{r_w^2} \frac{1}{(1 + (\tau/\tau_d))} \left(\frac{\tau^2}{\tau_f^2} + 1 - 2\frac{\tau}{\tau_f} \cos\alpha \right) \right]}{\left[N\left(1 + \frac{\tau}{\tau_d}\right) \sqrt{1 + \frac{r_w^2 \tau}{z_w^2 \tau_d}} \right]} \tag{8.21}$$

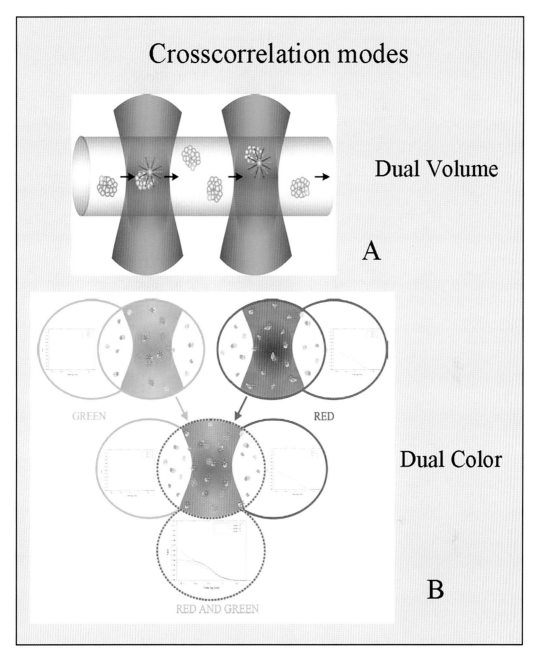

Figure 8.4 Two different cross-correlation modes have proved extremely effective during the past few years. (A) Spatial cross-correlation benefits from the fact that one molecule only correlates with itself, even after crossing the distance between the two measurement foci. Thus, the determination of flow velocities and directions becomes feasible. (B) An even more versatile technique is dual-color cross-correlation. In this case, the concomitant movement of two spectrally separable chromophores (e.g., green and red) can be monitored. Because the cross-correlation amplitude is directly proportional to the relative number of double-labeled species, this kind of cross-correlation analysis is mainly applied to association and dissociation reactions in which physical or chemical linkages are formed or cleaved.

Auto- vs. cross-correlation analysis

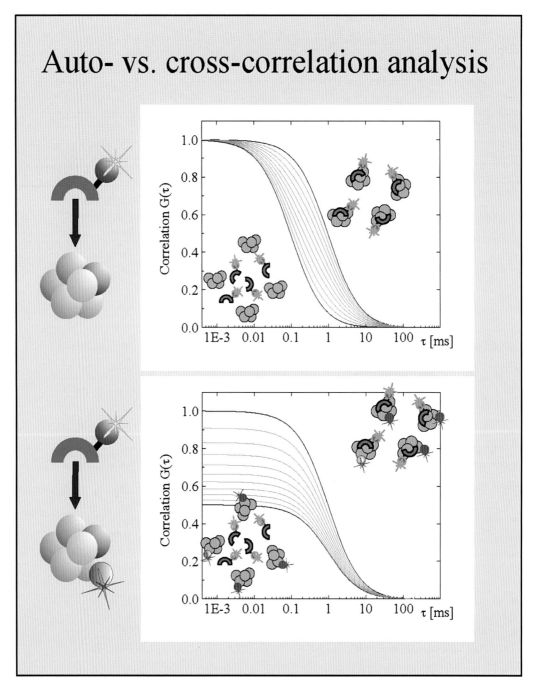

Figure 8.5 In principle, binding reactions can be monitored both by auto- and cross-correlation measurements. The binding of a small, labeled ligand to a bulky receptor, for example, will slow down its motion. Thus, the autocorrelation curves will be shifted toward longer timescales. Unfortunately, the increase in mass needs to be substantial to be seen on a logarithmic timescale. Therefore, it is much easier to monitor the association reaction of two differently labeled substances by dual-color cross-correlation. The amplitude of the cross-correlation curve is directly proportional to the number of double-labeled particles, irrespective of their mass or geometry.

Therefore, the detection efficiency for both channels has been assumed to be a three-dimensional Gaussian distribution $W_i(\underline{r})$, as in equation 8.3. The detection volumes are displaced by the distance R in the x/y plane and are assumed to be of equal size, with average diffusion-governed residence times of the fluorescent molecules of $\tau_d = r_w^2/4D$ (D being the diffusion coefficient). Because dual-volume fluorescence cross-correlation schemes are used most frequently to analyze directed flow or active transport processes, equation 8.21 also comprises a characteristic flow time τ_f in addition to the τ_d accounting for super imposed diffusion. The flow is assumed to be perpendicular to the z direction, enclosing the angle α with the connecting vector$\underline{R}$ between the two volume elements. Additionally, it is considered to be uniform inside the confocal detection volume and constant over the measurement time. The flow velocity (velocity vector $\underline{V}_{xy}$) can be derived easily from the flow time τ_f:

$$\tau_f = \left|\underline{R}\right|\Big/\left|\underline{V}_{xy}\right| = R\big/v \tag{8.22}$$

In contrast to fluctuation autocorrelation functions showing a steady decay, the cross-correlation function in equation 8.21 exhibits a local maximum at

$$\tau_{\max} = -\left(\tau_d + \frac{\tau_f^2}{2\tau_d}\left(\frac{R}{r_w}\right)^{-2}\right)$$
$$+ \sqrt{\left(\tau_d + \frac{\tau_f^2}{2\tau_d}\left(\frac{R}{r_w}\right)^{-2}\right)^2 + \tau_f^2\left(1 + \left(\frac{R}{r_w}\right)^{-2}\right) + 2\tau_d\tau_f \cos\alpha} \tag{8.23}$$

This value corresponds to the time that a molecule needs to travel the distance between the two centers of the focal spots. If diffusion can be neglected as a result of $\tau_f \gg \tau_d$, equation 8.21 can be simplified as follows:

$$G_{cc}(\tau) = \frac{1}{N}\exp\left[-\frac{R^2}{r_w^2}\left(\frac{\tau^2}{\tau_f^2} + 1 - 2\frac{\tau}{\tau_f}\cos\alpha\right)\right] \tag{8.24}$$

In this case, the maximum $\tau_{\max}$ approaches $\tau_f \cos\alpha$. If $\alpha = 0$ (i.e., if both detection volumes are aligned to the flow direction), the location of the maximum ($\tau_{\max} = \tau_f$) thus, in principle, allows a good estimation of the flow velocity for known R, without even fitting the cross-correlation curves.

In the case of only one volume element, the difference vector $\underline{R}$ is zero and with

$$\tau_f = r_w\big/v \tag{8.25}$$

the standard autocorrelation function for uniform translation with superimposed diffusion is obtained (Magde et al., 1978):

$$G_{ac}(\tau) = \left[N\left(1 + \frac{\tau}{\tau_d}\right)\sqrt{1 + \frac{r_w^2\tau}{z_w^2\tau_d}}\right]^{-1}\exp\left[-\left(\frac{\tau}{\tau_f}\right)^2\frac{1}{\left(1 + (\tau/\tau_d)\right)}\right] \tag{8.26}$$

The simplified expression for quasi-nondiffusion ($\tau_f \gg \tau_d$) is then identical to equation 8.19c:

$$G_{ac}(\tau) = \frac{1}{N} \exp\left(-\frac{\tau^2}{\tau_f^2}\right) \tag{8.27}$$

The previous correlation functions for diffusion and flow in one- and two-volume geometries were derived assuming one-photon excitation. In case of two-photon excitation, the characteristic diffusion time is $\tau_d = r_s^2/8D$, and the flow times τ_f forauto- and cross-correlation curve are defined as

$$\tau_{f,ac} = r_s/2v \tag{8.28}$$

$$\tau_{f,cc} = R/2v \tag{8.29}$$

In conclusion, although conventional autocorrelation is sufficient for describing the velocity of single molecules, dual-beam cross-correlation allows one to separate isotropic from anisotropic movements (e.g., to monitor flow directions).

Dual-Color Cross-Correlation

The use of dual-color fluorescence cross-correlation schemes was first proposed by Eigen and Rigler (1994) for ultrasensitive diagnostic applications. By cross-correlation, they improved the signal-to-background ratio in more or less complex reaction schemes with fluorescently labeled probes, in which, in favor of detecting extremely low concentrations of target molecules, a large excess of unbound probe ought to be suppressed. In a search for (infectious) particles at low concentrations, this can be achieved by designing probes of different color specifically directed to different binding sites of a common target molecule. By splitting the emission light into two separate channels and cross-correlating the two fluorescence signals arising from the same measurement volume, positive contributions to the cross-correlation function would then arise only if the different probes bind to their common target and thus form dual-labeled complexes.

Although diagnostic applications of FCS orr elated methods have indeed shown to be much improved with such a dual-color implementation (Bieschke et al., 2000), the most attractive and prominent applications of dual-color FCS to date are analyses of bimolecular association and dissociation processes in the nanomolar concentration range (Schwille et al., 1997a, 2001). To probe such interactions between two different molecular species, the dual-color cross-correlation mode has proved to be far superior to conventional FCS because of its inherent measurement selectivity. The underlying concept of using spectrally distinct dyes is to record coordinated spontaneous fluctuations specifically in the two respective detection channels, which unambiguously reflect the existence of physical or chemical linkages between the two fluorophores. Thus, dual-color cross-correlation analysis can be considered a dynamic analogue to colocalization techniques frequently used in fluorescence imaging, with the important advantage of considerably reduced false-positive signals resulting from the extremely low probability of coordinated fluctuations of two fully independent measurement parameters. Dual-color cross-correlation analysis has been applied to probe kinetics of irreversible association (Schwille et al., 1997a,b), the polymerization of DNA in polymerase chain reaction (Rigler et al., 1998), the early aggregation events in prion protein PrP^{Sc} rod formation (Bieschke and Schwille, 1998; Bieschke et al., 2000), and binding of DNA duplexes to transcription activator proteins (Rippe, 2000), and has proved particularly valuable in studying enzyme kinetics, in which the specific endonucleolytic cleavage of a double-labeled substrate by EcoRI (Kettling et al., 1998) and other nucleases (Koltermann et al., 1998) has been investigated. Recently, the first applications of dual-color cross-correlation to cell biological applications were demonstrated, mapping out endocytic pathways and dissociation of cholera toxin subunits aftercellularuptake (Bacia et al., 2002).

The theoretical concept can be outlined as follows. Generalizing equation 8.4, the fluctuating signals recorded in the two detection channels are given as

$$\delta F_1(t) = \int W_1(\underline{r}) \eta_1 \delta \left(C_1(\underline{r}, t) + C_{12}(\underline{r}, t) \right) dV$$

$$\delta F_2(t) = \int W_2(\underline{r}) \eta_2 \delta \left(C_2(\underline{r}, t) + C_{12}(\underline{r}, t) \right) dV$$

(8.30)

where $W_i(\underline{r})$ is the spatial intensity distribution of the fluorescence emission for species i (i = 1,2), $C_i(\underline{r}, t)$ is the concentration for the single-labeled species i (i = 1,2), and $C_{12}(\underline{r}, t)$ is the concentration of the double-labeled species.

The motion of the different components is supposed to be described by the term $M_i(\tau)$ (cf. eq. 8.19). Assuming ideal conditions, in which both channels have identical $W_i(\underline{r})$ (sharing the same effective volume element V_{eff}), fully separable emission spectra, and a negligible emission–absorption overlap integral, the following correlation curves can be derived (Schwille, 2001):

$$\text{Autocorrelation:} \quad G_i(\tau) = \frac{(\langle C_i \rangle M_i(\tau) + \langle C_{12} \rangle M_{12}(\tau))}{V_{eff}(\langle C_i \rangle + \langle C_{12} \rangle)^2} \quad with \quad i = 1, 2$$

$$\text{Cross-correlation:} \quad G_\times(\tau) = \frac{\langle C_{12} \rangle M_{12}(\tau)}{V_{eff}(\langle C_1 \rangle + \langle C_{12} \rangle)(\langle C_2 \rangle + \langle C_{12} \rangle)}$$

(8.31)

There is, however, one additional advantage of this technique in comparison with the autocorrelation mode: If there is no reaction-induced quenching or fluorescence enhancement, and no particle exchange in the sample, the amplitude of the cross-correlation function is directly proportional to the concentration of double-labeled particles. Knowing the amplitudes of the autocorrelation curves and, thus, the concentrations of both single-labeled species, the concentration $\langle C_{12} \rangle$ can be determined from equation 8.31 as follows:

$$\langle C_{12} \rangle = \frac{G_\times(0)}{G_1(0) \cdot G_2(0) \cdot V_{eff}}$$

(8.32)

A correct evaluation of concentrations measured by cross-correlation requires good prior knowledge of the system or a careful calibration procedure of the two basic parameters resulting from FCS analysis: the effective volume element and the lateral characteristic residence time.

EXPERIMENTAL REALIZATION

Standard Confocal Configurations

Autocorrelation Setup

The confocal FCS setup, which has already been mentioned briefly earlier, is illustrated schematically in figure 8.1. The exciting radiation provided by a laser beam is directed into a microscope objective via a dichroic mirror and is focused on the sample. Because the sample molecules are usually dissolved in aqueous solution, water immersion objectives with a high NA (ideally >0.9) are used. The fluorescence light from the sample is collected by the same objective and passed through the dichroic and the emission filter. The pinhole in the image plane (field aperture) blocks any fluorescence light not originating from the focal region, thus providing axial resolution. Afterward, the light is focused onto the detector, preferably an APD ora photomultiplierwith single-photon sensitivity (quantum efficiency of well more than 50%). For practical reasons of easier alignment, the pinhole is nowadays often replaced by the entrance slit of an optical fiber that is directly coupled to the detector. The losses induced by this fiber-coupled detection system can be suppressed to less than 10%, the adjustment gets much easierbecause only the fiberentr ance has to be positioned, and the "pinhole" size can be very easily varied by changing the fiber core diameters (typically between 25 and 200 μm).

Depending on the dye system, one may use argon or argon–krypton lasers, which allow the choice between multiple laser lines and thus provide a versatile system. Inexpensive alternatives are single-line HeNe lasers or even laser diodes. Depending on the beam quality and diameterof the laser, one might considerinser ting a beam expanderoran optical filter before the laser beam is coupled into the microscope. The larger the beam diameter and the higher the degree of filling of the objective's back aperture, the smaller the resulting focal volume. By overfilling the back aperture of the objective, the resolution limit for excitation is obtained, and a diffraction-limited focal volume of approximately 0.5 μm in diameter can be achieved in the visible spectral range.

The sample carrier depends on the respective application. For test measurements, a simple glass coverslip, on which a drop of sample solution can be placed, will usually be sufficient. More elaborate measurements, preventing evaporation or artifacts induced by air perturbations, can be performed in special sealable chambers or deep-well slides.

The signal-to-noise ratio of the FCS curves depends critically on the chosen filter system. First, there is the dichroic mirror serving basically as a wavelength-dependent beam splitter. It deflects excitation light and transmits the red-shifted fluorescence, but its blocking efficiency for the scattered laser light is usually very poor, less than OD3 (three orders of magnitude). Therefore, one or more additional selective emission filters are required. Bandpass filters adapted to the emission properties of the observed dye are recommended to guarantee high enough detection specificity at sufficient photon yields. Bandwidths of 30 to 50 nm allow suppression of both scattered laser light (Rayleigh scattering) and Raman scattering (usually of the solvent, because the concentration of the analyte is too low), which in wateris red-shifted 3380 cm^{-1} relative to the laser line.

The fluorescence signal from the detector is, in most setups, fed directly into a control PC where it is autocorrelated quasi online by a hardware correlator board (e.g., ALV-5000; ALV, Langen, Germany) for measurement times typically between 10 and 120 s. Both the fluorescence signal and the calculated curve can be displayed simultaneously on the monitorto facilitate adjustment and control of the setup. Data can be saved in ASCII format and imported in any math program for further analysis. Fitting routines using the Levenberg–Marquardt nonlinear least-square routine have proved very efficient.

Two-Volume Cross-Correlation

To create two parallel excitation beams for the dual-beam experiments, the excitation light is split and reunited slightly displaced by two polarizing beam splitters (Brinkmeier et al., 1999; Dittrich and Schwille, 2002). To equalize the intensity of both laser beams, a rotatable $\lambda/2$ plate (adapted either to visible or infrared light) is inserted in the beam path. Both parallel beams are now coupled into the microscope in the same way as for single-volume FCS, reflected by a dichroic mirror and focused by the same objective. The fluorescence light is again collected by the objective and, afterpassing an interference filterforeffi-cient suppression of Raman scattering light, is finally focused onto the apertures of two bundled optical fibers with a diameter of 100 μm and a centerdistance of approximately 120 μm. The use of optical fibers in the detection path is particularly advisable in this cor-relation mode, because the adjustment of two independent pinholes so close to each other would be extremely demanding. The fluorescence detected through each of the fibers is fed into a separate detector. The photon count signals are again correlated quasi online by the PC card.

Dual-Color Cross-Correlation

The experimental realization of a dual-color cross-correlation setup is technically rather challenging, because it requires exact spatial superposition of the two detection volumes (see fig. 8.6 for a comparison between different dual-color setups). In one-photon setups, in which two laserbeams have to be combined into the same focal spot of the objective, the alignment is highly critical and has to be controlled indirectly by FCS calibration measurements (Schwille et al., 1997a) or directly by measuring the illumination profiles with a specifically designed focus scanner(Rigleret al., 1998) offered by the designers of a commercial cross-correlation prototype (Zeiss ConfoCor, Jena, Germany). Although the determination of the full illumination point-spread functions at different colors is very attractive from an optical point of view, the calibration measurements that can be performed in every FCS setup are adequate for practical purposes to guarantee a sufficient overlap of the measurement volumes. The idea is to illuminate a dye solution with comparable excitation cross-sections at both wavelengths successively with both lasers without changing the detection pathway. If the amplitudes and decay times of the measured correlation functions are equal or at least similar, with deviations less than 10%, the alignment is close to perfect and sensible cross-correlation measurements are possible. De novo alignment of the two foci works in an equivalent way: First, the detection path is adjusted with one illuminating beam, then this first beam is blocked and the second beam is aligned to yield the same correlation curve without changing the detection optics. The procedure is harder to perform, but also works if the excitation cross-sections differ considerably.

To provide good spectral selectivity, the emission of the dyes and, even more crucial, the transmission spectra of the filters ideally should hardly overlap to minimize cross-talk.

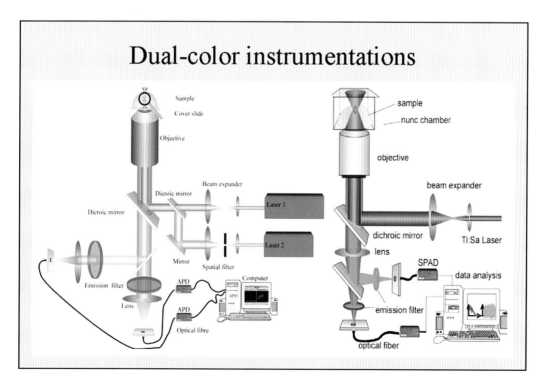

Figure 8.6 The "traditional" cross-correlation setup requires superimposing two laser lines for simultaneous excitation of the two different chromophores (left). A more recent approach benefits from the nonlinearity of two-photon excitation, which makes simultaneous excitation of two (carefully selected) spectrally different dyes by one pulsed infrared laser possible (right).

As a rule of thumb, the transmission maxima of the bandpass filters should be at least 100 nm apart, if possible. A critical feature for dual-color cross-correlation with such largely different emission spectra is, however, the chromatic aberration of the microscope lens. If only one pinhole is used and the wavelength selection is performed behind it, there is no possibility to correct against chromatic displacements of the focal spots in the z direction at different wavelengths by adjustment of the detection optics. The objective's correction properties get very critical in this case (Schwille et al., 1997a). An alternative setup with two pinholes, in which the wavelength selection can be additionally performed in a parallel beam path as used in commercial units (Kettling et al., 1998), offers slightly more flexibility although it cannot fully eliminate bad chromatic correction. Additionally, pinholes of different sizes are, in principle, required for the ideal case of equal effective detection volumes to cancel out the differences induced by the wavelength-dependent detection point-spread functions.

An interesting alternative to avoid alignment efforts for the illuminating laser beams is to use a laserin multiline mode. This has been successfully applied by Winkler et al. (1999). However, to achieve focal spots of the same size, selective filters that reduce the diameter of the green beam have to be used. Additionally, it has to be guaranteed that the microscope objective is perfectly corrected against chromatic aberration.

Two-Photon Auto- and Cross-Correlation Setup

In principle, two-photon FCS is rather easy to perform if one possesses a suitable laser system. Because the two-photon excitation process requires a very high instantaneous photon flux, the use of pulsed (best: femtosecond) lasersystems is recommendable to limit the overall intensities that have to be applied. The two-photon setup is, in principle, almost analogous to the one-photon setups (fig. 8.1), with the only difference being that the primary dichroic beam splitter (710DCSPXR; AHF Analysentechnik, Tübingen, Germany) before the objective lens is actually a short- rather than a long-pass filter, because the emission signal is shifted to shorter (visible) wavelengths compared with the excitation (infrared) beam. The two-photon experiments in our own lab are carried out with 80-MHz, 100-fs pulsed excitation of a mode-locked tunable Spectra Physics (Mountain View, California) Tsunami Titanium-Sapphire laser, using additional filters for efficient suppression of infrared excitation light. As a result of the absence of any Raman scattering in the spectral vicinity of the emission signal, very broad bandpass filters with very little signal loss can be used (e.g., D600/200; AHF Analysentechnik). The use of pinholes is usually not required because of the inherent spatial sectioning of the two-photon process, providing a square dependence of the excitation efficiency from the illumination intensity. However, the APDs that are commonly used in FCS (SPCM; Perkin Elmer, Waltham, Massachusetts) exhibit an active surface of only about 150 μm in diameterand practically do act as a pinhole. If wide-field detection is required, one should preferably work with sensitive photomultipliers that usually provide larger active surfaces. An advantage of missing pinholes is certainly the ease in alignment, because pinhole adjustment is often the most critical part of FCS setups. For bettercompatibility with one-photon excitation, ourgr oup usually works with the same pinhole-equivalent optical fibers for both excitation modes. It can be shown that large fiber diameters do not introduce additional restrictions to the effective two-photon probe volume, which is by itself much smallerthan in the one-photon case (Schwille et al., 1999a, Heinze et al., 2002).

The clearadvantages of two-photon excitation come into play fordual- ormulticolor applications such as cross-correlation. The broad excitation spectra for two-photon excitation allow one to access multiple dyes with largely different emission by one single illuminating laser. Although the concept of multicolor two-photon excitation has previously been utilized in confocal imaging applications (Xu et al., 1996), its suitability for single-molecule-based techniques, requiring the detection of two labels on a single molecule within the limited time frame of a molecule's dwelling in the focal spot, has so far not been demonstrated. Clearly, the choice of a proper dye system is crucial for such applications, because the chosen dyes should not only exhibit similarexcitation and distinct emission spectra, but also comparable photobleaching quantum yields at a given wavelength and intensity.

The setup for two-photon cross-correlation analysis (fig. 8.6) is only slightly modified compared with the two-photon autocorrelation case by inserting a dichroic beam splitter and anotherfiberentr ance forthe second APD. Although the same detection systems as for one-photon cross-correlation (fig. 8.6) are often used for better compatibility, the excitation and detection pathways can, in principle, be significantly simplified compared with confocal cross-correlation FCS geometries, because only one laser line is used for excitation and no pinholes are, in principle, required in the detection pathway. The critical task in two-photon dual-color cross-correlation is to find a system of suitable dyes that not only exhibit minimal spectral overlap in their emission characteristics, but also tolerate the same excitation intensities without considerable photobleaching. Thus, an optimization procedure of both parameters—wavelength and intensity—has to be performed to find a compromise of optical conditions forwhich the dyes exhibit similarper formance. Forthat purpose, the excitation wavelength is usually scanned throughout the range of the tunable infrared laser, and the fluorescence emission yield is recorded for both dyes independently.

Commercial Fluorescence Correlation Spectroscopy Instruments

Recently, a commercial instrument (ConfoCor II, C. Zeiss, Jena, Germany) has been made available that combines fluorescence microscopy, confocal laser scanning microscopy, and one-photon FCS in one setup. Other FCS modules from various optical or biotech companies are apparently awaiting to be released or are already available as custom-made products. In particular, for cell biological applications, a combination between imaging and FCS in a fully functional standard fluorescence or scanning microscope unit is very handy, because measurements can be taken at defined positions within the cell (Bacia et al., 2002). Although the auto- and cross-correlation of images taken by standard or two-photon scanning microscopy has been performed successfully to investigate very slow processes on cell surfaces (Wiseman et al., 2000), there is a strong need to increase the temporal resolution in these or similar applications that can, so far, only be achieved by limiting the observation volume (e.g., by parking the beam at defined positions after having acquired the images). To date, no two-photon equivalent of such a scanning FCS instrument can be commercially purchased despite the mutual dependence of the techniques of two-photon imaging and two-photon FCS, and the complementary information to be gained (Diaspro et al., 2001). However, because the technical requirements for a two-photon combination system are less restrictive than for one-photon systems as a result of the irrelevance of pinholes and the possibility of multicolorapplications with only a single illuminating laserline, we expect such a unit to be on the market in due time.

Appropriate Dye Systems

Standard One-Photon Dyes

Among the first dyes tested for their suitability in FCS measurements were fluorophores already known from fluorescence spectroscopy and microscopy (e.g., fluorescein), and laser dyes. However, the requirements imposed on a fluorophore in single-molecule techniques are a high quantum efficiency and large absorption cross-section, as in traditional spectroscopy. The most crucial property is photostability, which must be high enough for the dye to withstand the enormous powerin the laserfocus, which may be on the orderof several 100 kW/cm^2. Fluorescein, in standard applications being considered a rather good fluorophore, photobleaches at comparably low excitation powers, giving rise to unwanted artifacts in FCS and related methods.

Forsingle-molecule applications, it is thus now mostly substituted by the specifically engineered dye Alexa 488 (Molecular Probes, Eugene, Oregon) with similar absorption and emission characteristics and enhanced photostability. The Alexa dye family exhibits a very large selection of different colors. The absorption maxima range from 350 to 750 nm, covering more than the visible spectrum (cf. fig. 8.7). Other suitable dyes are rhodamines such as Rhodamine Green, tetramethylrhodamine (TMR), Rhodamine B and 6G, and cyanines (Cy2, Cy3, Cy5). It is also important to note that one- and two-photon properties of dyes can be extremely different. Although Cy5 is the red partner in most one-photon cross-correlation and FRET experiments, it is completely unsuitable for two-photon excitation and must be replaced by, for example, Alexa 633.

The majorcommon disadvantage of all these synthetic dyes consists of the fact that the biological system has to be specifically labeled prior to the experiments. This is often difficult and tedious, especially for intracellular measurements, in which the protein of interest has to be loaded back into the cell after labeling. An exciting alternative to such strategies is provided by the existence of fluorescent proteins—in other words, probes that can be cloned and genetically fused to the interesting target molecules. Today, there are two kinds of autofluorescent proteins in particular that are commonly used for such applications: GFP (Tsien, 1998) and its recently discovered homologue DsRed. Originally, GFP was produced by the jellyfish *Aequorea Victoria* and DsRed by the coral *Discosoma,* but both proteins and their mutants can be expressed in a manifold of different cells and organisms, because the expression does not require any cofactors. As a result of their large excitation cross-sections and quantum yields, and high photostabilities, both dyes are very well suited for single-molecule studies, as has been demonstrated in many applications to date (Dickson et al., 1997; Kubitscheck et al., 2000), and are regularly used for FCS measurements in vitro (Haupts et al., 1998; Schwille et al., 2000; Heikal et al., 2001; Jung et al., 2001; Malvezzi-Campeggi et al., 2001) and in vivo (Brock et al., 1999; Schwille et al., 1999a; Chen et al., 2002).

How to Choose Dyes for Two-Photon Excitation

Because two-photon excitation is, photophysically, quite a unique process involving different parity selection rules, the excitation spectra of fluorophores can differ considerably from their corresponding one-photon spectra even if the emission spectra are hardly distinguishable, as in most cases. Unfortunately, determining the two-photon excitation spectra of different dyes turns out to be quite difficult. The experimentally determined two-photon excitation spectra very often exhibit a significant blue shift relative to the one-photon spectra, indicating that the photophysical transition first occurs to a higher excited state. After internal relaxation, the system finally returns to the same excited state as for the one-photon process, and the emission properties are the same. This complicated, formally symmetry-forbidden absorption process in conjunction with the pulsed excitation may be responsible forthe fact that the maximum numberof photons that a dye molecule emits before undergoing photodestruction is often significantly less than that for the classic quantum mechanically allowed processes. Moreover, as a result of the quadratic intensity dependence, the range of applicable powers is much more narrow.

To determine the appropriate wavelength that maximizes the fluorescence yield of a certain dye species under two-photon excitation, one of the most crucial parameters forFCS, the average photon yield permolecule persecond, often denoted η, has to be considered (Schwille et al., 1999a). The value of η can be most comfortably determined from the measured autocorrelation curves. The amplitude of the normalized correlation curve G(0) is the inverse of the average number of molecules <N> simultaneously present

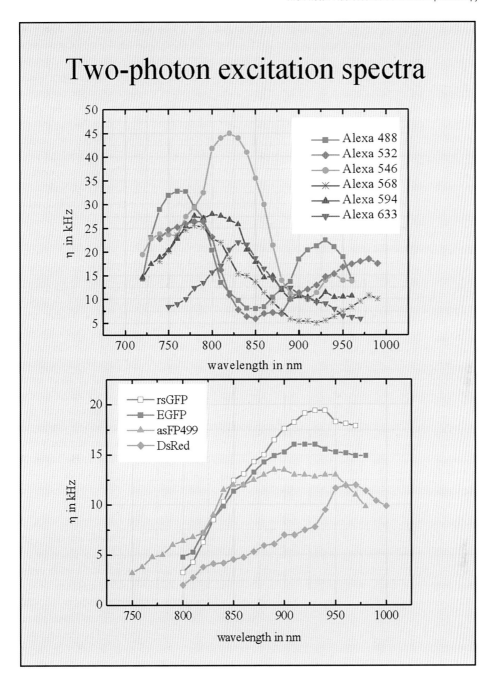

Figure 8.7 In contrast to one-photon excitation, there are no standard spectrophotometers for recording two-photon excitation spectra. Therefore, the data shown here were determined by FCS, giving the maximum number of detected photons per molecule and second (η) that could be achieved for excitation at the given wavelength. (A) Spectra for six different Alexa dyes (Molecular Probes, Carlsbad, California). (B) Spectra for four autofluorescent proteins.

in the effective volume element. Thus, the number of photons per molecule per second η is just the product of G(0) with the average count rate $<F>$ recorded by the detector. The determination of $\eta = G(0) < F >$ is important in two respects: (1) it yields absolute values of the detected fluorescence per molecule that are not affected by local changes in the fluorophore concentration and (2) it is closely related to the signal-to-noise ratio for the crucial fast timescale of the correlation curve (Koppel, 1974; Qian and Elson, 1991; Kask et al., 1997) and thereby provides a criterion for signal quality. The value of η depends directly on excitation rates and is thus also limited by the maximum allowable intensity values.

To determine the proper excitation wavelength of a certain dye for two-photon FCS, the most straightforward way to proceed is to scan the excitation wavelength in 10- to 20-nm steps throughout the appropriate regime (usually, between 700 and 1000 nm), record fluorescence correlation curves, and check the photon emission rates per molecule values η accordingly. In so doing, special care must be taken to monitor accurately both intensity and pulse width, but also the transmission characteristics of the optical system (in particular, the microscope objective), because they usually worsen as longer infrared wavelength are approached. To guarantee a fair comparison of two-photon excitation efficiencies throughout the covered wavelength range, the power of the pump laser has to be regulated to provide equivalent excitation features for each data point. Alternatively, another routine can be applied that is a little more strenuous on the experimenter, but provides an even better overview over optimum excitation conditions. Here, the intensity is varied independently for each wavelength until the maximum η for a certain wavelength is reached. Interestingly, there is a significant dependence of photobleaching on the excitation wavelength that limits the maximum values that can be achieved particularly at shorter (blue-shifted) wavelengths. Therefore, the η spectra obtained by the second routine often show slightly enhanced values in the long-wavelength range. Because FCS, as all single-molecule techniques, always strives for optimum photon count rates, these *maximum* η spectra are of higher practical value. Figure 8.7 exemplifies such spectra for the most common dyes used in our laboratory. In can be seen that the Alexa family (MolecularPr obes), as for one-photon excitation, can be used very well for two-photon applications. Fortunately, also the clonal probes GFP and DsRed are well excitable within the standard two-photon range. Recently, even two-photon FCS measurements of native proteins containing a multitude of tryptophan residues have been achieved (Lippitz et al., 2002). These results are very promising because two-photon excitation of ultraviolet transitions provides a comfortable alternative to standard ultraviolet illumination, which is always problematic in terms of transmission efficiency of standard microscope optics. However, the strong photobleaching propensity of tryptophan appears to be a severe impediment for measurements on the single-molecule level, allowing only very low η values (i.e., signal-to-noise ratios), even for proteins with high tryptophan content.

An interesting observation is that most of the two-photon spectra exhibit a very pronounced blue shift of their excitation maxima compared with the one-photon spectra. This can easily be evidenced for the dye Alexa 488, which is usually excited around 488 nm (hence its name) and now clearly shows a more pronounced peak at 760 nm (corresponding to 380 nm for one-photon excitation). As mentioned earlier, this dye-specific blue shift can, however, be used to excite two dyes simultaneously with different emission characteristics to perform two-photon dual-color cross-correlation experiments. Only one laser line is required for excitation, and the inherent axial resolution renders pinholes redundant, so that adjustment is greatly simplified.

APPLICATIONS

Concentration and Aggregation Measurements

As outlined earlier, the autocorrelation amplitude G(0) is, by definition, the normalized variance of the fluctuating fluorescence signal δ F(t). Using this definition and combining it with Poissonian statistics, which applies for ensembles of very few molecules, it becomes evident that G(0) equals the reciprocal number of molecules N in the effective volume element V_{eff} (eq. 8.14). Thus, in principle, absolute local concentrations can be determined very precisely if the size of the confocal volume element V_{eff} is known (Thompson, 1991; Eigen and Rigler, 1994; Schwille et al., 1997a). This, however, is not as easy as it may sound. Because FCS experiments are restricted to nanomolar concentrations, a variety of complications has to be dealt with. First, most protein molecules tend to adhere to charged glass or plastic surfaces like coverslips or the walls of reaction chambers. For more concentrated solutions, this effect passes unnoticed. In the nanomolar and subnanomolar range, a significant amount of the molecules contained in the sample may simply disappear that way, unless specially coated containers are used. Second, photophysical damage may reduce the number of detected molecules even further, so that an overall accuracy for N of more than 20% to 30% is hard to achieve in vitro. There is an alternative approach for calibrating the excitation volume that does not rely on the exact knowledge of one concentration for comparison. If the diffusion coefficient D of one molecular species has been very precisely determined by some other technique, it is possible to calibrate $r_{w(s)}$ and $z_{w(s)}$ by using the fit parameters for τ_d and r_w/z_w from the autocorrelation curve.

The determination of relative local concentrations is much more exact, because it does not depend on the size of the detection volume. When performing intracellular measurements, this can be useful to control expression levels of the intrinsically fluorescent proteins GFP or DsRed. Of course, it is also possible to load a cell with fluorescently labeled reagents (e.g., by microinjection or electroporation) and control the increase in concentration by FCS.

Another parameter of crucial importance is the molecular brightness η, which also was mentioned earlier. This parameter is calculated by dividing the average fluorescence count rate by the number of molecules within the illuminated region:

$$\eta = \frac{\langle F(t) \rangle}{N} = \langle F(t) \rangle \cdot G(0)$$

Although it is mainly used to quantify the performance of fluorescent probes or the quality of a particular setup, this brightness parameter is much more versatile: Knowing η, it is also possible to rate fluorescence quenching resulting from a changed environment of the chromophore, or the fluorescence enhancement of single particles resulting from aggregation effects (Chen et al., 2000).

Calculating not only the variance (i.e., the second moment), but also higher moments of the fluorescence signal can provide interesting additional information. Higher order autocorrelation analysis, as introduced by Palmer and Thompson (1989), can be used to identify subpopulations differing only by their respective molecular brightness values. Although most chemical reactions affect the fluorescence quantum efficiency, only very few cause mobility changes large enough for a separation by standard autocorrelation analysis. Despite its high susceptibility to noise, this kind of brightness distribution analysis has become the method of choice formany applications in biomolecularscr eening (Schaertl et al., 2000; Klumpp et al., 2001). Only recently has it been more and more substituted by photon-counting histogram analysis—a static method complementary to FCS that can be performed using basically the same setup. Fluorescence brightness histogram analysis was introduced almost simultaneously with two-photon (Chen et al., 1999) and one-photon excitation Fluorescence Intensity Distribution Analysis (FIDA) (Kask et al., 1999). Because the exact knowledge of the excitation intensity distribution is much more crucial for this method than for FCS, the two-photon variant, guaranteeing a Gaussian–Lorentzian intensity profile if pinholes are fully omitted, renders data analysis much simpler, whereas Kask et al. (1999) introduced the concept of generating functions for analysis of their one-photon FIDA data, a procedure that might appear more cumbersome for standard FCS users. Nevertheless, the practical use of both techniques could be demonstrated, because many fluorescent assays involving binding or even aggregation of probes do introduce severe changes in the brightness of fluorescent species, an effect that standard FCS is often unable to resolve as a result of the lack of knowledge about molecularquenching effects.

Mobility Analysis

Diffusion in Solution and Cellular Systems

The determination of mobility-related parameters of biologically relevant molecules is one of the primary goals of FCS analysis in aqueous solution, and one for which it is especially suited. The submicrometer spatial resolution also makes it a useful technique for intracellular measurements, because cellular compartments can be specifically addressed (fig. 8.8). When trying to figure out how biological processes work in detail, it is essential to distinguish between diffusion or active transport, anomalous subdiffusion, or even convection. Signal transduction or metabolic regulatory pathways can only be fully understood after the underlying transport mechanisms are revealed and well characterized. Because supporters of the FCS technique initially had to struggle with background suppression and unwanted photobleaching, the FRAP (fluorescence recovery after photobleaching) method has long been preferred for in vivo motility studies. The latter requires much higher dye concentrations and is thus less liable to suffer from the autofluorescence background (for a detailed discussion of FCS vs. FRAP, see Petersen and Elson [1986]). However, the temporal resolution is mainly limited to the millisecond timescale, so that FCS offers both higher dynamic performance and increased sensitivity. Low chromophore concentrations and laserpowersuffice forthis equilibrium measurement, which is definitely less stressful forthe cells underinvestigation.

However, because of this inherent sensitivity, the proper selection of dyes is crucial, because some synthetic dyes in particular seem to interfere with the monitored mechanism or introduce their own dynamics to the system. Many standard dyes, such as rhodamines and cyanines, are highly lipophilic and tend to associate with intracellular membranes, inducing severe deviations from free diffusion in the cytosol if they are not sufficiently shielded by their often much larger target molecule. In figure 8.8, a selection of intracellular measurements is depicted. It is quite obvious that the mobility of the fluorophores strongly depends on the cellular environment. The half-value decay time of the curves, which allows for a crude estimation of the mobility of the labeled molecules, varies by several orders of magnitude between the small dye in buffer and the large labeled receptor on the plasma membrane. The corresponding diffusion coefficients that can be calculated from the decay times using equations 8.13 and 8.13a range from 3×10^{-6} cm^2/s forthe free dye to 10^{-10} cm^2/s for the bulky receptor.

Anomalous Diffusion

It is known from other techniques such as FRAP or particle tracking that membrane-bound receptors often exhibit anomalous subdiffusion. As outlined earlier, the mean square displacement of the molecules is, in this case, not proportional to the measurement time, but exhibits a fractal time dependence (Bouchaud and Georges, 1990; Saxton, 1993). Among possible reasons for such a non-Brownian behavior are environmental heterogeneities (e.g., different lipid phases or rafts), nonspecific interactions with other particles, or local confinement. The underlying principles for nonhomogeneous diffusion can only be speculated on. It is quite astonishing that not only membrane-bound proteins exhibit such a strange motility, but also the lipids themselves show deviations from the normal homogeneous diffusion in natural cell membranes (Schwille et al., 1999b). To exclude any label-induced artifacts, measurements in single-phase model membranes of giant unilamellar vesicles have been performed. In these idealized model systems, the lipids indeed show a perfectly homogeneous two-dimensional diffusion. An altered membrane composition of the model

Intracellular FCS applications

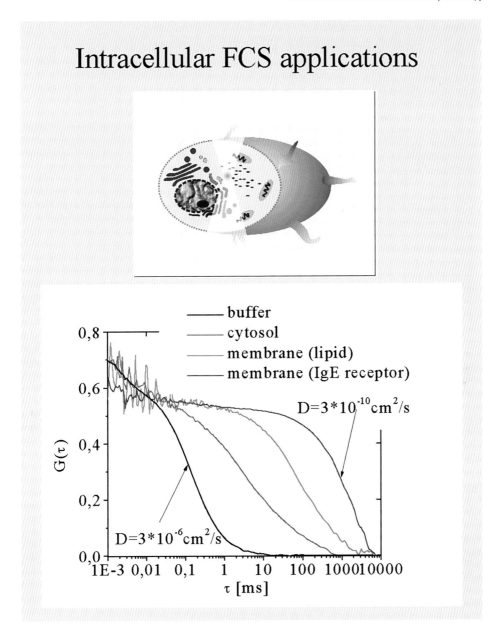

Figure 8.8 Fluorescence correlation spectroscopy can be measured at any position within the cell, with a spatial precision of ≈0.5 μm. Various autocorrelation curves recorded in different cellular compartments demonstrate the enormous difference in motility between free diffusion of dye in buffer solution and intracellular measurements. The slowest diffusion of single molecules so far observed by FCS was a fluorescently labeled immunoglobulin E (IgE) receptor in the cell membrane, with average residence times in the focal volume of about a second.

membranes, resulting in a distinctive phase separation, however, led to multiple diffusion coefficients for the labeled lipids (Korlach et al., 1999). Therefore, one might assume that environmental heterogeneities play a key role in the anomalous subdiffusion phenomenon.

The term *anomalous subdiffusion* is not restricted to intracellular phenomena. Quite to the contrary, it may be applied to any random walk phenomenon that cannot be described by homogeneous Brownian diffusion and forwhich no otherexplanation holds. Alternative sources of such nonnormal diffusion behavior are, for example, gradual (time-dependent) changes in the diffusion coefficient (e.g., during protein folding or unfolding). Because of the complex nature of biological systems, it often requires evidence from various measurements to rule out clearly any artifacts before motility can be termed truly anomalous. The cytosolic curve in figure 8.8 is an example of anomalous diffusion caused by the interactions of the dye with the cellular environment. The deviation from normal diffusion can be seen clearly in the more shallow decay of this curve in comparison with the two neighboring ones of which at least the solution-based curve exhibits true three-dimensional Brownian diffusion. The increased viscosity of the cytosol compared with the buffer solution mainly results in a parallel shift of the autocorrelation curve toward longer times. Here, however, the shape of the curve is changed as well, such that interactions with intracellular membranes must be suspected (Schwille et al., 1999a). If the dye is replaced by the rather inert GFP, normal diffusion in the cytoplasm can be observed (Dittrich and Schwille, 2001). The diffusion coefficient is reduced three to four times relative to buffer measurements, which corresponds to the increased viscosity. As has been shown recently, the diffusion of EGFP in the nucleus suddenly becomes anomalous as well, although it is perfectly homogeneous in the cytosol (Wachsmuth et al., 2000). This may justify the assumption that the topology of cellular organelles as well as molecular crowding have a nonnegligible effect on the diffusional properties.

Active Transport in Cells

When looking on a molecular scale, even such an apparently simple organism like a single cell is a highly complicated system. The many different substances needed for the cell's metabolism have to be exactly at the right place at the right time for the cell to sustain its biological function. There are various possibilities for a molecule to be delivered from the place it is produced to where it is needed. Obviously, the simplest means of transportation is diffusion along a concentration gradient. No complicated mechanism is required for this rather unspecific way. However, such a substance must be rather ubiquitous within the cell, because the majority of the molecules are serving only to maintain the concentration gradient. Thus, it is generally much more efficient for the cell to use directed transport along an internal (tubular) network, resembling the coverage of a country with highways. Much work is currently devoted to characterizing the endocytic and secretory trafficking pathways of proteins or vesicles, or receptor internalization in signal transduction. Fluorescence correlation spectroscopy is well suited for distinguishing between diffusive and oriented motions of dilute molecules on a submillisecond timescale. For very slow processes on timescales of several seconds to minutes, time-lapse imaging has been used successfully. Fascinating images were have been recorded showing GFP-labeled proteins being transported through the various compartments of the secretory pathway (Hirschberg et al., 1998).

By 1978, the theoretical background was laid for analyzing different active transport processes such as laminar flow or a combination of directed plug flow and diffusion (Magde et al., 1978). For the latter, the diffusion autocorrelation function must be expanded by an exponential term (eq. 8.26). Although there are examples describing FCS analysis of a laminarflow obeying Hagen–Poiseuille's law, orthe uniform electro-osmotic flow, measurements of active transport phenomena in living cells are still scarce.

As an example of flow analysis by two-photon FCS, the mobility of GFP targeted to the plastid stroma in tubular structures interconnecting separate plastids in vascularplants was investigated (Köhleret al., 2000). Because exchange of GFP between different interconnected plastids was obviously possible, the major task of FCS analysis consisted of revealing the nature of the extremely slow transport phenomenon. For this study, two-photon excitation appeared to be much better suited than one-photon FCS because of its improved background suppression and increased toleration by plant cells. Positioning the laserfocus directly on a plastid-connecting tubule, it could be shown that most of the GFP was contained within vesicular structures. Thus, it is indeed actively transported, although in rather large batches, and is superimposed on anomalous diffusion.

Molecular Interactions

Binding and Unbinding

In the previous sections, the potential of FCS to determine mobility coefficients with high spatial resolution was emphasized. The majority of FCS applications, however, has been concerned with reactions occurring on a much larger timescale than the short time window given by the residence time of a molecule within the measurement volume. Such reactions can be easily followed by continuous FCS monitoring (i.e., successive autocorrelation curves are recorded with the shortest integration time possible). From these curves, changes in accessible parameters such as the diffusion time, molecular brightness, or concentration can be determined. An ideal reaction system consists of one small, labeled probe or ligand, and a comparatively large, nonfluorescent target molecule. To induce significant changes in the diffusion time (fig. 8.5), the mass ratio should preferably be more than an order of magnitude, as a result of the approximate cube root dependence of the diffusion coefficient on molecularmass (discussed later). Analyzing the system is especially easy if only one fluorescent species (e.g., fully free or fully bound probe molecules) is present at the start and the end of the association process. During the reaction, the percentage of the complex will increase continuously, until chemical equilibrium is reached or one species has completely reacted. Therefore, it is possible to determine the diffusion time of the labeled probe from a reference measurement. The other curves can then be fitted with two diffusing components, fixing the diffusion time of the smaller component to the value determined beforehand (Schwille et al., 1997b). Kinjo and Rigler (1995) first developed this approach to measuring the binding kinetics of short fluorescent DNA probes to a longer DNA target; the same measurement scheme was further applied to a comparison of hybridization kinetics of DNA probes with different binding sites to folded RNA (Schwille et al., 1996). Mobility analysis has meanwhile proved to be an extremely powerful tool for a large variety of ligand–receptorsystems (Schüleret al., 1999; Wohland et al., 1999; Margeat et al., 2001). So far, no two-photon applications of this principle have been reported, which may be the result of the decreased volume elements under two-photon excitation (Schwille et al., 1999a): The smaller the probe volume and the shorter the average diffusion times, the smaller the dynamic range in distinguishing different mobility coefficients represented in a single correlation curve.

Another situation in which the mobility of molecules can be altered dramatically is when membrane binding is involved. When a molecule adheres to the membrane itself or to membrane proteins such as receptors, both the diffusion type (two-dimensional instead of three-dimensional) and the characteristic timescales are changed. Recently, first applications in cell cultures were reported, investigating the insertion of GFP-labeled palmitoylated Lyn analogs (Schwille et al., 1999a) or epidermal growth factor receptor (EGFR)–GFP fusion protein (Brock et al., 1999) into the plasma membrane, and the binding and displacement of proinsulin C-peptide (Rigler et al., 1999), and epidermal growth factor (EGF) (Pramanik et al., 2001) to and from cell membranes.

Cross-Correlation to Follow Enzyme Kinetics

The obvious solution to the dilemma of lacking specificity in resolving different species (e.g., reaction educts, intermediates, or products) is dual-color cross-correlation. Although the required setup, in particular for one-photon excitation, is considerably more elaborate because of the second required laser and detector, and also more difficult to adjust, dual-color cross-correlation is much more versatile, and data analysis can be significantly simplified (Schwille, 2001). In contrast to the autocorrelation applications described earlier, which focused mostly on analysis of the functional form of the correlation curves, the most important parameter is now simply the cross-correlation amplitude. As outlined in the theory section, this amplitude is a direct measure of the concentration of double-labeled particles diffusing through the focal volume. In principle, one simply focuses on the occurrence of coincident fluctuations in the two emission channels, induced by concerted motion of spectrally distinguishable labels. All kinds of reactions leading either to a separation oran association of the two labeled species can thus be monitored. Underideal conditions (i.e., no cross-talk between the detectors), the amplitude $G_x(0)$ is zero unless double-labeled particles are present in the sample. This makes feasible fast yes-or-no decisions based on this parameter. The tremendously enhanced detection specificity was first shown in association reactions of two small complementary DNA oligonucleotides carrying green and red labels, respectively (Schwille et al., 1997a). It was proved that the absolute concentration of the dimer could be directly monitored by cross-correlation.

On the basis of dual-color cross-correlation, Kettling et al. (1998) demonstrated a biologically very attractive approach to characterize enzyme kinetics at extremely low enzyme concentrations (> 1.6 pM). The assay understudy was the cleavage of double-labeled double-stranded DNA (dsDNA) by *Eco*RI restriction endonuclease. The inherently simple information provided by this technique, where yes/no decisions about enzyme activity could be made just by recording the cross-correlation amplitude, renders it very attractive forfast screening applications. The acquisition time persample could be reduced to less than a second, without substantially impairing reliability (Koltermann et al., 1998).

To demonstrate the powerof two-photon excitation forthese applications, the same biochemical assay based on cross-correlation has been subject to dual-color two-photon FCS (Heinze et al., 2000), simultaneously accessing two fluorescent species with minimal spectral overlap in their emission properties—Rhodamine Green and Texas Red—by a single infrared laser line. Figure 8.9 shows a plot of η forboth Rhodamine Green and Texas Red versus the two-photon excitation wavelength. During the scan, laser intensity and pulse width were controlled and kept constant at 30 mW and 100 fs, respectively. The optimal wavelength for the following cross-correlation experiments, in which both dyes are excited equally well, appeared to be 830 nm. Meanwhile, a large variety of otherpossible dye combinations fordual- and even triple-colorexcitation could be identified.

Recently, it was demonstrated that this kind of dual-color application can be expanded to assays fully based on genetically encoded probes such as GFP and DsRed (Kohl et al., 2002). Thus far, the development of an all-protein-based enzyme or protein binding assays forpossible use with dual-colorcr oss-correlation orotherkinds of single-molecule analysis in living cells has been limited by the absence of suitable pairs of fluorescent proteins that fulfill the necessary criteria for the assay (i.e., largely different emission spectra but similar photostabilities). We provided first proof of principle that state-of-the-art dual-color two-photon FCS can indeed be applied to genetically encoded fluorescent tags to perform kinetic enzyme analysis on the single-molecule level. To achieve this goal, a combined FRET /cross-correlation enzyme assay was created based on red-shifted green fluorescent protein (rsGFP) and DsRed connected by a short protein linker of 32 amino acid (aa) length, containing the target site of rTEV, a recombinant fragment from *Tobacco Etch Virus* protease. Upon protease cleavage, the separation of the fluorescent proteins from the reporter construct was monitored quantitatively, discriminating different enzyme concentrations within the nanomolar range. Fluorescence resonance energy transfer efficiency of the substrate was simultaneously observed during the protease digest by recording the molecular photon yield persecond η, providing an alternative means to monitor the cleavage reaction. Although, in theory, a FRET efficiency of 100% would seriously hamper cross-correlation

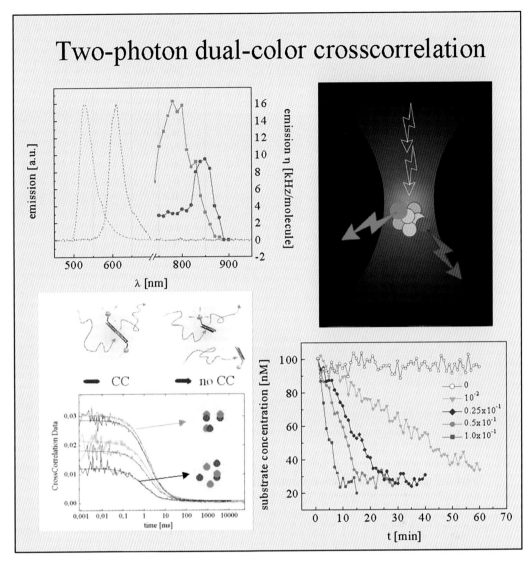

Figure 8.9 In 2000, the first dual-color cross-correlation experiment with two-photon excitation was performed by Heinze et al. In the top left picture, the (unaltered) emission spectra and the two-photon absorption spectra for the two dyes Rhodamine Green and Texas Red are shown. They were used to follow an enzymatic digestion reaction in which double-labeled DNA strands were cleaved by the restriction enzyme *Eco*RI. As a consequence, the monitored amplitude of the cross-correlation function decreased (bottom left). In the diagram on the bottom right, the decrease in substrate concentration (which is directly proportional to the cross-correlation amplitude) for different enzyme concentrations is depicted.

analysis, it was shown that transfer efficiencies of between 30% and 60% can be tolerated if properly corrected for in the data analysis.

Internal Dynamics

As a last example, the effects of fast fluctuations in the fluorescence emission on the autocorrelation curve are discussed, although the studies mentioned here have thus far exclusively been performed by standard one-photon FCS. Among the mechanisms leading to fast reversible changes in the fluorescence emission yield are intramolecular quenching resulting from electron transfer (Edman et al., 1996; Widengren et al., 1997), light-induced blinking resulting from a protonation reaction in several GFP mutants (Haupts et al., 1998; Schwille et al., 2000), and photoisomerization as observed for cyanine dyes (Widengren and Schwille, 2000). An interesting combination of FRET and FCS is the analysis of reversible intramolecular dynamics, such as structural "breathing" modes in proteins and nucleic acids with metastable ternary structures. Here, the changing proximity of the two labels introduces fluctuations in the FRET efficiency, and thus in the fluorescence brightness of both donor and acceptor. Several FRET/FCS studies on diffusing and fixed single molecules have been reported (Bonnet et al., 1998; Wallace et al., 2000; Kim et al., 2002), and although an extension of this principle to study the intermediate steps in protein folding processes could so far not be achieved, this fascinating outlook has often been advertised.

As already indicated earlier, there is a much more common phenomenon leading to blinking on fast timescales: At the high excitation rates applied for FCS measurements, the quantum mechanically forbidden transition between the first excited singlet and triplet state becomes increasingly probable. This so-called *intersystem crossing* in fact occurs in most dyes, as a result of their complex photophysical nature. The excited state lifetime for the triplet state is usually several orders of magnitude larger than the fluorescence lifetime. Mathematically, the corresponding fast-dynamics part of the autocorrelation function can be described by equation 8.17 (Widengren et al., 1995). The triplet-state parameters depend mainly on the excitation intensity, but the molecularenvir onment of the dye also plays a significant role. Molecular oxygen is one of the most common triplet-state quenchers, but some heavy metal ions have also been shown to alter the triplet-state kinetics (Widengren and Rigler, 1998). This sensitivity to ion concentrations could potentially be useful for probing the intracellular environment. Observing triplet dynamics by FCS thus opens up a fascinating perspective to map out the molecular microenvironment with submicrometer resolution.

However, not only ion concentrations can be determined in such a way; there is also a number of chromophores that exhibit pH-dependent emission characteristics, which are useful tools to probe the local pH. The most commonly known single-molecule pH meter is GFP, the clonal green fluorescent protein, being particularly suited for intracellular applications. Predominantly the long-wavelength mutants EGFP and YFP (yellow fluorescent protein) show an interesting emission behavior on timescales between micro- and milliseconds, which is easily accessible to FCS (Schwille et al., 2000). Haupts et al. (1998) identified a reversible protonation of the chromophore of EGFP to be responsible for the flickering dynamics. In bulk solution of EGFP, the average fluorescence emission is known to decrease to zero at low pH ($pK_a = 5.8$), paralleled by a decrease of the absorption at 488 nm and an increase at 400 nm. As a result of this spectral shift in the absorption spectrum, it is no longerpossible to excite the molecule with the selected laserline at 488 nm, and it appears dark. In the same way as for triplet blinking, the autocorrelation function shows an additional fast kinetic component, which can be described by the same mathematical formalism. The time constant of the protonation-induced flickering decreases with pH as expected, whereas the average fraction of molecules in the nonfluorescent (i.e., protonated) state increases to more than 80% at pH 4.5.

Interestingly, the most prevalent flickering phenomenon for one-photon FCS, triplet dynamics, does not appearto play a significant role fortwo-photon excitation (Dittrich and Schwille, 2001). Even at maximum achievable emission rates and at excitation intensities in which photobleaching can be already observed, there is hardly any clear blinking effect noticeable in the fast range of the autocorrelation curves. The reasons for this apparent difference, although far from being fully understood, certainly lay in the different photophysical nature of the excitation process, making it more likely to populate higher excited states that show a greater propensity as photobleaching precursors than for transitions to the triplet state. This issue is discussed next in a little more detail.

Fluorescence Correlation Spectroscopy in Microfluidic Systems

Microfluidics refers to the design, manufacturing, and practical use of microscale devices that handle very small volumes of fluids on the order of nano- to picoliters. The availability of chip-based channels or capillaries combined with miniaturized pumps, mixers, and valves opens up fascinating prospects to develop versatile total-analysis systems integrated on microscopic geometries (μTAS) (Mitchell, 2001). In recent years, a multitude of possible applications for microfluidic devices has been proposed and introduced. A particularly promising field is the downscaling of various chemical and biochemical processes, including the investigation of reaction kinetics, sensing and separation of reaction products (Kopp et al., 1998; Gao et al., 2001), as well as manipulation and sorting of cells and particles (Fu et al., 1999; Furlong et al., 2001). As one of the currently most important topics forbiochemical analytics, the detection, sizing, and sequencing of DNA fragments in flowing streams (Chou et al., 1999; Van Orden et al., 1999; Foquet et al., 2002) should be mentioned.

The increasing prevalence of microfluidic systems of various geometries and materials for the downscaling of chemical or biochemical processes raises a strong demand for adequate techniques to determine flow parameters precisely and to control fluid and particle manipulation. Of all readout parameters, fluorescence analysis of the fluid or suspended particles is particularly attractive, because it can be used without mechanical interference and with a sensitivity high enough to detect single molecules in aqueous environments. In a recent study, a precise determination of flow parameters such as velocity and direction in microstructured channels by one- and two-photon FCS was shown (Dittrich and Schwille, 2002). Figure 8.10 shows such microfluidic elements used and constructed in our laboratory, and some representative measurements of auto- and dual-beam cross-correlation on molecules flowing at different speeds. As outlined in the theory section, both the one- and two-photon excitation mode are well suited for simultaneous determination of flow velocity and direction with single-molecule sensitivity, and fordistinguishing otherfast dynamics superimposed on the transport process. However, certain advantages of two-photon excitation, such as highly restricted detection volumes and low scattering background, render it particularly valuable for measurements in tiny channel systems made of glass, polymethylmethacrylate (PMMA), or silicone. As an additional advantage of two-photon excitation, the pulsed excitation with an 80-MHz repetition rate allows for a delayed excitation between the two foci by $\approx$6 ns, which makes it particularly unlikely for molecules in the interfocal space to be sufficiently excited and reduces the autocorrelation cross-talk in dual-volume applications.

As outlined in the theory section, dual-beam cross-correlation provides a valuable means of controlling not only flow velocity but also monitoring flow directions. In many complex microfluidic structures with T or X junctions (fig. 8.10), designed to combine fluids or to manipulate immersed particles, the control of proper directionality, with respect to potential turbulence arising at intersections, is of crucial relevance. For example, in the rather extreme case of inverted flow direction, a flip-flop of the forward and backward cross-correlation curves can be observed.

Photobleaching in Two-Photon Fluorescence Correlation Spectroscopy

It is a well-known but not yet sufficiently understood fact that many standard fluorophores for single-molecule experiments exhibit significantly reduced photochemical lifetimes if

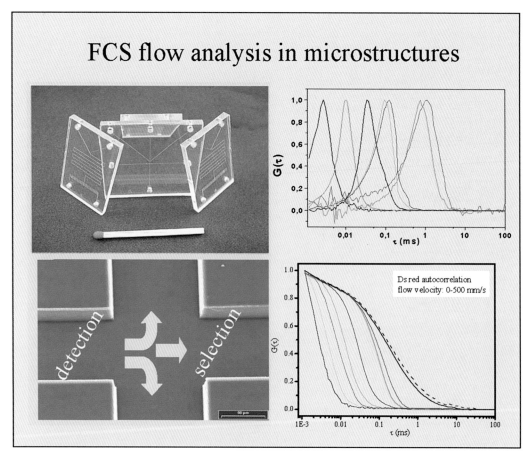

Figure 8.10 Depending on the application (i.e., sorting or flow measurements), a variety of different microfluidic channel systems can be designed. The chips are typically the size of a microscope slide. The rectangular channels, which are several tens of micrometers wide, can be seen in the top left as thin white lines resulting from scattering effects (electron microscope image below). On the right, cross- (top) and autocorrelation (bottom) curves of DsRed are shown, with the dye molecules being pumped at different speeds through the microchannels. With increasing flow velocity, the autocorrelation curves become steeper and are shifted toward faster timescales, because the average time of molecules spent in the focus is decreasing and becoming less disperse. Thus, it is possible to determine the flow velocity by comparing diffusion curves with flow curves. However, the cross-correlation curves show a much more pronounced change when monitoring flow processes: They peakat the characteristic passage time required for a particle to get from the first to the second focus (top right).

subject to two-photon excitation, resulting in lower achievable emission rates. Considering the many advantageous effects of two-photon excitation—in particular, with respect to single-molecule analysis in the cellularenvir onment (Schwille et al., 1999a) ormulticolor applications (Heinze et al., 2000)—this fact strongly calls for further elucidation of potential photobleaching mechanisms. The occurrence of photobleaching in FCS measurements is usually evidenced by the artificially decreased diffusion times of a dye molecule in a calibrated setup, if the intensity is increased over a certain threshold (fig. 8.11A). In an early study comparing the performance of one- and two-photon excitation (Schwille et al., 1999a), it could be demonstrated that the dramatic differences in achievable photon emission rates η, limiting the potential use of at least in vitro two-photon FCS, are in fact very dye-specific. Two dyes of particular interest for intracellular applications of FCS are TMR and EGFP, the enhanced red-shifted mutant of GFP. The first is probably the best studied and most applied dye in FCS measurements to date, with known photon emission rates per

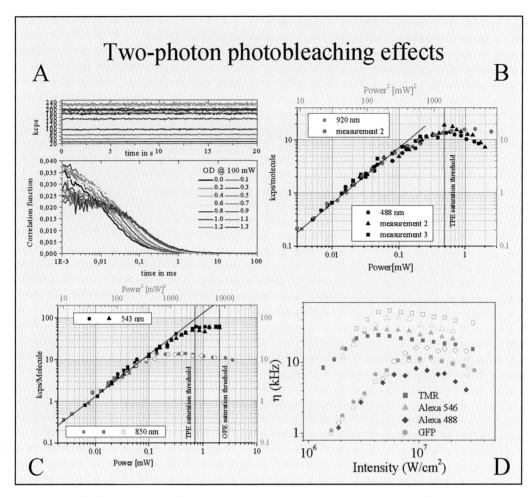

Figure 8.11 (A–D) When determining the optimal common wavelength to excite two dyes at the same wavelength, one also has to take into account different bleaching behavior. In A, intensity traces and the recorded autocorrelation curves are shown for different excitation intensities. Above a certain excitation power, the autocorrelation curves shift toward faster decay time scales, indicating that an increasing number of molecules is "opticuted" before leaving the focal volume by diffusion. At some point, even the detected fluorescence intensity is decreasing. Plotting η versus excitation power (or the square of excitation power, for two-photon excitation) on a double-logarithmic scale, the linear increase at low power is followed by a plateau, indicating saturation. At even higher power, the curve starts to decrease when photobleaching becomes prominent. (B) For GFP, one-photon excitation (OPE) and two-photon excitation (TPE) show similar levels of saturation and photo-bleaching, whereas for TMR (C) they exhibit different saturation levels and thus different photostabilities. Far higher η values can be obtained by one-photon excitation. (D) By adding ascorbic acid as a stabilizer to the solution, photobleaching can be dramatically reduced for a multitude of dyes (not for GFP), and maximum obtainable η can be increased by a factor of two to three. TMR, tetramethylrhodamine.

molecule η up to 200 kHz in well-designed setups with high detection efficiency, and with a large variety of labeling possibilities for many kinds of biologically active molecules. Green fluorescent protein mutants on the other hand, as already mentioned, have become the most powerful labeling tools in molecular biology (Tsien, 1998).

To compare the effects of photobleaching-limited emission rates, figure 8.11B shows an excitation powerdependence of η values forEGFP, and figure 8.11C shows it forTMR (Schwille et al., 1999a). Fora bettercompar ison, the effective detection volume elements were held the same size, which can be achieved by introducing very small (25-μm) pinholes in the one-photon setup. The y-axis parameter η in 10^3 photons/s (measured in kHz) is held the same for both excitation alternatives, whereas the excitation power axes are scaled for best overlap at low power. The two-photon power is squared to represent the underlying process. In the case of EGFP, the power dependencies of η forboth excitation methods can be brought to an almost perfect overlap, suggesting that at proper power levels, two-photon excitation allows an equivalent signal-to-noise ratio quality of FCS measurements over almost the entire range. The linear or quadratic, respectively, power dependence is observed up to values of $\eta \approx 5$ kcps/molecule, then a leveling off resulting from excited-state saturation and/or photobleaching can be seen. The maximal η that could be reached forEGFP is 15 kcps/molecule, farbelow the saturation expected forsinglet-state excitation. Because the limit is the same forcontinuous one-photon and pulsed two-photon excitation, slow processes such as intersystem crossing to the triplet state are assumed to be limiting the fluorescence output.

For TMR however, the two-photon curve deviates from the linear regime at much lower levels of η than the one-photon curve. Although the one-photon curve stays linear up to 20 kHz and reaches maximum values of 70 kHz, the two-photon curve shows a sudden plateau just above 13 to 15 kHz. The fact that strong differences between one- and two-photon excitation with respect to photobleaching and obtainable emission rates can be observed forthe synthetic dye TMR and other, forexample, Alexa dyes but not in GFP can be the result of at least two factors: First, it could be associated with the extremely blue-shifted two-photon absorption maximum of 850 nm for TMR (corresponding to 425 nm) relative to the one-photon maximum at 543 nm. These shifts are also quite pronounced in most of the Alexa dyes, but not in GFP. Second, the structure of GFP with a rigid β-sheet barrel enclosing and shielding the chromophoric unit from external solvent might be responsible for the better tolerance of intensive pulsed infrared light.

In a recent study (Dittrich and Schwille, 2001) aiming to address several factors that could be responsible for an enhanced dynamic bleaching, the roles of triplet states and radical formation within the solvent were particularly investigated. To determine the influence of triplet states, the oxygen concentration in the solution was varied. The potential role of radical formation was tested by adding stabilizing reagents, many of them known as *radical annihilators* (Eggeling et al., 1998). The results clearly show that in contrast to one-photon experiments, triplet states do not seem to provide the major pathway of photobleaching in two-photon excitation. According to the strong reduction of photochemical lifetimes without significant intersystem crossing, other mechanisms must be responsible for the severe limitation of photon emission yields. By addition of ascorbic acid known to be an antioxidant, a reduced bleaching rate can be observed, apparent in the much better preservation of τ_d compared with the pure water solution. In contrast to comparable one-photon measurements, the addition of ascorbic acid for many dye systems results in a shift of the saturation point I_{max} toward higher intensities, with dramatically enhanced photon yields (fig. 8.11D). The bleaching rates determined at intensities above I_{max} are lower than in pure water at the same intensities. Only the bleaching rate of GFP, with its chromophoric group being tightly shielded by a barrel of amino acid chains, is not influenced by ascorbic acid.

This observation strengthens the previously mentioned hypothesis that bleaching is, at least to a certain extent, mediated by photo-induced reactions within the solvent such as radical formation and annihilation. The relative stability of GFP toward two-photon excitation in general, as well as the fact that antioxidants do not enhance the maximum photon yields for this fluorophore, might be the result of the limitation of both: collision between the chromophore and reactive agents but also collision between the chromophore and ascorbic acid by the tight shielding of the protein barrel.

In conclusion, although the apparent limitations of two-photon fluorescence yield by photobleaching are far from being fully understood, and although the maximum performance of one-photon measurements in vitro, with photon count rates up to several hundred kilo-Hertz per molecule, seem out of reach by two-photon excitation, there are a multitude of reasons to prefer this excitation mode for certain FCS applications. Intracellular measurements, requiring by far lower excitation intensities, and measurements in confined geometries particularly profit from the inherent sectioning of the excitation process and from the much lower scattering background. In addition, the perspectives of dual- or multicolor experiments for both laser scanning microscopy and FCS are very promising in the experimentally simplified two-photon mode in which only a single excitation line may be utilized.

REFERENCES

Aragón, S. R., Pecora, R. Fluorescence correlation spectroscopy as a probe of molecular-dynamics. *J. Chem. Phys.* 64, 1791 (1976).

Bacia, K., Majoul, I., Schwille P. Probing the endocytic pathway in live cells using dual-color fluorescence cross-correlation analysis. *Biophys. J.* 83, 1184 (2002).

Berland, K. M., So, P. T. C., Gratton, E. 2-photon fluorescence correlation spectroscopy - method and application to the intracellular environment. *Biophys. J.* 68, 694 (1995).

Berne, B. J., Pecora, R. Laser light-scattering from liquids. *Annu. Rev. Phys. Chem.* 25, 233 (1974).

Bieschke, J., Giese, A., Schulz-Schaeffer, W., Zerr, I., Poser, S., Eigen, M., Kretzschmar, H. Ultrasensitive detection of pathological prion protein aggregates by dual-color scanning for intensely fluorescent targets. *Proc. Natl. Acad. Sci. USA* 97, 5468 (2000).

Bieschke, J, Schwille, P. *Fluoresc. Microsc. Fluoresc. Probes* 2, 81 (1998).

Bonnet, G., Krichevsky, O., Libchaber, A. Kinetics of conformational fluctuations in DNA hairpin-loops. *Proc. Natl. Acad. Sci. USA* 95, 8602 (1998).

Bouchaud, J. P., Georges, A. Anomalous diffusion in disordered media: Statistical mechanisms, models ,and physical applications. *Phys. Rep.* 195, 127 (1990).

Brinkmeier, M., Dörre, K., Stephan, J., Eigen, M. Two beam cross correlation: A method to characterize transport phenomena in micrometer-sized structures. *Anal. Chem.* 71, 609 (1999).

Brock, R., Vàmosi, G., Vereb, G., Jovin, T. M. Rapid characterization of green fluorescent protein fusion proteins on the molecular and cellular level by fluorescence correlation microscopy. *Proc. Natl. Acad. Sci. USA* 96, 10123 (1999).

Chen, Y., Müller, J. D., Ruan, Q. Q., Gratton, E. Molecular brightness characterization of EGFP in vivo by fluorescence fluctuation spectroscopy. *Biophys. J.* 82, 133 (2002).

Chen, Y., Müller, J. D., So, P. T. C., Gratton, E. The photon counting histogram in fluorescence fluctuation spectroscopy. *Biophys. J.* 77, 553 (1999).

Chen, Y., Müller, J. D., Tetin, S. Y., Tyner, J. D., Gratton, E. Probing ligand protein binding equilibria with fluorescence fluctuation spectroscopy. *Biophys. J.* 79, 1074 (2000).

Chou, H. P., Spence, C., Scherer, A., Quake, S. A microfabricated device for sizing and sorting DNA molecules. *Proc. Natl. Acad. Sci. USA* 96, 11 (1999).

Denk, W., Strickler, J. H., Webb, W. W. 2-photon laser scanning fluorescence microscopy. *Science* 248, 73 (1990).

Diaspro, A., Chirico, G., Federici, F., Cannone, F., Beretta, S., Robello, M. Two-photon microscopy and spectroscopy based on a compact confocal scanning head. *J. Biomed. Opt.* 6, 300 (2001).

Dickson, R. M., Cubitt, A. B., Tsien, R. Y., Moerner, W. E. On/off blinking and switching behavior of single molecules of green fluorescent protein. *Nature* 388, 355 (1997).

Dittrich, P., Malvezzi-Campeggi, F., Jahnz, M., Schwille, P. Accessing molecular dynamics in cells by fluorescence correlation spectroscopy. *Biol. Chem.* 382, 491 (2001).

Dittrich, P. S., Schwille, P. Photobleaching and stabilization of fluorophores used for single-molecule analysis with one- and two-photon excitation. *Appl. Phys. B.* 73, 829 (2001).

Dittrich, P.S.; Schwille, P.. Spatial two-photon fluorescence cross-correlation spectroscopy for controlling molecular transport in microfluidic structures. *Anal. Chem.* 74; 4472–4479 (2002).

Edman, L., Mets, Ü., Rigler, R. Confirmational transitions monitored for single molecules in solution. *Proc. Natl. Acad. Sci. USA* 93, 6710 (1996).

Eggeling, C., Fries, J. R., Brand, L., Guenther, R., Seidel, C. A. M. Monitoring conformational dynamics of a single molecule by selective fluorescence spectroscopy. *Proc. Natl. Acad. Sci. USA* 95, 1556 (1998).

Ehrenberg, M., Rigler, R. Rotational brownian-motion and fluorescence intensity fluctuations. *Chem. Phys.* 4, 390 (1974).

Eigen, M., Rigler, R. Sorting single molecules: Application to diagnostics and evolutionary biotechnology. *Proc Natl. Acad. Sci. USA* 91, 5740 (1994).

Elson, E. L., Magde, D. Fluorescence correlation spectroscopy. 1. Conceptual basis and theory. *Biopolymers* 13, 1 (1974).

Fahey, P. F., Barak, L. S., Elson, E. L., Koppel, D. E., Wolf, D. E., Webb, W. W. Lateral diffusion in planar lipid bilayers. *Science* 195 305 (1977).

Feder, T. J., Brust-Mascher, I., Slattery, J. P., Baird, B., Webb, W. W. Constrained diffusion or immobile fraction on cell surfaces: A new interpretation. *Biophys. J.* 70, 2767 (1996).

Foquet, M., Korlach, J., Zipfel, W., Webb, W. W., Craighead, H. G. DNA fragment sizing by single molecule detection in submicrometer-sized closed fluidic channels. *Anal. Chem.* 74, 1415 (2002).

Fu, A. Y., Spence, C., Scherer, A., Arnold, F. H., Quake, S. R. A microfabricated fluorescence-activated cell sorter. *Nat. Biotechnol.* 17, 1109 (1999).

Furlong, E. E. M., Profitt, D., Scott, M. P. Automated sorting of live transgenic embryos. *Nat. Biotechnol.* 19, 153 (2001).

Gao, J., Xu, J., Locascio, L. E., Lee, C. S. Integrated microfluidic system enabling protein digestion, peptide separation, and protein identification. *Anal. Chem.* 73, 2648 (2001).

Gennerich, A., Schild, D. Fluorescence correlation spectroscopy in small cytosolic compartments depends critically on the diffusion model used. *Biophys. J.* 79, 3294 (2000).

Haupts, U., Maiti, S., Schwille, P., Webb, W. W. Dynamics of fluorescence fluctuations in green fluorescent protein observed by fluorescence correlation spectroscopy. *Proc. Natl. Acad. Sci. USA* 95, 13573 (1998).

Heinze, K. G., Koltermann, A., Schwille, P. Simultaneous two-photon excitation of distinct labels for dual-color fluorescence crosscorrelation analysis. *Proc. Natl. Acad. Sci. USA* 97, 10377 (2000).

Heinze, K. G., Rarbach, M., Jahnz, M., Schwille, P. Two-photon fluorescence coincidence analysis: Rapid measurements of enzyme kinetics. *Biophys. J.* 83: 1671–1681 (2002).

Heikal, A. A., Hess, S. T., Webb, W. W. Multiphoton molecularspectr oscopy and excited-state dynamics of enhanced green fluorescent protein (EGFP): Acid-base specificity. *Chem. Phys.* 274, 37 (2001).

Hirschberg, K., Miller, C. M., Ellenberg, J., Presley, J. F., Siggia, E. D., Phair, R. D., Lippincott-Schwartz, J. Kinetic analysis of secretory protein traffic and characterization of Golgi to plasma membrane transport intermediates in living cells. *J. Cell Biol.* 143, 1485 (1998).

Jung, G., Bräuchle, C., Zumbusch, A. Two-color fluorescence correlation spectroscopy of one chromophore: Application to the E222Q mutant of the green fluorescent protein. *J. Chem. Phys.* 114, 3149 (2001).

Kask, P., Gunther, R., Axhausen, P. Statistical accuracy in fluorescence fluctuation experiments. *Eur. Biophys. J.* 25(3), 163 (1997).

Kask, P., Palo, K., Ullmann, D., Gall K. Fluorescence-intensity distribution analysis and its application in biomolecular detection technology. *Proc. Natl. Acad. Sci. USA* 96, 13756 (1999).

Kettling, U., Koltermann, A., Schwille, P., Eigen, M. Real-time enzyme kinetics monitored by dual-color fluorescence cross-correlation spectroscopy. *Proc. Natl. Acad. Sci. USA* 95, 14116 (1998).

Kim, H. D. Nienhaus, G. U., Ha, T., Orr, J. W., Williamson, J. R., Chu, S. Mg2+-dependent conformational change of RNA studied by fluorescence correlation and FRET on immobilized single molecules. *Proc. Natl. Acad. Sci. USA* 99, 4284 (2002).

Kinjo, M., Rigler, R. Ultrasensitive hybridization analysis using fluorescence correlation spectroscopy. *Nucl. Acids Res.* 23, 1795 (1995).

Klumpp, M., Scheel, A., Lopez-Calle, E., Busch, M., Murray, K. J., Pope, A. J. Gand binding to transmembrane receptors on intact cells or membrane vesicles measured in a homogeneous 1-microliter assay format. *J. Biomolec. Screening* 6, 159 (2001).

Kohl, T., Heinze, K. G., Kuhlemann, R., Koltermann, A., Schwille, P. A protease assay for two-photon crosscor-relation and FRET analysis based solely on fluorescent proteins. *Proc. Natl. Acad. Sci. USA* 99:12161–12166 (2002).

Köhler, R. H., Schwille, P., Webb, W. W., Hanson, M. Active protein transport through plastid tubules: Velocity quantified by fluorescence correlation spectroscopy. *J. Cell Sci.* 113, 3921 (2000).

Koltermann, A., Kettling, U., Bieschke, J., Winkler, T., Eigen, M. Rapid assay processing by integration of dual-color fluorescence cross-correlation spectroscopy: High throughput screening for enzyme activity. *Proc. Natl. Acad. Sci. USA* 95, 1421 (1998).

Kopp, M. U., de Mello, A. J., Manz, A. Chemical amplification: Continuous-flow PCR on a chip. *Science* 280, 1046 (1998).

Koppel, D. E. Statistical accuracy in fluorescence correlation spectroscopy. *Phys. Rev. A* 10, 1938 (1974).

Korlach, J., Schwille, P., Webb, W. W., Feigenson, G. F. Characterization of lipid bilayer phases by confocal microscopy and fluorescence correlation spectroscopy. *Proc. Natl. Acad. Sci. USA* 96, 8461 (1999).

Kubitscheck, U., Kuckmann, O., Kues, T., Peters, R. Imaging and tracking of single GFP molecules in solution. *Biophys. J.* 78, 2170 (2000).

Lippitz, M., Erker, W., Decker. H., van Holde, K. E., Basche, T. Two-photon excitation microscopy of tryptophan-containing proteins. *Proc. Natl. Acad. Sci. USA* 99, 2772 (2002).

Magde, D., Elson, E. L., Webb, W. W. Thermodynamic fluctuations in a reacting system: measurement by fluorescence correlation spectroscopy. *Phys. Rev. Lett.* 29, 705 (1972).

Magde, D., Elson, E. L., Webb, W. W. Fluorescence correlation spectroscopy .2. Experimental realization. *Biopolymers* 13, 29 (1974).

Magde, D., Elson, E. L., Webb, W. W. Fluorescence correlation spectroscopy .3. Uniform translation and laminar-flow. *Biopolymers* 17, 361 (1978).

Malvezzi-Campeggi, F., Jahnz, M., Heinze, K. G., Dittrich, P., Schwille, P. Light-induced flickering of DsRed provides evidence for distinct and interconvertible fluorescent states. *Biophys. J.* 81, 1776 (2001).

Margeat, E., Poujol, N., Boulahtouf, A., Chen, Y., Müller, J. D., Gratton, E., Cavailles, V., Royer, C. A. E human estrogen receptoralpha dimerbinds a single SRC-1 coactivatormolecule with an affinity dictated by agonist structure. *J. Mol. Biol.* 306, 433 (2001).

Mitchell, P. Turning the spotlight on cellular imaging: Advances in imaging are enabling researchers to track more accurately the localization of macromolecules in cells. *Nat. Biotechnol.* 19, 1013 (2001).

Palmer, A. G., Thompson, N. L. Theory of sample translation in fluorescence correlation spectroscopy. *Biophys. J.* 51, 339 (1987).

Palmer, A. G., Thompson, N. L. High-order fluorescence fluctuation analysis of model protein clusters. *Proc. Natl. Acad. Sci. USA* 86, 6148 (1989).

Petersen, N. O. Elson, E. L. Measurements of diffusion and chemical-kinetics by fluorescence photobleaching recovery and fluorescence correlation spectroscopy. *Methods Enzymol.* 130, 454 (1986).

Pramanik, A., Rigler, R. Ligand-receptor interactions in the membrane of cultured cells monitored by fluorescence correlation spectroscopy. *Biol. Chem.* 382, 371 (2001).

Qian, H., Elson, E. L. Analysis of confocal laser-microscope optics for 3-d fluorescence correlation spectroscopy. *Appl. Opt.* 30, 1185 (1991).

Rigler, R., Földes-Papp, Z., Meyer-Almes, F. J., Sammet, C., Völcker, M., Schnetz, A. Fluorescence cross-correlation: A new concept for polymerase chain reaction. *J. Biotechnol.* 63, 97 (1998).

Rigler, R., Mets, U., Widengren, J., Kask, P. Fluorescence correlation spectroscopy with high count rate and low-background: Analysis of translational diffusion. *Eur. Biophys. J.* 22, 169 (1993).

Rigler, R., Pramanik, A., Jonasson, P., Kratz, G., Jansson, O. T., Nygren, P. A., Stahl, S., Ekberg, K., Johansson, B. L., Uhlen, S., Uhlen, M., Jornvall, H., Wahren, J. Specific binding of proinsulin C-peptide to human cell membranes. *Proc. Natl. Acad. Sci. USA* 96, 13318 (1999).

Rigler, R., Widengren, J. *Bioscience* 3, 180 (1990).

Rippe, K. Simultaneous binding of two DNA duplexes to the NtrC-enhancercomplex studied by two-colorfluor escence cross-correlation spectroscopy. *Biochemistry* 39, 2131 (2000).

Saxton, M. J. Lateral diffusion in an archipelago - single-particle diffusion. *Biophys. J.* 64, 1766 (1993).

Schaertl, S., Meyer-Almes, F. J., Lopez-Calle, E., Siemers, A., Kramer, J. A novel and robust homogeneous fluorescence-based assay using nanoparticles for pharmaceutical screening and diagnostics. *J. Biomolec. Screening* 5, 227 (2000).

Schüler, J., Frank, J., Trier, U., Schäfer-Korting, M., Saenger, W. A. Interaction kinetics of tetramethylrhodamine trans-ferrin with human transferrin receptor studied by fluorescence correlation spectroscopy. *Biochemistry* 38, 8402 (1999).

Schwille, P. In *Fluorescence Correlation Spectroscopy: Theory and Applications*. Elson, E. L., Rigler, R., eds. Springer, Berlin, (2001), p. 360.

Schwille, P., Bieschke, J., Oehlenschläger, F. Kinetic investigations by fluorescence correlation spectroscopy: The analytical and diagnostic potential of diffusion studies. *Biophys. Chem.* 66, 211 (1997a).

Schwille, P., Haupts, U., Maiti, S., Webb, W. W. Molecular dynamics in living cells observed by fluorescence correlation spectroscopy with one- and two-photon excitation. *Biophys. J.* 77, 2251 (1999a).

Schwille, P., Korlach, J., Webb, W. W. Fluorescence correlation spectroscopy with single-molecule sensitivity on cell and model membranes. *Cytometry* 36, 176 (1999b).

Schwille, P., Kummer, S., Heikal, A. A., Moerner, W. E., Webb, W. W. Fluorescence correlation spectroscopy reveals fast optical excitation-driven intramolecular dynamics of yellow fluorescent proteins. *Proc. Natl. Acad. Sci. USA* 97, 151 (2000).

Schwille, P., Meyer-Almes, F.- J., Rigler, R. Dual-color fluorescence cross-correlation spectroscopy for multicomponent diffusional analysis in solution. *Biophys. J.* 72, 1878 (1997b).

Schwille, P., Oehlenschläger, F., Walter, N. Quantitative hybridization kinetics of DNA probes to RNA in solution followed by diffusional fluorescence correlation analysis. *Biochemistry* 35, 10182 (1996).

Thompson, N. L. In *Topics in Fluorescence Spectroscopy.* Lakowicz, J. R., ed. Plenum Press, New York (1991), vol. 1, p. 337.

Tsien, R. Y. The green fluorescent protein. *Annu. Rev. Biochem.* 67, 509 (1998).

Van Orden, A., Cai, H., Goodwin, P. M., Keller, R. A. Efficient detection of single DNA fragments in flowing sample streams by two photon fluorescence excitation. *Anal. Chem.* 71, 2108 (1999).

Wachsmuth, M., Waldeck, W., Langowski, J. Anomalous diffusion of fluorescent probes inside living cell nuclei investigated by spatially-resolved fluorescence correlation spectroscopy. *J. Mol. Biol.* 298, 677 (2000).

Wallace, M. I., Ying, L., Balasubramanian, S., Klenerman, D. Ratiometric analysis of single-molecule fluorescence resonance energy transfer using logical combinations of threshold criteria: A study of 12-mer DNA. *J. Phys. Chem. B* 104, 11551 (2000).

Widengren, J., Dapprich, J., Rigler, R. Fast interactions between Rh6G and dGTP in water studied by fluorescence correlation spectroscopy. *Chem. Phys.* 216, 417 (1997).

Widengren, J., Mets, Ü., Rigler, R. Fluorescence correlation spectroscopy of triplet-states in solution: A theoretical and experimental-study. *J. Chem. Phys.* 99, 13368 (1995).

Widengren, J., Rigler, R. Review - Fluorescence correlation spectroscopy as a tool to investigate chemical reactions in solutions and on cell surfaces. *Cell. Mol. Biol.* 44, 857 (1998).

Widengren, J., Schwille, P. Characterization of photoinduced isomerization and back-isomerization of the cyanine dye Cy5 by fluorescence correlation spectroscopy. *J. Phys. Chem. A* 104, 6416 (2000).

Winkler, T., Kettling, U., Koltermann, A., Eigen, M. Confocal fluorescence coincidence analysis: An approach to ultra high-throughput screening. *Proc. Natl. Acad. Sci. USA* 96, 1375 (1999).

Wiseman, P. W., Squier, J. A., Ellisman, M. H., Wilson, K. R. Two-photon image correlation spectroscopy and image cross-correlation spectroscopy. *J. Microsc.* 200, 14 (2000).

Wohland, T., Friedrich, K., Hovius, R., Vogel, H. Study of ligand-receptor interactions by fluorescence correlation spectroscopy with different fluorophores: Evidence that the homopentameric 5-hydroxytryptamine type 3(As) receptor binds only one ligand. *Biochemistry* 38, 8671 (1999).

Xu, C., Zipfel, W., Shear, J. B., Williams, R. M., Webb, W. W. Multiphoton fluorescence excitation: New spectral windows forbiological nonlinearmicr oscopy. *Proc. Natl. Acad. Sci. USA* 93, 10763 (1996).

9

Nanoscopy: The Future of Optical Microscopy

STEFAN W. HELL AND ANDREAS SCHÖNLE

In 1873, Ernst Abbe discovered that the resolution of a focusing light microscope is limited by diffraction. This physical insight became one of the most prominent paradigms in the natural sciences, with paramount importance in biology. Although the advent of confocal and multiphoton fluorescence microscopy facilitated three-dimensional imaging (Sheppard and Kompfner, 1978; Wilson and Sheppard, 1984; Denk et al., 1990), the resolution issue remained a fundamental physical limitation of the method (Pawley, 1995). It is only in recent years that methods with the ability to push resolution beyond this limit have appeared (Hell, 1997). In this chapter we discuss how the performance of optical microscopes operating with visible light can be pushed to its ultimate limit as predicted by Abbe, but in all three directions of space, and, more important, how the *diffraction barrier* can be broken altogether. We review a new class of fluorescence microscopes that has no theoretical resolution limit and has already led to a substantial resolution increase in practice. We believe that the introduction of these concepts mark the advent of optical nanoscopy (Hell, 2003).

The resolution of a focusing light microscope can be assessed by the FWHM of the focal spot, which is commonly referred to as the *point-spread function*. Loosely speaking, this focal spot is the pen, with which the image of the object is drawn and, therefore, if identical molecules are within the FWHM distance, the molecules cannot be separated in the image. Hence, improving the resolution is largely equivalent to narrowing the point-spread function or features thereof. In a conventional microscope, the FWHM of the point-spread function is about $\lambda/(2n \sin \alpha)$, with λ denoting the wavelength, n the refractive index, and α the semiaperture angle of the lens. Without changes to the principal method, the spot size can only be decreased by using shorter wavelengths and larger aperture angles (Abbe, 1873; Born and Wolf, 1993), but this strategy must face a dead end for biological applications, because wavelengths λ smallerthan 350 nm are not compatible with live-cell imaging and the lens half-aperture is technically limited to 70 deg. In the best case, established focusing microscopes resolve 180 nm in the focal plane (x,y) and merely 500 to 800 nm along the optic axis (z) (Pawley, 1995).

One possible strategy to defeat the fundamental limit imposed by diffraction is to abandon focusing altogether (Pohl and Courjon, 1993). Near-field microscopes use ultrasharp tips or tiny apertures to localize the interaction of the light with the object to subdiffraction dimensions. However, this approach limits imaging to surfaces and care has to be taken to avoid imaging artifacts (Hecht et al., 1997).

Clearly, it is preferable to improve resolution fundamentally without giving up one of the most prominent advantages of light microscopy: the ability to obtain three-dimensional resolved images from within the specimen. In combination with modern fluorescence techniques, this provides the basis for functional imaging in living cells and therefore the observation and analysis of dynamic biological processes in their natural environment. Although this problem has challenged many physicists (Toraldo di Francia, 1952; Lukosz, 1966), it did not really stimulate feasible proposals in the past.

Only recently have methods emerged that are promising candidates to open the nanoscale to optical light microscopy. In the first part of this chapter we explore how optical resolution can be pushed to its ultimate limit in fluorescence microscopy. The use of two opposing lenses brings axial resolution to the level of its transverse counterpart, yielding almost uniform resolution in all three directions. This is a major improvement over conventional and even confocal microscopy. Different methods based on this principle are outlined, analyzed, and compared.

However, all attempts to improve the resolution by merely changing the optical components of the microscope remain fundamentally limited by diffraction. This is best understood in the frequency world. Here, the fluorescence spot is described as being composed of spatial frequencies transmitted by the microscope. The optical properties of a device are described by the optical transfer function (OTF), which describes the strength with which these frequencies are transferred to the image (Goodman, 1968; Wilson and Sheppard, 1984). Hence, the ultimate resolution limit is given by the highest frequency passed. Point-spread function and OTF are intertwined by Fourier mathematics: The sharper the point-spread function, the broader the OTF.

The highest frequency k transmitted is limited by the wavelength, $\lambda = 2\pi/k$, of the light passing through the setup. The dependence of the excitation probability on the square of the electrical field leads to convolutions in frequency space, allowing frequencies of up to $2k$ to be transmitted. Finally, resolving optics can be used both in the illumination and the detection path. The multiplication of the excitation and the detection probability of a fluorophore is equivalent to another convolution in frequency space, and therefore the optimum cutoff achievable is at $4k$. This implies that all microscopes relying solely on improvements of the optical instrument have an ultimate resolution limit of just under 100 nm in all directions, if visible light is used and all transmitted frequencies can be restored in the image. This theoretical mark is indeed almost reached by the 4Pi-confocal microscope described in more detail later (Hell et al., 1997; Egner et al., 2002).

To overcome this fundamental limit and obtain what is usually called *superresolution,* higher frequencies have to be created. It was soon realized that this can be achieved by introducing nonlinearities in the interaction of the excitation light with a dye. Therefore, a longstanding popular notion was that superresolution is readily attained by the cooperative absorption of many photons (Göppert-Mayer, 1931; Bloembergen, 1965; Sheppard and Kompfner, 1978). However, it turned out that m-photon excitation ($m > 1$) of a fluorophore has not pushed the resolution limit and is unlikely to do so in the future. Although it is true that m-photon absorption occurs mainly at the center of the spot, the concomitant narrowing of the effective spot is spoiled by the fact that m-photon excitation usually means that the excitation energy is split between the photons (Denk et al., 1990). Consequently the photons have m times lowerenergy, thus an m times longerwavelength, which results in m times larger focal spots to start with. Besides this, $m > 3$ requires very high intensities (Xu et al., 1996), leading to complicated setups and possible damage to the sample.

Therefore, *multiphoton* concepts, in which the detection of a photon occurs only after the consecutive absorption of multiple excitation photons, have been proposed. In this case, high intensities are not required and photon energy subdivision does not occur (Hänninen et al., 1996; Schönle and Hell, 1999; Schönle et al., 1999). However, they require specific conditions that are probably hard to meet in practice.

As a matter of fact, it was not until the mid 1990s when the first viable concepts appeared to break the diffraction barrier (Hell and Wichmann, 1994; Hell and Kroug, 1995). They all share a common principle: They switch markers on or off in well defined, subdiffraction sized regions of the sample. To this end they exploit reversible saturated optically linear (fluorescence) transition between two molecular states (RESOLFT). Because there is no physical limit to the degree of saturation, there is no longer a theoretical limit to the resolution. Thus, a whole family of RESOLFT microscopy methods that can achieve nanoscale resolution in *all* directions has been established (Hell, 1997; Hell et al., 2003; Heintzmann et al., 2003).

The full potential of this approach is currently being explored, but it is obvious that these methods represent a fundamentally new approach to resolution increase without theoretical limit. The second part of this chapter gives an introduction to the general idea and outlines strategies, implementations, and initial applications of RESOLFT-based superresolution microscopes. Encouraging results have recently been reported, such as the first demonstration of spatial resolution of $\lambda/25$ with focused light and with regular lenses (Westphal et al., 2003b).

Last, a potential road map toward imaging with nanometer resolution in live cells is discussed. Bridging the gap between electron and current light microscopy, a *nanoscope* working with focused light should be a powerful tool for unraveling the relationship between structure and function in cell biology.

AXIAL RESOLUTION IMPROVEMENT WITH TWO LENSES

A simple explanation for the suboptimal optical resolution of a confocal fluorescence microscope is that the focusing angle of the objective lens is not a full solid angle of 4π. If it were, the focal spot would be spherical and the axial resolution would be as good as its lateral counterpart. Therefore, an obvious way to decrease the axial spot close to its optimal value is to synthesize a focusing angle that is as close to 4π as possible. As mentioned earlier, the focusing angle of a single lens is technically limited, but a larger wavefront can be achieved by using two opposing lenses (Hell, 1990; Hell and Stelzer, 1992b; Gustafsson et al., 1995).

Of course, this requires that the light from both sides is coherent. In other words, the two beams need to interfere in the focal spot where a standing wave will be created. In case of constructive interference, this will produce a maximum with an axial FWHM of about $\lambda/4$ in the focal point, with several side maxima at $\sim m\lambda/2$, with m = 0, 1, 2, 3. The method will work if the axial resolution is dominated by the central maximum.

The simplest way to achieve this interference is implemented in the standing wave microscope (SWM) (Lanni, 1986; Bailey et al., 1993). The basic idea in standing wave microscopy is to produce a *flat* standing wave of laserlight to create a set of excitation nodal planes in the sample along the optic axis. It uses wide-field detection on a camera, like any conventional fluorescence microscope. A substantial increase of axial resolution has initially been claimed by this technique, which has been experimentally realized and applied to cellular imaging. Standing wave microscopy produced stacks of multiple, thin interference layers in a sample (Lanni, 1986), but has not unambiguously resolved features stretching over an axial range that is larger than half the wavelength (0.2–0.4 μm) (Krishnamurthi et al., 1996; Freimann et al., 1997). Although standing wave microscopy might prove useful for more specialized applications, it cannot deliver extended axial images of arbitrary objects with improved axial resolution (Nagorni and Hell, 2001a,b). We give a more thorough explanation for this later. An intuitive reason is the fact that all side maxima equal the maximum in the focal spot in height, leading to strong artifacts in the image, which cannot be removed without a priori knowledge about the sample.

It can be safely stated that, although the accurate alignment of two lenses posed initial challenges, the real physical problem in the development of microscopes with opposing lenses was the avoidance of lobe-induced artifacts. Hence, it was not until the advent of spot-scanning 4Pi-confocal (Hell, 1990; Hell and Stelzer, 1992a) and wide-field I^5M microscopy (Gustafsson et al., 1995) that the axial resolution in three-dimensional imaging was significantly improved through interference (Schrader and Hell, 1996; Hell et al., 1997; Gustafsson, 1999; Gustafsson et al., 1999). Both use lenses with high NAs and implement one or several of the following three mechanisms to suppress further the side lobes in the point-spread function:

1. Confocalization (Hell, 1990; Hell and Stelzer, 1992b)
2. Multi-photon excitation (Hell and Stelzer, 1992a)
3. Exploiting excitation/fluorescence wavelength disparities (Hell and Stelzer, 1992a,b; Gustafsson et al., 1995)

The latter is particularly efficient if both wavefront pairs are brought to interfere in the sample and at the detector (Hell and Stelzer, 1992b; Gustafsson et al., 1995), respectively, because the respective side lobes no longer coincide in space. A single mechanism may be sufficient; however, to date, the implementation of at least two mechanisms has proved to be more reliable. After initial demonstration (Schrader and Hell, 1996), superresolved axial separation with two-photon 4Pi-confocal microscopy has been applied to fixed cells (Hell et al., 1997). Image quality can be further improved by applying nonlinear restoration (Holmes, 1988; Carrington et al., 1995; Holmes et al., 1995). Under biological imaging conditions, this typically improves the resolution up to a factor of two in both the transverse and the axial directions. Therefore, in combination with image restoration, two-photon 4Pi-confocal microscopy has resulted in a resolution of ≈ 100 nm in all directions, as first witnessed by the imaging of filamentous actin (Hell et al., 1997) and immunofluorescently labeled microtubule (Hell and Nagorni, 1998; Nagorni and Hell, 1998) in mouse fibroblasts.

The very effective lobe-reducing measures of (1) confocalization and (2) two-photon excitation are to some extent restrictive. Clearly, nonconfocal wide-field detection and regular illumination would make 4Pi microscopy more versatile. Therefore, the related approach of I^5M (Gustafsson et al., 1995, 1996; Gustafsson, 1999; Gustafsson et al., 1999) confines itself to using mechanism 3: the simultaneous interference of both the excitation and the (Stokes-shifted) fluorescence wavefront pairs. Impressive work has demonstrated this method to yield three-dimensional images of actin filaments with slightly better than 100-nm axial resolution in fixed cells (Gustafsson et al., 1999). To remove the side lobe artifacts, I^5M-recorded data are deconvolved off-line.

The benefits of I^5M are readily cited: single-photon excitation with arguably less photobleaching, an additional 20% to 50% gain in fluorescence signal, and lower cost. However, the relaxation of the side lobe suppression comes at the expense of increased vulnerability to sample-induced aberrations, especially with nonsparse objects (Nagorni and Hell, 2001a,b). Thus, I^5M imaging, which has so far relied on oil immersion lenses, has required mounting of the cell in a medium with a refractive index of $n = 1.5$ (Gustafsson et al., 1999). Live cells inevitably necessitate aqueous media ($n = 1.34$). Moreover, water immersion lenses have a poorer focusing angle and therefore larger lobes to begin with (Bahlmann et al., 2001). Potential strategies for improving the tolerance of I^5M are the implementation of a nonlinearexcitation mode, as well as the combination with pseudoconfocal orpatter ned illumination (Gustafsson, 2000). Although these measures again add physical complexity, they may have the potential to render I^5M more suitable for live cells.

To understand the respective benefits and limitations of SWM, I^5M, and 4Pi microscopy, we have to go beyond the comparison of their technical implementation (Gustafsson et al., 1999). To assess underwhich conditions these microscopes will be able to deliver three-dimensional resolved images with superior resolution, their point-spread functions and optical transfer functions (OTFs) have to be analyzed and compared in detail. A very useful and explanatory comparison of the OTF *supports* has been published (Gustafsson. 1999). However, as we see in the following paragraph, the support alone is insufficient for evaluation. The success of increasing the axial resolution with coherent counterpropagating beams not only depends on the achievable optical bandwidth, but on whetherand how they transferobject frequencies *within* this bandwidth. Specifically, we will address the importance of gaps and weak parts that occur in some systems along the optic axis of the OTF (Krishnamurthi et al., 1996; Gustafsson et al., 1999). They will be quantified fora particularoptical setting and it will be shown that they are intimately connected with the optical arrangement. It will then be demonstrated that gaps in the OTF may render the removal of artifacts impossible, so that a genuine increase in axial resolution becomes impossible as well.

COMPARISON OF THE POINT-SPREAD FUNCTIONS

Because the absolute phase of the illumination light is lost upon absorption, fluorescence image formation can be treated as an incoherent process, for which the product of the excitation and detection point-spread function gives the effective point-spread function (Wilson and Sheppard, 1984). The difference between the concepts is reflected by the different structure of the excitation and detection point-spread function of these microscopes, and is ultimately determined by the substantial differences in the effective point-spread function governing the image formation (fig. 9.1).

Three major types of 4Pi-confocal microscopy have been reported (Hell and Stelzer, 1992b). They differ on whether the spherical wavefronts are coherently added for illumination, for detection, or for both simultaneously, and are referred to as types A, B, and C, respectively. To reveal how these microscopes compare with standard techniques, figure 9.1 also displays the point-spread function of the conventional epifluorescence and the standard confocal microscope. The epifluorescence microscope features uniform illumination intensity throughout the sample volume. In contrast, the SWM and the I^5M are illuminated by characteristic flat standing wave patterns.

To obtain a practically relevant comparison between their performance, the point-spread function for NA = 1.35 and oil immersion with refractive index $n = 1.51$ were calculated. The assumption of NA = 1.35 rather than 1.4 is in good agreement with what is found in experiments. For single-photon excitation, an excitation and detection wavelength of 488 nm and 530 nm, respectively, is assumed; for two-photon excitation, 800 nm was chosen. The finite size of the confocal pinhole and the camera pixels was neglected in the detection point-spread function because, when correctly adjusted, these parameters do not affect the general conclusions of this study. The point-spread functions were numerically computed in a volume of $128 \times 128 \times 512$ pixels in the x, y, and z directions, respectively, for cubic pixels with 20 nm length. The numberand size of the pixels facilitated the numerical calculation of the transfer function by Fourier transformation; of the 512 pixels in the z direction, only data based on the central 256 pixels are shown in figure 9.1.

The excitation point-spread function of the confocal microscope, as well as the *detection* point-spread function of the wide-field, confocal, SWM, and 4Pi-type microscopes are regular intensity point-spread functions. Assuming excitation of and emission from an arbitrarily oriented dipole transition of the dye, the excitation and emission point-spread function of a single lens is well approximated by the absolute square of the electrical field in the focal region:

$$h = \left| \vec{E}_1(z, r, \phi) \right|^2 \qquad (9.1)$$

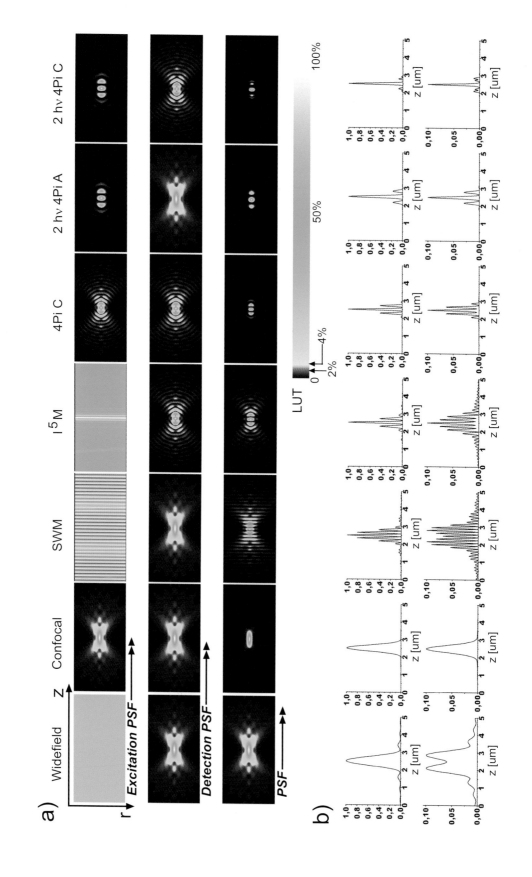

It depends on the axial and radial distance to the geometric focus (z and r, respectively) and the polarangle ϕ. It was calculated using the vectorial theory presented by Richards and Wolf (1959) assuming unpolarized light.

For the excitation point-spread function of the 4Pi-confocal microscopes (types A and C) and the detection point-spread function of the I^5M and the 4Pi-confocal microscopes (types B and C), one considers constructive interference between the two spherical wavefronts. The point-spread function is therefore given by the coherent addition of two beams

$$h = \left| \vec{E}_1(z, r, \phi) + \vec{E}_2(z, r, \phi) \right|^2 \tag{9.2}$$

The field of the opposing lens is given by

$$\vec{E}_2(\vec{r}) = \mathbf{M}\vec{E}_1\left(\mathbf{M}^{-1}\vec{r}\right) \tag{9.3}$$

where $\mathbf{M}$ is the coordinate transform from the system of lens number 2 to lens number 1 and is a diagonal matrix that inverts the z component for a triangular cavity and the y and z components fora rectangularcavity (Bahlmann and Hell, 2000). In the case of two-photon excitation (Denk et al., 1990), the effective excitation point-spread function is obtained by squaring the point-spread function at the excitation wavelength.

The excitation intensity in the SWM is fundamentally different from that of a 4Pi-confocal microscope (fig. 9.1A) because it is given by a plane standing wave along the optic axis:

$$h = I_0 \cos^2(kz) \tag{9.4}$$

where I_0 denotes a constant, k is the wave number, and we assume constructive interference at the common geometric focus.

While the 4Pi-confocal microscope uses a spatially coherent, pointlike laser illumination, wide-field illumination with a lamp or a laser, normally in Köhler mode, is used in the I^5M. The physical consequences of this difference are best explained as follows. If one decomposes the two spherical wavefronts of the 4Pi illumination into partial plane waves incident from different angles corresponding to different points of the illumination apertures, *all* partial plane waves of the aperture interfere with each other. In the I^5M, the illumination light is not spatially coherent throughout the aperture. Therefore only *pairs* of plane waves originating from corresponding points of the illumination apertures are mutually coherent, forming a plane wave with the wave vector scaling with the cosine of the azimuth angle θ. The excitation point-spread function of the I^5M can be calculated by

Figure 9.1 (facing page) Comparison of the intensity point-spread function of the conventional wide-field, confocal, standing wave, I^5M, 4Pi-confocal type C, two-photon 4Pi-confocal type A, and the two-photon 4Pi-confocal type C microscopes. (A) Excitation, detection, and effective point-spread function are shown in the upper, center, and bottom panels, respectively. The color lookup table (LUT) emphasizes weakregions of the point-spread function. The aliasing effect in the excitation point-spread function of the SWM is the result of the LUT. (B) Axial profiles along the optic axis (upper row) are compared with axial profiles through a laterally offset point characterized by an intensity of 10% of that found at the geometric focal point (lower row). Although in the conventional, standing wave, and I^5M microscopes the lateral defocus is associated with pronounced changes in the axial profiles, in the confocalized systems the shape of the profiles is largely unaltered, so that their effective point-spread function can be factored into a radial and axial function.

adding the intensity of *plane* standing waves. Assuming uniform intensity throughout the exit pupil of the lens, the point-spread function is given by

$$h(z) = I_0 \int d\phi \int d\theta \sin(\theta) \cos^2(k_0 z \cos \theta) = 2\pi I_0 \int d\theta \sin(\theta) \cos^2(k_0 z \cos \theta) \quad (9.5)$$

The integral is from zero to the half-aperture angle of the lens.

If the I^5M elects to illuminate the sample critically, which may not be recommended because of potential nonuniformities in the illumination, its excitation point-spread function is found by integrating a 4Pi excitation point-spread function over the field of view in the focal plane. The integration accounts for the averaging resulting from the wide-field illumination. This leads to a result that is not very much different from that predicted by equation 9.5 and hence not further discussed. Another version of this approach also not discussed is the incoherent illumination interference (I^3M) and the excitation field synthesis. The I^3M is easier to align inasmuch as it does not require coherent detection through both objective lenses, but offers far less favorable imaging conditions for improving the resolution. The excitation point-spread function of the SWM and the I^5M are displayed in figure 9.1A next to that of the 4Pi (type C), which is fundamentally different.

Figure 9.1A systematically compares the x-z sections along the optic axis of the point-spread function of the wide-field conventional, confocal, standing wave, I^5M, 4Pi-confocal type C, two-photon excitation 4Pi-confocal type A, and two-photon excitation 4Pi type C microscopes (Hell and Stelzer, 1992a). The color lookup table has been chosen so that the regions of weak signal are emphasized. This reveals subtle but important differences among the point-spread function. Areas of low but nonnegligible intensity are important because they cover a large volume and substantially contribute to the image formation.

When scrutinizing the point-spread functions (fig. 9.1A), the differences become apparent. As a result of the incoherent addition of the standing wave spectrum, the local minima in the excitation point-spread function of the I^5M are not zero, as is the case in the SWM. Still, although the excitation modes of these microscopes are similar, those of the 4Pi-confocal microscopes are different. The key difference is that as a result of focusing, the 4Pi point-spread functions are confined in the lateral direction so that contributions from outer lateral parts of the focal region are reduced. The confinement has important consequences. Although the 4Pi-confocal microscope, especially its two-photon version, exhibits two rather low lobes, the I^5M and even more so the SWM feature a multitude of lobes and fringes on either side of the focal plane. The second consequence is that the 4Pi-confocal point-spread function, to good approximation, can be separated into an axial and a radial function (Hell et al., 1995; Schrader et al., 1998):

$$h(r, z) \cong c(r) \, h_l(z) \quad (9.6)$$

This feature can be recognized in figure 9.1B, where the axial profiles through the focal point are plotted for all microscopes (top), including the conventional and the confocal, next to laterally offset profiles (bottom). The lateral offset was chosen such that the point-spread function had dropped to 10% of its maximum. The key to separability is the suppression of out-of-focus regions by an effective point-spread function that scales quadratically, cubically, orhigherwith the local intensity distribution, as is the case in a confocalized or multiphoton excitation system. The separability is, therefore, a particular feature of the 4Pi-confocal and multiphoton arrangements, and we see later that it is the prerequisite for simple online removal of artifacts introduced by the side lobes. The additional use of two-photon excitation leads to a further suppression of the outer parts of the excitation focus and thus of the side lobes of the 4Pi illumination mode. Figure 9.1 also reveals that, together with coherent detection (type C), two-photon excitation 4Pi-confocal microscopy features an almost lobe-free point-spread function.

In the SWM and the I^{5}M, the number and relative height of the lobes increase dramatically when moving away from the focal point, because an effective suppression mechanism is missing. In the SWM, the lobes become even higherthan the central peak itself. In the I^{5}M, the secondary maxima are as high as the first maxima of the single-photon 4Pi-confocal microscope of type C.

COMPARISON AND SIGNIFICANCE OF THE OPTICAL TRANSFER FUNCTIONS

We mentioned earlier that the point-spread function can be considered the pen with which the image is drawn. Features that are finer than its width will not appear in the image. Mathematically, the image (I) is the convolution of the object O (e.g., the distribution of fluorescence dye in the sample) and the point-spread function h:

$$I(\vec{r}) = [O \otimes h](\vec{r}) \tag{9.7}$$

To understand the significance of the OTF, we transform this equation to the Fourier domain using the convolution theorem:

$$\hat{I}(\vec{k}) = \hat{O}(\vec{k}) \cdot \hat{h}(\vec{k}) \tag{9.8}$$

The hat denotes the Fourier transform, $\hat{h}$ is the OTF, and the convolution becomes a multiplication. This is the mathematical equivalent to the fact that the OTF determines which spatial frequencies propagate to the image. If the OTF were nonzero everywhere, we could divide the Fourier transform of the image by it, Fourier back-transform, and obtain our object. In practice, the OTF is bandwidth limited (fig. 9.2) and has weak regions. In regions where the OTF is small, such a division leads to a strong amplification of noise and, consequently, to artifacts.

Linear deconvolution is based on this division approach, but it introduces a special treatment for frequencies not transmitted by the OTF. It is only capable of restoring frequencies where the OTF is not weak or zero. The foremost advantage of a contiguous and generally strong OTF is, therefore, the possibility to apply such a linear deconvolution.

In the case of missing frequencies, a correct representation of the object in the image can be given only if these frequencies are extracted from a priori knowledge of the object. This extraction is mathematically more complex and often not viable. Linear deconvolution is computationally facile and fast. Speed is of particular importance because the interference artifacts are ideally removed online so that the final image is immediately accessible.

The comparison of the effective OTF (fig. 9.2) highlights the severe gaps in the SWM, rendering linear deconvolution impossible. Linear deconvolution is reportedly possible in I^{5}M (Gustafsson et al., 1999). However, because the gaps in I^{5}M are filled with rather low amplitudes, in the presence of noise, linear deconvolution may not be straightforward and must render artifacts for objects that are either not sparse or do contain spatial frequencies that are transferred only weakly. As a result of the continuity of the OTF and the strong amplitudes throughout its support, 4Pi-confocal microscopy fulfills the preconditions for lineardeconvolution. In fact, lineardeconvolution along the axial direction based on the separability of the point-spread function has been described and successfully applied for the removal of interference artifacts in complex objects such as in dense filamentous actin (Schrader et al., 1998) and microtubular networks (Nagorni and Hell, 1998). This important procedure will now be placed on a more general basis, both in the spatial as well as in the frequency domains.

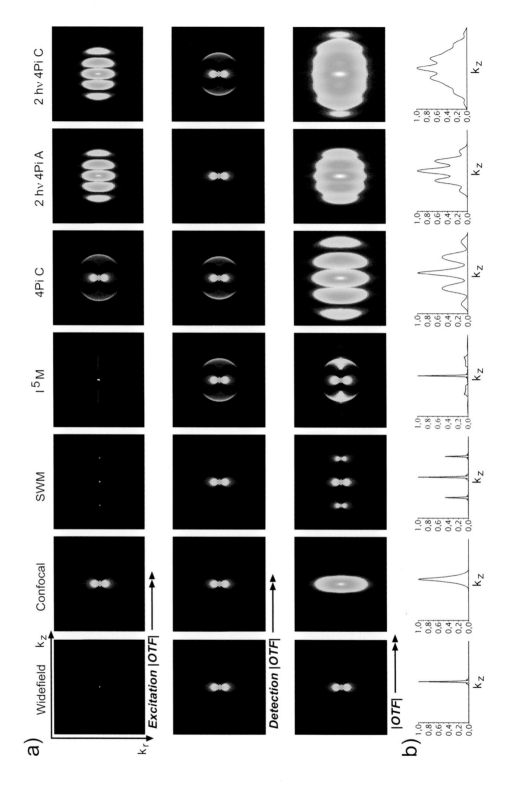

DECONVOLUTION IN THE SPATIAL DOMAIN

The 4Pi-confocal, but not the SWM and I^{5}M OTF, allows one to perform a *one-dimensional* linear deconvolution that just eliminates the effects of the lobes in the image. The axial factor of the point-spread function (eq. 9.6) is decomposed into the convolution of a function $h_p(z)$, which describes the shape of a single peak, and a lobe function $l(z)$ containing the information about the position and relative height of the lobes:

$$h_l(z) \cong h_p(z) \otimes l(z) \tag{9.9}$$

From the viewpoint of image processing, h_p plays the role of a blurring function and l is the replication function that is responsible for the "ghost images" noticed in unprocessed 4Pi images. Therefore, to remove the interference multiple-maxima effect, only the elimination of the effect of the lobe function l is required. This can be done by algebraic inversion.

We use a discrete notation, in which each lobe is represented by a component l_i of the vector $\mathbf{l}$, with l_0 denoting the strength of the central lobe and the index running from $-n$ to n. Negative indices denote lobes left of the central one. If the lobe distance in pixels is denoted by d, the values of the object along the line are given by O_j and those of the image are given by I_j, the convolution is given by

$$I_j = \sum_k l_k O_{j-dk} \tag{9.10}$$

We are looking for an inverse filter $\mathbf{l}^{-1}$ inverting this convolution. Therefore we seek

$$O_j = \sum_s l_s^{-1} I_{j-ds} = \sum_s \sum_k l_s^{-1} l_k O_{j-ds-dk} \tag{9.11}$$

for all possible objects. The inverse filter therefore needs to fulfill the condition

$$\sum_s l_s^{-1} l_{j-s} = \delta_{j0} \tag{9.12}$$

At this point we can arbitrarily choose the length of the inverse filter and assume the index running from $-m$ to m. The index j in equation 9.12 can take values from $-m - n$ to $m + n$ and we have, therefore, $2(m + n) + 1$ equations with $2m + 1$ unknowns, which usually cannot be solved. An approximation is found by only considering the equations for $j = -m \ldots m$. The problem is then equivalent to solving a linear Toeplitz problem, with the Toeplitz matrix given by the vector $\mathbf{l}$ (Press et al., 1993; Nagorni and Hell, 2001a). The approximation is good if the edges of the inverse filters are small, because the remaining equations are nearly satisfied. In practice, this is the case if the first-order lobes are less than 45%. In this case, the error is practically not observable. A typical length of the inverse

Figure 9.2 (facing page) Comparison of the modulus of the OTF of the conventional wide-field, confocal, standing wave, I^{5}M, 4Pi-confocal type C, two-photon 4Pi-confocal type A, and the two-photon 4Pi-confocal type C microscopes, calculated numerically by Fourier transforming the point-spread function in figure 9.1. (A) Excitation OTF (upper panels), detection OTF (center panels), and effective OTF (bottom panels). The zero frequency point is in the center of each panel. The maximal frequency displayed in the r and z directions is $2\pi/80$ nm^{-1}. The lookup table is the same as in figure 9.1. (B) Axial profiles through the center of the OTF. Despite the fact that the OTF of the interference-based microscopes all feature a larger maximum bandwidth in the z direction, they fundamentally differ in contiguity and strength within the region bordered by the highest frequency. Note the pronounced frequency gaps in the SWM and the depressions in the I^{5}M.

filter is 11. In conjunction with a lobe height of 45%, the edges of the inverse filter feature a modulus less than 1% of the filter maximum. For lobes of 35% relative height, this value drops to a value of only about 0.08%. Thus, using this technique, the separability of the 4Pi-confocal point-spread function allows a fast way to find an image equivalent to imaging with a single main maximum. This method is referred to as *point deconvolution* because the inverse filter is discrete and nonzero only at a few points.

It should be noted that the lobes of the point-spread function correspond to suppressed regions along the optic axis of the OTF and therefore this method tries to restore suppressed frequencies by linear deconvolution even though the actual calculation takes place entirely in the spatial domain. If the lobes become too high (the depressed regions of the OTF too low and wide), the method fails and produces artifacts because the approximation for the inverse filter becomes inaccurate.

DECONVOLUTION IN THE FREQUENCY DOMAIN

As outlined earlier, linear deconvolution in the frequency domain basically relies on inverting equation 9.8, but has to deal with the vanishing regions of the OTF. An estimate of the object's frequency spectrum is obtained using the formula

$$\hat{E}(k) = \hat{I}(k)\hat{h}^*(k) \Big/ \left(|\hat{h}(k)|^2 + \mu\right) \tag{9.13}$$

The regularization parameter μ (Bertero et al., 1990) sets a lower bound on the denominator to avoid amplification of frequencies in which the modulus of the OTF is small and therefore the frequency spectrum of the image is dominated by noise. If the OTF is a convex function, the effect of regularization is similar to smoothing. In most cases considered in this chapter, however, the OTF is not convex and the situation is more complicated because small values are found not only at the boundaries of the OTF, but also in central parts of its support, for example in the vicinity of the minima or at the frequency gaps (fig. 9.2). If the lobes are too high, the level of the minima of the OTF is comparable or smaller than the noise level. This is definitely the case in the SWM, but also in the presence of slight aberrations in the I^5M, and in a more aberrated 4Pi-confocal setting. In this case, an adequate μ to avoid noise artifacts renders an estimate with lobe artifacts.

A more precise analysis is possible when the point-spread function is separable into a peak function and a lobe function. It also allows us to approach the problem of lobe removal in a more rigorous way than in the spatial domain:

$$h\left(\vec{r}\right) = h_p\left(\vec{r}\right) \otimes l\left(z\right) \tag{9.14}$$

Please note that this method does not rely on the ability to separate the point-spread function in a radial and an axial part, and that, different from the treatment in the spatial domain, the peak function now contains the radial dependence. As mentioned earlier, the lobe function is described by the relative height of the lobes and is given by

$$l\left(z\right) = \sum_s l_s \delta\left(z - ds\right) \tag{9.15}$$

with d now being the lobe distance in units of length. The frequency spectrum of the image is then given by

$$\hat{I}(\vec{k}) = \hat{O}(\vec{k}) \cdot \hat{h}(\vec{k}) = \hat{O}(\vec{k}) \cdot \hat{h}_p(\vec{k}) \cdot \hat{l}(k_z) \tag{9.16}$$

And if the lobe function is symmetric

$$\hat{l}(k_z) = 1 + \sum_{s>0} 2l_s \cos\left(\pi s d \cdot k_z\right) \tag{9.17}$$

It is straightforward to see that a critical height for the first-order lobes is 50%. Assume that the point-spread function consists of a main maximum and two primary lobes of half the strength. The right-hand side of equation 9.17 is then equal to zero for axial frequencies corresponding to the distance d. In other words, if the lobes are 50% (and higher), the frequency represented by the lobes is not transferred by the OTF. In practice, the critical lobe height at which the OTF first contains zeros is shifted to a value slightly more than 50%. The reason is the effect of the secondary lobes in the point-spread function that is neglected in the previous example. Still, the 50% threshold is an excellent rule of thumb that applies to SWM, I^{5}M, and 4Pi-confocal microscopy as well, for fundamental reasons.

On the other hand, a closer look at equation 9.17 reveals that the lobes can be removed from the image by direct Fourier inversion if the Fourier transform of the lobe function is much larger than the noise level everywhere. The frequency spectrum of the lobe-free image is then obtained by dividing the frequency spectrum of the image by the Fourier transform of the lobe function.

In a typical 4Pi setup of type A, an avalanche photodiode (APD) was used as a detector with a typical dark count rate less than 1 count/pixel that is negligible (Hell et al., 1997). Hence, the only significant source of noise is the Poisson noise of the photon counting process. The Poisson noise manifests itself in the data as white noise that does not depend on the frequency. Because the primary lobes are experimentally well below 50%, the first minima of the OTF are at 19%, which is well above the noise level of typically 0.5% to 1%, and direct lobe removal is therefore possible. The necessity for nonvanishing amplitudes becomes apparent again. If the 4Pi OTF had regions close to zero, as was theoretically described for the I^{5}M, the multiplication with a high number in this region would result in strong amplification of noise and therefore compromise the results. In the SWM, this removal of interference artifacts is virtually impossible.

Lobe removal by direct inversion also works with complex objects, as demonstrated with the fluorescently labeled microtubular network of a mammalian cell. The linear deconvolution in the spatial domain and the corresponding direct inversion via the frequency domain lead to virtually identical results. In figure 9.3, just one resulting image is shown because the two images do not exhibit noteworthy differences in the hard copy. Both final 4Pi-confocal images are free of artifacts, thus underlining the ruggedness and self-consistency of the linear deconvolution. A comparison with the confocal image reveals the fourfold improved axial resolution of the 4Pi-confocal technique.

DISCUSSION OF THE VARIOUS CONCEPTS

Although standing wave, I^{5}M, and 4Pi-confocal microscopy all use interfering counterpropagating beams, a detailed study of their respective imaging properties disclosed fundamental differences. The differences are most notable in the effective point-spread function and OTF of these microscopes, as shown in figures 9.1 and 9.2. The major discrepancy between the effective point-spread function of the SWM and I^{5}M on the one hand, and the 4Pi-confocal on the other, is the fact that in the 4Pi-confocal microscope the point-spread function is spatially more confined. The point-spread function in I^{5}M and even more so in SWM exhibit pronounced interference fringes of higher order in their outer region. In contrast to this, the point-spread function of a two-photon excitation 4Pi microscope falls off very steeply

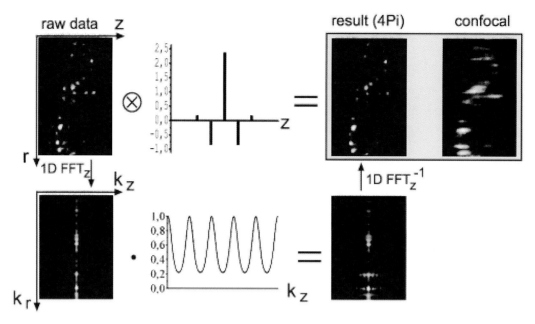

Figure 9.3 4Pi-confocal microscopy allows deconvolution by direct linear inversion. Raw x/z image data of microtubules of a mouse fibroblast cell obtained by two-photon 4Pi-confocal microscopy of type A. Linear deconvolution in the spatial domain (upper row) and in the frequency domain (lower row) yield almost identical results, shown in the boxed panel on the left-hand side. The comparison with its confocal counterpart (boxed image, right) reveals a fourfold improved axial resolution over the 4Pi-confocal microscope.

with departure from the focal point, so that the otherwise three-dimensional deconvolution is reduced to a one-dimensional linear problem.

More importantly, in the SWM and I^5M, the fringes of the point-spread function are not only manifested along the optic axis but also in the lateral direction. Therefore, the removal of artifacts, if possible at all, requires a full three-dimensional deconvolution with the three-dimensional point-spread function. This also implies the acquisition of a complete three-dimensional data stack and a more elaborate off-line deconvolution. Because of its spatial confinement, the 4Pi-confocal point-spread function can be factored to a lateral function and an axial function, which in turn enables a fast (point) deconvolution that removes the interference fringe structures and renders a final image during data acquisition. Therefore, care should be taken when comparing the speed of imaging of these microscopes. The scanning procedure in a 4Pi-confocal microscope is certainly slower than the wide-field imaging in SWM and I^5M. However, in 4Pi-confocal microscopy, for the axial separation of objects, it is sufficient to perform a *single* axial scan. Likewise, the lineardeconvolution is carried out online, which renders the final image immediately.

This comparison has also revealed that the linear deconvolution is effectively performed in the "classical way"—namely, by direct Fourier inversion. This is accomplished in genuine object data and without the requirement for regularization. The superiority of the 4Pi-confocal microscope with respect to the suitability for deconvolution is also reflected in the OTF. It was shown that the OTF of an SWM exhibits gaps along the optic axis that make the removal of the interference artifacts virtually impossible. The OTF of the I^5M is superior in this respect, because it is nonzero throughout its support. Still, it is very weak over a considerable region when compared with the strong zero-frequency peak. Hence, to benefit from its continuity, I^5M data must be recorded with a very high signal-to-noise-ratio and it may be applicable only to sparse, nonextended objects such as points or sparse fine lines. The use of cooled CCD cameras with low intrinsic noise is mandatory in these microscopes. The OTF of the 4Pi-confocal microscopes are contiguous. In the critical regions, the 4Pi-confocal OTF (fig. 9.2) exhibits significant values in the 19% to 32% range. This feature is of key relevance to the removal of the interference artifacts and hence for unambiguous, object-independent three-dimensional microscopy with improved axial resolution.

The laterally broader excitation OTF of the 4Pi-confocal microscope leads to a better filling of its effective OTF. The broader OTF is rooted in the fact that the excitation light is *focused*. Hence, although scanning with a focused beam inevitably reduces imaging speed, it also results in fundamentally better imaging properties of the microscope. An intuitive explanation for the superiority of the spherical waves is that their coherent addition is the only mode that truly increases the total aperture of the microscope. We will see later that the reduced imaging speed can be improved by parallelizing image acquisition. This will show that the beam-scanning approach of the 4Pi microscope can be well suited for video-rate imaging at much improved resolution.

Flat field standing wave excitation inherent to the I^5M and SWM trades off collected frequencies. The loss of optical frequencies is so significant (fig. 9.2) that the ability to provide unambiguous axial resolution is put at risk. This OTF study corroborates previous studies (Krishnamurthi et al., 1996) that in the SWM it is almost impossible to distinguish unambiguously two axially separated objects, unless the object is thinner than half the wavelength (Freimann et al., 1997). The insights gained earlier explain why this technique has neverpr ovided evidence forunambiguous axial imaging of axially extended objects, despite its early success in producing thin-stapled layers of excitation. Although both standing wave techniques and 4Pi microscopy have been successfully applied to the ultraprecise measurement of axial distances (Schmidt et al., 2000; Schneider et al., 2000; Albrecht et al., 2002) and object sizes (Egner et al., 2002; Failla et al., 2002), we are therefore convinced that images such as those in figure 9.4 and published elsewhere (Hell et al., 1997; Hell and Nagorni, 1998; Nagorni and Hell, 1998; Gustafsson et al., 1999) cannot be accomplished with the SWM forfundamental physical reasons.

However, similar axial resolution improvement and imaging is reportedly possible with I^5M combined with off-line image restoration (Gustafsson et al., 1999). This advantage over standing wave microscopy is ultimately rooted in the properties of the OTF discussed earlier. It has also been suggested that by combining I^5M with fringe pattern illumination and subsequent image restoration, one can alleviate the problem of the zero-frequency peak singularity in the I^5M (Heintzmann and Cremer, 1998; Gustafsson, 2000). In principle, one can also increase the lateral resolution up to that of (restored) confocal microscopy with this fringe pattern illumination wide-field scheme. Leaving aside the technical complexity of controlling the interference patterns of typically four pairs of beams, this suggestion just confirms that an unambiguous axial resolution requires the use of a wider angular spectrum. In fact, the spherical beams in a 4Pi-confocal microscope can be regarded as a complete spectrum of interfering plane waves coming from all angles available. Hence, from the standpoint of imaging theory, the combination of I^5M with fringe pattern illumination is a modification of I^5M toward a scheme that is more similar to the 4Pi arrangement; the improvements of the OTF are gained by conditions that are more similar to the *focusing* conditions found in the 4Pi-confocal microscope.

In real samples, the OTF of the microscopes is compromised by aberrations that were not included in this comparison of concepts. Residual misalignments induced by slight variations of the refractive index in the sample will play a role. However, successful 4Pi-confocal imaging of the mouse fibroblast cytoskeleton (Nagorni and Hell, 1998) and the recent I^5M imaging of similar structures (Gustafsson et al., 1999) have revealed that aberration effects are surmountable. Still, it is expected that an important benefit of a filled OTF is ruggedness of operation and lower sensitivity to potential misalignment.

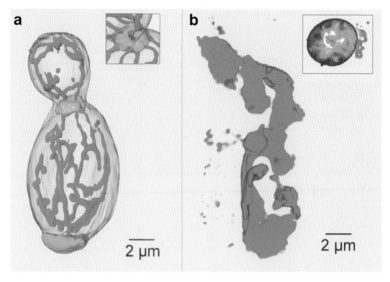

Figure 9.4 (A) Green fluorescent protein-labeled mitochondrial compartment of live *Saccharomyces cerevisiae*. The organelle displays strong tubular ramification of a single large body that is exclusively located beneath the plasma membrane (counterstained in blue). (Inset) A mitochondrial tubule can be followed through the thickened cell wall at the budding site (Egner et al., 2002). (B) Golgi apparatus as represented by the GalTase-EGFP expression in a live *Vero* cell (Enger et al., 2004). Note the convoluted structure of the Golgi apparatus featuring ribbons and fractionated stacks, as well as smaller tubular and vesicular subcompartments. (Insert) Epifluorescence overview image of the same cell colocalizes the organelle with the nucleus counterstained in blue. (Data in A are adapted from Egner et al. [2002]; data in B are from Egner et al. [2004].)

It is evident that image deconvolution requires prior knowledge of the point-spread function and hence its explicit determination with a pointlike object. This is particularly important because the point-spread function, and the complex OTF depend on the relative phase of the two counterpropagating wavefronts. It has been shown that in 4Pi-confocal microscopy, the particular mode of the interference—constructive, destructive, or anything in between—is of lesser importance (Hell and Nagorni, 1998). The same may also apply for the other concepts. So far, the structure of the point-spread function has been determined by measuring the response of test objects, such as fibers or fluorescent beads. However, it has been shown that the relative phase can be extracted directly from the image data (Blanca et al., 2002). This is of great importance in the case when the imaged object itself alters the relative phase of the interfering beams.

The SWM and the I^{5}M are therefore not wide-field versions of the 4Pi-confocal microscope, and vice versa; the latter is not a scanning version of the first. Basic differences between the contiguity and strength of their OTFs determine whether a particular system is able to superresolve axially at all. These optical properties are so critical that they outweigh by far the relevance of the more obvious technical differences.

In fact, the ruggedness of the 4Pi approach allowed its application to superresolved imaging of specimen in aqueous media using waterimmer sion lenses (Bahlmann et al., 2001). This is particularly noteworthy because water lenses feature NAs of not more than 1.2, thus increasing side lobes and making the problem of missing frequencies in the OTF more severe. Because imaging in water is one of the prerequisites of live-cell imaging—one of the most important applications of modern fluorescence microscopy—it can be concluded that for realistic aperture angles, the inherent lack of contiguity of the OTF in the SWM most likely renders the removal of interference ambiguities absolutely impossible. The I^5M becomes more viable through filling the frequency gaps present in the SWM, albeit with values that are weak with respect to the zero-frequency components. Both systems have yet to prove their applicability to live-cell imaging.

APPLICATIONS

As mentioned earlier, the reduced imaging speed of a beam-scanning microscope compared with wide-field imaging can be improved by parallelizing image acquisition. Recently, such a multifocal variant, termed *MMM-4Pi* (Hell et al., 1997), has indeed translated the 100-nm three-dimensional resolution into live-cell imaging. This method has provided superior three-dimensional images of the reticular network of GFP-labeled mitochondria in live budding yeast cells (fig. 9.4A). Cell-induced phase changes proved more benign than anticipated, but they are likely to confine these methods to the imaging of individual cells or thin cell layers. The deep modulation of the focal spot provided a new tool to measure the thickness of cellular constituents in the 50- to 500-nm range with a precision of a few nanometers (Egner et al., 2002). This property has been used to detect changes of ≈ 20 nm in the diameter of mitochondrial tubules upon a change of growth conditions.

The recent development of sample chambers with appropriate air–carbon dioxide conditions allows one to sustain cell viability over periods up to 48 hours and enabled 4Pi imaging in live mammalian cells. By imaging the Golgi-resident proteins UDP-galactosyltransferase and heparansulfate-2-O-sulfotransferase as EGFP fusion proteins, this work resulted in the first three-dimensional representation of the Golgi apparatus of a live mammalian cell at ≈ 100 nm resolution in all directions (Egner et al., 2004) (fig. 9.4B). The results indicate that ≈ 100-nm three-dimensional resolution can be obtained in the imaging of protein distributions in the cytosol and probably also in the nucleus. Extending the technique to multicolor detection will improve the microscope's ability to colocalize axially differently tagged proteins by the same factor of three to seven, which is likely to become important in protein interaction studies.

Up until now, live-cell imaging has been the prerogative of two-photon 4Pi-confocal microscopy. In its most recent version, confocalization, two-photon excitation, and the use of excitation/fluorescence disparities have been synergistically implemented in a compact 4Pi unit that was firmly interlaced with a state-of-the-art confocal scanning microscope (Leica TCS-SP2, Mannheim, Germany). Consequently, a sevenfold improved axial resolution (80 nm) overconfocal microscopy has been achieved in live-cell imaging with a rugged system (Gugel et al., 2004).

BREAKING THE DIFFRACTION BARRIER: A ROAD MAP TO FLUORESCENCE NANOSCOPY

As outlined earlier, confocal and related imaging modalities like I^5M and 4Pi microscopy may surpass the diffraction barrier, *but they do not break it.* Breaking implies the potential of featuring an infinitely sharp focal spot, or an infinitely large OTF bandwidth.

In 1994, a theoretical paper appeared that detailed a concept to eliminate the resolution-limiting effect of diffraction without eliminating diffraction itself (Hell and Wichmann, 1994). It was termed *stimulated emission depletion* (STED) microscopy, because the quenching of the molecular fluorescent state through stimulated emission was utilized. It was shortly followed by the proposal of ground-state depletion microscopy as a further concept with molecular resolution potential (Hell and Kroug, 1995). Both share the same principle to break Abbe's barrier: A focal intensity distribution with a zero point in space

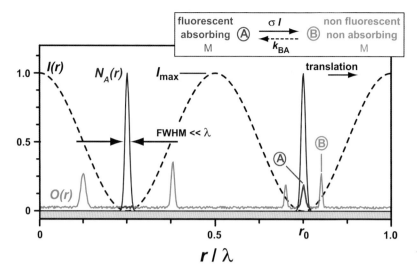

Figure 9.5 The principle of achieving diffraction-unlimited spatial resolution by saturating a transition from a detectable state (A) to a nondetectable state (B). The object is illuminated with light of wavelength λ, which drives molecules from state A to state B. After saturation, the population $N_A(r)$ of state A (blue) is confined to small regions around the nodal points of the incident intensity distribution $I(r)$ (dotted). If state A but not state B fluoresce, only those parts of an object $O(r)$ that lie within these regions (red) contribute to the signal. No signal is detected from other parts of the object (green). The width of the peaks of $N_A(r)$ depend on the saturation factor and can be infinitely narrow in theory.

effects a *saturated depletion* of a molecular state that is essential to the fluorescence process (Hell and Kroug, 1995; Hell, 1997). After the depletion the state can be populated again—that is, the saturated depletion is reversible. Because any reversible transition that can be saturated is a potential candidate (Hell, 1997; Dyba and Hell, 2002; Hell et al., 2003), the choice of the utilized process is solely determined by practical conditions, such as the required intensities, available light sources, photobleaching, and, with respect to applications in cell biology, compatibility with live-cell imaging. To reveal the essentials as well as the potential of this radically different concept that we called *RESOLFT,* we now review the breaking of the diffraction barrier in a general way.

The basic idea underlying stimulated emission and ground-state depletion microscopy can be generalized by considering a molecule with two arbitrary states A and B, between which the molecule can be switched (fig. 9.5). In fluorophores, typical examples for these states are the ground state and an excited state, conformational and isomeric states. The transition A $\rightarrow$ B is induced by light, but no restriction is made about the transition B $\rightarrow$ A. It may be spontaneous, but it could also be induced by light, heat, orany othermechanism. The only further assumption is that at least one of the states is critical to the generation of the signal. In fluorescent microscopy, this means that the dye can fluoresce only (or much more intensively) in state A.

The manner in which diffraction-unlimited resolution in fluorescence imaging (or any kind of manipulation, probing, or so forth, that depends on one of the states) is achieved in such a simple setting is illustrated in figure 9.5. First the sample is illuminated with light that drives the transition from A to B. Its intensity distribution, $I(r)$, features one or several naughts (i.e., positions with zero intensity). After illumination, the probability, $N_A(r)$, of finding molecules in state A depends on $I(r)$ and has peaks where $I(r)$ is zero. If we choose the maximum intensity I_{max} in such a way that it is many times (ζ times, $\zeta = I_{max}/I_S$ is called the *saturation factor*) higherthan the saturation intensity I_S (the threshold at which 50% of the molecules are switched into state B) of the transition A → B, then molecules will be almost exclusively in state B even at positions where $I(r)$ is only a small fraction ($\approx 5/\zeta$) of I_{max}. This means that while molecules remain in state A in the naughts, they are switched to state B even at places very close to them, and the peaks of $N_A(r)$ become very sharp. Therefore, upon probing the fluorescence, we can be sure that it stems exclusively from the immediate vicinity of the positions where $I(r)$ is zero. To image the sample $O(r)$, we scan the position of the naughts and record the amount of fluorescence detected from each position. The resolution of this process is determined by the FWHM of the peaks in $N_A(r)$, which in turn is solely determined by the saturation factor and can become infinitely narrow in theory.

Although figure 9.5 and the following calculations are presented in one dimension for clarity, this idea is readily extended to all directions in space, and hence to three-dimensional imaging. Conventional camera-based detection is possible if the nodes are farther apart than the classic resolution limit of the microscope, because fluorescence can then be assigned to the respective nodes.

Please note that complete depletion of A (orcomplete darkness of B) is *not* required. It is sufficient that the nonnodal region features a constant, notably lower, probability to emit fluorescence, so that it can be distinguished from its sharp counterpart. Even if not A, but B is the brighter state, one can read out B and may obtain the same superresolved image after subtraction. It should be noted that all these alterations do not interfere with the theoretically unlimited resolution, but will usually lead to a compromised signal-to-noise ratio because the light from nonnodal regions invariably contributes to the photon shot noise.

The nonlinear intensity dependence brought about by saturation is radically different from the nonlinearity connected with m-photon excitation, m−th harmonics generation, coherent anti-Stokes-Raman scattering (Sheppard and Kompfner, 1978; Shen, 1984), and so forth. In the lattercases, the nonlinearsignal stems from the simultaneous action of *more than one photon* at the sample, which would only work at high focal intensities. By contrast, the nonlinearity brought about by saturation and depletion stems from a change in the *population* of the involved states, which is effected by a *single*-photon process— namely, stimulated emission. Therefore, unlike in m-photon processes, strong nonlinearities are achieved at comparatively low intensities.

To clarify the principle method and get a rough estimate of the relationship between saturation factor and resolution, let us briefly consider a simple example. We denote the rates of A → B and B → A with k_{AB} and k_{BA}, respectively. The time evolution of the normalized populations of the two states n_A and n_B is then given by

$$dn_A/dt = -k_{AB}n_A + k_{BA}n_B = -dn_B/dt \qquad (9.18)$$

Independent of its initial state, afteran illumination time

$$t \geq (k_{BA} + k_{AB})^{-1} \qquad (9.19)$$

when the equilibrium is approximately reached, the population of state A is given by

$$N_A = k_{BA}/(k_{BA} + k_{AB}) \qquad (9.20)$$

The rate at which state A is depleted is given by $k_{AB} = \sigma I$, where σ denotes the molecular cross-section, and the intensity is written as photon flux per unit area. Hence, the equilibrium population is given by

$$N_A(r) = k_{BA}/(\sigma I(r) + k_{AB}) \qquad (9.21)$$

And $N_A = 1/2$ for the saturation intensity

$$I_s = k_{BA}/\sigma \tag{9.22}$$

From equation 9.22 we see that where $I(r) \gg I_s$, all molecules end up in B. Thus, if we choose

$$I(r) = I_{max}f(r) \tag{9.23}$$

with $I_{max} \gg I_s$, molecules in state A are only found in the nodes of the diffraction-limited distribution function $f(r)$. As an example we choose a standing wave

$$f(x) = \sin^2(2\pi x/\lambda) \tag{9.24}$$

for illumination. A simple calculation shows that the FWHM of the peaks of N_A and hence the resolution of the microscope is then given by

$$\Delta x = \lambda \pi^{-1} \arcsin\left(\sqrt{k_{BA}/\sigma I_{max}}\right) \approx \lambda \Big/ \left(\pi\sqrt{\zeta}\right) \tag{9.25}$$

A saturation factor of $\zeta = 1000$ yields $\Delta x \approx \lambda/100$, but in principle the spot of A molecules can be continuously squeezed by increasing ζ.

Stimulation Emission Depletion Microscopy

Stimulation emission depletion microscopy (STED) produces subdiffraction resolution and subdiffraction-size fluorescence volumes in exactly this way, by saturating the depletion of the fluorescent state of the dye. The physical conditions, a setup, and a typical focal spot are displayed in fig. 9.6. The fluorophore in the fluorescent state S_1 (state A) is switched to the ground state S_0 (state B) with a doughnut-shaped beam. The saturated depletion of S_1 confines fluorescence to the central naught. With typical saturation intensities ranging from 1 to 100 MW/cm^2, saturation factors of up to 120 have been reported (Klar et al., 2000, 2001). This should yield a 10-fold resolution improvement over the diffraction barrier, but imperfections in the doughnut have limited the improvement to five- to sevenfold in experiments (Klar et al., 2001). Utilizing STED wavelengths of $\lambda = 750$ to 800 nm, a lateral resolution of as low as 28 nm has been achieved in experiments with single molecules (Westphal et al., 2003b).

 As stated earlier, light microscopy resolution can be described either in real space or in spatial frequencies. In real space, the resolution is assessed by the FWHM of the focal spot. Figure 9.7A shows the measured profile of the point-spread function in the focal plane (x) for a conventional fluorescence microscope along with its sharper subdiffraction STED fluorescence counterpart. Note the 5.5-fold improvement of resolution with STED.

 Figure 9.7B shows the OTF of a conventional microscope along with the approximate fivefold enlarged OTF of the STED fluorescence microscope. The marked bandwidth increase signifies a fundamental breaking of Abbe's diffraction barrier in the focal plane in the case of STED. The measurements were carried out with an excitation wavelength of $\lambda = 635$ nm, an oil immersion lens with an NA of 1.4, and with the smallest possible probe: a single fluorescent molecule (Westphal et al., 2003b).

 The FWHM of the point-spread function or the bandwidth of the OTF of the microscope are just estimates; a thorough description of the resolution requires the complete functions. Moreover, knowing these functions in full also enables improving the resolution by mathematical deconvolution. Note that in figure 9.7B the OTF falls off with larger spatial frequencies. Provided that, in the image, these frequencies are not swamped by noise, they can be artificially elevated by multiplication (see arrows). Mathematically, this amounts

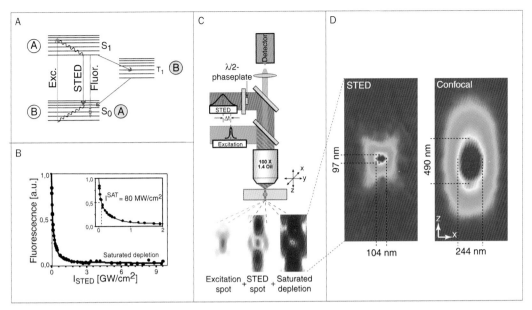

Figure 9.6 Physical conditions, setup, and typical focal spot for STED. (A) Energy diagram of an organic fluorophore. Molecules in the excited state S_1 return to the ground state S_0 by spontaneous fluorescence emission. Return to S_0 may also be enforced by light through stimulated emission (Einstein, 1917)—a phenomenon with the same cross-section and intensity dependence as normal absorption. To prevail over the spontaneous return, STED currently uses intense light pulses with a duration of a fraction of the S_1 lifetime. Tuning the STED wavelength to the red edge of the emission spectrum prevents reexcitation by the same pulses. T_1 is a darktriplet state that can be accessed through S_1 and then returns to S_0 within 1 to $10^4 \mu$s. (B) Saturated depletion of S_1 with increasing STED pulse intensity I_{STED}, as measured by the remaining fluorescence of an organic fluorophore. Depletion of S_1 saturates with increasing I_{STED} and therefore establishes a nonlinear relationship between the fluorescence and the intensity applied for STED. The saturation is essential in breaking the diffraction barrier because it ensures that the ability of fluorophores to emit photons is effectively switched off outside the immediate vicinity of the focal naught. (C) Sketch of a point-scanning STED microscope. Excitation and STED are accomplished with synchronized laser pulses focused by a lens into the sample, sketched as green and red beams, respectively. A detector registers the fluorescence. The panels below outline the corresponding spots at the focal plane: the excitation spot (left) is overlapped with the STED spot, featuring a central naught (center). Saturated depletion by the STED beam reduces the region of excited molecules (right) to the very zero point, leaving a fluorescent spot of subdiffraction dimensions shown in D. (D) Fluorescent spot in the STED and in the confocal microscope. Note the doubled lateral and fivefold-improved axial resolution. The reduction in dimensions (x, y, z) yields ultrasmall volumes of subdiffraction size—here, 0.67 attoliter (Klar et al., 2000)—corresponding to an 18-fold reduction compared with its confocal counterpart. The spot size is not limited on principle grounds but by practical circumstances, such as the quality of the naught and the saturation factor of depletion.

to a deconvolution in real space. Because the higher frequencies are responsible for small details in the image, deconvolution results in a further image sharpening.

As an example of linearly deconvolved STED microscopy, two molecules at a 62-nm distance are distinguished in full by two sharp peaks (fig. 9.7C) (Westphal et al., 2003b). The individual peaks are sharper (33 nm) than the initial peak of 40 nm, as a result of deconvolution. The effective OTF afterdeconvolution is slightly augmented at lowerfr equencies, as indicated by the arrows in figure 9.7B.

Subdiffraction images with threefold axial and doubled lateral resolution have been obtained with membrane-labeled bacteria and live budding yeast cells (Klar et al., 2000). Although there is preliminary evidence for increased nonlinear photobleaching of certain markers with the elevated intensities (Dyba and Hell, 2003), there is no indication that the intensities applied currently would hamper the imaging of live cells. This is not surprising because the intensities are by two to three orders of magnitude lower

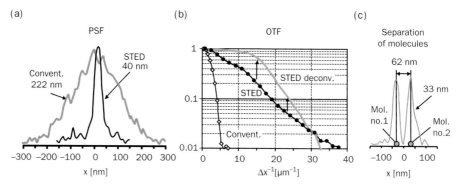

Figure 9.7 (A) The intensity profile of the effective point-spread function (PSF) quantifies the focal blur in the microscope. Identical fluorescent objects that are closer than the FWHM of the PSF cannot be distinguished. (B) The optical transfer function (OTF) is an equivalent representation of the resolution, giving the bandwidth of the spatial frequencies passed to the image; the broader the OTF, the better the resolution. The data plotted in A and B are gained by probing the fluorescent spot of a scanning microscope with a single molecule of the fluorophore JA 26, both in the conventional (Abbe–type) mode and with STED. Conditions: $n = 1.5$, $\alpha = 67$ deg, wavelengths λ: 635 nm (excitation), 650 to 720 nm (fluorescence collection), and 790 nm (STED). Note the 5.5-fold sharper PSF (A) and the equally broader OTF (B) of STED compared with the diffraction-limited conventional microscope. (C) Subdiffraction resolution with STED microscopy. Two identical molecules located in the focal plane that are only 62 nm apart can be entirely separated by their intensity profile in the image. A similar clear separation by conventional microscopy would require the molecules to be at least 300 nm apart. (Data adapted from Westphal et al. [2003b].).

than those used in MPM (Denk et al., 1990). Moreover, STED has proved to be single-molecule sensitive, despite the proximity of the STED wavelength to the emission peak. In fact, individual molecules have been switched on and off by STED upon command (Westphal et al., 2003a).

The power of STED and 4Pi microscopy has been synergistically combined to demonstrate for the first time an axial resolution of 30 to 40 nm in focusing light microscopy. Superior x/z images of membrane-labeled bacteria have been shown as a result (Dyba and Hell, 2002). More recently, STED 4Pi microscopy has been extended to immunofluorescence imaging. A spatial resolution of $\approx$50 nm has been demonstrated in the imaging of the microtubular meshwork of a mammalian cell (fig. 9.8) (Dyba et al., 2003). These results indicate that the basic physical hurdles have been taken toward attaining a three-dimensional resolution of the order of a few tens of nanometers. Because the samples were mounted in an aqueous buffer(Dyba and Hell, 2002; Dyba et al., 2003), the results indicate that the optical conditions for obtaining subdiffraction resolution are met under the physical conditions encountered in live-cell imaging.

Such ultrasmall detection volumes are also critical to a number of sensitive bioanalytical techniques. For example, fluorescence correlation spectroscopy (FCS) (Magde et al., 1972) relies on small focal volumes to detect rare molecular species or interactions in concentrated solutions (Eigen and Rigler, 1994; Elson and Rigler, 2001). Although volume reduction can be obtained by nanofabricated structures (Levene et al., 2003), STED may prove instrumental in attaining spherical volumes at the nanoscale. Published results imply the possibility of a further decrease of the volume by another order of magnitude (Dyba and Hell, 2002; Westphal et al., 2003b). First applications may be hampered by the requirement of an additional pulsed laser that is tuned to the red edge of the emission spectrum of the dye. Nevertheless STED is, so far, the only known method to squeeze a fluorescence volume to the zeptoliter scale without mechanical contact. Creating ultrasmall volumes,

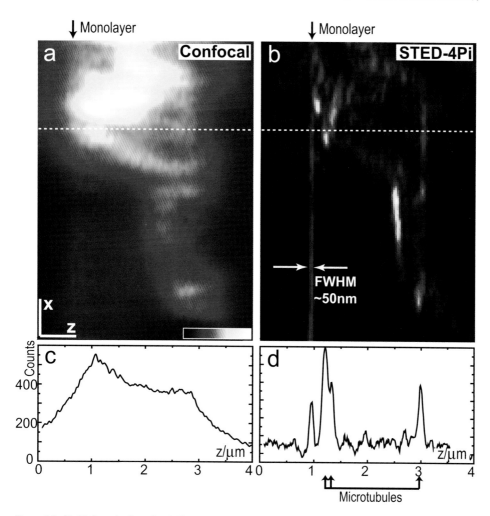

Figure 9.8 (A, B) Standard confocal (A) and STED 4Pi (B) x/z sections from the same site of the microtubular network of a human embryonic kidney (HEK) cell labeled by immunofluorescence. The straight vertical line serving as a resolution reference stems from a monomolecular fluorescent layer on the coverslip. The STED 4Pi image was linearly filtered to remove the effect of side lobes. Note the fundamentally improved clarity in B. (C, D) The profiles of the image data along the marked lines quantify an ≈15-fold improved axial resolution of the STED 4Pi microscope over its confocal counterpart. The profiles of the microtubules (FWHM of 60–70 nm) are broader than the response to the monolayer (≈50 nm). The STED 4Pi microscope is able to distinguish spatially dense features and to reveal weak objects next to bright clusters. Because the cell was mounted in an aqueous buffer and recorded with water immersion lenses, the results indicate that the optical conditions for obtaining subdiffraction resolution can be met in live cells as well. (Data from Dyba et al. [2003].)

tens of nanometers in diameter, by STED may be a pathway to improving the sensitivity of fluorescence-based bioanalytical techniques (Weiss, 2000; Laurence and Weiss, 2003).

Nevertheless, STED microscopy is still at an early stage of development. The suitability of laserdiodes both for excitation and for depletion has been demonstrated (Westphal et al., 2003a), but further efforts are required to implement STED into fast scanning systems. The lack of compact pulsed light sources in the visible range has thus far confined STED investigations to red-emitting dyes. However, as more efficient light sources become available, both visible fluorophores and fluorescent proteins will be interesting candidates for saturated depletion (Gryczynski et al., 1994; Lakowicz and Gryczynski, 1997). Shorter wavelengths will also lead to higher spatial resolution. A further increase in the intensity might be barred in aqueous media by intolerable photobleaching. Although STED pulses more than 300 ps recently improved dye photostability (Dyba and Hell, 2003), saturation factors $\zeta > 200$ might not be readily attainable.

Variations of Reversible Saturable Optically Linear (Fluorescence) Transition between Two Molecular States Microscopy

Fortunately, this limitation can be counteracted. The required saturation intensity can be lowered by choosing systems in which k_{BA} is small. It has therefore been proposed to deplete the ground state (now state A) by targeting an excited state (B) with a comparatively long lifetime (Hell and Kroug, 1995; Hell, 1997), such as the metastable triplet state T_1 (fig. 9.6A). In many fluorophores, T_1 can be reached through S_1 with a quantum efficiency of 1% to 10% (Lakowicz, 1983). Being a forbidden transition, the relaxation of T_1 is 10^3 to 10^5 times slower than that of S_1, thus giving $I_s = 0.1$ to 100 kW/cm^2. The signal to be measured (from the naught) is the fluorescence of the molecules that remain in the singlet system, through a synchronized further excitation (Hell and Kroug, 1995). The disadvantage here is the involvement of T_1 in photobleaching. Potential alternatives are metastable states of rare earth metal ions that are fed through chelates.

Another option is to deplete S_0 by saturating S_1 (now B), as has been proposed recently (Heintzmann et al., 2002). This is perhaps the simplest realization of saturated depletion, because it requires just excitation wavelength matching. However, as the fluorescence emission maps the spatially extended "majority population" in state B, the superresolved images (represented by state A) are hidden under a bright signal from B. Hence, gaining these images requires computational extraction, which makes this approach prone to noise, unless the sample is very sparse. Nevertheless, simplicity of raw data acquisition may render it attractive for the imaging of fixed cells. The saturation intensity is of the same order as in STED microscopy, because the saturation of fluorescence also competes against the spontaneous decay of S_1. Therefore attaining $\zeta > 200$ might involve similar photostability issues.

One possible solution to the quest for large saturation factors at low intensities should be compounds with two (semi) stable states (Dyba and Hell, 2002; Hell et al., 2003). If the rate k_{BA} (and the spontaneous rate k_{AB}) almost vanish, large saturation factors are attained at very low intensities. The lowest useful intensity is set by the concomitant increase in switching time. In the ideal case, the marker is a bistable fluorescent compound that can be photoswitched, at separate wavelengths, from a fluorescent state A to a dark state B, and vice versa. Recently, a photoswitchable coupled molecular system, based on a photochromic diarylethene derivative and a fluorophore, has been reported (Irie et al., 2002). Using equation 9.25, one can assess that focusing less than 100 μW of deep-blue "switch-off light" to an area of 10^{-8} cm^2 for 50 μs should yield better than 5 nm of spatial resolution. Targeted optimization of photochromic or other compounds toward fatigue-free switching and visible light operation could therefore open up radically new avenues in microscopy and data storage (Hell et al., 2003).

For live-cell imaging, fluorescence proteins are more advantageous. Any fluorescent protein that can be pushed to a dark state (Hell, 1997; Hell et al., 2003) (and vice versa), and with a lifetime longer than 10 ns may result in larger saturation factors. Even more attractive are, however, fluorescent proteins that can be "switched on and off" at different wavelengths (Hell et al., 2003). An example is as FP595 (Lukyanov et al., 2000). According to the published data, it allows saturated depletion of the fluorescence state with intensities of less than a few watts per square centimeter. Under favorable switching conditions, these or similar fluorescent proteins should allow a spatial resolution of better than 10 nm (Hell et al., 2003). The involved intensities should also enable parallelization of saturation through an array of minima or dark lines. Initial realization of very low-intensity depletion microscopes may, however, be challenged by switching fatigue (Irie et al., 2002) and overlapping action spectra (Lukyanov et al., 2000). Nevertheless, the prospect of attaining nanoscale resolution with regular lenses and focused light is an incentive to surmount these challenges by strategic fluorophore modification (Hell et al., 2003).

CONCLUSION

Although most textbooks still portray light microscopy as resolution limited, during the past few years, concepts have emerged that are poised to change this view radically. The coherent use of two opposing lenses can greatly improve the suboptimal axial resolution of a confocal microscope. However, the mere generation of an interference pattern is insufficient. The key to axial resolution improvement is the creation of conditions that allow the elimination of the effect of multiple interference fringes. The comparative study of SWM, I^5M, and (two-photon) 4Pi-confocal microscopy from the standpoint of imaging theory reveals that this problem has been solved in the latter.

Although this constitutes a major step forward by yielding almost uniform resolution of around 100 nm along the axial and the transverse directions, diffraction still limits the resolution. In contrast, the approaches based on switching between molecular states outlined in the latter part of this chapter lead to the fundamental breaking of the diffraction barrier, as demonstrated in a number of imaging experiments (Hänninen, 2002). Although the new methods exploit nonlinearities in the interaction between light and dye, they are radically different from methods relying on higher order transitions in the dye, such as, for example, m-photon excitation, $m-$th harmonics generation, or coherent anti-Stokes-Raman scattering (Sheppard and Kompfner, 1978; Shen, 1984). In the latter, the nonlinear signal stems from the action of *more than one photon* at the sample at the same time, and hence high focal intensities are needed as a result of the small interaction cross-sections involved. In contrast, the new approach introduces nonlinear dependence on light intensity by saturating a linear optical transition, avoiding this problem and its spatial resolution is only limited by the attainable saturation level.

Stimulated emission depletion microscopy probably is the most promising realization of the new concept to date. It has been applied to single fluorescent molecules, as well as to simple fixed and live biological specimens, and has demonstrated resolution improvements by a factor of as much as six and has resulted in the smallest fluorescent volumes that have ever been created with focused light. Combined with 4Pi microscopy, STED resulted in the first demonstration of immunofluorescence imaging with a resolution of 30 to 50 nm in the axial direction. Considering the fact that the method is very young, and its theoretical potential, this value is not the limit but the starting point for further developments. Future research on spectroscopy conditions and on practical aspects (Stephens and Allen, 2003) will undoubtedly lead to even more impressive resolving power.

The saturation level in STED is limited as a result of the fast relaxation of the excited state. In contrast, meta-stable switchable dyes and fluorescent proteins should allow high levels of saturation at low light intensities (Hell et al., 2003). This detail is expected to be essential to opening up the cellularnanoscale to imaging with visible light and regularlenses. Although first candidates have been named, dedicated synthesis or protein engineering might uncover a whole new range of suitable markers.

Without a doubt, physics greatly contributed to the advances in biology and biotechnology in the past. Even more intriguing, it now might be up to biotechnology to topple a longstanding alleged paradigm of physics.

Note added in proof: Nanoscopy is a rapidly evolving field, and this chapter reflects the research status prior to publication. For a review describing the most recent advances, please see Hell, 2007.

ACKNOWLEDGMENTS

This chapter is largely based on two previous publications: Nagorni and Hell (2001a) and Hell (2003). We thank M. Dyba, A. Egner, S. Jakobs, J. Jethwa, L. Kastrup, J. Keller, and I. Lein for valuable discussions and for contributing to the work summarized in this chapter.

REFERENCES

Abbe, E. Beiträge zur Theorie des Mikroskops und der mikroskopischen Wahrnehmung. Arch. f. Mikroskop. Anat., 1873;9:413–420.

Albrecht, B. F., A. V. Schweizer, A. C. Cremer. Spatially modulated illumination microscopy allows axial distance resolution in the nanometer range. Appl. Opt., 2002;41(1):80–86.

Bahlmann, K., and S. W. Hell. Polarization effects in 4Pi confocal microscopy studied with water-immersion lenses. Appl. Opt., 2000;39(10):1653–1658.

Bahlmann, K., S. Jakobs, and S. W. Hell. 4Pi-confocal microscopy of live cells. Ultramicroscopy, 2001;87:155–164.

Bailey, B., D. L. Farkas, D. L. Taylor, and F. Lanni. Enhancement of axial resolution in fluorescence microscopy by standing-wave excitation. Nature, 1993;366:44–48.

Bertero, M., G. J. Brakenhoff, F. Malfanti, and H. T. M. Van der Voort. Three-dimensional image restoration and super-resolution in fluorescence confocal microscopy. J. Microsc., 1990;157:3–20.

Blanca, C. M., J. Bewersdorf, and S. W. Hell. Determination of the unknown phase difference in 4Pi-confocal microscopy through the image intensity. Opt. Commun., 2002;206:281–285.

Bloembergen, N. *Nonlinear Optics.* 1965. New York: Benjamin.

Born, M., and E. Wolf. *Principles ofOptics.* 6th ed. 1993. Oxford: Pergamon Press.

Carrington, W. A., R. M. Lynch, E. D. W. Moore, G. Isenberg, K. E. Fogarty, and F. S. Fay. Superresolution in three-dimensional images of fluorescence in cells with minimal light exposure. Science, 1995;268:1483–1487.

Denk, W., J. H. Strickler, and W. W. Webb. Two-photon laser scanning fluorescence microscopy. Science, 1990; 248:73–76.

Dyba, M., and S. W. Hell. Focal spots of size l/23 open up far-field fluorescence microscopy at 33 nm axial resolution. Phys. Rev. Lett., 2002;88:163901.

Dyba, M., and S. W. Hell. Photostability of a fluorescent marker under pulsed excited-state depletion through stimulated emission. Appl. Opt., 2003;42(25):5123–5129.

Dyba, M., S. Jakobs, and S. W. Hell Immunofluorescence stimulated emission depletion microscopy. Nat. Biotechnol., 2003;21(11):1303–1304.

Egner, A., S. Jakobs, and S. W. Hell. Fast 100-nm resolution 3D-microscope reveals structural plasticity of mitochondria in live yeast. Proc. Natl. Acad. Sci. USA, 2002;99:3370–3375.

Egner, A., S. Verrier, A. Goroshkov, H.- D. Söling, and S. W. Hell. 4Pi-microscopy of the Golgi apparatus in live mammalian cells. J. Struct. Biol., 2004;147(1):70–76.

Eigen, M., and R. Rigler. Sorting single molecules: Applications to diagnostics and evolutionary biotechnology. Proc. Natl. Acad. Sci. USA, 1994;91:5740–5747.

Einstein, A. Zur Quantentheorie der Strahlung. Physik. Zeitschr., 1917;18:121–128.

Elson, E. L., and R. Rigler, eds. *Fluorescence Correlation Spectroscopy: Theory and Applications.* 2001. Berlin: Springer.

Failla, A. V., U. Spoeri, B. Albrecht, A. Kroll, and C. Cremer. Nanosizing of fluorescent objects by spatially modulated illumination microscopy. Appl. Opt., 2002;41(34):7275–7283.

Freimann, R., S. Pentz, and H. Hörler. Development of a standing-wave fluorescence microscope with high nodal plane flatness. J. Microsc., 1997;187(3):193–200.

Goodman, J.W. *Introduction to Fourier Optics.* 1968. New York: McGraw Hill.

Göppert-Mayer, M. Über Elementarakte mit zwei Quantensprüngen. Ann. Phys. Lpzg., 1931;9:273–295.

Gryczynski, I., V. Bogdanov, and J. R. Lakowicz. Light quenching and depolarization of fluorescence observed with laser pulses: A new experimental opportunity in time-resolved fluorescence spectroscopy. Biophys. Chem., 1994; 49:223–232.

Gugel, H., J. Bewersdorf, S. Jakobs, J. Engelhardt, R. Storz, and S. W. Hell. Combining 4Pi excitation and detection delivers seven-fold sharper sections in confocal imaging of live cells. Biophys. J. 2004;87;4146–4152.

Gustafsson, M. G. L. Extended resolution fluorescence microscopy. Curr. Opin. Struct. Biol., 1999;9:627–634.

Gustafsson, M. G. L. Surpassing the lateral resolution limit by a factor of two using structured illumination microscopy. J. Microsc., 2000;198(2):82–87.

Gustafsson, M. G. L., D. A. Agard, and J. W. Sedat. Sevenfold improvement of axial resolution in three-dimensional widefield microscopy using two objective lenses. Proc. SPIE, 1995;2412:147–156.

Gustafsson, M. G., D. A. Agard, and J. W. Sedat. Three-dimensional widefield microscopy with two objective lenses: Experimental verification of improved axial resolution. Prof. SPIE, 1996;2655:62–66.

Gustafsson, M. G. L., D. A. Agard, and J. W. Sedat. I5M: Three-dimensional widefield light microscopy with better than 100 nm axial resolution. J. Microsc., 1999;195:10–16.

Hänninen, P. Beyond the diffraction limit. Nature, 2002;419:802.

Hänninen, P. E., L. Lehtelä, and S. W. Hell. Two- and multiphoton excitation of conjugate dyes with continuous wave lasers. Opt. Commun., 1996;130:29–33.

Hecht, B., H. Bielefledt, Y. Inouyne, D. W. Pohl, and L. Novotny. Facts and artifacts in near-field optical microscopy. J. Appl. Phys., 1997;81:1492–2498.

Heintzmann, R., and C. Cremer. Laterally modulated excitation microscopy: Improvement of resolution by using a diffraction grating. SPIE Proc., 1998;3568:185–195.

Heintzmann, R., T. M. Jovin, and C. Cremer. Saturated patterned excitation microscopy: A concept for optical resolution improvement. J. Opt. Soc. Am. A, 2002;19(8):1599–1609.

Hell, S. W. *Double-Scanning Confocal Microscope.* 1990. European patent.

Hell, S. W. Far-field optical nanoscopy. Science, 2007;316:1153–1158.

Hell, S. W. Increasing the resolution of far-field fluorescence light microscopy by point-spread-function engineering. In *Topics in Fluorescence Spectroscopy*, J. R. Lakowicz, ed. 1997. New York: Plenum Press, p. 361–422.

Hell, S. W. Toward fluorescence nanoscopy. Nat. Biotechnol., 2003;21(11):1347–1355.

Hell, S. W., S. Jakobs, and L. Kastrup. Imaging and writing at the nanoscale with focused visible light through saturable optical transitions. Appl. Phys. A, 2003;77:859–860.

Hell, S. W., and M. Kroug. Ground-state depletion fluorescence microscopy: A concept for breaking the diffraction resolution limit. Appl. Phys. B, 1995;60:495–497.

Hell, S. W., and M. Nagorni. 4Pi confocal microscopy with alternate interference. Opt. Lett., 1998;23(20):1567–1569.

Hell, S. W., M. Schrader, P. E. Hänninen, and E. Soini. Resolving fluorescence beads at 100–200 distance with a two-photon 4Pi-microscope working in the near infrared. Opt. Commun., 1995;117:20–24.

Hell, S. W., M. Schrader, and H. T. M. van der Voort. Far-field fluorescence microscopy with three-dimensional resolution in the 100 nm range. J. Microsc., 1997;185(1):1–5.

Hell, S. W., and E. H. K. Stelzer. Fundamental improvement of resolution with a 4Pi-confocal fluorescence microscope using two-photon excitation. Opt. Commun., 1992a;93:277–282.

Hell, S., and E. H. K. Stelzer. Properties of a 4Pi-confocal fluorescence microscope. J. Opt. Soc. Am. A, 1992b;9:2159–2166.

Hell, S. W., and J. Wichmann. Breaking the diffraction resolution limit by stimulated emission: Stimulated emission depletion microscopy. Opt. Lett., 1994;19(11):780–782.

Holmes, T. J. Maximum-likelihood image restoration adapted for non-coherent optical imaging. JOSA A, 1988;5(5):666–673.

Holmes, T. J., S. Bhattacharyya, J. A. Cooper, D. Hanzel, V. Krishnamurthi, W. Lin, B. Roysam, D. H. Szarowski, and J. N. Turner. Light microscopic images reconstruction by maximum likelihood deconvolution. In *Handbook of Biological Confocal Microscopy,* J. Pawley, ed. 1995. New York: Plenum Press, p. 389–400.

Irie, M., T. Fukaminato, T. Sasaki, N. Tamai, and T. Kawai. A digital fluorescent molecular photoswitch. Nature, 2002;420(6917):759–760.

Klar, T. A., E. Engel, and S. W. Hell. Breaking Abbe's diffraction resolution limit in fluorescence microscopy with stimulated emission depletion beams of various shapes. Phys. Rev. E, 2001;64:1–9.

Klar, T. A., S. Jakobs, M. Dyba, A. Egner, and S. W. Hell. Fluorescence microscopy with diffraction resolution limit broken by stimulated emission. Proc. Natl. Acad. Sci. USA, 2000;97:8206–8210.

Krishnamurthi, V., B. Bailey, and F. Lanni. Image processing in 3-D standing wave fluorescence microscopy. Proc. SPIE, 1996;2655:18–25.

Lakowicz, J. R. *Principles of Fluorescence Spectroscopy.* 1983. New York: Plenum Press.

Lakowicz, J. R., and I. Gryczynski. Fluorescence quenching by stimulated emission. In *Topics in Fluorescence Spectroscopy,* J. R. Lakowicz, ed. 1997. New York, NY: Plenum Press, p. 305–355.

Lanni, F. *Applications of Fluorescence in the Biomedical Sciences*, ed. D. L. Taylor. 1986. New York: Liss, p. 520–521.

Laurence, T. A., and S. Weiss. How to detect weak pairs. Science, 2003;299(5607):667–668.

Levene, M. J., J. Korlach, S. W. Turner, M. Foquet, H. G. Craighead, and W. W. Webb. Zero-mode waveguides for single-molecule analysis at high concentrations. Science, 2003;299:682–686.

Lukosz, W. Optical systems with resolving powers exceeding the classical limit. J. Opt. Soc. Am., 1966;56: 1463–1472.

Lukyanov, K. A., A. F. Fradkov, N. G. Gurskaya, M. V. Matz, Y. A. Labas, A. P. Savistky, M. L. Markelov, A. G. Zaraisky, X. Zhao, Y. Fang, W. Tan, and S. A. Lukyanov. Natural animal coloration can be determined by a nonfluorescent green fluorescent protein homolog. J. Biol. Chem., 2000;275(34):25879–25882.

Magde, D., E. L. Elson, and W. W. Webb. Thermodynamic fluctuations in a reacting system: Measurement by fluorescence correlation spectroscopy. Phys. Rev. Lett., 1972;29(11):705–708.

Nagorni, M., and S. W. Hell. 4Pi-confocal microscopy provides three-dimensional images of the microtubule network with 100- to 150-nm resolution. J. Struct. Biol., 1998;123:236–247.

Nagorni, M.. and S. W. Hell. Coherent use of opposing lenses for axial resolution increase in fluorescence microscopy. I. Comparative study of concepts. J. Opt. Soc. Am. A, 2001a;18(1):36–48.

Nagorni, M., and S. W. Hell. Coherent use of opposing lenses for axial resolution increase in fluorescence microscopy. II. Powerand limitation of nonlinearimage restoration. J. Opt. Soc. Am. A, 2001b;18(1):49–54.

Pawley, J., ed. *Handbook ofBiological Confocal Microscopy*. 1995. New York: Plenum Press.

Pohl, D. W., and D. Courjon. *Near Field Optics*. 1993. Dordrecht: Kluwer.

Press, W. H., B. P. Flannery, S. A. Teukolsky, and W. T. Vetterling. *Numerical Recipes in C*. 2nd ed. 1993. Cambridge: Cambridge University Press, p. 1020.

Richards, B., and E. Wolf. Electromagnetic diffraction in optical systems II. Structure of the image field in an aplanatic system. Proc. R. Soc. Lond. A, 1959;253:358–379.

Schmidt, M., M. Nagorni, and S. W. Hell. Subresolution axial distance measurements in far-field fluorescence microscopy with precision of 1 nanometer. Rev. Sci. Instrum., 2000;71:2742–2745.

Schneider, B., B. Albrecht, P. Jaeckle, D. Neofotistos, S. Söding, T. Jäger, and C. Cremer. Nanolocalization measurements in spatially modulated illumination microscopy using two coherent illumination beams. Proc. SPIE, 2000; 3921:321–330.

Schönle, A., P. E. Hänninen, and S. W. Hell. Nonlinear fluorescence through intermolecular energy transfer and resolution increase in fluorescence microscopy. Ann. Phys. (Leipzig), 1999;8(2):115–133.

Schönle, A.. and S. W. Hell. Far-field fluorescence microscopy with repetitive excitation. Eur. Phys. J., 1999;6:283–290.

Schrader, M., K. Bahlmann, G. Giese, and S. W. Hell. 4Pi-confocal imaging in fixed biological specimens. Biophys. J., 1998;75:1659–1668.

Schrader, M., and S. W. Hell. 4Pi-confocal images with axial superresolution. J. Microsc., 1996;183:189–193.

Shen, Y. R. *The Principles ofNonlinear Optics*. 1st ed. 1984. Hoboken, NJ: Wiley.

Sheppard, C. J. R., and R. Kompfner. Resonant scanning optical microscope. Appl. Opt., 1978;17:2879–2882.

Stephens, D. J., and V. J. Allen. Light microscopy techniques for live cell imaging. Science, 2003;300:82–91.

Toraldo di Francia, G. Supergain antennas and optical resolving power. Nuovo Cimento Suppl., 1952;9:426–435.

Weiss, S. Shattering the diffraction limit of light: A revolution in fluorescence microscopy? Proc. Nat. Acad. Sci. USA, 2000;97(16):8747–8749.

Westphal, V., C. M. Blanca, M. Dyba, L. Kastrup, and S. W. Hell. Laser-diode-stimulated emission depletion microscopy. Appl. Phys. Lett., 2003a;82(18):3125–3127.

Westphal, V., L. Kastrup, and S. W. Hell. Lateral resolution of 28 nm (lambda/25) in far-field fluorescence microscopy. Appl. Phys. B, 2003b;77(4):377–380.

Wilson, T., and C. J. R. Sheppard. *Theory and Practice ofScanning Optical Microscopy*. 1984. New York: Academic Press.

Xu, C., W. Zipfel, J. B. Shear, R. M. Williams, and W. W. Webb. Multiphoton fluorescence excitation: New spectral windows forbiological nonlinearmicr oscopy. P.N.A.S. USA, 1996;93:10763–10768.

10
Fluorescence Imaging in Medical Diagnostics

STEFAN ANDERSSON-ENGELS, KATARINA SVANBERG,

AND SUNE SVANBERG

Humans orient themselves in the world largely utilizing their vision. The physical data recording in the retina of the eye is merged with extremely advanced data processing by the brain, by means of which it is possible to recognize objects and make sense of observations. Human vision largely works on the principle of imaging reflectance spectroscopy, during which the spectroscopic content is retrieved from the different color sensitivity of the retinal cones. However, the eye is sensitive in a very small region of the electromagnetic spectrum: from 400 (violet) to 700 nm (dark red). Obviously, a lot of information is available from other wavelength bands, and such data are now being utilized by spectroscopic and imaging systems largely extending the capability of the eye.

It is also well-known that the fluorescence phenomenon can reveal properties now invisible to the naked eye. By inducing fluorescence using invisible ultraviolet light, visible emitted light carries information additional to the one obtained in reflected light. Well-known applications of fluorescence techniques are found in the forensic science, art inspection, and microscopy fields. Passports, monetary bills, and credit cards all carry numbers, letters, and images only visible in fluorescence.

Medical diagnostics strongly relies on "the trained doctor's eye" (i.e., reflectance information processed to a high level of sophistication using maybe decades of professional experience). It is not unexpected that fluorescence could add diagnostic capability, especially if combined with electronic data recording and processing. This has also been found to be the case, and the current chapter reviews the field of *fluorescence imaging in medical diagnostics.*

The organization of this chapter is as follows. We first give a medical background and clinical motivation to the rather large efforts now going on in the field. In particular, we address the question in which medical subfields the new diagnostic techniques of fluorescence imaging can have a particular impact. In the following section we give a general overview of pertinent concepts of fluorescence spectroscopy and imaging, with special emphasis on the efficient harvesting of data. We then provide an overview of the different approaches and implementation strategies taken by different actors in the field. Although we have made a good effort to be comprehensive, we apologize for incompleteness and potential unfortunate omissions of important work. This section is also accompanied by several examples of clinical imagery. Last, we consider possible future directions the field might take.

MEDICAL BACKGROUND AND CLINICAL MOTIVATION: DIAGNOSTIC TECHNIQUES FOR TUMOR VISUALIZATION

Malignant Disease and Its Detection

Early diagnosis of malignancies remains the most important prognostic factor in the treatment of cancer. Worldwide, more than half the number of malignant disease cases appears in developing countries, according to statistics from the World Health Report of the World Health Organization. Approximately 75% of these malignancies are found to be incurable as a result of delayed diagnosis (Garfinkel, 1991). Although there is a variety of diagnostic procedures to be utilized, there is obviously a lack of early and efficient detection modalities. This might be the result of economic reasons for the population in the developing world, but it might also be the result of other factors, such as inadequate techniques for precise tissue characterization. Even if biopsies are acquired, the exact spot forthe early tumormight be missed, orthe endoscopic procedure in the examination may overlook the diseased tissue. Irritation or inflammation may confuse early tumor monitoring, and the changed surface might falsely be stated as nonmalignant.

There is a variety of diagnostic procedures to be utilized for localizing malignancies (MacKay, 1984; Robb, 1994; Guy, 1996). Conventional X-ray examination based on ionizing radiation has a long tradition, and the use of this technique continues to be of great value in a variety of diagnostic schemes. Another X-ray-based technique is computerized tomography for visualizing sections of the body in axial, coronal, or sagittal cuts. Concerns with these techniques are the risks related to the X-ray doses delivered to the patients (Hall, 1994). A typical chest examination is estimated to deposit 200 μGy to the lung parenchyma and approximately the same dose to the mammary glandular tissue. A urethrocystography deposits higherabsor bed doses, with typically 3000 μGy to the bone marrow and 20,000 μGy and 15,000 μGy to the testes and ovaries, respectively. A carotid angiography delivers 15,000 μGy to the bone marrow. By the introduction of digital plates, the absorbed dose to the patient is reduced by approximately a factor of 10.

Techniques that do not involve ionizing radiation are magnetic resonance imaging and ultrasound. These imaging techniques are claimed to be absolutely safe because no X-rays are involved. This relies on an assumption that the high magnetic fields involved do not cause any biological effect. The risks have, however, not been evaluated in similar detail as forthese forX-r ay-based methods.

The use of radioisotopes fortumordetection belongs to the field of nuclearmedicine. The technique is based on the tracer principle with the administration of a radionuclide to the patient. Various radionuclides are known to be accumulated preferably in certain tissue types, such as ^{123}I forthe thyroid gland and ^{99}Tc forosseous metastases. Examples of radionuclide detection are single-photon emission computed tomography (or SPECT) and positron emission tomography (or PET).

Direct visual inspection with or without magnification is performed in various clinical situations, such as in the oral cavity, the glottic area, or the outer surface of the genital tract of females. For the investigation of hollow organs in the body, endoscopic techniques are used. An endoscope is eithera flexible ora rigid one forvisualization of interiorwalls. The investigation is performed with the illumination of a white lamp and the detection is based on what the reflected light reveals.

To motivate the introduction of new optically based techniques, it is important to evaluate the detection potential of existing methods—in particular, the possibility for early detection of cancerorpr ecancerin a comparison with the new methods. From a clinical point of view, it is also important to restrict the arenas of action to those where optical characterization can be evaluated and compared with the existing gold standard techniques for tissue characterization: histopathology.

Some Statistical Concepts

The diagnostic impact of a technique is evaluated by the ability of the technique to detect cancer or precancer accurately. Statistical analysis refers to *sensitivity, specificity, positive* and *negative predicted values,* and *accuracy* forthe technique used. In the search of malignant transformation areas, the localization of the starting site of a cancer (a precancer) is of highest importance to achieve high sensitivity. A positive result from this type of examination means that the test indicates a lesion, whereas a true-positive and a false-positive result means that the test indicates a lesion that is later verified or not verified by a gold standard method (often histopathology), respectively. True and false-negative results are defined correspondingly; no lesion is correctly or incorrectly indicated by the test. The *sensitivity* of a diagnostic method is widely defined as its ability to identify correctly those patients who have the disease or, expressed differently, the proportion of patients with the disease and those who also have positive test results:

$$Sensitivity = \frac{true\ positives}{true\ positives + false\ negatives} \tag{10.1}$$

Specificity in a statistical sense is the ability of an investigation to identify correctly those patients who do not have the disease, or the proportion of patients without disease who also have negative results:

$$Specificity = \frac{true\ negatives}{false\ positives + true\ negatives} \tag{10.2}$$

Accuracy lumps together the true positives and the true negatives, and is therefore of less value in the evaluation of a technique and is not considered further here. The *positive predicted value* indicates the probability of whether the disease is actually present if the test is positive:

$$Positive\ predicted\ value = \frac{true\ positives}{true\ positives + false\ positives} \tag{10.3}$$

The *negative predicted value* defines whetherthe disease is likely to be absent if the result is negative:

$$Negative\ predicted\ value = \frac{true\ negatives}{true\ negatives + false\ negatives} \tag{10.4}$$

The sensitivity and specificity of a test are generally independent of the prevalence of the disease and are therefore often called *intrinsic operating characteristics.* On the other hand, the predicted positive and negative values and the accuracy are highly dependent on the prevalence of the disease in the examined group or population and, for example, regional variations may influence the result. It can also be noted that a very high specificity of a diagnostic method is required in applications with low prevalence to provide valuable diagnostic information.

Optical techniques such as fluorescence imaging are of particular interest for surface visualization of various epithelial tissues such as the skin or the mucosa of interior hollow organs including easily reachable areas, such as the oral cavity or genital tract, besides endoscopically accessible organs. For tumor detection in solid organs, such as liver, kidney, and breast parenchyma, other techniques are applicable. Lesions located deep in the tissue are more difficult to visualize using optical techniques because of poor penetration and multiple scattering of light in tissue. In addition, for the detection of metastases to osseous tissue, fluorescence will most likely never serve as a diagnostic tool.

Spectroscopically Based Optical Detection

The main technique based on spectroscopical analysis of tissue is laser- or light-induced fluorescence (LIF). Optical detection, and in particular LIF, exhibits various characteristics, which indicate, when fully developed, that the method could become important in many clinical situations:

- Noninvasiveness
- Real-time capability
- Endoscopic compatibility
- Point monitoring (scientific mode) and imaging (ultimate goal) implementation

An obvious advantage with fluorescence detection is the real-time characterization aspect. This means that LIF can be utilized interactively during the procedure and can give updated information during the diagnostic procedure. The property of fluorescence to change early in the development of malignant lesions and the utility for identification of premalignant lesions are of particular clinical interest.

During an endoscopic diagnostic procedure, the epithelium of the hollow organ is visualized and the surface is judged from its appearance in the reflected light illumination. Signs that are registered by the endoscopist are the variations in color and the structure of the mucosa. If the mucosa is reddened, the most probable reason for this is a chronic irritation of some origin, such as inflammation of the tissue. The clinical challenge is, of course, to differentiate benign lesions from early malignancies. By their clinical appearance, these conditions may look very similar, and in this situation a biopsy is acquired.

The goal with the development of LIF is to bring the technique to the status of being so accurate that the biopsy sampling can be guided by the fluorescence signal (*smart biopsy*), as indicated in figure 10.1. Thus, blind biopsy could be avoided, sparing healthy tissue and increasing the accuracy of tissue sampling. The closest to *immediate* diagnosis in conventional medicine is the interactive process of performing frozen-section evaluation. The term is synonymous for *intraoperative microscopic consultation*. This can usually be performed within a time span of 30 to 45 minutes for each biopsy sample. This process is usually not used in connection with early tumor detection, but rather it is used as an interactive process in connection with surgical procedures—for example, Moh's surgery for aggressive skin malignancies or during neurosurgical procedures for glioma resection.

The term *smart biopsy* includes guidance for the biopsy procedure both to avoid blind tissue sampling and to avoid acquiring unnecessary biopsies. This is of particular importance in certain organs, such as the vocal cords or other sensitive structures, like the cervical area in the female gynecological tract.

Spectroscopically based characterization is based on very early biochemical as well as morphological changes in the tissue. Because biomolecular changes occur very early in the transformation from fully normal to diseased tissue, even precancerous lesions might be visualized. When the malignant transformation proceeds, structural changes in the cells appear, which might be visualized during an endoscopic investigation. These changes guide the endoscopist in sampling tissue for further histopathological characterization. However,

Figure 10.1 One of the main goals with fluorescence examinations is to guide *smart biopsies*.

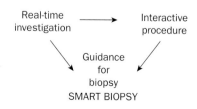

Endoscopic investigation
Visual inspection of hollow organs with the option of tissue sampling

CLINICAL CHALLENGE!

Differentiate inflammation/chronic irritation from neoplasia
visualize small malignancies–noninvasive cancer or precancer

Guidance for tissue sampling

Avoid
blind
biopsy

Minimize
or
avoid biopsies

Figure 10.2 The clinical challenge during an endoscopic examination in a case of a suspected tumor.

these changes might also be related to other conditions, such as inflammation, and, although suspicious, may only contain benign tissue. The clinically relevant challenge is summarized in figure 10.2.

Fluorescence is normally excited with ultraviolet, near-ultraviolet, or violet light, which results in shallow tissue penetration. This is not a limitation, but it is rather of high clinical importance because early changes occur in the thin epithelium. Although near-infrared excited fluorescence diagnostics may be a powerful tool for identifying certain deep lesions, in this chapterwe confine ourdiscussion to superficial lesions. Fully developed fluorescence detection—in particular, in an imaging mode—can obviously play an important role as a complementary method for detection of early, noninvasive malignancies or precancerous lesions in various clinical specialties. Point-monitoring detection may be of clinical value for certain applications as well as to build up a spectral "handbook" of various tissue types in different organs. A correlation between the spectral signal and histopathology has to be performed during the developmental and evaluation phases. A precise spatial resolution must be achieved based on the traditional biopsy sampling procedure, including full histologic investigation. During this process, a histopathological "tissue map" is built up for correlation with the fluorescence signatures in an imaging mode.

Clinical Challenges

The main clinical challenge in many specialties is to differentiate benign conditions: inflammation and early noninvasive carcinoma or precancer. Inflammation is best defined as the local reaction of vascularized tissue to injury. Inflammation can be caused by living organisms, such as bacteria or virus, but also by physical agents (radiation, trauma) and chemical agents (smoke). The classic local signs of inflammation are heat, redness, swelling, pain, and, in some cases, loss of function. With acute inflammation there is always a hemodynamic effect, with vasodilation causing increased blood flow in connection with exudative processes resulting from changes in the permeability of the vessels. The clinical signs that are observed are changed appearance of the epithelium, mainly in terms of colorand thickness of the mucosa, which might be similarto what is seen during early malignant transformation. An early cancer or precancer might indicate ulcer

Table 10-1. Applications to various clinical specialties

Location	Clinical Challenge
Urinary bladder	Delineate areas of dysplasia and/or carcinoma in situ with/without concomitantly occurring, easily recognizable malignant papillary tumor growth
	Visualize and characterize tissue type of the tumor "bottom" after resection of, for example, papillary tumors to judge the surgical radicality for minimizing recurrences postoperatively
Bronchus	Visualize areas of early malignancies in cases in which signs of malignant cells are revealed (e.g., clinical symptoms or positive sputum cytology [occult lung cancer or precancer])
	Identify early recurrence at the bronchus end after, for example, pulmonectomy or lobectomy
	Delineate areas of carcinoma in situ in the surrounding of clearly recognizable tumor
Female genital tract	Characterize lesions in the vulva and cervix uteri—in particular, the transformation zone of the portio
	Detect noninvasive cancer or cancer in situ in connection with the treatment planning of a known higher stage malignant tumor
Skin	Characterize various skin lesions
	Differentiate between malignant and benign lesions
	Delineate tumor margins interactively in connection with surgery (e.g., Moh's surgery for morphea-growing basal cell carcinoma)
Head and neck	Identify dysplasia with or without a connection with leukoplakia (oral cavity, naso/pharynx glottis)
	Delineate tumor borders on, for example, the vocal cords (implications for treatment regimen)
Gastrointestinal tract	Characterize various types of colonic polyps
	Differentiate between neoplastic (adenomatous, villous, and, in particular, flat, sessile polyps) and nonneoplastic polyps (hyperplastic or metaplastic)

as well as inflammation. Often, a precancer or a cancer is also associated with inflammation. Therefore, the clinical signs are not specific regarding whether it is a benign inflammatory or a premalignant process in the mucosa. This applies for many organs, such as the urinary bladder, the female genital tract, the ear–nose–throat region, the bronchus, and the esophagus. Beside this more general consideration, each organ system exhibits particular indications for a need for more specific possibilities for visualization of diseased areas. An overview of clinical challenges in different organs is given in Table 10.1.

In addition to this overview of various organ systems, a more detailed description is presented here to illuminate and define clinical areas and situations in which the diagnostic procedures could be improved by introducing fluorescence as a complementary technique.

Lower Urinary Tract

Although imbedded in the deep part of the pelvis, the urinary bladder is fully accessible for endoscopic investigation. The close relationship of the female genital tract to the urinary bladder is of considerable significance. There are three major types of malignancies in the urinary bladder, out of which transitional cell carcinomas represent approximately 90%. From a morphological point of view, the malignancies can be divided into a few groups in which the papillary tumors are easily recognizable, with their exophytic growth pattern often attached by a stalk to the mucosa. Besides these tumors, there are the flat lesions that grow only as a plaquelike thickening of the mucosa without a well-defined papillary structure. These tumors often tend to be more aggressive and anaplastic compared with papillary malignancies. As in many other specialties, it is of high importance to recognize these flat tumors, which usually have a more severe biological natural history. These tumors may also grow in conjunction with papillary tumors, which is an even more threatening situation. The less aggressive tumors may be treated and the more malignant ones may be left unrecognized.

Papillary tumors are often treated by cutting the tumors at the stem bottom by means of laser coagulation or electrical loop strangulation. It is known that there is a tendency for recurrences after these procedures. One obvious reason for this might be that complete surgical resection is not achieved and residual tumor is left at the bottom.

Bronchus

Bronchoscopy is a well-established diagnostic procedure, with growing importance with the increasing incidence of lung cancer. Among lung malignancies, bronchogenic carcinomas are the preeminent ones. More than 90% of the carcinomas arise from the epithelium in the bronchus. Chronic bronchitis, usually generated by the paramount etiological influence of cigarette smoking, causes atypical metaplasia and dysplasia of the respiratory epithelium. This in turn provides a possible base for cancerous transformation. An investigation of the sputum may reveal atypical cells, but because the epithelium overall shows bronchitis with hyperemia and swelling of the mucosa, the site of interest may be difficult to reveal. The morphological changes are most pronounced in the smaller bronchi but may also extend to the proximally located bifurcation of the trachea.

Carcinoma in situ and dysplasia of various grades can also occur in the neighborhood of invasive tumor and should be included in the treatment to provide as much of a radical treatment as possible.

Female Genital Tract

The main area of interest in the female genital tract is the portio of the cervix uteri. It is extremely common with lesions of the cervix. Most of them represent nonspecific acute or chronic cervicitis. The transformation zone or the squamocolumnar junction is the area of activity, with an intermingling of squamous and columnar epithelium. The transformation zone has dilated capillaries but with normal architecture. This is also the site of squamous metaplasia, a physiological condition characterized by temporarily hosting a keratinized mucosa, although in normal conditions nonkeratinized stratified squamous mucosa is noted. The transformation zone is also the site where the glands open up and the Nabothian cysts are found. More than 90% of cervical lesions appear in this zone.

Conventional diagnostic procedures include a vaginal smear for cytological examination of the exfoliated cells, followed by colposcopy if abnormal cells are revealed. However, the cytological test may be misinterpreted. The cells in metaplasia can be so active that they are mistaken fordysplasia. On the otherhand, cells in malignancies can be of nonvital origin and may not reveal the malignancy. The colposcopy procedure is performed with an acetic acid wash and a green filter for revealing the blood vessel structures. At least in 30% of cases, cervicitis cannot be separated from cervical intraepithelial neoplasias with various degree of dysplasia, including carcinoma in situ with severe dysplasia. This problem is further illustrated by the fact that approximately 75% of all biopsies taken in the cervical area result in a benign diagnosis.

Skin

Most patients with skin lesions are seen by a nondermatologically trained physician, which results in an unnecessarily large number of excisions and biopsies. Pigmented as well as nonpigmented skin lesions can be divided into benign and malignant types. It is of particular importance to notice that, forexample, a malignant melanoma should neverbe sampled, because the diagnostic procedure may be a hazard for the future prognosis of the patient. With an increasing incidence of skin lesions, the clinical problem of differentiating benign and malignant skin lesions is accelerating.

There is a certain number of recurrences in the treatment of malignant skin lesions, and many of these cases can certainly be explained by including too-small margins in the treatment procedure. In aggressive morphea-growing basal cell carcinomas, a very time-consuming treatment procedure is sometimes used with an interaction between pathology and surgery. This may result in a full-day surgery.

Head and Neck

Oral cancer accounts for approximately 5% of all malignant tumors in humans. Because the oral cavity is easily reachable for visualization, cancers should be detected at an early stage. If biopsy procedures are performed at the right spot, the accuracy in the histopathological evaluation is close to 100%. Surface scraping of the lesions is mostly used for screening tests. However, exfoliative examination of the oral cavity secretion is not reliable, resulting from the fact that the containing proteolytic enzymes rapidly destroy the cells, with false-negative results.

Malignancy of the vocal cords is more common in men than in women. Cigarette smoking and most probably also other environmental factors play an important role for cancer development. Very often the tumor starts from hyperkeratotic and thickened mucosa as a sign of chronic inflammation, in which the cells change into metaplastic and dysplastic ones. Malignant tumors of the vocal cords are 95% squamous cell carcinomas and all start as thin in situ lesions, laterwith a thickening of the epithelium and the development of ulcers, which sometimes show secondary infections, which shield the malignant tumor and disturb the diagnosis.

Gastrointestinal Tract

Carcinomas and adenomas of the colon are among the most common forms of neoplasias in men and women. Adenomas are also called *epithelial polyps* because they protrude into the lumen of the large intestine. In a population of olderadults they appearwith a frequency of 5% to 25%, oreven as high as up to 50% in autopsy studies. The polyps are divided into two main groups: neoplastic or nonneoplastic hyperplastic polyps. The hyperplastic polyps arise from the cells deep in the crypts of the colonic mucosal glands. With a certain degree of mitotic activity and loss of growth control, the polyps are built up and migrate upward, forming a hyperplastic benign polyp. Further loss of growth control is thought to result in neoplastic polyps. These polyps include increased cell division also from the upper part of the colonic glandularcr ypts.

Neoplastic polyps are divided into two main groups: adenomatous polyps or polypoid adenomas, and villous adenomas. Adenomatous polyps are mainly pedunculated lesions with a well-defined stalk that occasionally appears as sessile hemi- or spherical lesions sitting directly on the colonic mucosa. These flat, sessile lesions are the ones most commonly seen during colonoscopy. Villous adenomas are most often of the sessile type. Carcinoma in situ without invasion into the colonic wall is seen in approximately 10% of villous adenomas.

The major significance relates to the fact that these polyps are precancerous lesions. Therefore, every adenomatous polyp represents a diagnostic challenge because there is a certain risk of cell atypia harbored among the cells in the polyps. There is evidence for the adenoma–carcinoma sequence, which explains a certain number of colon cancer cases. It is a well-established fact that patients with adenomas are at an increased risk of developing colon cancer.

The diagnostic problem in the clinic is that approximately 90% of the polyps are hyperplastic and only 10% of the polyps are at risk for the patients. The clinical challenge is to characterize the tissue type and differentiate between the two main types in real time, and thus avoid many unnecessary excisions.

CONCEPTS OF FLUORESCENCE SPECTROSCOPY AND IMAGING

Basic Considerations

Fluorescence imaging can be performed with systems of different degrees of complexity. The high performance and selectivity of advanced and complex systems have to be put in relation to low price and ease of operation that may pertain for more simple systems.

Imaging is always based on the spectroscopic properties in individual object points, and thus the starting point for imaging must be a detailed study of the spectral and temporal properties of the fluorescence from different types of tissue, which must be performed in close connection to classic pathological identification. Imaging is based on the data obtained from thousands of individual points. Clearly, a presentation of all these individual spectra does not provide information comprehensible to a clinician or to anybody else for that matter. Rather, the data have to be compressed into an easily interpretable two-dimensional image, in which abnormal tissue is noted in a predetermined way. This is the purpose of a spectroscopic imaging system, which can be implemented in a number of ways. Here we discuss the concept of *multispectral imaging,* which is a very general class of techniques pertinent to fields ranging from astronomy and satellite imagery to medical imaging and microscopy. The extraction of data can be performed using an *optimized contrast function* orusing *multivariate techniques.* The basic idea behind the use of contrast functions is to compress a spectrum into a single number, which becomes large for the pixels with selected properties (i.e., is malignant) and low for all other pixels. Then a threshold can be set and contrast function pixel values above the threshold are displayed as *fulfilling a cancer criterion.* Multivariate techniques include many varieties, such as partial least squares and principal component analysis (PCA). The latter technique reduces large amounts of data into a small number of basic functions (principal components), and an individual spectrum can be localized in, for example, a two-dimensional diagram, where, for instance, the second principal component (PC2) value (score) is plotted versus the first principal component (PC1) score.

Fluorescence investigations of tissue can focus on the emission properties of the naturally occurring chromophores of tissue (*autofluorescence* studies) oron artificially introduced fluorescent substances, frequently referred to as *fluorescent markers.* Out of the numerous biomolecular species occurring in normal and carcinogenic tissue, elastin, collagen, carotene, tryptophan, and NADH can be mentioned. Typical fluorescence spectra of such substances excited with a nitrogen laser at 337 nm are shown in figure 10.3 (Andersson-Engels et al., 1992).

As seen in figure 10.3, all the substances exhibit a bluish fluorescence, but with slightly different peak wavelengths. Clearly, a combination of such spectra into a tissue composite signature will also exhibit a bluish fluorescence light distribution. It might be surprising that the spectral appearance can still be sufficiently different between normal and malignant tissue to allow tumor discrimination based on autofluorescence only, as discussed in more detail later. Fluorescent tumor markers accumulate preferentially with malignant

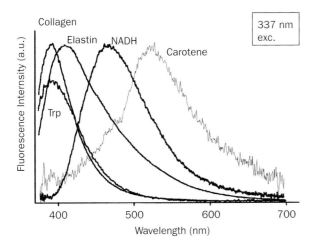

Figure 10.3 Fluorescence curves for a number of compounds present in human tissue. The spectra have been recorded for 337 nm excitation and are normalized to the same maximum intensity. (From Andersson-Engels et al. [1992].)

cells, with a relative selectivity ranging from 2 up to perhaps 10. The tumor markers used in fluorescence diagnostics are chosen to provide a specific fluorescence in the red spectral region, where autofluorescence is weak and unstructured. Tumor markers are also frequently selected to perform photodynamic therapeutic procedures, which are then called *photosensitizers*.

Point monitoring of tissue fluorescence is most conveniently performed with fiber optically coupled laserfluor osensors, in which a single fiberis used both forlaunching the excitation light and for receiving the fluorescence used. The data are obtained by reading out the full fluorescence spectrum captured with a CCD detector placed in the image plane of a spectrograph that disperses the fluorescence brought to its entrance slit by the optical fiber. The development of clinically adapted point-monitoring instrumentation is illustrated in figure 10.4, where three systems, developed at the Lund University Medical Laser Center,

Figure 10.4 Photographs of three fiber-based fluorosensors for pointwise tissue diagnostics. (top) System with nitrogen laser excitation and an intensified optical multichannel analyzer system (Andersson-Engels et al., 1991, 1992). (bottom) System of "carry-on-luggage" size comprising a nitrogen–dye laser combination and a compact fluorescence spectrometer with fast spectral gating allowing time-resolved spectra to be recorded (Klinteberg et al. 2005). (facing page) System with violet diode laser excitation and spectral recording in an integrated minispectrometer with CCD readout (Gustafsson et al., 2000).

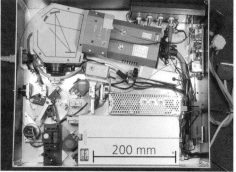

Figure 10.4 (*Continued*)

are shown (Andersson-Engels et al., 1991; Gustafsson et al., 2000; Klinteberg, 2005). The sequence represents a change in volume from 1 m^3 to about 1 dm^3, and a change in weight from 100 kg to 1 kg.

Basically, fluorescence imaging can be performed based on *spectrally resolved* or *temporally resolved* data. Again, point recording best serves to describe the concepts. Recordings in the spectral and temporal domains are shown in figures 10.5 and 10.6. Each figure contains data from malignant as well as normal tissue. Spectrally resolved data show the fluorescence intensity emitted from the sample as a function of wavelength. The data are integrated over a certain time. What this means is very clear for the case of continuous excitation. In the case of short-pulse (nanosecond) laser excitation, fluorescence is released with a certain decay time, frequently on the order of 1 to 20 ns, corresponding to the excited state lifetime. In a spectrally resolving system, the total integrated emission is normally recorded. On the contrary, a time-resolving system normally observes the fluorescence as a function of time in a selected wavelength band. Hybrid concepts also exist in which spectral and temporal resolution are combined.

Figure 10.5 (Svanberg et al., 1998) shows the spectrally resolved fluorescence intensity for diseased and normal tissue of a patient who had previously received delta aminolevulinic acid (ALA). The production of protoporphyrin IX (PpIX) is enhanced in malignant cells, whereas the surrounding tissue only shows traces of the substance, which is characterized by sharp peaks at 635 nm and a less prominent one at 705 nm. The strength of the PpIX signal can be enhanced by exciting at a wavelength of 405 nm, where the molecules have their absorption maxima, instead of at 337 nm. However, this shorter wavelength better excites the autofluorescence, which produces a broad bluish hump in the spectrum. It is a general observation that blue fluorescence is reduced in tumor tissue compared with normal tissue. At least two factors may be responsible for this observation: Tumors have an increased blood flow, which reduces ultraviolet light penetration into the tissue and thus fluorescence production. Furthermore, the balance between the strongly fluorescent substance NADH and the more weakly fluorescing form NAD^+ may be shifted in malignant tissue (Andersson-Engels et al., 1990a,b).

In figure 10.6, temporal fluorescence decay curves for short-pulse ultraviolet excitation are shown for normal and malignant tissue in a Wistar-Furth rat injected intravenously with

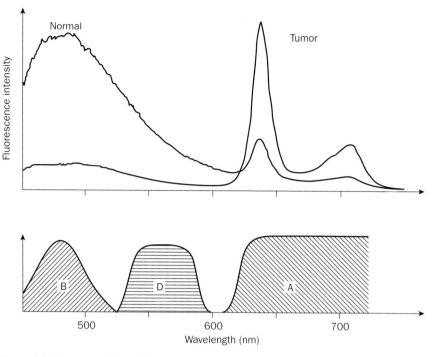

Figure 10.5 Examples of tissue fluorescence spectra recorded for 337 nm excitation. δ-Aminolevulinic acid had been administered 4 hours earlier, leading to protoporphyrin IX formation in tumor tissue and to a lesser extent in normal surrounding tissue as well. Three fluorescence bands are selected as being useful for imaging recordings (Svanberg et al., 1998).

Photofrin. The fluorescence is detected in a spectral band around the 630-nm porphyrin emission peak present in tissue. It can be seen that a long-lived major component is present in the tumor. There is also a more short-lived decay component resulting from the background autofluorescence at which the red peak is located.

It should be noted that for multispectral as well as multitemporal data, recording more than one excitation wavelength can be utilized to enhance further the diagnostic capability of the system. Foreach excitation wavelength, a set of data is obtained and data sets can then be merged in computerized contrast enhancement for the unambiguous identification of diseased tissue.

Data Recording and Handling

Spectral and temporal detections are presented in figures 10.5 and 10.6. The selected bands suggest an integration of the spectral or temporal signal falling into the particular band. This is done to facilitate data handling and to avoid fully resolving systems. Although it goes beyond the simplest implementation of fluorescence imaging, it is highly desirable to record data in more than one spectral or temporal band. This is related to the desire to perform imaging in some *dimensionless quantity*. The simplest type of such a quantity is a *ratio*—forexample, the ratio of the intensities $I(\lambda_1)$ and $I(\lambda_2)$ recorded in spectral bands centered around λ_1 and λ_2. The attractiveness of such dimensionless quantities is related to the fact that they are immune to

- Distance changes between the tissue and measurement equipment
- Variations in the angle of incidence of radiation on tissue
- Fluctuations in the illumination source and detection system efficiency

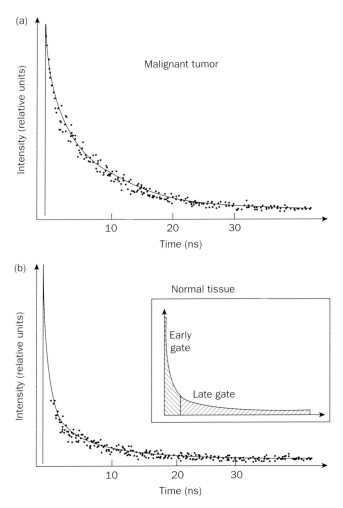

Figure 10.6 Temporally resolved fluorescence decay curves obtained with short-pulse excitation. Two time gates, an early one (prompt) and a delayed one (afterglow), are indicated. Clearly, the late fluorescence intensity divided by the early intensity yields very different results for tumor and normal tissue. (Adapted from Andersson-Engels et al. [1990a].)

This is very valuable, because many of these aspects are uncontrollable in clinical practice. The dimensionless quantity (e.g., a ratio signal) is basically only sensitive to the intrinsic chemical properties of the tissue. The *shape* of the spectrum or the *temporal decay time* is studied rather than the fluorescence photon yield. In connection with figure 10.5, we noted that a tumor is characterized by a red peak increase accompanied by a blue decrease. Thus, we see that by dividing the red intensity by the blue, a contrast enhancement is obtained at the same time that a dimensionless quantity is formed. This also constitutes a *contrast function*, as defined earlier, with desirable properties. Clearly, the recording and utilization of many bands allows a description of more complicated shapes in *multispectral* or *multitemporal imaging*.

The desire to record the data in more than one spectral or temporal band immediately brings up an additional aspect to consider: Are data to be recorded *sequentially* or *simultaneously?* Sequential recording may be technically simpler but may not be allowable for moving objects. These aspects are discussed next in connection with a description of different available systems.

Because fluorescence, on which medical imaging currently under consideration is based, is a weak phenomenon, the question of achieving a sufficient *signal-to-noise ratio* and the related *background rejection* aspects need adequate attention. Fluorescence induced by a continuous wave (CW) source is always minute in comparison with daylight or typical operating-lamp backgrounds. Such excitation thus requires the recording in, essentially, darkness. This can automatically be provided in endoscopic applications, but otherwise only by switching off the illumination—a step that is not desirable in routine health care. These inconveniences can be avoided by pulsed excitation combined with *gated detection* using, for example, a microchannel plate image intensifier in front of a CCD detector. With typical excitation times of 10 ns and a gate time of 100 ns, the induced fluorescence normally overwhelms the background radiation. Still, a sufficient excitation pulse energy is needed to provide a sufficient image quality in view of the shot noise limitations in the detection.

Fluorescence Imaging Modes

An image can be recorded in a number of different ways. For a stationary object, image points can be interrogated one by one sequentially in a *point-scanning* (whisk broom) mode. It is also possible to image a full line simultaneously using so-called *pushbroom scanning geometry,* commonly used in satellite imagery. Last, simultaneous *full two-dimensional recording* can be accomplished. With reference to existing prototype systems we can also identify at least six different fluorescence imaging modes, designated as modes A through F in the following sections.

Mode A: Single-Band Imaging

Single-band imaging is the simplest way of fluorescence imaging, which is also compatible with visual inspection. A CW illumination source (e.g., a mercury lamp with a strong line at 365 nm and anotherone at 405 nm) can typically be used as an excitation source. An additional strong line is available at 254 nm, but should not be used for medical imaging because there is a 325-nm wavelength limit under which the photons are energetic enough to split DNA molecules. The excitation line of interest is selected with a proper filter, which can be a colored glass filter, an interference filter, a liquid filter, or a combination thereof. At the detection site, the excitation light is effectively blocked by a colored glass filter, leaving only the induced fluorescence to be recorded. All fluorescence can be recorded, or selected by a bandpass filter. The human eye is sensitive in the range 400 (violet) to 700 nm (dark red), whereas normal video cameras extend the region somewhat both toward the ultraviolet and the infrared. Because only one band is used, the recording becomes very sensitive both to nonuniform illumination and surface topography. The surface can be intermittently observed also in white-light illumination to orient fluorescence features to the area under investigation. It should be noted that the human brain is a unique "online computer" for image evaluation, because it can interpret features that remain stationary although the field of view and the illumination is changed. The Storz system described later is an example of this first type of simple fluorescence imaging systems.

Mode B: Dual-Band Imaging

If the fluorescence of the tissue is observed in two spectral bands, the advantages of dimensionless imaging and contrast function formation can be utilized. Such a system can be implemented either by shifting bandpass filters in front of a CCD camera or, better, by using two separately filtered cameras imaging the same scene, as schematically illustrated in figure 10.7. For efficient photon economy, a dichroic beam splitter should be used for dividing the fluorescence. By using two separate cameras, it is possible to monitor both images exactly simultaneously, which eliminates motion artifacts that may confuse the

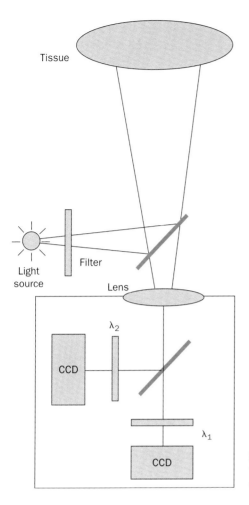

Figure 10.7 Principal arrangement of a dual-band fluorescence imaging system operating with two cameras.

data from a single-camera system. Full digital data processing is performed in a system as described by the Lausanne group (Wagnières et al., 1999). A variety of the dual-band two-camera concept was developed in the Santa Barbara/Xillix system (Novadaq Technologies, Mississauga, Canada, formerly Xillix; discussed later), in which the outputs from the two cameras individually drive two channels of a standard RGB color monitor. Changes in false-color hues, as tumors are encountered, are interpreted by the "onboard computer" of the operator based on experience. Because the images are not digitized, the advantages of dimensionless imaging are not realized.

Mode C: Multicolor Imaging

When more than two spectral bands are used, we denote this as *multicolor imaging*. We have already given the motivation for an extended wavelength coverage for being able to construct more advanced and powerful contrast functions. Because biological tissue fluorescence, in contrast to, for example, gaseous samples, exhibits only slow wavelength variations, a limited number of spectral bands will basically characterize the full spectrum, because nothing but a smooth intensity transition can occur between the selected wavelength bands. Observation of the spectra given in figure 10.5 gives some indication of how many bands might be needed for a reasonable characterization of the full spectrum. Four bands clearly would provide a fair description. In particular, it is important to monitor the background signal before the isolated emission peak of a fluorescent tumor marker to be able to assess the intensity of a truly fluorescent marker-dependent intensity at the nominal peak wavelength. Even in the complete absence of a fluorescent marker-related peak, a filter at that wavelength would see the background intensity of the autofluorescence signal. Clearly, for low fluorescent marker concentrations, when the specific signal might only be manifested as a minute deviation from the spectrally smoothly falling-off autofluorescence intensity, prepeak monitoring is needed. If this is achieved, the signals on the peak I_{red}, just before the peak I_{yellow} and in the blue spectral region I_{blue} can be combined with the contrast function

$$F_c = (I_{red} - cI_{yellow})/I_{blue} \qquad (10.5)$$

where the constant c is selected to give a zero numerator for normal tissue without any fluorescent marker uptake. In this way, the true porphyrin signal divided by the autofluorescence is monitored as the contrast function.

A practical implementation of multicolor imaging based on early work at the Lund Medical LaserCenter(Montán   et al., 1985; Andersson et al., 1987) is discussed later. A special feature of this system is that it allows simultaneous monitoring of four individually filtered fluorescence images followed by digital processing in near real time. A schematic diagram of this concept is shown in figure 10.8.

Regarding the choice of the limited number of wavelengths available in a multicolor imaging system, further considerations than those just discussed for sensitized tissue pertain. If autofluorescence only is used for tissue characterization, it can be particularly valuable to choose wavelength pairs for which the absorption from blood are the same for the two wavelengths included in each pair (Andersson-Engels et al., 1989b). Blood reabsorption with peaks at 405, 540, and 580 nm frequently makes a strong imprint on the fluorescence

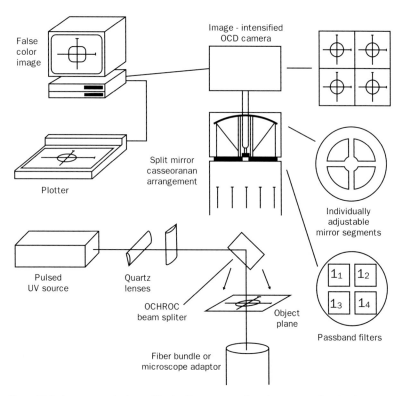

Figure 10.8 Arrangement of a multicolor fluorescence imaging system for simultaneous image recording in four selected bands. (From Andersson-Engels et al. [1990a].)

spectrum, creating, for example, artificial peaks, which correspond to the edges of blood absorption. If a function such as

$$F_c = (I(440 \text{ nm}) - I(360 \text{ nm}))/(I(520 \text{ nm}) - I(600 \text{ nm})) \qquad (10.6)$$

is monitored, its value is dimensionless and at the same time independent of blood reabsorption. These considerations are particularly relevant when imaging tissues with occasional bleeding.

Mode D: Pushbroom Hyperspectral Imaging

In this type of imaging sensor, all points along a line are illuminated by short-wavelength radiation, and the induced fluorescence is for each spatial location dispersed into a spectrum that is picked up by a CCD detector. The general principle is shown in figure 10.9. The line on the tissue under study is imaged along the entrance slit of a spectrometer. The grating characteristics and the slit width determine the spectral resolution. Although in airborne or satellite configurations the measurement line advancement in the passive pushbroom imager is provided by the flight by the platform, the scan using a stationary system is provided by a sweep mirror. Such a geometry was used in an imaging fluorescence lidar system (Edneret al., 1994) and is used in an advanced medical fluorescence system developed by STI Inc. (discussed later).

The advantages of a pushbroom hyperspectral imager (also called an *imaging spectrometer*) are that the appropriate bands can be selected and channels can be appropriately "binned" using software, because all the information on the target is available to be appropriately selected for image processing. The physical scan of the excitation line over the area to be imaged takes some time, and obviously the object must be fixed during this time.

Mode E: Fourier Transform Spectroscopic Imaging

An imaging Fourier transform spectrometer as schematically illustrated in figure 10.10 is used in another type of fluorescence imaging in which the whole spectrum is recorded in the object points, but now they are all in parallel without any scanning across the target. However, a different type of scanning still has to be performed: the displacement of the mirror in the imaging interferometer, which can be of the Michelson or the Sagnac type, as illustrated in figure 10.10 (Pham et al., 2001). Thus, the same demands for a stationary object as for the imaging spectrometer just discussed pertains to this type of system. For each pixel in the detector matrix a signal directly connectable to a particular spot on the object is recorded as a function of mirror displacement, and for each spatial point

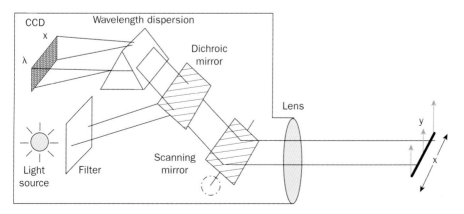

Figure 10.9 Schematic view of a pushbroom hyperspectral imaging system.

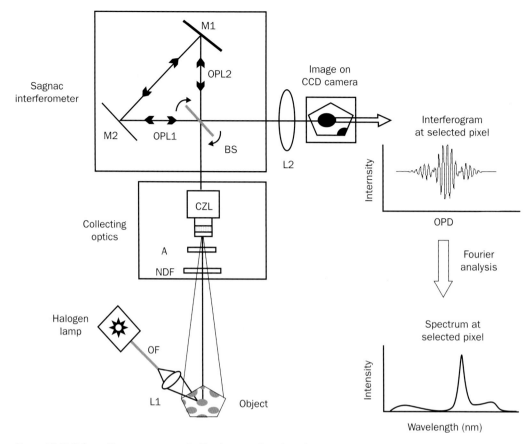

Figure 10.10 Schematic arrangement of a Fourier transform imaging system. (Adapted from Pham et al. [2001].)

the corresponding spectrum can be calculated using a fast Fourier transform algorithm. Because the whole object is illuminated simultaneously, the signal-to-background situation is worse than for the pushbroom hyperspectral imager, in which the excitation source photons at a single occasion are concentrated on a line, helping them to overcome the background photons better. The Fourier transform imager thus becomes particularly sensitive to background.

Mode F: Time-Gated Imaging

All systems discussed so far have used time-integrated fluorescence intensities without any consideration of characteristic decay times of the excited states populated by the illumination source. For implementing the characteristics shown in figure 10.6 into an imaging scheme, the object can be illuminated by a short ultraviolet pulse of a few nanosecond duration, and the fluorescence selected in a bandpass filter impinges on the entry face of a "gateable" image intensifier. With a gate time of typically 2 ns, a "prompt" image and different delayed images can be captured by sweeping the delay of the intensifier gate. A schematic diagram of this technique, implemented at the Politecnico di Milano and discussed later, is shown in figure 10.11 (Cubeddu et al., 1997b). By evaluating the image frames pixel by pixel for the different delay times, an effective lifetime value for the fluorescence in each spatial pixel can be evaluated and displayed in false colors, providing tumor demarcation as explained in connection with figure 10.6. It should be noted that decay time can be a very useful contrast function, as discussed earlier.

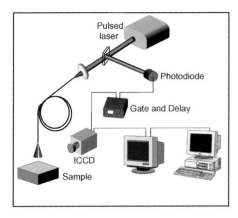

Figure 10.11 Schematic arrangement of a temporally resolving fluorescence imaging system. (From Cubeddu et al. [1997b].)

Hybrid Imaging Modes

By combining a spectrometer with a streak camera, time-resolved decay curves can be recorded simultaneously for each fluorescence light wavelength following short-pulse excitation. To create an image, the point under study must be sequentially selected much as is done in a whisk broom scanner used in air- and spaceborne multispectral imaging. If such a study is made for many excitation wavelengths, maximal information on the fluorescence characteristics of the tissue under study is gained, clearly at the expense of complexity and long scanning times, and corresponding demands for a stationary patient.

It should also be noted that fluorescence imaging data can be combined advantageously with reflectance data recorded with the same system, but now using a white-light illumination source. Here well-known phenomena such as reddening of tissue, normally ocularly evaluated by the trained doctor's eye, can be put on a more objective footing. In a more trivial implementation, a normal color video representation of the tissue under study can be recorded and mixed into the processed data from the fluorescence imaging system to provide a convenient orientation to help in the malignancy demarcation process.

FLUORESCENCE IMAGING SYSTEMS

In this section, various types of fluorescence imaging systems for tissue characterization are exemplified. The aim is not to cite all publications presented within this field of research, but rather to describe the principles in examples and discuss the potential applications of the various designs. As discussed earlier, there is always a trade-off between instrumental complexity and the amount of diagnostic information gained by a measurement. Small, compact, simple systems can be low-cost and easy to operate, whereas more complex systems may be better at filtering out the diagnostic information of interest. Before the current status of systems operating in the different modes is discussed, let us look briefly at the historical development of fluorescence imaging for tissue diagnostics. More thorough reviews on this subject can be found, for example, in Urbach et al. (1976), Daniell and Hill (1991), Andersson-Engels and Wilson (1992), Andersson-Engels et al. (1997), and Wagnières et al. (1998).

Our review focuses on fluorescence imaging for in vivo identification of malignant tumors. We limit the scope to macroscopic imaging, because microscopic techniques are covered in other chapters in this volume. Other primary fields of application of in vivo fluorometry include tissue metabolic studies (Tamura et al., 1989; Horvath et al., 1994; Cordeiro et al., 1995), cardiovascular diagnosis (Perk et al., 1991, 1993; Deckelbaum, 1994; Deckelbaum et al., 1995; Papazoglou, 1995; Warren et al., 1995), and ophthalmology

(Docchio, 1989). Most of the in vivo fluorescence imaging studies published, however, deal with oncological applications to identify early malignant lesions fordefining tumorextent and spread to adjacent tissues and as a guide to optimize localized treatments of solid tumors.

Historical Introduction

Only a long time after Stokes discovered the phenomenon of fluorescence in 1852, was it considered for biomedical applications. In 1911, Stübel reported that animal tissues emitted visible fluorescence light after ultraviolet illumination. This was tissue autofluorescence. Approximately a decade earlier, another important finding was made by Raab, a student of von Tappeiner in Munich, when investigating toxic properties of the fluorescent substance acridine, a coal tar derivative, to the paramecium (Raab, 1900). He observed that the toxicity required both light and the fluorescent substance. von Tappeiner later found that oxygen was also required for the toxic reaction to take place, and coined the term *photodynamic therapy* (PDT) to describe this phenomenon (von Tappeiner and Jodlbauer, 1907). The two fields of fluorescence diagnostics using exogenous tumor markers and PDT have since been tightly connected. Policard (1924) was the first to observe spontaneous fluorescence from experimental tumors illuminated by a Wood lamp, and explained this by selective accumulation of porphyrins in tumors. Later, Auler and Banzer (1942), as well as Figge et al. (1948), investigated the accumulation and retention of porphyrins. As a result of these studies, Figge et al. (1948) proposed that hematoporphyrin fluorescence could be used for cancer diagnostics (Rasmussen-Taxdal et al., 1955). Winkelman and Rasmussen-Taxdal (1960) performed the first quantitative measurements of fluorescence in vivo utilizing exogenous fluorophores in 1960. They compared the results of their fluorescence measurement with those of spectrophotometry of porphyrins chemically extracted from tissue. In the same year, Lipson and Baldes (1960) reported on a derivative of hematoporphyrin, which they called *hematoporphyrin derivative* (HpD), as a fluorescent tumormar kerand photosensitizerforPDT. Many studies followed with this substance during the 1960s and '70s, suggesting that it might be possible to use HpD fluorescence to detect a number of endoscopically reachable tumors clinically (Lipson and Blades, 1960; Lipson et al., 1961, 1964a,b; Gregorie and Green, 1965; Leonard and Beck, 1971; Kelly and Snell, 1976). At the end of the 1970s, the interest in using HpD as a fluorescent tumor marker drastically increased as a result of a breakthrough of HpD PDT of malignant tumors (Dougherty et al., 1975). At the same time, it was also recognized that there existed a need to develop instruments allowing quantitative detection of the fluorescence, rather than visual inspection (Sanderson et al., 1972; Carpenter et al., 1977; Profio et al., 1977). During the 1980s, intense research was conducted to improve the understanding of the precise mechanisms of the interaction of photosensitizers for PDT as well as fluorescent tumor markers with tissue, unraveling the interesting properties of these substances. Second-generation photosensitizers with improved properties were also developed. The next majorstep toward clinical acceptance fora technique using exogenous agent-induced fluorescence for diagnostics of malignant tumors was the introduction of ALA (Malik and Lugaci, 1987; Kennedy et al., 1990; Kennedy and Pottier, 1992). In parallel, diagnostics based on tissue autofluorescence alone attracted increasing interest, because it could be used for screening purposes without any potential side effects resulting from an exogenous tumor marker. During the past decade, many studies have been performed to develop and evaluate commercial prototypes of fluorescence imaging systems for specific indications. Clinical and regulatory approvals have been achieved for several diagnostic procedures in many countries.

Current Status

Recent reviews of the use of fluorescence imaging for various clinical indications are presented in Pfau and Sivak (2001), Kennedy et al. (2000), Sutedja et al. (2001b), and

Hirsch et al. (2001). A numberof systems have been developed fora variety of clinical indications using different algorithms to differentiate malignant tissue from normal tissue. The clinical applications examined in most detail are identification of early-stage lung, urinary, and laryngeal tumors, and tumors in the gastrointestinal tract. The systems used differ both in terms of excitation source and detection schemes.

We first discuss some aspects of importance for the excitation source in fluorescence imaging. The excitation source determines which fluorophores can be excited, the depth probed by the investigation, the overall excitation power available, and the option of suppressing background light. The optics used to illuminate the area investigated also add some requirements to the light source. Different clinical applications have very different requirements in these respects. For example, delineation of the borders of a skin tumor are very different from fluorescence guidance in resecting a brain tumor, which again will differ from an endoscopic examination of lung tumor.

Anothermajortask in selecting an excitation source fora specific application is to identify the best type: laser, lamp, or light-emitting diode (LED). The price and complexity of the system depends drastically on this. Lamps are often used for white-light illumination. Much can be gained if the same source can be used for the fluorescence examination as well. There are, however, a few limitations in using lamps as an excitation source for fluorescence imaging. The first is to meet the power requirements for fluorescence detection of a large surface. Lamps most often have a broad spectral output and they thus need spectral filtering to suppress light at the detected fluorescence wavelengths. Reflected excitation light will otherwise pass the detection filters and be superimposed on the fluorescence signal, yielding a high background. Also, the power of the remaining light may be insufficient, especially if the light has to be guided through fiber optics, because the coupling of the light from a lamp into fiber optics is not always very efficient as a result of the relatively low brightness of lamps. The low power and brightness also limit the use of LEDs. For direct surface illumination applications, for which fiber delivery is not necessary, one can, however, use a matrix of LEDs to excite fluorophores. An advantage with LEDs is that they emit within a relatively narrow band, so they may not need as much spectral filtering as lamps. Lasers are mainly interesting because of their narrow spectral emission profile and their high brightness, making it possible to focus the light efficiently into thin optical fibers. A drawback with lasers, aside from the price and added complexity, may be speckle patterns formed in the illuminated area resulting from the long coherence length of the light.

The excitation wavelength is anotheressential parameterin defining the diagnostic information revealed. The wavelength determines how well different fluorophores are excited in the tissue and thus the information available for diagnostics (Eker, 1999). The excitation wavelength also determines the tissue depth probed by the diagnostic procedure, because the penetration depth of light in tissue is strongly wavelength dependent. The depth may range from a few hundred micrometers for near-ultraviolet light to several millimeters for red light (depending on detection geometry). Shallow penetration is favorable in detecting thin lesions, such as carcinoma in situ, thus avoiding fluorescence contributions from underlying healthy tissue. Other factors to consider in selecting excitation wavelength are light-induced tissue damage (especially below about 325 nm, where DNA starts to absorb) and technological aspects, such as transmission of optical components. The poor transmission of conventional endoscopes forlight below 400 nm drastically limits the possibilities for short-wavelength fluorescence endoscopy.

As mentioned earlier, pulsed or modulated light may be used to suppress ambient background light and to make it possible to subtract any such contributions from the fluorescence signal. This may prove important for procedures that cannot be performed without white-light guidance. This is typically the case forall areas examined without an endoscope, such as during open surgical procedures and skin tumor examinations. In endoscopic applications, the only light present in the examination area is guided through the endoscope.

If the white examination light is turned off, the procedure may be continued in the fluorescence examination view. If both the conventional white-light and fluorescence modes are necessary for the examination, it is necessary to be able to switch modes quickly.

With regard to the detection scheme, the systems can be classified according to the modes described earlier. In the following sections we exemplify the different types of systems with instruments described in the literature and discuss their potential for clinical diagnostics.

Mode A: Single-Band Imaging

Mode A is the simplest and most straightforward fluorescence imaging method. Systems belonging to this group are characterized by fluorescence detection through a filter, selecting out the most important fluorescence wavelength. Typically the fluorescence from a fluorescent tumor marker is visualized. Fluorescence diagnostic imaging started to develop using this technique with HpD as a fluorescent tumor marker. Profio and Balchum performed much of the early work in developing fluorescence bronchoscopy to identify and delineate early-stage tumors in the lungs. All early findings in in vivo fluorescence imaging were obtained using violet mercury-lamp light excitation and ocular inspection of HpD fluorescence. A first improvement of the technique was to replace the eyes with electronic detection to allow simple data storage and image processing. Later, lasers replaced lamps as excitation sources, as a result of the better coupling efficiency of this light to fiber optics, making the images clearer and less noisy (Doiron et al., 1979; Profio and Doiron, 1979). It was then also required that one modified the distal end of the source fiber with a microlens to illuminate the entire area viewed by the endoscope (Profio, 1984). The currently existing mode A systems are quite similar to the early systems.

The simplest possible mode A system fordelineation of basal cell carcinoma of the skin was recently described (Wennberg et al., 1999). The diagnostic potential of the system, based on filtered lamp excitation and filtered CCD camera detection of PpIX fluorescence, was evaluated in a small clinical study. Forskin lesions, no fiberoptics are necessary for guiding the light back and forth to the lesion. Neither is it of importance to have white-light examination possibilities incorporated in the system. The measurement geometry can also be fixed and standardized to reduce interrecording variability. Thus, a simple system may provide reliable diagnostic information.

In general, however, the restriction to a single wavelength seriously limits the diagnostic potential, because no spectroscopic information is provided. Nor is the measured fluorescence intensity at a single wavelength necessarily reliable, because the detected fluorescence signal depends on many factors of no diagnostic relevance, such as detection geometry, intensity of excitation light, and so forth. A third aspect, especially apparent in imaging using a tumor marker through an endoscope, is that the fluorescence signal often gives insufficient information for guiding the instrument during the procedure. In the fluorescence mode, most of the image contains information from normal tissue with a weak fluorescence signal from the tumor marker and it can be difficult to identify the tissue area to examine. This problem can be solved in several ways without making the detection system very complex. The first solution was developed by Profio et al. (1983) when they developed a rapid switching between white light and fluorescence mode. All endoscopic true mode A instruments used today utilize this feature to allow examination both in conventional white-light reflection as well as in fluorescence. A system developed for investigations of the upper gastrointestinal tract is conceptually very similar to early systems by Profio, although it uses a pulsed laser source and gated CCD camera to allow simultaneous white-light and fluorescence imaging (Sukowski et al., 2001).

An interesting application of mode A instruments is for short-wavelength excitation fluorescence endoscopy. Short-wavelength endoscopy may be of particular interest for diagnosing gastrointestinal lesions based on tissue autofluorescence (Kapadia et al., 1990;

Rava et al., 1991; Richards-Kortum et al., 1991; Bottiroli et al, 1992; Marchesini et al., 1992; Schomackeret al., 1992; Bottiroli et al., 1995; Yang et al., 1995; Wang et al., 1996; Zonios et al., 1998; Chwirot et al, 1999; Eker et al., 1999). As mentioned previously, endo-scopes exhibit low transmission at short wavelengths. The poor transmission is not a problem if the fluorescence light is detected with a small CCD camera at the distal end of the endo-scope. Feld et al. developed a digital fluorescence endoscope to examine autofluorescence from polyps in the colon (Wang et al., 1996, 1999). For this application, it is not trivial to use more than one detection band, because the filtering must be at the distal end of the endoscope. They used ultraviolet laser excitation delivered through a fiber in one of the open instrument channels. The CCD camera is only sensitive above 400 nm and thus suppressed scattered laser light while detecting the emitted blue fluorescence. The same CCD camera was used to capture the white-light images. For this instrument they used sophisticated image processing to filter out much of the geometric artifacts present in the recorded images.

Another novel concept to correct for geometric and illumination artifacts in the image is used in the Storz system as described by Kriegmair et al. (1995) and in a similar system from Olympus, as described by Ehsan et al. (2001). Such a system, illustrated schematically in figure 10.12, is in the twilight zone between modes A and C. The system is designed to visualize PpIX fluorescence through an endoscope after systemic or local administration of ALA. In this system, the sole detection filteris selected in such a way that not only is the tumor marker fluorescence detected, but also some scattered excitation light and tissue autofluorescence. The filter is used to balance these signals so they all can be visualized simultaneously. The detector is, in this case, a normal RGB video camera, yielding some spectroscopic information. The scattered excitation light thus appears blue; the autofluo-rescence, green; and the tumor marker fluorescence, red. It is therefore the mixed color rather than the intensity of the presented image that yields the diagnostic information. In this way, the system offers much diagnostic information despite its simplicity. It is also one of the most widely used systems, evaluated in clinical studies mainly forbladdertumor s (Filbeck et al., 1999; Hartmann et al., 1999; Koenig et al., 1999; Kriegmair et al., 1999; De Dominicis et al., 2001; Ehsan et al., 2001; Frimberger et al., 2001; Riedl et al., 2001; Zaak et al., 2001), but also for tumors in the oral cavity (Leunig et al., 1996; Betz et al., 2002), and cervical (Hillemanns et al., 2000), gastrointestinal (Gahlen et al., 1999; Orth et al., 2000), and breast tumors (Ladner et al., 2001). An example of an invasive tumor of the vocal fold of a 73-year-old man is illustrated in figure 10.13.

Mode B: Dual-Band Imaging

As just mentioned, many advantages can be gained if two fluorescence signals are recorded instead of one (Ankerst et al., 1984; Montán et al., 1985; Profio and Balchum, 1985; Profio et al., 1986; Baumgartner et al., 1987; Wagnières et al., 1990; Lam et al., 1991; Palcic et al., 1991; van den Bergh, 1994; Wagnières et al., 1997; Lazarev et al., 1999; Scott et al., 2000; Fischeret al., 2001). If two ormore independent signals are obtained, it is possible to evaluate the fluorescence as a dimensionless quantity by forming a ratio of intensities. In this way, the measurement will be less dependent on measurement geometry and the spatial profile of the illumination light, as discussed earlier. This is of particular importance in fluorescence endoscopy, in which it is difficult to control the measurement geometry precisely. One can also better correct for any background superimposed on the image by subtracting signals from each other. This is of great interest when imaging a weak tumor marker fluorescence emission superimposed on a high autofluorescence background, as is the case for low-dose porphyrin (especially weakly fluorescent photofrin) fluorescence endoscopy. It is obviously also possible to obtain more spectroscopic information if two, instead of one, wavelength bands are measured.

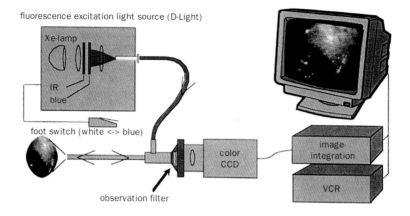

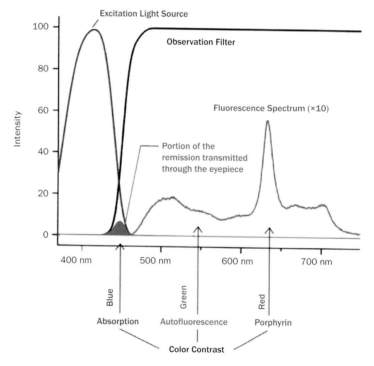

Figure 10.12 Schematic illustration of the Storz system. (From Baumgartner et al. [1999] and Brand et al. [1999].)

The signals can be measured in two fluorescence emission bands (most common) using two excitation sources at one fluorescence emission band, or at two different delays after pulsed excitation. Here we consider the latter group to belong to mode D (time-resolved imaging) and discuss it later. The two images can be measured either sequentially or in parallel. Parallel imaging is sometimes required to avoid movement artifacts in the presented processed image. Examinations benefiting from parallel imaging include gastrointestinal endoscopy and bronchoscopy, during which the examined tissue is constantly moving.

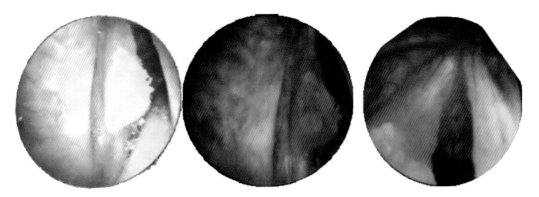

Figure 10.13 Images of an invasive tumor on the vocal fold. To the left a white-light image is shown. The tumor in this case is clearly visible. About the same view is shown in the middle in the blue-light autofluorescence mode. To the right, another view clearly illustrates the ability to localize the border of the lesion with this system. (Photos courtesy of Dr. Roland Rydell.)

Parallel detection is only possible in the fluorescence emission mode, which is part of the reason why this is the most frequently used concept.

The most widely used mode B system today is a commercial product called the LIFE system constructed by Xillix Technology, Inc., from the concepts of Profio et al. and Lam et al. The system is developed both for endoscopic examinations of the lung, gastrointestinal, and ear–nose–throat tracts. It is schematically illustrated in figure 10.14. The system utilizes a CW light source in the violet wavelength region, either a high-pressure mercury lamp at 405 and 436 nm, or a He:Cd laser at 442 nm. The diagnostic capability of this system is based on the ratio between the red and the green fluorescence emission bands. The presentation of the recorded images is very similar to that of the Storz system: A color image is displayed that yields a balance between red and green fluorescence. The systems differ, however, in two important ways: The LIFE system is based on pure autofluorescence detection and does not involve any tumor marker, and the diagnostics are based on two fluorescence images recorded at optimally selected bandpasses rather than a color CCD camera. Based on these differences, it is clear that an examination with the LIFE system does not require the administration of a tumor-marking agent. This may be of great importance for certain applications because the risk of possible side effects is minimized and it is easier to include patients with a low risk of tumor growth. The increased complexity with two recorded images yields an added freedom to select the wavelength bands of diagnostic interest. This is of importance for this system, because it is based on relatively small but statistically significant differences in the tissue autofluorescence, rather than the very large differences in fluorescence emission when tumor markers are used. The diagnostic information is interpreted from an image displayed on a video screen. The two recorded images are both fed to the same monitor as the red and green video signals. Normal tissue is seen as greenish, whereas malignant lesions appear brownish.

The LIFE system has been widely evaluated and is now used at many centers for diagnostic imaging in combination with bronchoscopy (Lam et al., 1992a,b, 1993a,b; Lam and Becker, 1996; Khanavkar et al., 1998; Kurie et al., 1998; Lam et al., 1998; Lam and Palcic, 1999; Lam et al., 1999; Palcic et al., 1999; Zeng et al., 1999; Kusunoki et al., 2000; Lam et al., 2000; Kennedy et al., 2001; Sutedja et al., 2001a; van Rens et al., 2001), endoscopy of the gastrointestinal tract (DaCosta et al., 1999; Haringsma and Tytgat, 1999; Izuishi et al., 1999; Marcon, 1999; Zent et al., 1999; Abe et al., 2000; Haringsma et al., 2001; Kobayashi et al., 2001) and diagnostic procedures in the ear–nose–throat region (Zargi et al., 1997a,b; Kulapaditharom and Bookitticharoen, 1998; Zargi et al., 2000). Typical examples of images produced by the system are shown in figure 10.15. From the literature, it is obvious that the

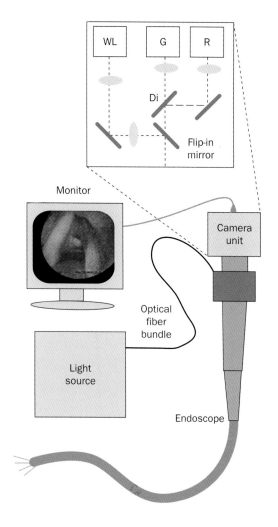

Figure 10.14 A schematic of the optical arrangements in the LIFE system. (Adapted from Palcic et al. [1991].)

system requires a learning period for the endoscopist before it provides valuable diagnostic information, but this training period clearly improves diagnostic precision when used in combination with conventional procedures. One of the main difficulties in interpreting the images is that a small change in colormay be difficult to observe because of a high variation in intensity within the images. These variations are much the result of geometric effects acquired during recording (some regions in a recorded image are much closer to the endoscope, yielding a higher intensity than regions farther away).

Mode C: Multicolor Imaging

The mode A and B systems have been evaluated in a large number of patients in several disciplines. They have proved to provide information of diagnostic interest for these applications. This is very promising and makes it clear that fluorescence can provide new information about tissue not available from conventional diagnostic procedures, and this information can be obtained using relatively simple instruments. However, there is, for many applications, a need to improve further the precision in the diagnostic procedure before it will be fully accepted as a routine procedure in the clinic. In particular, it would be valuable if the sometimes relatively high false-positive rate reported could be lowered (Kennedy et al., 2000; Pfau and Sivak, 2001). Today, it can in some cases be difficult to distinguish inflammation from high-grade dysplasia and cancer. In addition, new applications, not yet evaluated in detail, in part because of inconclusive fluorescence patterns in initial

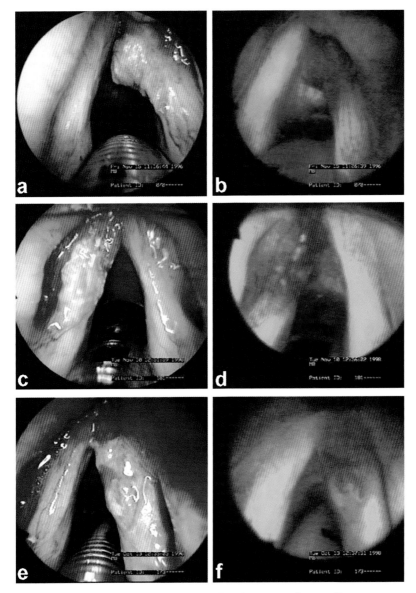

Figure 10.15 (A–F) Microlaryngoscopic (A, C, E) and corresponding autofluorescence (B, D, F) views of two cancerous lesions (A–D) and one precancerous lesion (E, F) using the LIFE system. The lesions are seen as darker and more reddish than the surrounding more normal vocal fold. (Adapted from Zargi et al. [2000].)

studies, may also be possible with more sophisticated diagnostic techniques. As many fluorophores and chromophores contribute to the spectroscopic content (by reabsorbing some of the fluorescence light in its transport to the detector), the diagnostic potential may be improved if more than two fluorescence signals are recorded. This would make it possible to separate better the contributions from the different fluorophores and chromophores, and it can be performed using multispectral imaging techniques. The following list presents some applications for which multispectral imaging might be valuable:

- It is obviously of interest to utilize ratio imaging better than in the previously described techniques to compensate fully for geometric effects in recorded fluorescence signals (Andersson-Engels et al., 1994, 1995; Tassetti et al., 1997; Svanberg et al., 1998; Andersson-Engels et al., 2000). Such effects yield intensity variations of no relevance in the classification of tissue type. For some applications, especially if no white-light image is captured in parallel, this intensity information could, however, be of interest to identify easily the area investigated. This is, forinstance, the case in both the Storz and LIFE systems.

- Autofluorescence imaging provides usually relatively small differences between tissue types to be characterized. A multispectral approach might yield more robust diagnostic procedure to utilize fully the information available (Chwirot et al., 1999), with a lowernumberof false-positive results.

- One application requiring a multispectral approach is imaging in the presence of strongly absorbing blood (the blood content may be of little interest for the diagnostics) (Andersson-Engels et al., 1989a, 1990a). By forming a ratio between fluorescence light in two wavelength bands with the same absorption in blood, a criterion based on the fluorescence signal but almost independent of blood absorption can be used, as discussed earlier.

- Another similar application is to record a fluorescence signal independent of the optical properties of the examined tissue. A powerful double ratio technique was developed by Sterenborg et al. (1996) for this purpose. This is of importance in, for example, attempts to measure absolute concentrations of fluorophores in tissue, such as for dosimetry purposes in connection with PDT (Sinaasappel and Sterenborg, 1993; Saarnak et al., 1998; Stefanidou et al., 2000). It would also be of interest for diagnostic purposes in highly absorbing tissue, such as malignant melanomas (Sterenborg et al., 1996; Bogaards et al., 2001), in addition to attempts to stage more precisely cervical intraepithelial neoplasia, for example (Bogaards et al., 2002).

Multispectral systems tend to become complex, because several images at different wavelengths are recorded. It is thus an important issue for most applications to reduce the complexity as much as possible without reducing the signal quality or content. One can distinguish two different approaches to obtain multiple images without needing several cameras: sequential and parallel imaging. In sequential imaging, images at different wavelengths of the same tissue are recorded one by one, shifting emission filters in between measurements. A limitation with sequential imaging is possible tissue movement, making it difficult to correlate the images in the subsequent images, resulting in edge effects in areas with varying fluorescence intensity. Sequential imaging is sometimes preferred to avoid the need of multiple cameras making the systems complex, and is also necessary if several excitation wavelengths are used for the multispectral approach (Sinaasappel and Sterenborg, 1993; Bogaards et al., 2001; Ramanujam et al., 2001). An alternative is to use special optics to fit a number of images on the same camera. Normally four images are captured, one on each quadrant of the detector. In this way only one camera is needed and all images can still be recorded in parallel (Montán et al., 1985; Andersson et al., 1987; Andersson-Engels et al., 1994, 1995; Svanberg et al., 1998; Andersson-Engels et al., 2000; Hewett et al., 2001; Ramanujam et al., 2001). The drawback is the slightly reduced spatial resolution with such a system. An illustration of the design of such a system and an image produced is illustrated in figure 10.16.

The four images recorded in parallel are computer processed for viewing in a contrast function optimized foreach particularapplication. An advantage with this system is the use of a pulsed light source and a gated detector. This concept allows the simultaneous use of fluorescence imaging and normal white-light endoscopic examination. Many practitioners trained using normal white-light examinations appreciate the potential of adding

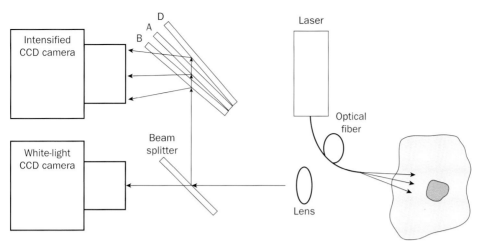

Figure 10.16 Illustration of the optical arrangement of the multicolor imaging system currently used in Lund.

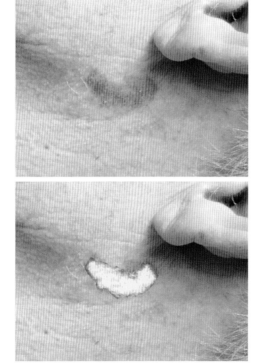

Figure 10.17 Example of a processed multicolor fluorescence image of a basal cell carcinoma after topical application of ALA. (top) White-light image. (bottom) Digitally processed image. (From Klinteberg et al. [2002].)

fluorescence information to the information they are used to working with, rather than replacing the white light with a fluorescence examination, or sequential examinations. Because as many as fourdetection wavelengths can be utilized, this system is more flexible in optimizing the diagnostic information extractable for various clinical applications. The system could work for tissue autofluorescence as well as for a combination of autofluorescence and fluorescence from an exogenous marker. An example of an image produced from a basal cell carcinoma of the skin after application of ALA is shown in figure 10.17 (Klinteberg et al., 2002).

Mode D: Pushbroom Hyperspectral Imaging

The probably most advanced system suggested for biomedical diagnostic imaging to date is based on hyperspectral fluorescence emission imaging. Even more spectral information is obtained in such a system, in which a large number of spectral bands are imaged. Such data are three-dimensional in character, a function of two spatial coordinates and one spectral coordinate. This means that the entire fluorescence emission spectrum is recorded for each position in the image. The data are often recorded using a two-dimensional CCD camera scanning one of the dimensions. Two majorclasses of systems can be identified: eithera pushbroom type of scanning system or a system in which the spectra are recorded with a Fourier transform instrument. The two types of systems each have their advantages and limitations.

The enormous amount of data contained in each type of image sets a high demand on the evaluation routine. For the necessary data reduction in the evaluation, one could utilize the same type of multivariate analysis frequently used for fluorescence point spectroscopy for tissue diagnostics. It is, however, also of interest to utilize the context of the image in the analysis. Advanced routines developed forsatellite imaging could be utilized forthis purpose.

Photos of hyperspectral diagnostic imaging systems developed by Science and Technology International Inc. are shown in figure 10.18 (Gustafsson et al., 2000; DeWeert et al., 2003; Gustafsson et al., 2003a,b). The device has been developed fordetection of cervical neoplasia. This instrument is typically operated with a cube size (x, y, λ) of $150 \times 150 \times 50$ (fig. 10.19). Although the instruments can operate at higher resolution, this combination provides a good balance of performance with speed. Both white-light and fluorescence scans can be collected, and the instrument is equipped with a digital RGB spotting camera that collects colposcopic images, as illustrated in the typical image produced by the system in figure 10.20. The typical area imaged measures 35×35 mm.

Mode E: Fourier Transform Spectroscopic Imaging

This mode provides the same information as a mode D instrument, but in this instrument the spectral dimension is scanned rather than one of the spatial dimensions. The same discussion just presented is thus valid in this case as well. With full spectral resolution provided with a system like this, it is possible to study very subtle changes in the fluorescence emission. It is thus possible, forexample, to detect slight changes in fluorescence emission of a fluorophore resulting from changes in the microenvironment. Protoporphyrin bound to various subcellular compartments could thus be differentiated. This might be of specific interest for fluorescence tissue diagnostics (Garini et al., 1996). A typical processed image clearly

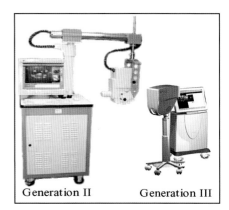

Generation II   Generation III

Figure 10.18 Hyperspectral diagnostic imaging cervical Generation II device, and the Generation III device. The images are to the same scale.

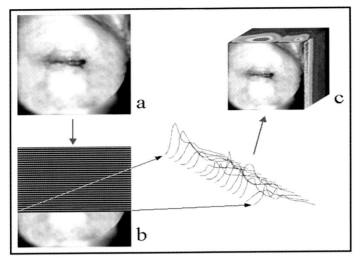

Figure 10.19 Schematic of hyperspectral diagnostic imaging imaging.
(A) Ordinary colposcopic image of the cervix uteri. (B) Pushbroom
hyperspectral scan. (The example shown is a white-light reflectance scan.)
(C) The final product is a hyperspectral data "cube" typically measuring 150
lines × 150 samples × 50 bands. Together with ancillary data, each cube
represents about 2 MB of data.

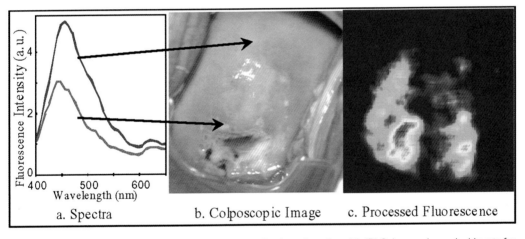

Figure 10.20 (A) Fluorescence spectra recorded after application of acetic acid. (B) Colposcopic cervical image for
a case of cervical intraepithelial neoplasia CIN I. (C) Processed fluorescence image showing probable pathology.
The colors range from blue for normal tissue to yellow and red for various degrees of pathology.

visualizing the borders of a basal cell carcinoma of the skin is illustrated in figure 10.21
(Roth et al., 2003).

Mode F: Time-Gated Imaging

Another approach to increase the information content in fluorescence imaging is to record
information of fluorescence lifetimes. As discussed earlier, diagnostics are based on mapping the relative concentrations of a few fluorophores. Fluorescence lifetime measurements
provide another dimension to separate contributions from various fluorophores with more
or less overlapping fluorescence emission profiles. Some fluorophores may thus be easier
to distinguish using the fluorescence lifetime properties than the emission spectra. Unique

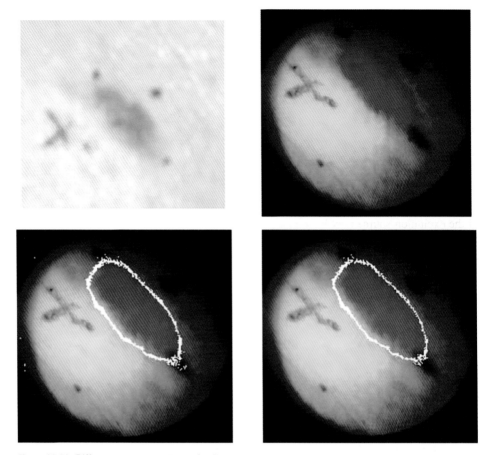

Figure 10.21 Different representations of a fluorescence measurement of a basal cell carcinoma shown in the upper left corner (a photograph of the lesion taken prior to any treatment). The black points and crosses are those drawn prior to measurement to identify regions of interest when viewed in the Fourier transform imaging spectrometer FTIS objective. The upper right view shows the image reconstructed from hyperspectral data. The similarity mapping procedure, evaluating the image using model spectra for the lesion and healthy tissue averaged from the other three chosen sample images, was used to create the white border in the image to the lower left. The white pixels, intended to show the border between the lesion and healthy tissue, had a score between −0.2 and 0 point on a scale on which 1 point is most similar to the lesion and −1 point is most similar to healthy tissue. The white threshold pixels in the lower right image represent pixels; the partial least squares (PLS) score was between 0.4 and 0.5 point on a scale where 0 point represents healthy tissue and 1 point represents lesion. (From Roth et al. [2003].)

properties for fluorescence lifetime measurements, which may be important in certain applications, are the independence of measurement geometry and robustness to interference from nonfluorescent tissue chromophores. A recorded fluorescence lifetime does not depend on the signal level, and thus not the distance between the tissue and detectorororothergeometr ic factors in the measurement, as discussed earlier. Also, other tissue chromophores may alter the detected fluorescence emission intensity and spectrum as a result of reabsorption of part of the fluorescence emission in its path through the tissue to the detector. In fluorescence lifetime measurements, the signal level will be decreased as a result of this reabsorption, but the shape of the decay curve and the lifetime will again be unaltered (as long as the lifetime is so long that the time for the propagation of the light will not influence the recorded lifetime). Thus, time-resolved techniques may be preferable, even for some applications in which the contributions from two fluorophores can be distinguished equally well using fluorescence emission spectroscopy or time-resolved measurements.

The fluorescence lifetime information is not utilized in most fluorescence imaging systems because of the fast electronics needed, making the systems more complex and expensive. Fluorescence lifetime systems require a pulsed excitation source and a gated detector. Most fluorophores important for fluorescence diagnostics of tissue exhibit a multiexponential decay. The decay can often be fitted to a curve consisting of three exponentials (Andersson-Engels et al., 1990b). The lifetimes are often between 100 ps and 10 ns. Porphyrins may have a slightly longer lifetime. Often it is sufficient to measure an effective lifetime in which a single exponential curve approximates the fluorescence decay. The effective lifetime of tissue is often a few nanoseconds. This means that the time resolution of a detection system should be at least on the order of a nanosecond or better . On the camera side, this is possible with fast image-intensified CCD cameras. The requirement of the excitation source may be more difficult for a clinically adapted system. A nitrogen laser or solid-state laser with picosecond pulses can provide the excitation light needed. Such systems tend, however, to be somewhat bulky and/or complex for clinical measurements. A system also requires high-speed electronics for triggering and gating purposes. The technique is also based on a number of recorded images from sequential laser pulses, each image with a different delay between laser pulse and gate of the detector to evaluate the fluorescence lifetimes. Such a recording is thus sensitive to tissue movements during the recording.

A smart, simplified scheme of this technique is to use time-gated imaging with a long delay to enhance the relative contribution from the long-lived fluorophores. This could, for instance, be utilized to visualize an exogenous tumor marker with a long fluorescence lifetime, such as a porphyrin. Such a technique would thus not require sequential imaging, making it insensitive to movement artifacts. It will, however, be sensitive to measurement geometry, if no spectral ratio is formed.

Promising results from feasibility studies have been published (Cubeddu et al., 1993a, b; Kohl et al., 1993; Cubeddu et al., 1997a,b,c; Dowling et al., 1997; Cubeddu et al., 1998; Dowling et al., 1998a,b, 1999; Jones et al., 1999; Andersson-Engels et al., 2000; Balas, 2001). An amazingly good tumor demarcation capability has been demonstrated with very low concentrations of fluorescent tumor markers in animal studies (Cubeddu et al., 1995). An image of a basal cell carcinoma after topical application of ALA produced with the systems illustrated in figure 10.11 is shown in figure 10. 22. This technique may be interesting also in utilizing the pure tissue autofluorescence as the diagnostic information.

Hybrid Imaging Mode

There are many combinations of the previously mentioned techniques that may provide better sensitivity and specificity for specific clinical applications than the techniques described here. Also, combinations with other spectroscopic techniques can offer interesting diagnostic capabilities. A full utilization of such information is, to our knowledge, not yet implemented in any system. However, based on the potential of such systems, we believe that better systems will be developed in the near future , providing enhanced diagnostic capabilities.

FUTURE DIRECTIONS

Early detection of malignant disease is a key to the successful abatement of one of the major plagues of mankind. Fluorescence imaging is a new diagnostic modality that is now coming of age. Although the use of exogenous tumor-seeking fluorescent agents has been quite successful in many contexts, such as for bladder cancer detection,   it is even more

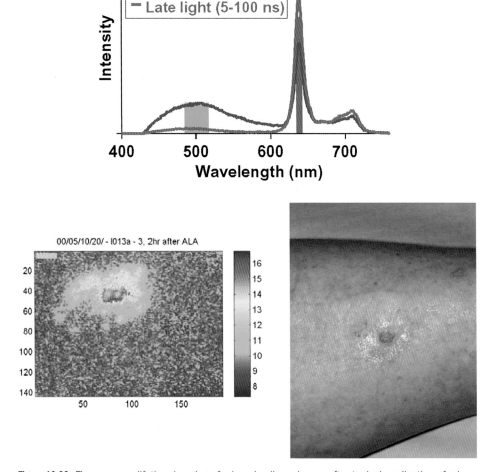

Figure 10.22 Fluorescence lifetime imaging of a basal cell carcinoma after topical application of a low dose of ALA. (From Andersson-Engels et al. [2000].)

challenging to develop techniques that fully rely on natural tissue chromophores only. This is because of increased practical convenience, and because of an easier and more clear-cut regulatory situation. The future clinical role of fluorescence imaging will also greatly depend on how well the instruments can be adapted to the practical situation. This will largely determine how the technique will be accepted by clinicians. Achieving a simple and intuitive user interface and data presentation will most likely be very important issues in this context. Clearly, the effort to achieve powerful instrumentation will be influenced by the market, with large-scale applications being more attractive than very specialized ones demanded in a few clinics only. The development of reliable low- and medium-cost instruments will accelerate the expansion of the market. In this context, the recent progress made in violet and ultraviolet diode lasers constitutes an important development, as do the advances in compact spectrometerand detectortechnology. The possibilities combining powerful cancer imagery and interactive treatment modalities based on photodynamic and thermal therapy—perhaps involving robotics—will provide further impetus for a dynamic development of the field.

REFERENCES

Abe, S., K. Izuishi, H. Tajiri, T. Kinoshita, and T. Matsuoka. Correlation of in vitro autofluorescence endoscopy images with histopathologic findings in stomach cancer. *Endoscopy* **32**, 281–286 (2000).

Andersson, P. S., S. Montán, and S. Svanberg. Multi-spectral system for medical fluorescence imaging, *IEEE J. Quant. Electr.* **QE-23**, 1798–1805 (1987).

Andersson-Engels, S., R. Berg, K. Svanberg, and S. Svanberg. Multi-colour fluorescence imaging in combination with photodynamic therapy of d-amino levulinic acid (ALA) sensitised skin malignancies. *Bioimaging* **3**, 134–143 (1995).

Andersson-Engels, S., G. Canti, R. Cubeddu, C. Eker, C. A. F. Klinteberg, A. Pifferi, K. Svanberg, S. Svanberg, P. Taroni, G. Valentini, and I. Wang. Preliminary evaluation of two fluorescence imaging methods for detection of basal cell carcinomas of the skin. *Lasers Surg. Med.* **26**, 76–82 (2000).

Andersson-Engels, S., Å Elner, J. Johansson, S.- E. Karlsson, L. G. Salford, L.- G. Strömblad, K. Svanberg, and S. Svanberg. Clinical recording of laser-induced fluorescence spectra for evaluation of tumour demarcation feasibility in selected clinical specialties. *Lasers Med. Sci.* **6**, 415–424 (1991).

Andersson-Engels, S., A. Gustafsson, J. Johansson, U. Stenram, K. Svanberg, and S. Svanberg. Laser-induced fluorescence used in localizing atherosclerotic lesions. *Lasers Med. Sci.* **4**, 171–181 (1989a).

Andersson-Engels, S. A. Gustafsson, J. Johansson, U. Stenram, K. Svanberg, and S. Svanberg. Investigation of possible fluorophores in human atherosclerotic plaque. *Lasers Life Sci.* **5**, 1–11 (1992).

Andersson-Engels, S. J. Johansson, U. Stenram, K. Svanberg, and S. Svanberg. Diagnosis by means of fluorescent light emission from tissue. U.S. patent no. 5,115,137, (1989b).

Andersson-Engels, S., J. Johansson, U. Stenram, K. Svanberg, and S. Svanberg. Malignant tumor and atherosclerotic plaque diagnosis using laser-induced fluorescence. *IEEE J. Quant. Electr.* **26**, 2207–2217 (1990a).

Andersson-Engels, S., J. Johansson, and S. Svanberg. The use of time-resolved fluorescence for diagnosis of atherosclerotic plaque and malignant tumours. *Spectrochim. Acta* **46A**, 1203–1210 (1990b).

Andersson-Engels, S., J. Johansson, and S. Svanberg. Medical diagnostic system based on simultaneous multispectral fluorescence imaging. *Appl. Opt.* **33**, 8022–8029 (1994).

Andersson-Engels, S., C. A. F. Klinteberg, K. Svanberg, and S. Svanberg. In vivo fluorescence imaging for tissue diagnostics. *Phys. Med. Biol.* **42**, 815–824 (1997).

Andersson-Engels, S., and B. C. Wilson. *In vivo* fluorescence in clinical oncology: Fundamental and practical issues. *J. Cell Pharmacol.* **3**, 48–61 (1992).

Ankerst, J., S. Montán, K. Svanberg, and S. Svanberg. Laser-induced fluorescence studies of hematoporphyrin derivative (HpD) in normal and tumor tissue of rat. *Appl. Spectrosc.* **38**, 890–896 (1984).

Auler, H., and G. Banzer. Untersuchungen über die Rolle der Porphyrin bei Geschwulstkranken Menschen und Tieren. *Z. Krebsforsch.* **53**, 65–68 (1942).

Balas, C. A novel optical imaging method for the early detection, quantitative grading, and mapping of cancerous and precancerous lesions of cervix. *IEEE Trans. Biomed. Eng.* **48**, 96–104 (2001).

Baumgartner, R., H. Fisslinger, D. Jocham, H. Lenz, L. Ruprecht, H. Stepp, and E. Unsöld. A fluorescence imaging device for endoscopic detection of early stage cancer: Instrumental and experimental studies. *Photochem. Photobiol.* **46**, 759–763 (1987).

Betz, C., S., H. Stepp, P. Janda, S. Arbogast, G. Grevers, R. Baumgartner, and A. Leunig. A comparative study of normal inspection, autofluorescence and 5-ALA-induced PPIX fluorescence for oral cancer diagnosis. *Int. J. Cancer* **97**, 245–252 (2002).

Bogaards, A., M. C. Aalders, A. J. L. Jongen, E. Dekker, and H. J. C. M. Sterenborg. Double ratio fluorescence imaging for the detection of early superficial cancers. *Rev. Sci. Instrum.* **72**, 3956–3961 (2001).

Bogaards, A., M. C. G. Aalders, C. C. Zeyl, S. de Blok, C. Dannecker, P. Hillemanns, H. Stepp, and H. J. C. M. Sterenborg. Localization and staging of cervical intraepithelial neoplasia using double ratio fluorescence imaging. *J. Biomed. Opt.* **7**, 215–220 (2002).

Bottiroli, G., A. C. Croce, D. Locatelli, R. Marchesini, E. Pignoli, S. Tomatis, C. Cuzzoni, S. Di Palma, M. Dal Fante, and P. Spinelli. Natural fluorescence of normal and neoplastic human colon: A comprehensive "ex vivo" study. *Lasers Surg. Med.* **16**, 48–60 (1995).

Bottiroli, G., M. Dal Fante, R. Marchesini, A. C. Croce, C. Cuzzoni, S. Di Palma, E. Pignoli, and P. Spinelli. Naturally occurring fluorescence of adenocarcinomas, adenomas and non-neoplastic mucosa of human colon: Endoscopic and microscopic spectrofluorometry. In *Photodynamic Therapy and Biomedical Lasers,* eds. P. Spinelli, M. Dal Fante, and R. Marchesini, pp. 898–902 (Elsevier Science Publishers, Amsterdam, 1992).

Brand, S., H. Stepp, T. Ochsenkuhn, R. Baumgartner, G. Baretton, J. Holl, C. von Ritter, G. Paumgartner, M. Sackmann, and G. Baumgartner. Detection of colonic dysplasia by light-induced fluorescence endoscopy: A pilot study. *Int. J. Colorectal Dis.* **14**, 63–68 (1999).

Carpenter, R. J., R. J. Ryan, H. B. Neel, and D. R. Sanderson. Tumour fluorescence with haematoporphyrin derivative. *Ann. Otol. Rhinol. Laryngol.* **86**, 661–666 (1977).

Chwirot, B. W., M. Kowalska, N. Sypniewska, Z. Michniewicz, and M. Gradziel. Spectrally resolved fluorescence imaging of human colonic adenomas. *J. Photochem. Photobiol. B* **50**, 174–183 (1999).

Cordeiro, P. G., R. E. Kirschner, Q. Y. Hu, J. J. C. Chiao, H. Savage, R. R. Alfano, L. A. Hoffman, and D. A. Hidalgo, Ultraviolet excitation fluorescence spectroscopy: A noninvasive method for the measurement of redox changes in ischemic myocutaneous flaps. *Plast. Reconstr. Surg.* **96**, 673–680 (1995).

Cubeddu, R., G. Canti, A. Pifferi, P. Taroni, and G. Valentini. A real time system for fluorescence lifetime imaging. *Proc. SPIE* **2976**, 98–104 (1997a).

Cubeddu, R., G. Canti, A. Pifferi, P. Taroni, and G. Valentini. Fluorescence lifetime imaging of experimental tumors in hematoporphyrin derivative-sensitized mice. *Photochem. Photobiol.* **66**, 229–236 (1997b).

Cubeddu, R., G. Canti, P. Taroni, and G. Valentini. Time-gated fluorescence imaging for the diagnosis of tumors in a murine model. *Photochem. Photobiol.* **57**, 480–485 (1993a).

Cubeddu, R., G. Canti, P. Taroni, and G. Valentini. Tumour visualization in a murine model by time-delayed fluorescence of sulphonated aluminium phthalocyanine. *Lasers Med. Sci.* **12**, 200–208 (1997c).

Cubeddu, R., A. Pifferi, P. Taroni, G. Valentini, and G. Canti. Tumor detection in mice by measurement of fluorescence decay time matrices. *Opt. Lett.* **20**, 2553–2555 (1995).

Cubeddu, R., A. Pifferi, P. Taroni, G. Valentini, G. Canti, C. Lindquist, S. Andersson-Engels, S. Svanberg, I. Wang, and K. Svanberg. Multispectral and lifetime imaging for the detection of skin tumors. *Proc. OSA Trends in Optics and Photonics.* **22**, 106–109 (1998).

Cubeddu, R., P. Taroni, and G. Valentini. Time-gated imaging system fortumordiagnosis. *Opt. Eng.* **32**, 320–325 (1993b).

DaCosta, R. S., H. Andersson, B. C. Wilson, S. Cirocco, M. Hassaram, and N. E. Marcon. *Correlative Studies ofAutof luorescence, Immuno/histopathology and Ultrastructures ofex Vivo Tissues and Primary Cultured Epithelial Cells from Human Normal, Preneoplastic and Neoplastic Colorectal Mucosa.* Presented at the CLEO/Europe—EQEC Focus Meetings, München, Germany (1999).

Daniell, M. D., and J. S. Hill. A history of photodynamic therapy. *Aust. N. Z. J. Surg.* **61**, 340–348 (1991).

Deckelbaum, L. I. Cardiovascularapplications of lasertechnology. *Lasers Surg. Med.* **15**, 315–341 (1994).

Deckelbaum, L. I., S. P. Desai, C. Kim, and J. J. Scott. Evaluation of a fluorescence feedback system for guidance of laser angioplasty. *Lasers Surg. Med.* **16**, 226–234 (1995).

De Dominicis, C., M. Liberti, G. Perugia, C. De Nunzio, F. Sciobica, A. Zuccala, A. Sarkozy, and F. Iori. Role of 5-aminolevulinic acid in the diagnosis and treatment of superficial bladder cancer: Improvement in diagnostic sensitivity. *Urology* **57**, 1059–1062 (2001).

DeWeert, M. J., J. Oyama, E. McLaughlin, E. Jacobson, J. Håkansson, G. S. Bignami, U. Gustafsson, P. Troy, V. Poskiene, K. Kriukelyte, R. Ziobakiene, A. Vaitkuviene, S. Pålsson, M. Soto Thompson, U. Stenram, S. Andersson-Engels, S. Svanberg, and K. Svanberg. Analysis of the spatial variability in hyperspectral imagery of the uterine cervix in vivo. *Proc. SPIE* **4959**, 67–76 (2003).

Docchio, F. Ocular fluorometry: Principles, fluorophores, instrumentation, and clinical applications. *Lasers Surg. Med.* **9**, 515–532 (1989).

Doiron, D. R., E. Profio, R. G. Vincent, and T. J. Dougherty. Fluorescence bronchoscopy for detection of lung cancer. *Chest* **76**, 27–32 (1979).

Dougherty, T. J., G. B. Grindey, R. Fiel, K. R. Weishaupt, and D. G. Boyle. Photoradiation therapy II. Cure of animal tumors with haematoporphyrin and light. *J. Natl. Cancer Inst.* **55**, 115–119 (1975).

Dowling, K., M. J. Dayel, S. C. W. Hyde, J. C. Dainty, P. M. W. French, P. Vourdas, M. J. Lever, A. K. L. Dymoke-Bradshaw, J. D. Hares, and P. A. Kellett. Whole-field fluorescence lifetime imaging with picosecond resolution using ultrafast 10-kHz solid-state amplifier technology. *IEEE J. Sel. Top. Quant. Electr.* **4**, 370–375 (1998a).

Dowling, K., M. J. Dayel, S. C. W. Hyde, P. M. W. French, M. J. Lever, J. D. Hares, and A. K. L. Dymoke-Bradshaw. High resolution time-domain fluorescence lifetime imaging for biomedical applications. *J. Mod. Opt.* **46**, 199–209 (1999).

Dowling, K., M. J. Dayel, M. J. Lever, P. M. W. French, J. D. Hares, and A. K. L. Dymoke-Bradshaw. Fluorescence lifetime imaging with picosecond resolution for biomedical applications. *Opt. Lett.* **23**, 810–812 (1998b).

Dowling, K., S. C. W. Hyde, J. C. Dainty, P. M. W. French, and J. D. Hares. 2-D Fluorescence lifetime imaging using a time-gated image intensifier. *Opt. Comm.* **135**, 27–31 (1997).

Edner, H., J. Johansson, S. Svanberg, and E. Wallinder. Fluorescence lidar multicolor imaging of vegetation. *Appl. Opt.* **33**, 2471–2479 (1994).

Ehsan, A., F. Sommer, G. Haupt. and U. Engelmann. Significance of fluorescence cystoscopy for diagnosis of superficial bladdercancerafterintr avesical instillation of delta aminolevulinic acid. *Urol. Int.* **67**, 298–304 (2001).

Eker, C. *Optical Characterization ofTissue for Medical Diagnostics.* PhD diss., Lund Institute of Technology, Lund, Sweden (1999).

Eker, C., S. Montán, E. Jaramillo, K. Koizumi, C. Rubio, S. Andersson-Engels, K. Svanberg, S. Svanberg, and P. Slezak. Clinical spectral characterisation of colonic mucosal lesions using autofluorescence and d aminolevulinic acid sensitization. *Gut* **44**, 511–518 (1999).

Figge, F. H. J., G. S. Weiland, and L. O. J. Manganiello. Cancer detection and therapy: Affinity of neoplastic, embryonic and traumatised tissues for porphyrins and metalloporphyrins. *Proc. Soc. Exp. Med. Biol.* **68**, 640–641 (1948).

Filbeck, T., W. Roessler, R. Knuechel, M. Straub, H. J. Kiel, and W. F. Wieland. 5-Aminolevulinic acid-induced fluorescence endoscopy applied at secondary transurethral resection after conventional resection of primary superficial bladdertumor s. *Urology* **53**, 77–81 (1999).

Fischer, F., E. F. Dickson, R. H. Pottier, and H. Wieland. An affordable, portable fluorescence imaging device for skin lesion detection using a dual wavelength approach for image contrast enhancement and aminolaevulinic acid-induced protoporphyrin IX. Part I. Design, spectral and spatial characteristics. *Lasers Med. Sci.* **16**, 199–206 (2001).

Frimberger, D., D. Zaak, H. Stepp, R. Knuchel, R. Baumgartner, P. Schneede, N. Schmeller, and A. Hofstetter. Autofluorescence imaging to optimize 5-ALA-induced fluorescence endoscopy of bladder carcinoma. *Urology* **58**, 372–375 (2001).

Gahlen, J., R. L. Prosst, M. Pietschmann, M. Rheinwald, T. Haase, and C. Herfarth. Spectrometry supports fluorescence staging laparoscopy after intraperitoneal aminolaevulinic acid lavage for gastrointestinal tumours. *J. Photochem. Photobiol. B* **52**, 131–135 (1999).

Garfinkel, L. Cancer statistics and trends. In *American Cancer Society Textbook ofClinical Oncology,* eds. A. I. Holleb, D. J. Fink, and G. P. Murphy, pp. 1–6 (American Cancer Society, Atlanta, Ga., 1991).

Garini, Y., N. Katzir, D. Cabib, R. Buckwald, D. G. Soenksen, and Z. Malik. Spectralbio-imaging. In *Fluorescence Imaging Spectroscopy and Microscopy,* eds. X. F. Wang and B. Herman, pp. 88–124 (John Wiley, New York, 1996).

Gregorie, H. B., and J. F. Green. Haematoporphyrin derivative fluorescence in malignant neoplasms. *J. S. C. Med. Assoc.* **61**, 157–164 (1965).

Gustafsson, U., E. McLaughlin, E. Jacobson, J. Håkansson, P. Troy, M. J. DeWeert, S. Pålsson, M. Soto Thompson, S. Svanberg, A. Vaitkuviene, and K. Svanberg. Fluorescence and reflectance monitoring of human cervical tissue in vivo: A case study. *Proc. SPIE* **4959**, 100–110 (2003a).

Gustafsson, U., E. McLaughlin, E. Jacobson, J. Håkansson, P. Troy, M. J. DeWeert, S. Pålsson, M. Soto Thompson, S. Svanberg, A. Vaitkuviene. and K. Svanberg. In-vivo fluorescence and reflectance imaging of human cervical tissue. *Proc. SPIE* **5031**, 521–530 (2003b).

Gustafsson, U., S. Pålsson, and S. Svanberg. Compact fibre-optic fluorosensor using a continuous wave violet diode laser and an integrated spectrometer. *Rev. Sci. Instrum.* **71**, 3004–3006 (2000).

Guy, C. N. The second revolution in medical imaging. *Contemp. Phys.* **37**, 15–46 (1996).

Hall, E. *Radiobiology for the Radiologist* (J.B. Lippincott, Philadelphia, 1994).

Haringsma, J., and G. N. Tytgat. Fluorescence and autofluorescence. *Baillieres Best Pract. Res. Clin. Gastroenterol.* **13**, 1–10 (1999).

Haringsma, J., G. N. Tytgat, H. Yano, H. Iishi, M. Tatsuta, T. Ogihara, H. Watanabe, N. Sato, N. Marcon, B. C. Wilson, and R. W. Cline. Autofluorescence endoscopy: Feasibility of detection of GI neoplasms unapparent to white light endoscopy with an evolving technology. *Gastrointest. Endosc.* **53**, 642–650 (2001).

Hartmann, A., K. Moser, M. Kriegmair, A. Hofstetter, F. Hofstaedter, and R. Knuechel. Frequent genetic alterations in simple urothelial hyperplasias of the bladder in patients with papillary urothelial carcinoma. *Am. J. Pathol.* **154**, 721–727 (1999).

Hewett, J., V. Nadeau, J. Ferguson, H. Moseley, S. Ibbotson, J. W. Allen, W. Sibbett, and M. Padgett. The application of a compact multispectral imaging system with integrated excitation source to in vivo monitoring of fluorescence during topical photodynamic therapy of superficial skin cancers. *Photochem. Photobiol.* **73**, 278–282 (2001).

Hillemanns, P., H. Weingandt, R. Baumgartner, J. Diebold, W. Xiang, and H. Stepp. Photodetection of cervical intraepithelial neoplasia using 5-aminolevulinic acid-induced porphyrin fluorescence. *Cancer* **88**, 2275–2282 (2000).

Hirsch, F. R., W. A. Franklin, A. F. Gazdar, and P. A. J. Bunn. Early detection of lung cancer: Clinical perspectives of recent advances in biology and radiology. *Clin. Cancer Res.* **7**, 5–22 (2001).

Horvath, K. A., K. T. Schomacker, C. C. Lee, and L. H. Cohn. Intraoperative myocardial ischemia detection with laser-induced fluorescence. *J. Thorac. Cardiovasc. Surg.* **107**, 220–225 (1994).

Izuishi, K., H. Tajiri, T. Fujii, N. Boku, T. Ohtsu, T. Ohnishi, M. Ryu, T. Kinoshita, and S. Yoshida. The histological basis of detection of adenoma and cancer in the colon by autofluorescence endoscopic imaging. *Endoscopy* **31**, 511–516 (1999).

Jones, R., K. Dowling, M. J. Cole, D. Parsons-Karavassilis, M. J. Lever, P. M. W. French, J. D. Hares, and A. K. L. Dymoke-Bradshaw. Fluorescence lifetime imaging using a diode-pumped all-solid-state laser system. *Elec. Lett.* **35**, 256–258 (1999).

Kapadia, C. R., F. W. Cutruzzola, K. M. O'Brien, M. L. Stetz, R. Enriquez, and L. I. Deckelbaum. Laser-induced fluorescence spectroscopy of human colonic mucosa: Detection of adenomatous transformation. *Gastroenterology* **99**, 150–157 (1990).

Kelly, J. F., and M. E. Snell. Haematoporphyrin derivative: A possible aid in the diagnosis and therapy of carcinoma in the bladder. *J. Urol.* **115**, 150–151 (1976).

Kennedy, J. C., and R. H. Pottier. Endogenous protoporphyrin IX: A clinically useful photosensitizer for photodynamic therapy, *J. Photochem. Photobiol. B* **14**, 275–292 (1992).

Kennedy, J. C., R. H. Pottier, and D. C. Pross. Photodynamic therapy with endogenous protoporphyrin IX: Basic principles and present clinical experience. *J. Photochem. Photobiol. B* **6**, 143–148 (1990).

Kennedy, T. C., S. Lam, and F. R. Hirsch. Review of recent advances in fluorescence bronchoscopy in early localization of central airway lung cancer. *Oncologist* **6**, 257–262 (2001).

Kennedy, T. C., Y. Miller, and S. Prindiville. Screening for lung cancer revisited and the role of sputum cytology and fluorescence bronchoscopy in a high-risk group. *Chest* **117**, 72S–79S (2000).

Khanavkar, B., F. Gnudi, A. Muti, W. Marek, K. M. Muller, Z. Atay, T. Topalidis, and J. A. Nakhosteen. [Basic principles of LIFE–autofluorescence bronchoscopy. Results of 194 examinations in comparison with standard procedures for early detection of bronchial carcinoma—overview]. *Pneumologie* **52**, 71–76 (1998).

Klinteberg, C. A. F., Andreasson M, Sandström O, Andersson-Engels S, Svanberg S. Compact medical fluorosensor for minimally invasive tissue characterization. *Rev. Sci. Instrum* **76**, 034303–034309 (2005).

Klinteberg, C. A. F., I. Wang, I. Karu, T. Johansson, N. Bendsoe, K. Svanberg, S. Andersson-Engels, S. Svanberg, G. Canti, R. Cubeddu, A. Pifferi, P. Taroni, and G. Valentini. *Diode Laser-Mediated ALA-PDT Guided by Laser-Induced Fluorescence Imaging.* Report, Lund Reports on Atomic Physics LRAP-287 (Lund Institute of Technology, Lund, Sweden, 2002).

Kobayashi, M., H. Tajiri, E. Seike, M. Shitaya, S. Tounou, M. Mine, and K. Oba. Detection of early gastric cancer by a real-time autofluorescence imaging system. *Cancer Lett* **165**, 155–159 (2001).

Koenig, F., F. J. McGovern, R. Larne, H. Enquist, K. T. Schomacker, and T. F. Deutsch. Diagnosis of bladder carcinoma using protoporphyrin IX fluorescence induced by 5-aminolaevulinic acid. *BJU Int.* **83**, 129–135 (1999).

Kohl, M., J. Neukammer, U. Sukowski, H. Rinneberg, D. Wöhrle, H.- J. Sinn, and E. A. Friedrich. Delayed observation of laser-induced fluorescence for imaging of tumors. *Appl. Phys. B* **56**, 131–138 (1993).

Kriegmair, M., H. Stepp, P. Steinbach, W. Lumper, A. Ehsan, H. G. Stepp, K. Rick, R. Knuchel, R. Baumgartner, and A. Hofstetter. Fluorescence cystoscopy following intravesical instillation of 5-aminolevulinic acid: A new procedure with high sensitivity for detection of hardly visible urothelial neoplasias. *Urol. Int.* **55**, 190–196 (1995).

Kriegmair, M., D. Zaak, H. Stepp, R. Baumgartner, R. Knuechel, and A. Hofstetter. Transurethral resection and surveillance of bladder cancer supported by 5-aminolevulinic acid-induced fluorescence endoscopy. *Eur. Urol.* **36**, 386–392 (1999).

Kulapaditharom, B., and V. Boonkitticharoen. Laser-induced fluorescence imaging in localization of head and neck cancers. *Ann. Otol. Rhinol. Laryngol.* **107**, 241–246 (1998).

Kurie, J. M., J. S. Lee, R. C. Morice, G. L. Walsh, F. R. Khuri, A. Broxson, J. Y. Ro, W. A. Franklin, R. Yu, and W. K. Hong. Autofluorescence bronchoscopy in the detection of squamous metaplasia and dysplasia in current and former smokers. *J. Natl. Cancer Inst.* **90**, 991–995 (1998).

Kusunoki, Y., F. Imamura, H. Uda, M. Mano, and T. Horai. Early detection of lung cancer with laser-induced fluorescence endoscopy and spectrofluorometry. *Chest* **118**, 1776–1782 (2000).

Ladner, D. P., R. A. Steiner, J. Allemann, U. Haller, and H. Walt. Photodynamic diagnosis of breast tumours after oral application of aminolevulinic acid. *Br. J. Cancer* 33–37 (2001).

Lam, S., and H. D. Becker. Future diagnostic procedures. *Chest Surg. Clin. North Am.* **6**, 363–380 (1996).

Lam, S., J. Y. Hung, S. M. Kennedy, J. C. LeRiche, S. Vedal, B. Nelems, C. E. Macaulay, and B. Palcic. Detection of dysplasia and carcinoma in situ by ratio fluorometry. *Am. Rev. Respir. Dis.* **146**, 1458–1461 (1992a).

Lam, S., J. Hung, and B. Palcic. Mechanism of detection of early lung cancer by ratio fluorometry. *Lasers Life Sci.* **4**, 67–73 (1991).

Lam, S., T. Kennedy, M. Unger, Y. E. Miller, D. Gelmont, V. Rusch, B. Gipe, D. Howard, J. C. LeRiche, A. Coldman, and A. F. Gazdar. Localization of bronchial intraepithelial neoplastic lesions by fluorescence bronchoscopy. *Chest* **113**, 696–702 (1998).

Lam, S., C. MacAulay, J. Hung, J. LeRiche, A. E. Profio, and B. Palcic. Detection of dysplasia and carcinoma in situ with a lung imaging fluorescence endoscope device. *J. Thorac. Cardiovasc. Surg.* **105**, 1035–1040 (1993a).

Lam, S., C. MacAulay, B. Jaggi, P. Eng, and B. Palcic. Early detection of lung cancer by fluorescence imaging: The Vancouver experience. Prevention, early diagnosis and treatment of endobronchial lung cancer. (1992b).

Lam, S., C. MacAulay, J. C. LeRiche, and B. Palcic. Detection and localization of early lung cancer by fluorescence bronchoscopy. *Cancer* **89**, 2468–2473 (2000).

Lam, S., C. MacAulay, and B. Palcic. Detection and localization of early lung cancer by imaging techniques. *Chest* **103**, 12S–14S (1993b).

Lam, S., and B. Palcic. Autofluorescence bronchoscopy in the detection of squamous metaplasia and dysplasia in current and former smokers. *J. Natl. Cancer Inst.* **91**, 561–562 (1999).

Lazarev, V. V., R. A. Roth, Y. Kazakevich. and J. Hang. Detection of premalignant oral lesions in hamsters with an endoscopic fluorescence imaging system. *Cancer* **85**, 1421–1429 (1999).

Leonard, J. R., and W. L. Beck. Haematoporphyrin fluorescence: An aid in diagnosis of malignant neoplasms. *Laryngoscope* **81**, 365–372 (1971).

Leunig, A., K. Rick, H. Stepp, R. Gutmann, G. Alwin, R. Baumgartner, and J. Feyh. Fluorescence imaging and spectroscopy of 5-aminolevulinic acid induced protoporphyrin IX for the detection of neoplastic lesions in the oral cavity. *Am. J. Surg.* **172**, 674–677 (1996).

Lipson, R. L., and E. J. Baldes. The photodynamic properties of a particular haematoporphyrin derivative. *Arch. Dermatol.* **82**, 508–516 (1960).

Lipson, R. L., E. J. Baldes, and A. M. Olsen. Hematoporphyrin derivative: A new aid for endoscopic detection of malignant disease. *J. Thorac. Cardiovasc. Surg.* **42**, 623–629 (1961).

Lipson, R., L. E. J. Baldes, and A. M. Olsen. Further evaluation of the use of haematoporphyrin derivative as a new aid forthe endoscopic detection of malignant disease. *Dis. Chest* **46**, 676–679 (1964a).

Lipson, R. L., J. H. Pratt, E. J. Baldes, and M. B. Dockerty. Hematoporphyrin derivative for detection of cervical cancer. *Obstet. Gynecol.* **24**, 78–84 (1964b).

MacKay, R. S. *Medical Images and Displays: Comparison ofNuclear Magnetic Imaging, Ultrasound, X-rays and Other Modalities* (Wiley, New York, 1984).

Malik, Z., and H. Lugaci. Destruction of erythroleukaemic cells by photoinactivation of endogenous porphyrins. *Br. J. Cancer* **56**, 589–595 (1987).

Marchesini, R., M. Brambilla, E. Pignoli, G. Bottiroli, A. C. Croce, M. Dal Fante, P. Spinelli, and S. Di Palma. Light-induced fluorescence spectroscopy of adenomas, adenocarcinomas and non-neoplastic mucosa in human colon. I. In vitro measurements. *J. Photochem. Photobiol. B* **14**, 219–230 (1992).

Marcon, N. E. Is light-induced fluorescence better than the endoscopist's eye? *Can. J. Gastroenterol.* **13**, 417–421 (1999).

Montán, S., K. Svanberg, and S. Svanberg. Multi-color imaging and contrast enhancement in cancer tumor localization using laser-induced fluorescence in hematoporphyrin derivative (HpD)-bearing tissue. *Opt. Lett.* **10**, 56–58 (1985).

Orth, K., D. Russ, R. Steiner, and H. G. Beger. Fluorescence detection of small gastrointestinal tumours: Principles, technique, first clinical experience. *Langenbecks Arch. Surg.* **385**, 488–494 (2000).

Palcic, B., S. Lam, J. Hung, and C. MacAulay. Detection and localization of early lung cancer by imaging techniques. *Chest* **99**, 742–743 (1991).

Papazoglou, T. G. Malignant and atherosclerotic plaque diagnosis: Is laser induced fluorescence spectroscopy the ultimate solution? *J. Photochem. Photobiol. B* **28**, 3–11 (1995).

Perk, M., G. J. Flynn, S. Gulamhusein, Y. Wen, C. Smith, B. Bathgate, J. Tulip, N. A. Parfrey, and A. Lucas. Laser-induced fluorescence identification of sinoatrial and atrioventricular nodal conduction tissue. *PACE* **16**, 1701–1712 (1993).

Perk, M., G. J. Flynn, C. Smith, B. Bathgate, J. Tulip, W. Yue, and A. Lucas. Laser-induced fluorescence emission: I. The spectroscopic identification of fibrotic endocardium and myocardium. *Lasers Surg. Med.* **11**, 523–534 (1991).

Pfau, P. R., and M. V. J. Sivak. Endoscopic diagnostics. *Gastroenterology* **120**, 763–781 (2001).

Pham, T. H., C. Eker, A. Durkin, B. J. Tromberg, and S. Andersson-Engels. Quantifying the optical properties and chromophore concentrations of turbid media by chemometric analysis of hyperspectral, diffuse reflectance data collected using a Fourier interferrometric imaging system. *Appl. Spectrosc.* **55**, 1035–1045 (2001).

Policard, A. Etudes sur les aspects offerts par des tumeur experimentales examinée à la lumière de Woods. *CR. Soc. Biol.* **91**, 1423 (1924).

Profio, A. E. Laser excited fluorescence of hematoporphyrin derivative for diagnosis of cancer. *IEEE J. Quant. Electr.* **QE-20**, 1502–1507 (1984).

Profio, A. E., and O. J. Balchum. Fluorescence diagnosis of cancer. In *Methods in Porphyrin Photosensitization,* ed. D. Kessel, pp. 43–50 (Plenum Publishing, New York, 1985).

Profio, A. E., O. J. Balchum, and F. Carstens. Digital background subtraction for fluorescence imaging. *Med. Phys.* **13**, 717–721 (1986).

Profio, A. E., and D. R. Doiron. Laser fluorescence bronchoscope for localization of occult lung tumors. *Med. Phys.* **6**, 523–525 (1979).

Profio, A. E., D. R. Doiron, O. J. Balchum, and G. C. Huth. Fluorescence bronchoscopy for localization of carcinoma in situ. *Med. Phys.* **10**, 35–39 (1983).

Profio, A. E., D. R. Doiron, and G. C. Huth. Fluorescence bronchoscope for lung tumour localization. *IEEE Trans. Nucl. Sci.* **NS-24**, 521–524 (1977).

Raab, O. Über die Wirkung fluoreszierenden Stoffe auf Infusoria. *Z. Biol.* **39**, 524–546 (1900).

Ramanujam, N., J. Chen, K. Gossage, R. Richards-Kortum, and B. Chance. Fast and noninvasive fluorescence imaging of biological tissues in vivo using a flying-spot scanner. *IEEE Trans. Biomed. Eng.* **48**, 1034–1041 (2001).

Rasmussen-Taxdal, D. S., G. E. Ward, and F. H. J. Figge. Fluorescence of human lymphatic and cancer tissues following high doses of intravenous haematoporphyrin. *Cancer* **8**, 78–81 (1955).

Rava, R. P., R. Richards-Kortum, M. Fitzmaurice, R. Cothren, R. Petras, M. Sivak, H. Levin, and M. S. Feld. Early detection of dysplasia in colon and bladdertissue using laserinduced fluorescence. *Proc. SPIE* **1426**, 68–78 (1991).

Richards-Kortum, R., R. P. Rava, R. E. Petras, M. Fitzmaurice, M. Sivak, and M. S. Feld. Spectroscopic diagnosis of colonic dysplasia. *Photochem. Photobiol.* **53**, 777–786 (1991).

Riedl, C. R., D. Daniltchenko, F. Koenig, R. Simak, S. A. Loening, and H. Pflueger. Fluorescence endoscopy with 5-aminolevulinic acid reduces early recurrence rate in superficial bladder cancer. *J. Urol.* **165**, 1121–1123 (2001).

Robb, R. A. *Three-dimensional Biomedical Imaging: Principle and Practice* (VCH, New York, 1994).

Roth, J. E., S. Andersson-Engels, N. Bendsoe, and K. Svanberg. Hyperspectral fluorescence and reflectance imaging of basal cell carcinomas following topical administration of 5-aminolevulinic acid. (2003).

Saarnak, A. E., T. Rodrigues, J. Schwartz, A. L. Moore, T. A. Moore, D. Gust, M. J. C. van Gemert, H. J. C. M. Sterenborg, and S. Thomsen. Influence of tumour depth, blood absorption and autofluorescence on measurements of exogenous fluorophores in tissue. *Lasers Med. Sci.* **13**, 22–31 (1998).

Sanderson, D. R., R. S. Fontana, R. L. Lipson, and E. J. Baldes. Haematoporphyrin as a diagnostic tool: A preliminary report of new techniques. *Cancer* **30**, 1368–1372 (1972).

Schomacker, K. T., J. K. Frisoli, C. C. Compton, T. J. Flotte, J. M. Richter, T. F. Deutsch, and N. S. Nishioka. Ultraviolet laser-induced fluorescence of colonic polyps. *Gastroenterology* **102**, 1155–1160 (1992).

Scott, M. A., C. Hopper, A. Sahota, R. Springett, B. W. Mcllroy, S. G. Bown, and A. J. MacRobert. Fluorescence photodiagnostics and photobleaching studies of cancerous lesions using ratio imaging and spectroscopic techniques. *Lasers Med. Sci.* **15**, 63–72 (2000).

Sinaasappel, M., and H. J. C. M. Sterenborg. Quantification of the hematoporphyrin derivative by fluorescence measurement using dual-wavelength excitation and dual-wavelength detection. *Appl. Opt.* **32**, 541–548 (1993).

Stefanidou, M., A. Tosca, G. Themelis, E. Vazgiouraki, and C. Balas. In vivo fluorescence kinetics and photodynamic therapy efficacy of delta-aminolevulinic acid-induced porphyrins in basal cell carcinomas and actinic keratoses: Implications foroptimization of photodynamic therapy. *Eur. J. Dermatol.* **10**, 351–356 (2000).

Sterenborg, H. J. C. M., A. E. Saarnak, R. Frank, and M. Motamedi. Evaluation of spectral correction techniques for fluorescence measurements on pigmented lesions in vivo. *J. Photochem. Photobiol. B* **35**, 159–165 (1996).

Stübel, H. Die Fluoreszenz tierischer Gewebe in ultravioletten Licht. *Pflügers Arch.* **142**, 1 (1911).

Sukowski, U., B. Ebert, K. Zumbusch, K. Müller, B. Fleige, H. Lochs, and H. Rinneberg. Endoscopic detection of early malignancies in the upper gastrointestinal tract using laser-induced fluorescence imaging. *Proc. SPIE* **4156**, 255–261 (2001).

Sutedja, T. G., H. Codrington, E. K. Risse, R. H. Breuer, J. C. van Mourik, R. P. Golding, and P. E. Postmus. Autofluorescence bronchoscopy improves staging of radiographically occult lung cancer and has an impact on therapeutic strategy. *Chest* **120**, 1327–1332 (2001a).

Sutedja, T. G., B. J. Venmans, E. F. Smit, and P. E. Postmus. Fluorescence bronchoscopy for early detection of lung cancer: A clinical perspective. *Lung Cancer* **34**, 157–168 (2001b).

Svanberg, K., I. Wang, S. Colleen, I. Idvall, C. Ingvar, R. Rydell, D. Jocham, H. Diddens, S. Bown, G. Gregory, S. Montán, S. Andersson-Engels, and S. Svanberg. Clinical multi-colour fluorescence imaging of malignant tumours: Initial experience. *Acta Radiol.* **39**, 2–9 (1998).

Tamura, M., O. Hazeki, S. Nioka, and B. Chance. In vivo study of tissue oxygen metabolism using optical and nuclear magnetic resonance spectroscopies. *Annu. Rev. Physiol.* **51**, 813–834 (1989).

Tassetti, V., A. Hajri, M. Sowinska, S. Evrard, F. Heisel, L. Q. Cheng, J. A. Mielé, J. Marascaux, and M. Aprahamian. In vivo laser-induced fluorescence imaging of a rat pancreatic cancer with pheophorbide-a. *Photochem. Photobiol.* **65**, 997–1006 (1997).

Urbach, F., P. D. Forbes, R. E. Davies, and D. Berger. Cutaneous photobiology: Past, present and future. *J. Invest. Dermatol.* **67**, 209–224 (1976).

van den Bergh, H. Photodynamic therapy and photodetection of early cancer in the upper aerodigestive tract, the tracheobronchial tree, the oesophagus and the urinary bladder. In *Hardrontherapy in Oncology*, eds. U. Amaldi and B. Larsson, pp. 577–621 (Elsevier, Amsterdam, 1994).

van Rens, M. T., F. M. Schramel, J. R. Elbers, and J. W. Lammers. The clinical value of lung imaging fluorescence endoscopy for detecting synchronous lung cancer. *Lung Cancer* **32**, 13–18 (2001).

von Tappeiner, H., and A. Jodlbauer. *Die sensibiliserende Wirkung fluorescierender Substanzer: Gasammette Untersuchungen über die photodynamische Erscheinung* (FCW Vogel, Leipzig, 1907).

Wagnières, G., C. Depeursinge, P. Monnier, M. Savary, P. Cornaz, A. Châtelain, and H. van den Bergh. Photodetection of early cancer by laser induced fluorescence of a tumour-selective dye: Apparatus design and realization. *Proc. SPIE* **1203**, 43–52 (1990).

Wagnières, G. A., W. M. Star, and B. C. Wilson. *In vivo* fluorescence spectroscopy and imaging for oncological applications. *Photochem. Photobiol.* **68**, 603–632 (1998).

Wagnières, G. A., A. P. Studzinski, and H. E. van den Bergh. An endoscopic fluorescence imaging system for simultaneous visual examination and photodetection of cancer. *Rev. Sci. Instrum.* **68**, 203–212 (1997).

Wang, T. D., J. M. Crawford, M. S. Feld, Y. Wang, I. Itzkan, and J. van Dam. In vivo identification of colonic dysplasia using fluorescence endoscopic imaging. *Gastrointest. Endosc.* **49**, 447–455 (1999).

Wang, T. D., J. van Dam, J. M. Crawford, E. A. Preisinger, Y. Wang, and M. S. Feld. Fluorescence endoscopic imaging of human colonic adenomas. *Gastroenterology* **111**, 1182–1191 (1996).

Warren, S., K. Pope, Y. Yazdi, A. J. Welch, S. Thomsen, A. L. Johnston, M. J. Davis, and R. Richards-Kortum. Combined ultrasound and fluorescence spectroscopy for physico-chemical imaging of atherosclerosis. *IEEE Trans. Biomed. Eng.* **42**, 121–132 (1995).

Wennberg, A. M., F. Gudmundson, B. Stenquist, A. Ternesten, L. Mölne, A. Rosén, and O. Larkö. In vivo detection of basal cell carcinoma using imaging spectroscopy. *Acta Dermatol. Venereol.* **79**, 54–61 (1999).

Winkelman, J., and D. S. Rasmussen-Taxdal. Quantitative determination of porphyrin uptake by tumor tissue following parenteral administration. *Bull. Johns Hopkins Hosp.* **107**, 228–233 (1960).

Yang, Y., G. C. Tang, M. Bessler, and R. R. Alfano. Fluorescence spectroscopy as a photonic pathology method for detecting colon cancer. *Lasers Life Sci.* **6**, 259–276 (1995).

Zaak, D., M. Kriegmair, H. Stepp, R. Baumgartner, R. Oberneder, P. Schneede, S. Corvin, D. Frimberger, R. Knuchel, and A. Hofstetter. Endoscopic detection of transitional cell carcinoma with 5-aminolevulinic acid: Results of 1012 fluorescence endoscopies. *Urology* **57**, 690–694 (2001).

Zargi, M., I. Fajdiga, and L. Smid. Autofluorescence imaging in the diagnosis of laryngeal cancer. *Eur. Arch. Otorhinolaryngol.* **257**, 17–23 (2000).

Zargi, M., L. Smid, I. Fajdiga, B. Bubnic, J. Lenarcic, and P. Oblak. Detection and localization of early laryngeal cancer with laser-induced fluorescence: Preliminary report. *Eur. Arch. Otorhinolaryngol.* **254**, S113–S116 (1997a).

Zargi, M., L. Smid, I. Fajdiga, B. Bubnic, J. Lenarcic, and P. Oblak. Laser induced fluorescence in diagnostics of laryngeal cancer. *Acta Otolaryngol.* **527**, 125–127 (1997b).

Zeng, H., C. MacAulay, S. Lam, and B. Palcic. Light induced fluorescence endoscopy (LIFE) imaging system for early cancer detection. *Proc. SPIE* **3863**, 275–282 (1999).

Zonios, G., R. Cothren, J. M. Crawford, M. Fitzmaurice, R. Manoharan, J. van Dam, and M. S. Feld. Spectral pathology. *Ann. N Y Acad. Sci.* **838**, 108–115 (1998).

11

Fluorescence and Spectroscopic Markers of Cervical Neoplasia

INA PAVLOVA, REBEKAH DREZEK, SUNG CHANG,
DIZEM ARIFLER, KONSTANTIN SOKOLOV,
CALUM MACAULAY, MICHELE FOLLEN,
AND REBECCA RICHARDS-KORTUM

CLINICAL PERSPECTIVE ON FLUORESCENCE AND REFLECTANCE SPECTROSCOPY FOR DETECTION OF CERVICAL DYSPLASIA

The effective screening, diagnosis, and treatment of cervical intraepithelial neoplasia in the developed world reduce the mortality resulting from cervical cancer enormously. This is an area of intensive undertaking. Tragically, in the developing world, where these resources are not available, this disease is the leading cause of cancer death among women. Inexpensive, real-time diagnostic tools based on optical spectroscopy have the potential to reduce mortality associated with cervical cancer. Because detection of cancer is provided in real time, optical methods could enable combined detection and treatment in a single visit. The difficulties of follow-up contact in many developing countries makes the implementation of a see-and-treat methodology the only logical choice. As part of the new managed health reality in the developed world, there is an important need for improved screening and detection methods for cervical intraepithelial neoplasia that are both sensitive and cost-effective. Recent developments in optical diagnostic technologies have the potential to address both of these issues. Many groups have demonstrated that techniques based on quantitative optical spectroscopy have the potential to fulfill this need by providing accurate, objective, and instantaneous point-of-care diagnostic and screening tools (Alfano et al., 1987; Hung et al., 1991; Richards-Kortum et al., 1991; Schomacker et al., 1992; Lam et al., 1993a,b; Mahadevan et al., 1993; Ramanujam et al., 1994a,b; Richards-Kortum, 1994; Zuclich et al., 1994; Mourant et al., 1995; Vo-Dinh et al., 1995; Cothren et al., 1996; Ramanujam et al., 1996a,b,c; Zangaro et al., 1997; Lam et al., 1998; Perelman et al., 1998; Pogue et al., 1998; Tumer et al., 1998; Zuluaga et al., 1998; Brookner et al., 1999; Hornung et al., 1999; Utzinger et al., 1999; Harries, 1995; Mitchell, 1998). However, the connections among these optical signatures and the underlying morphology and biology are not well understood.

Recent work has begun to elucidate the biological and morphologic basis for differences in optical spectra of normal and precancerous cervix (Ramanujam et al., 1994b; Hornung et al., 1999; Brookner et al., 2000; Drezek et al., 2001a). These novel clinical studies indicate that spectral changes in dysplasia can be attributed to (1) an increase in epithelial NAD(P)H and a decrease in epithelial FAD fluorescence (Brookner et al., 2000; Drezek et al., 2001a), (2) a decrease in stromal collagen cross-link fluorescence,

(3) an increase in stromal hemoglobin absorption, and (4) an increase in epithelial scattering (Drezek et al., 2001b). Thus, spectroscopic measurements are sensitive to changes in epithelial metabolism, angiogenesis, nuclear morphology, tissue architecture, and epithelial/stromal interactions. Spectra could be analyzed to yield quantitative information about these features, which are intimately related to the dysplasia–carcinoma sequence. In this chapter we review recent progress to identify the spectroscopic markers of cervical neoplasia and to develop systems to assess them noninvasively in vivo.

Overview of Fluorescence and Reflectance Spectroscopy

Although of relatively recent interest in the medical field, optical spectroscopy has long been an important tool in analytical chemistry. Reflectance and fluorescence spectroscopy are routinely used to determine analyte concentrations and to monitor chemical kinetics in complex reactions. No other nondestructive means available at comparable cost can equal optical spectroscopy in sensitivity. Extensive methodologies have been developed to determine the chemical composition of dilute solutions from an analysis of optical spectra. Concentrations of metabolites as low as 10^{-9} M are routinely determined. The emergence of fiber optic technology has enabled development of remote sensing and monitoring systems based on spectroscopy. This, coupled with the large number of biologically important molecules with distinct optical spectra, has led to increasing interest in development of optical sensors to detect and monitor pathology.

When measuring optical spectra of tissue, a fiber optic probe provides illumination light at certain illumination wavelength, λ_x. The light impinges on the tissue where it interacts; light remitted through the tissue front surface after interaction is collected and sent to a sensitive detector (fig. 11.1). In this chapter we describe two primary types of optical spectra: reflectance and fluorescence. In reflectance, the illumination photon energy (wavelength) is not changed during the interaction, and the intensity of light remitted from the tissue surface at this wavelength is measured. A reflectance spectrum is a plot of the intensity of reflected light as a function of wavelength. In fluorescence spectroscopy, the photon energy (wavelength) changes during the interaction, and the intensity of light emitted from the tissue surface is detected at the emission wavelength, λ_m. There are three main types of fluorescence spectra. A fluorescence emission spectrum is a plot of the intensity of fluorescent light as a function of emission wavelength, produced when the tissue is illuminated at a particular excitation wavelength, λ_x. An excitation spectrum is obtained when the emission wavelength is fixed and the excitation wavelength is scanned. The fluorescence intensity and line shape of complicated optical samples such as human tissue are both a function of excitation and emission wavelengths. A complete fluorescence characterization of such

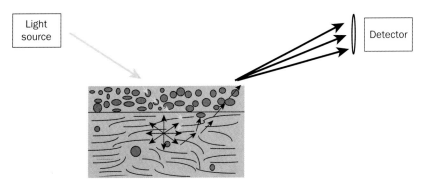

Figure 11.1 Interaction between tissue and light, illustrating reflectance and fluorescence spectroscopy.

samples requires the recording of a fluorescence excitation emission matrix, in which the fluorescence intensity is recorded as a function of both excitation and emission wavelengths.

When analyzing tissue fluorescence and reflectance spectra, it is important to consider in more detail the interactions that can take place between the photons and tissue progressing from illumination to measurement. A spectroscopist views tissue as a collection of chromophores—constituents that interact with light in some manner. In optical spectroscopy, three types of interaction are important—absorption, scattering, and fluorescence—and chromophores are classified as absorbers, scatterers, and/or fluorophores. During a scattering interaction, the direction of light travel is changed as a result of microscopic fluctuations in the tissue index of refraction, but the light intensity remains the same. Absorption interactions reduce the intensity of the light as photon energy is transferred to the absorbing molecule. Fluorescence can occur after absorption as the chromophore releases some of this absorbed energy in the form of fluorescent light at the emission wavelength. Most tissues are highly scattering, with the probability of a scattering interaction exceeding that of absorption by at least an order of magnitude. Most tissues are weakly fluorescent, with a small probability that fluorescence will occur after absorption. Thus, when light of a certain wavelength impinges on tissue, it is typically multiply scattered, after which absorption and perhaps fluorescence can occur. Further scattering and absorption can occur before light exits the tissue surface where it can be detected (fig. 11.1). In reflectance spectroscopy, light has undergone a combination of absorption and scattering events before reaching the detector, and therefore interrogates both tissue scatterers and absorbers. Similarly, the remitted light in fluorescence spectroscopy contains contribution from all three types of chromophores.

CLINICAL USE OF IN VIVO FLUORESCENCE AND REFLECTANCE SPECTROSCOPY

Algorithm Performance Using Fluorescence Spectroscopy

Fluorescence spectroscopy has recently been explored as a diagnostic tool for identifying disease in several human pathological models (Richards-Kortum and Sevick-Muraca, 1996; Wagnieres et al., 1998; Ramanujam, 2000). For the past few years, our group has developed optical techniques based on fluorescence spectroscopy to diagnose cervical precancer. The use of fluorescence spectroscopy is based on the hypothesis that the optical properties of normal and dysplastic tissue differ significantly, because the concentration and distribution of chromophores change with normal to precancerous conversion. For example, precancerous cells are known to have a modified rate of metabolic activity and proliferation in addition to impaired communication with other tissue components. These modifications are likely to affect the optical properties of tissue and to be detected by fluorescence spectroscopy. To record in vivo fluorescence spectra we used a system consisting of three main components: (1) a light source that provides monochromatic excitation light, (2) a fiber optic probe that delivers excitation light to the tissue and collects the remitted fluorescence signal, and (3) a polychromator coupled to a sensitive detector. A statistical algorithm was developed to test the diagnostic content of the in vivo acquired spectra, and its diagnostic capabilities were compared with the standard of care (Ramanujam et al., 1996a,b,c; Tumer et al., 1998; Brookner et al., 1999; Utzinger et al., 1999). Figure 11.2 shows this system in clinical use and illustrates typical fluorescence emission spectra at 337 nm excitation recorded with this device.

Initially we focused on determining which excitation wavelength could yield the greatest differences between normal and precancerous cervical tissue. We measured fluorescence excitation emission matrices of normal and abnormal cervical biopsies in vitro and determined that the optimal excitation wavelengths fordiagnosis are at 337, 380, and 460 nm

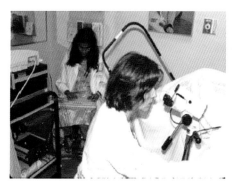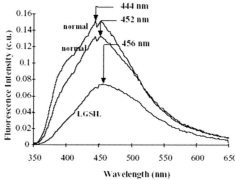

Figure 11.2 Typical spectra (right) measured from a normal and a dysplastic cervical site from the same patient using the system shown at left. LGSIL, low-grade squamous intraepithelial lesion.

(Mahadevan et al., 1993). Figure 11.2 shows emission spectra of normal and precancerous cervical sites obtained at 337 nm excitation from one patient. As the tissue progresses from normal to dysplastic, the fluorescence intensity decreases and the peak emission wavelength shifts to the red. In addition, results indicate that the fluorescence intensity and line shape vary substantially from patient to patient. Despite this variability, if spectra are obtained at multiple excitation wavelengths, good discrimination between normal cervix and pre-cancercan be achieved. We have developed an algorithm based on statistical methods to extract the spectral information that is most diagnostically relevant (Ramanujam et al., 1996a,b,c). Preprocessing methods, such as normalization, mean scaling, and a combination of mean scaling and normalization, corrected for interpatient and intrapatient variations in the spectra. Principal component analysis was used to reduce dimensionally each type of preprocessed spectral data with minimal information loss. Last, a probability-based classi-fication algorithm was developed using logistic discrimination (Ramanujam et al., 1996c). We measured spectra from 500 cervical sites in 95 patients and these data were divided randomly into a training set used to develop a diagnostic algorithm, and into a valida-tion set used to test the algorithm. The performance of the fluorescence-based algorithm (sensitivity, 86%; specificity, 74%) exceeds that of the Papanicolaou smear(sensitivity, 60%; specificity, 60%) at many clinical sites and is comparable with that of colposcopy in expert hands (sensitivity, 96%; specificity, 48%) (Tumer et al., 1998; Brookner et al., 1999; Utzingeret al., 1999).

In addition to the improved specificity, fluorescence spectroscopy allows measure-ments to be made in near real time, and results are available immediately. Because fluorescence measurement is not painful and does not require tissue removal, the entire epithelial volume can be interrogated, potentially reducing sampling error. Analysis of flu-orescence data is accomplished using algorithms implemented in software, so the need for clinical expertise is reduced. Thus, because optical spectra can be recorded remotely in nearr eal time without the need fortissue removal, and data analysis can be automated, we believe that optical diagnosis affords many important advantages over traditional tech-niques, including the potential to reduce the need for clinical expertise, reduce the number of unnecessary biopsies, enable combined diagnosis and therapy in those patients who might most benefit, and the potential to reduce health care costs (Cantor et al., 1998).

Algorithm Performance Using Reflectance Spectroscopy

Other studies have introduced the use of reflectance spectroscopy as a noninvasive diag-nostic tool for early cancer detection (Mourant et al., 1995; Perelman et al., 1998;

Hornung et al., 1999). In diffuse reflectance spectroscopy, light remitted from tissue has undergone a combination of scattering and absorption events before reaching the detector, providing information about absorbers and scatterers within the tissue. The use of this technique is based on the preposition that many significant tissue pathologies are accompanied by local architectural changes at the cellular and subcellular levels. These changes are hypothesized to affect the absorption and scattering properties of the overall tissue. A small percentage of the reflected light that reaches the detector comes from singly backscattered photons that originate from cell nuclei in the epithelial layers of tissue. Light-scattering spectroscopy uses this singly scattered light to provide information about the size and numberdistr ibution of cell nuclei. The size of epithelial cell nuclei is a major indicator for precancerous transformation of tissue, with smaller, denser nuclei seen in dysplastic cells.

Early investigations reported initial success aiding diagnosis of neoplasia in the bladder, the gastrointestinal tract, and the cervix with reflectance spectroscopy, and analyzed reflectance spectra empirically to yield a tissue diagnosis (Copperson et al., 1994; Mourant et al., 1995; Mourant et al., 1996). Copperson et al. (1994) developed a fiber optic probe (which was known as the Polarprobe) that measures tissue reflectance at four wavelengths in the visible and near infrared. In tests of 77 volunteers, the Polarprobe diagnosis agreed with colposcopy and histology in 85% to 99% of measurements, depending on tissue type. In addition, our group has explored the capability of spatially resolved in vivo reflectance spectroscopy to detect cervical precancer in a large diagnostic trial of 161 patients (328 sites) (Mirabal et al., 2002). Reflectance spectra were obtained with a spectroscopic system utilizing a probe with nine optical fibers placed in direct contact with the cervix. One of the fibers provided broadband illumination, and eight collection fibers at four different source–detector separations collected diffusely reflected light. As the separation between source and detectors increased, greater tissue depth was probed. Average reflectance spectra for each normal and precancerous tissue category at four different source–detector separations were calculated and are shown in figure 11.3. Spectra have valleys resulting from hemoglobin absorption at 420, 542, and 577 nm. Results indicate that the average spectra for each diagnostic class differ at all source–detector separations, with the greatest differences occurring at the closest distance between source and detector. The reflectance data were also used to develop and evaluate an algorithm based on Principal Component Analysis (PCA) and Mahalanobis distance forall combinations of source–detectorsepar ations. The diagnostic performance was high when only a single separation was used and did not increase significantly when data from additional separations was included. The algorithm distinguished normal squamous tissue from high-grade dysplastic tissue with a sensitivity of 81% and a specificity of 73%. Discrimination of normal columnar tissue and high-grade dysplastic tissue was also very good, with a sensitivity of 75% and specificity of 89%.

Recent work using elastic light scattering spectroscopy (LSS) has utilized models to separate signals resulting from absorption and scattering (Perelman et al., 1998; Tromberg, 1998). Tromberg (1998) presented a pilot study of 10 patients that indicated that differences in the tissue absorption and scattering coefficients in the near-infrared spectral regions can be used to discriminate normal and dysplastic cervical epithelium (Hornung et al., 1999). They calculated tissue absorption and scattering coefficients from frequency domain measurements of diffuse reflectance at multiple source–detector separations at 674, 811, 849, and 956 nm. Using these data they estimated tissue blood volume, watercontent, and oxygen saturation. Precancers showed up to a 15% decrease in absorption and scattering relative to normal tissue, reflecting an increase in epithelial thickness and water content. Because of the large source–detector separations used in these experiments, changes in scattering likely represent changes in stromal optical properties.

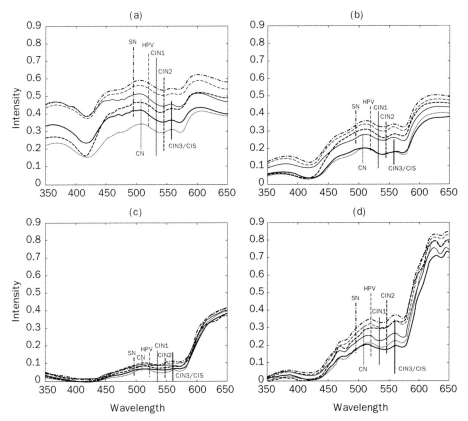

Figure 11.3 Average reflectance spectra by tissue diagnostic classification for different source–detector separations. (A) Position 0: 200 μm separation. (B) Position 1: 1 mm separation. (C) Position 2: 2 mm separation. (D) Position 3: 3 mm separation. Tissue spectra have been normalized by spectra from a standard solution of 1.05-μm-diameter polystyrene microspheres. Diagnostic classifications: squamous normal (SN), columnar normal (CN), low-grade dysplasia (human papillomavirus [HPV], cervical intraepithelial neoplasia [CIN I]), and high-grade dysplasia (CIN 2, CIN 3, carcinoma in situ [CIS]).

Perelman et al. (1998) presented a model to separate the contributions of single and multiply scattered light. Singly scattered light originates predominantly from the epithelium, whereas multiply scattered light is produced primarily in the stroma. A fine structure component in the singly backscattered light was attributed to Mie scattering from epithelial cell nuclei. Assuming that nuclei could be represented as homogeneous spheres, analysis of this fine structure yielded the nuclear diameter of epithelial cells. Tests of this technique have shown promise for the detection of dysplasia in the colon, esophagus, bladder, and oral cavity (Backman et al., 2000). Later work showed that polarization of the incident light could enable direct detection of the singly backscattered component (Sokolov et al., 1999; Gurjar et al., 2001) and that imaging approaches could extend this to large regions of surface epithelium (Gurjar et al., 2001).

Algorithm Performance Using Both Fluorescence and Reflectance Spectroscopy

Several groups have developed systems capable of measuring both fluorescence and reflectance spectra from the same tissue site. Because reflectance and fluorescence spectroscopy provide complementary information about biochemical and morphological properties of tissue, a combination of these techniques can lead to an improvement in diagnostic ability. The company, MediSpectra has carried out a clinical trial examining the combination of diffuse reflectance spectroscopy with fluorescence emission spectroscopy at 355 nm excitation (Nordstrom et al., 2001). Using a device that can measure fluorescence and diffuse reflectance spectra from 120 locations uniformly distributed over the cervix, they measured spectra from 41 women undergoing colposcopy with biopsy of suspicious areas. A multivariate algorithm based on the Mahalanobis distance was trained using half the data set. Fluorescence at this excitation wavelength could discriminate normal squamous tissue from high-grade dysplasia (high-grade squamous intraepithelial lesion [HGSIL]) with a sensitivity of 91% and a specificity of 93%; however, poor results were found for discriminating between metaplastic tissue and HGSIL. In contrast, reflectance spectra could be used to discriminate metaplasia and HGSIL with a sensitivity of 77% and a specificity of 76%. They suggest that future trials should investigate the combination of fluorescence and reflectance spectra.

A newly developed technique, called *trimodal spectroscopy* (orTMS), uses the combined effect of three spectroscopic techniques to detect and diagnose cervical precancer (Georgakoudi et al., 2002). Fluorescence and reflectance spectra from 44 patients were obtained simultaneously and analyzed to provide information about intrinsic (undistorted by absorption and reflectance events) fluorescence, diffuse reflectance, and light scattering from all measured sites. Intrinsic fluorescence provides qualitative information about the biochemical composition of tissue, whereas diffuse reflectance interrogates the absorbers and scatterers in the deeper layers of cervical tissue. In addition, light scattering was used to determine nuclear size of epithelial cells. An algorithm based on logistic regression and cross-validation was developed and validated using data from all three techniques. Results indicate that TMS can distinguish from normal and precancerous cervical tissue with a sensitivity of 92% and a specificity of 71%.

Effects of Biographical and Patient-to-Patient Covariates

Thus, diagnostic algorithms based on optical spectroscopy can classify tissue samples as diseased or nondiseased. To improve the performance of these diagnostic algorithms, the effects of patient-to-patient variations and biographical variables on tissue fluorescence spectra must be examined. During our early clinical studies, it was noted that there is great variability in the fluorescence spectra collected from different patients, even within a single histopathological category. For example, peak fluorescence intensities of normal tissues can vary by more than a factor of five from patient to patient, but within a single patient the standard deviation is usually less than 25% of the average value. Because of this large variation between patients, early data analysis was performed in a paired manner, which required that a normal and abnormal site be measured for each patient. Subsequent data analysis methods removed the need for paired data, but current diagnostic algorithms still require normalization and mean scaling of the data as part of the preprocessing to reduce the effects of interpatient variations.

To examine the effects of biographical variables on tissue fluorescence spectra, an analysis using data collected from two previously published clinical trials was performed; one study measured spectra from 395 sites in 95 patients referred to a colposcopy clinic with abnormal Papanicolaou smears, and the second study measured spectra from 204 sites in 54 patients self-referred for screening and expected to have a normal Papanicolaou smear. A diagnostic algorithm was developed and has been described in detail (Ramanujam et al., 1996a). For this analysis, data about age, race, menstrual cycle, menstrual status (pre-, peri-, post-), and smoking were collected. Principal component analysis on normalized and nonnormalized data was compared. An analysis of variance (ANOVA) was performed based on age, menopausal status, race, and smoking. There are clear intensity differences observed with menopausal status; postmenopausal patients exhibit higher emission intensities (fig. 11.4). Age appears to also influence intensity, with higher emission intensities seen with greater age. Caucasian women had slightly increased emission intensities compared with African American women, and smokers had slightly increased emission intensities compared with nonsmokers. Significant differences were observed in the principal components, which describe spectral data such as the patient's age or menopausal status, even when the pathological diagnosis of the measured tissue is the same. Current preprocessing techniques—normalization and mean scaling—are limiting because normalization ignores intensity differences between spectra and mean scaling requires an equal number of normal and abnormal sites per patient. Alternative preprocessing methods, which could account for the differences resulting from patient age or menopausal status without normalization, for example, could offer improvements in algorithm performance and applicability.

Effects of Menstrual Cycle on Fluorescence Spectra

Several findings (Gorodeski et al., 1989; Ferenczy and Wright, 1994) have shown that hormones affect the metabolism of epithelial cells in the cervix. Because these hormones are regulated throughout the menstrual cycle, we examined the effect of the menstrual cycle on in vivo fluorescence of the cervix (Change et al., 2001). Fluorescence spectra from 10 patients were measured for 30 consecutive days. For each patient, measurements were recorded from three sites, with positions that were fixed throughout the study. We applied PCA to the 900 sets of fluorescence spectra to investigate any correlation between the daily fluorescence signal and the menstrual cycle. Of the first five principle component scores from the emission spectra at 340 to 380 nm excitation, one was strongly correlated with the menstrual cycle in most of the patients, which accounted for less than 15% of the variance in the data. The corresponding principal component revealed two peaks corresponding to NADH and FAD emission maxima. The shape and score of the principal component show that NADH fluorescence increases during the proliferating phase of the cycle and decreases during the secretory phase. The opposite results were obtained for FAD fluorescence. To investigate whether the variations observed throughout the menstrual cycle have an effect on the diagnostic algorithm performance, we applied the data set from this study to an algorithm developed in a previous study (Ramanujam, 1996a). Specificity was 94.3%, which is higher than that reported previously (Tumer et al., 1998; Brookner et al., 1999; Utzinger et al., 1999). Thus, although the menstrual cycle does have an effect on cervical fluorescence, it seems that the day of the cycle is not a critical factor in developing a diagnostic algorithm.

Extending to Multispectral Optical Imaging

The majority of fluorescence and reflectance spectroscopy measurements described earlier were made through fiber optic probes that interrogate a small portion of the cervical epithelium. Multiple sites can be interrogated by manually scanning the probe across the tissue surface. Many groups are developing imaging systems to take advantage of the ability of optical systems to interrogate the entire cervical epithelium.

Based on a fluorescence spectroscopy algorithm (Ramanujam et al., 1996a,b), LifeSpex, a medical startup, developed a system to measure fluorescence images from the entire cervical epithelium at multiple excitation emission wavelength pairs. LifeSpex has successfully completed clinical studies with its cervical cancer detection device, Cerviscan. Sixty-seven patients were evaluated with this device, the liquid-based Papanicolaou smear, and colposcopy with biopsy. Results from all three tests were available from 52 women;

(a)

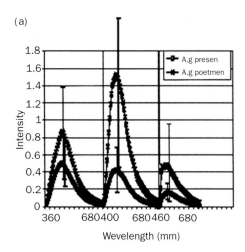

(b)

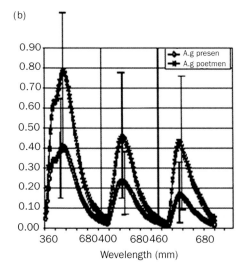

(c)

Figure 11.4 Three emission average spectra of pre- and postmenopausal women. (A) Normal sites from the screening setting. (B) Normal sites from the referral setting. (C) Dysplastic sites from the referral setting for (left to right) 337 nm, 380 nm, and 460 nm excitation. Error bars represent one standard deviation. The same set of arbitrary units is used in each graph so that intensity information can be compared.

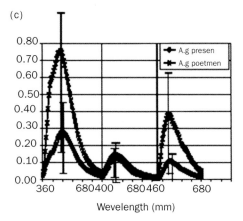

data from the first 228 sites in 42 patients were used to train the algorithm and data from 70 sites in 10 women were used as an independent validation set. These initial data show that the device discriminates precancerous cervical cancer lesions from normal tissue with a sensitivity and specificity of 84% and 93%, respectively, compared with colposcopically directed biopsy (Dattamajumdar et al., 2001a,b). The Cerviscan image correctly resolved five of seven atypical cells of undetermined significance (ASCUS) + low-grade squamous intraepithelial lesion calls made by liquid-based cytology and detected two patients with precancerous lesions that had been missed by liquid-based cytology.

A similar device was developed by a company SpectRx, which incorporates the ability to measure both reflectance and fluorescence. This device was used to measure the colposcopically visible cervical epithelium in 136 patients in the colposcopy setting (Ferris et al., 2001). An algorithm was derived to recognize cervices with cervical intraepithelial neoplasia 2 or greater. Encouraging sensitivities and specificities were reported (97% and 70%, respectively). In this study, algorithm results are reported from the same data set used to derive the algorithm; thus, estimates of sensitivity and specificity may be overly optimistic as a result of an overtraining bias.

Pogue et al. (1998, 2000) have developed a wavelength tunable digital imaging colposcope, capable of measuring reflectance images of the cervical epithelium. This device can retrieve spectral information from the acquired images. The reflectance images can be digitally stored and postprocessed. Preliminary images have been obtained at wavelengths of 400, 500, 515, 560, 577, 600, 700, 760, and 800 nm (Pogue et al., 1998). This group is beginning pilot studies to determine whether additional spectroscopic information and digital image processing can enhance colposcopic detection of cervical intraepithelial neoplasia (Pogue et al., 2000). They found that when images are separated into their RGB channels, the Eulernumbercan improve separation between metaplastic tissue and HGSIL, and the blue channel is most correlated with epithelial cell changes. An interesting modification of this concept was developed, in which polarized light is used for illumination and two polarized reflected light images are captured: one parallel and one perpendicular to the polarization of the illumination light (Pogue et al., 1998). The singly scattered light, generated mainly in the epithelium, maintains the illumination polarization and is the major contribution to the parallel image. Light in the perpendicular image is primarily light that has undergone multiple scattering events. Analysis of these two images can separate effects of scattering and absorption from superficial and deep tissues. Preliminary results appear promising for detection of basal cell carcinoma, even in highly pigmented tissues. Balas (2001; Balas et al., 1999) has developed a multispectral imaging system that has been used to obtain reflectance images of cervical tissue at serial time points after application of acetic acid. Linearly polarized white light illuminates the cervix, and reflected light is viewed by a CCD camera through a liquid crystal tunable filter (LCTF) to provide detection at 10 different wavelength ranges. Images were obtained before and at 10-s intervals for 10 minutes after application of acetic acid in a group of 16 women. Maximum contrast was reported at 525 nm (Balas, 2001) and it was shown that different lesion grades have different time courses of aceto-whitening, with higher grade lesions showing more persistent aceto-whitening (Balas et al., 1999).

CHROMOPHORES PRESENT IN EPITHELIAL TISSUE

To obtain an optical signal from tissue, the presence of an endogenous or exogenous chromophore is required. We have already mentioned that chromophores can be divided into scatterers, fluorophores, and absorbers. Furthermore, because we are investigating the native optical properties for tissue, endogenous chromophores (or naturally occurring fluorophores) are of special interest. There are several main biomolecules that are known

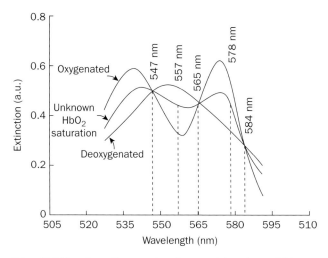

Figure 11.5 The absorption spectra of oxy- and deoxyhemoglobin.

to cause the majority of signals in fluorescence and reflectance spectroscopy, and their properties are summarized in this section.

Absorbers

Endogenous chromophores, such as oxy- and deoxyhemoglobin, melanin, myoglobin, and water, are primarily responsible for absorption of light in tissue. Hemoglobin is of special importance because it is the main absorber in the visible range of the spectrum and is present in vascularized tissues. Hemoglobin has a strong Soret band absorption near 420 nm; two smaller absorption bands are present in oxyhemoglobin near 540 and 580 nm. A single band is found in deoxyhemoglobin, near 560 nm. Reabsorption of fluorescence by these chromophores produces valleys in the emitted fluorescence spectra (fig. 11.5).

Fluorophores

Endogenous fluorescence provides an important tool in optical assessment of tissue metabolic status. The pyridine nucleotides and the flavins play an important role in cellular energy metabolism (Stryer, 1988). Figure 11.6 shows the fluorescence excitation and emission spectra of oxidized and pyridine nucleotide and flavoprotein. Nicotinamide adenine dinucleotide is the major electron acceptor; its reduced form is NADH, and the reduced nicotinamide ring is fluorescent with an excitation maximum near 360 nm and an emission maximum near 450 nm, whereas the oxidized form is not fluorescent (Lakowicz, 1985). Flavin adenine dinucleotide is the other major electron acceptor; the oxidized form, FAD, is fluorescent whereas the reduced form, $FADH_2$, is not (Masters and Chance, 1993). Thus, fluorescence spectroscopy can be used to probe changes related to oxidative respiration by monitoring the fluorescence of these cofactors.

   Another potential source of autofluorescence in tissue is protein fluorescence. The aromatic amino acids tryptophan, tyrosine, and phenylalanine contribute to protein fluorescence in the ultraviolet region of the spectrum (Lakowicz, 1985). Significant autofluorescence has been noted in the structural proteins collagen (Eyre and Paz, 1984) and elastin (Blomfield and Farrar, 1969). Collagen autofluorescence is associated with cross-links (Fujimoto, 1977), and elastin autofluorescence is also suspected to be associated with cross-links (Thornhill, 1975). In elastin, two major types of cross-links are the unique amino acids desmosine and isodesmosine (Eyre and Paz, 1984). Although it was initially believed

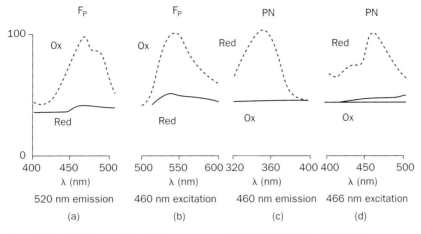

Figure 11.6 (A–D) Flavoprotein (Fp) fluorescence excitation and emission spectra, respectively (A, B). Oxidized and reduced pyridine nucleotide (PN) fluorescence excitation and emission spectra, respectively (C, D).

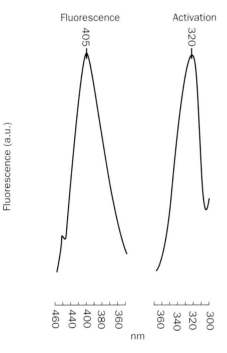

Figure 11.7 The excitation and emission spectra of the elastin cross-link. The emission (fluorescence) spectrum was measured at 320 nm excitation; the excitation (activation) spectrum was measured at 405 nm emission. (Adapted from Deyl et al. [1980].)

that desmosine is responsible for elastin autofluorescence, Thornhill (1975) showed that desmosine could be separated from the fluorescent material in elastin. Further work has indicated that the fluorescent material in elastin is the result of a tricarboxylic triamino pyridinium derivative, which is very similar to the fluorophore in collagen (Deyl et al., 1980). The excitation and emission maximum of this elastin cross-link is shown in figure 11.7, with an excitation maximum at 325 nm and an emission maximum at 400 nm.

There are two major mechanisms of collagen cross-links formation. Enzymatically formed cross-links (fig. 11.8) were believed to be the main fluorophores in collagen at 325 nm excitation (Fujimoto, 1977). Recently, a second mechanism of intermolecular cross-linking of collagen has been reported, which is age related and occurs via glycation

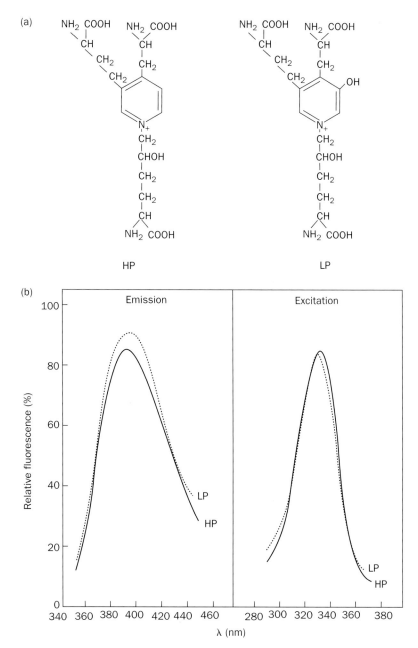

Figure 11.8 (A) Enzymatically formed cross-links: hydroxylysyl pyridinoline (HP), lysyl pyridinoline (LP). (B) Both HP and LP are autofluorescent with an excitation maximum at 325 nm and an emission maximum at 400 nm. (Adapted from Eyre and Paz [1984].)

(Bailey et al., 1998). Several types of fluorescent cross-links can result, including pentosidine (excitation maximum, 335 nm; emission maximum, 385 nm; accounts for25% to 40% of total collagen fluorescence), vesperlysine (370 nm, 440 nm, 5% of total fluorescence), crossline (380 nm, 460 nm), and argpyrimidine (320 nm, 380 nm) (Baynes, 1991; Odetti et al., 1994; Bailey et al., 1998). Collagen is the main structural protein (Kucharz, 1992) of cervical stroma and changes in the fluorescent properties of this fluorophore and specific cross-links can be indicators of normal to dysplastic conversion. During precancerous transformations, the communication between epithelial cells and underlying stroma changes, which could lead to oversecretion of matrix metalloproteinases. These enzymes are known to participate in the degradation of the stromal matrix (Parks and Mecham, 1998), thus indirectly affecting collagen fluorescence.

Scatterers

Because of the intense scattering of tissues, visible light can propagate from 0.5 to 1.0 mm into tissue, enabling one to extract information noninvasively throughout the entire epithelial thickness. The elastic tissue scattering arises as a result of the microscopic heterogeneities of refractive indices between extracellular, cellular, and subcellular components. Light scattering from individual cells is difficult to model because of the complexity of the cell structure and changes in the index of refraction. Because of this complexity, light propagation in tissue has traditionally been modeled using the assumption that cells can be described as perfectly homogeneous spheres. This is an unrealistic assumption; recent electromagnetic models have overcome this limitation by calculating the angular distribution of scattered light (phase function) from individual cells of arbitrary shape and dielectric structure. These models are based on a numerical solution of Maxwell's equations using a finite-difference time domain technique (Dunn and Richards-Kortum, 1996). With this method, the cell is "discretized" into a grid on which the permissiveness and conductivity of the different regions of the cell are specified, and the electrical and magnetic fields are successively updated at each grid point. This tool has been used to calculate the angular dependence of the scattered electromagnetic fields and intensities. Results show that, for amelanotic epithelial cells, which have a relatively low volume fraction of mitochondria, fluctuations in the nuclear index of refraction play an important role in determining the high-angle (backward) scattering, whereas the nuclear diameter plays an important role in determining forward scattering (Dunn et al., 1997). As the grade of dysplasia increases, our simulations predict that the scattering cross-section significantly increases, resulting from increases in nuclear size, optical density, and texture (Drezek et al., 2001b). Furthermore, the wavelength dependence of this scattering can be calculated using finite-difference time domain simulations (fig. 11.9) (Drezek et al., 2000). Other studies confirm increased scattering from tumorigenic cells from 500 to 790 nm (Mourant et al., 1998). Analysis of the wavelength dependence of this scattering indicates that the average size of the scatterers in these cells is on the of order 0.3 μm (much smallerthan the average nuclear diameter), and is larger in tumorigenic cells than in nontumorigenic cells (Mourant et al., 1998).

Thus, tissues contain a number of chromophores that can be interrogated using optical methods to provide information about the biochemical and structural composition. Because of the complex interactions among absorption, scattering, and fluorescence, and the heterogeneous structure of human tissue, our understanding of the relationship among measured tissue spectra and the tissue biochemistry, morphology, and architecture is rudimentary at best, despite a numberof pilot clinical studies that show diagnostic promise. One challenge of biomedical optics is to analyze measured tissue spectra to yield the contributions from various chromophores that are influenced by the disease process. This analytical approach requires advances in two key areas: (1) physical models relating measured spectra to changes in the physical structure of tissue and the optical properties of tissue, and (2) knowledge of the spatial variations of tissue's optical components and how these are modified by normal physiological processes such as aging, or by disease processes such as neoplasia. In the next section we review work designed to assess the spectral variations in living normal and precancerous tissue.

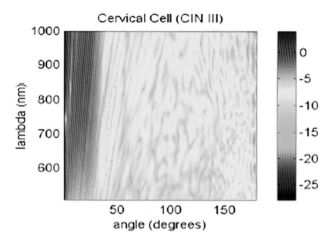

Figure 11.9 Wavelength and angular dependence of the log of the intensity of light scattered from a single cell.

MODULATION OF CHROMOPHORE CONCENTRATIONS AND OPTICAL PROPERTIES WITH DYSPLASIA

Models Based on Fresh Cervical Tissue Slices

We developed a model system to directly visualize and interpret living cervical tissue fluorescence. In particular, we used this model system to explore the origins of normal cervical tissue fluorescence and the biological basis for differences in the fluorescence of normal and precancerous cervix (Brookner et al., 2000; Drezek et al., 2001a). This goal was achieved through a multistep process involving the following: (1) cervical biopsy collection from volunteers, (2) fresh cervical tissue section preparations from biopsies, (3) imaging with fluorescence microscopy, (4) image analysis, and (5) histologic staining and diagnosis of the cervical biopsies. A brief description of each step can be found in the work of Brookner et al. (2000). Two major studies were performed, one involving only samples from normal biopsies and another in which paired normal and precancerous biopsies from the same patient were examined.

Based on the fluorescence pattern, images from normal cervical biopsies (Brookner et al., 2000) were divided into three categories: group 1, slices with bright epithelial fluorescence that was generally brighter than the stromal fluorescence; group 2, slices with similar fluorescence intensities in the epithelium and stroma; and group 3, slices with weak epithelial fluorescence and strong stromal fluorescence. A strong correlation was observed between age and the fluorescence pattern of the normal biopsies. The average age of group 1 patients is 30.9 years, and only 1 of the 12 patients is postmenopausal. The average age of the women in group 2 is 38.0 years, and three of the seven patients are postmenopausal. In group 3, the average age is 49.2 years and three of the four women are postmenopausal. Differences in the average age of patients in each of the groups are statistically significant at orbelow $P = .05$.

When comparing the fluorescence patterns of paired normal and precancerous cervical slices at 380 nm excitation (Drezek et al., 2001a), it is evident that fluorescence intensity increases in the epithelium of the dysplastic tissue relative to the normal tissue, whereas the fluorescence in the stroma significantly decreases (fig. 11.10). In particular, it was found that the average epithelial fluorescence increased in the dysplastic samples (109.3 ± 41.4) relative to the normal samples (85.8 ± 32.4). The means were found to be statistically different using a paired two-tailed Student's t test ($P = .036$). In contrast, the average stromal fluorescence

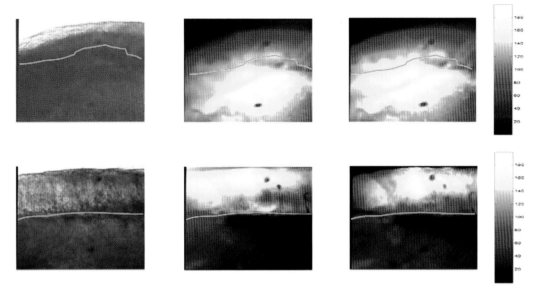

Figure 11.10 Bright-field images (left) and fluorescence micrographs at 380 nm excitation (middle) and 460 nm excitation (right) of tissue slices of normal (top row) and dysplastic (bottom row) cervical biopsies. As dysplasia develops, the NADH fluorescence of the epithelium increases and the collagen fluorescence of the stroma decrease. The green line indicates the basement membrane, and divides epithelium from stroma.

Figure 11.11 (Left) Effects of increasing age on tissue fluorescence. (Right) Effects of dysplasia on tissue fluorescence.

was markedly reduced in the dysplastic samples (104.2 ± 37.2) relative to the normal samples (160.7 ± 42.6). Between the normal and abnormal samples from the same patients, the intensities of fluorescence in the stroma were statistically different ($P = .001$ using a paired two-tailed Student's t test). In addition, the ratio of mean epithelial fluorescence to mean stromal fluorescence was computed for all paired samples at 380 nm excitation. The average ratios (± standard deviation) were found to be 0.55 ± 0.21 and 1.06 ± 0.23 for the normal and abnormal samples, respectively. Using a paired Student's t test, these ratios were found to be statistically different ($P = .0002$ using a two-tailed paired Student's t test). A summary of the results from the two short-term tissue culture studies is shown in figure 11.11, which illustrates how fluorescence changes with age and dysplasia at 380 nm excitation.

The significant epithelial fluorescence at 380 nm excitation observed in most of the tissue slices is most likely the result of NAD(P)H. Supporting this hypothesis, fluorescence measurements of ectocervical cells from primary cultures and two cervical cancer cell lines showed fluorescence consistent with NAD(P)H (Brookner et al., 2000). Presuming that NAD(P)H is the dominant epithelial fluorophore at 380 nm excitation, the increased

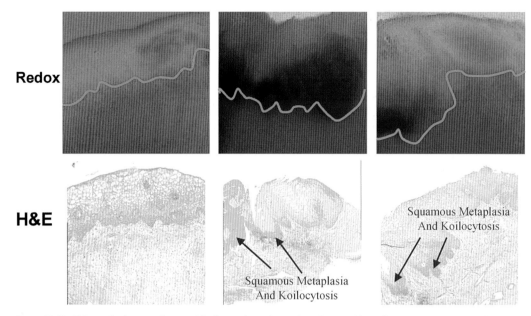

Redox

H&E

Squamous Metaplasia
And Koilocytosis

Squamous Metaplasia
And Koilocytosis

Figure 11.12 False color images (top row) indicate the redox ratio calculated from fluorescence images of cervical tissue slices of normal (left) and dysplastic (middle and right) cervix. High redox ratios are indicated in orange; low values in black. The green line indicates the basement membrane. The bottom row shows H&E-stained sections of the same samples. Note that areas of reduced redox ratio correspond directly with areas of dysplasia in the H&E sections.

fluorescence found in dysplastic samples relative to normal samples from the same patient suggests a higher NAD(P)H concentration in the areas of dysplasia. Moreover, an increased concentration in the reduced electron carrier NAD(P)H and a decreased concentration in the oxidized electron carrier FAD (Davis and Canessa-Fischer, 1965) indicate a higher metabolic rate and it is characteristic of dysplastic tissue. To determine whether the fluorescence images might contain evidence of metabolic changes between normal and precancerous tissue, we calculated redox images by dividing the 460-nm excitation image by the sum of the 380-nm and 460-nm excitation images (fig. 11.12). Marked changes in redox in the precancerous tissue were observed in one third of the samples. In these cases, redox ratios in the regions of dysplasia (0.1–0.2) were 17% to 40% of the average epithelial redox ratio of the normal tissue (0.5–0.6), suggesting an increased metabolic rate.

The main source of stromal fluorescence is believed to be the result of collagen cross-links. Because the density of these cross-links in collagen increases with age, an older person is expected to have a higher stromal fluorescence than a younger one. Results from the normal tissue slices study confirm this hypothesis. The weaker stromal fluorescence seen in abnormal samples is attributed to a decrease in the density of the stromal matrix. In particular, it is believed that dysplastic epithelial cells may interact with stromal collagen, destroying collagen cross-links and reducing stromal fluorescence (Hong and Sporn, 1997).

We also used a model of freeze-trapped biopsies to characterize the fluorescence intensity distribution within the epithelium and stroma of frozen human cervical tissues at the following excitation–emission wavelength pairs: 440, 525 nm (FAD) and 365, 460 nm (NAD[P]H) (Ramanujam et al., 2001). This model is different from the tissue culture model in that it preserves the vasculature and therefore the oxygen concentration of the tissue sample. A comparison of the redox ratios between the epithelia of dysplastic and nondysplastic tissues indicates that this ratio is lowest in severely dysplastic tissues, and is consistent with observations in the short-term tissue slices model described earlier.

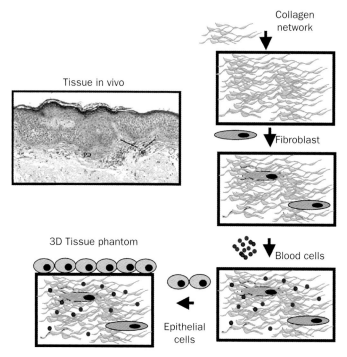

Figure 11.13 A schematic illustration of step-by-step multilayer reconstruction of an epithelial tissue using components characteristic of the human epithelium.

Results from our models are in contrast to data collected in previous studies performed with frozen-thawed tissue. In a study of frozen-thawed sections of cervical tissue at 380 nm excitation, Mahadevan (1998) found that the epithelium exhibited little autofluorescence. Lohmann et al. (1989) also imaged transverse cryosections of cervical tissue using 340 to 380 nm excitation and reported that neither normal nor dysplastic tissue exhibited significant fluorescence from the epithelial layer. It is likely that frozen-thawed tissue does not provide a realistic model of in vivo fluorescence. The redox state may be altered by the oxidation that occurs during cryosectioning or microscopic examination, causing NAD(P)H to be oxidized to NAD+, which is nonfluorescent.

The results from the short-term tissue culture and the freeze-trapped tissue biopsy studies suggest that these models can answer questions about the nature and distribution of major optical constituents in cervical tissue. Most important, these models indicate that there is a biological basis for the differences in the fluorescence of normal and abnormal cervical tissue.

Models Based on Three-Dimensional Tissue Phantoms

In addition to the fresh cervical tissue models, we also developed three-dimensional tissue phantoms that can help to understand the optical properties of the individual components of normal and neoplastic epithelial tissue, how they combine to produce the aggregate tissue spectrum, and the interactions between the epithelium and the stroma throughout the dysplasia-to-carcinoma sequence (Sokolov et al., 2002). Our approach, based on techniques developed in the field of tissue engineering (Parkhurst and Saltzman, 1992; Langer and Vacanti, 1993; Kuntz and Saltzman, 1997; Riesle et al., 1998), is to create phantoms based on a step-by-step reconstruction of epithelial tissue (fig. 11.13).

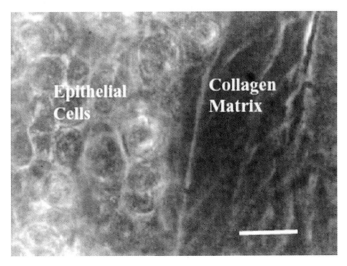

Figure 11.14 Phase contrast photograph of multiple layers of epithelial cells on top of collagen matrix. Scale bar, ≈20 μm.

We began with the stromal layer, which predominantly consists of a network of collagen bundles, and then progressively increased the model complexity by successively adding epithelial cells on top of the collagen matrix. In the framework of this model, different stages of epithelial cancer can be simulated using cell lines characteristic of normal, precancerous, and cancerous epithelium.

Phantoms consisting of a collagen matrix alone were created by using type I collagen from rat tail tendon (Sokolov et al., 2002). Comparison of bright-field and fluorescence images of the collagen gel and a stromal layer of a normal cervical biopsy revealed that the main morphological features in images of both specimens are extended linear fibers with bright autofluorescence at the 380-nm excitation wavelength. These fibers consist of many self-assembled collagen molecules with intra- and intermolecular cross-links stabilizing the structure.

Building on the results obtained with collagen gels, we developed procedures for preparation of multilayer tissue phantoms. Figure 11.14 shows an example of a two-layer tissue phantom with multiple layers of epithelial cells on top of a collagen matrix. This model system provides an ideal testbed for the forward and inverse models proposed here. To our knowledge, it is the only well-controlled model system that accurately reflects the optical properties of the epithelium and stroma, and their interactions throughout the development of dysplasia.

PUTTING IT TOGETHER: UNDERSTANDING IN VIVO SPECTRAL CHANGES IN TERMS OF BIOLOGICAL CHANGES

In the previous section, methods for visualizing and understanding cervical tissue fluorescence were discussed. The ultimate goal of these studies is to detect which endogenous fluorophores contribute to in vivo fluorescence spectra and to elucidate the biochemical basis for the observed differences in spectra of normal and abnormal cervix. To understand in vivo fluorescence quantitatively, mathematical models should be developed that relate the information from our tissue models to properties of clinically measured spectra. We used a method based on Monte Carlo modeling to calculate tissue fluorescence spectra of various tissue types (Drezek et al., 2001c).

Monte Carlo methods can be used to simulate the random walk of photons in a turbid medium and are readily adapted to predict remitted fluorescence, as described by Welch et al. (1997). It is necessary to know the absorption and scattering coefficients, the location and relative density of fluorophores within the tissue, as well as the intrinsic line shape of each fluorophore as input to the model. As an initial step, we modeled cervical tissue fluorescence spectra at 380 nm excitation. At this wavelength, our experiments with the fresh tissue slices indicate that fluorescence in the epithelium is dominated by NAD(P)H, and fluorescence in the stroma is dominated by collagen; thus, the cervix was modeled as two infinitely wide layers: the first representing the cervical epithelium and containing NADH fluorescence, and the second representing the stroma and containing collagen fluorescence. Each layer is assumed to have a homogeneous distribution of fluorophores, absorbers, and scatterers. Optical properties (scattering coefficient, absorption coefficient, anisotropy, and refractive index) for each layer are based on a combination of references in the literature (Zijlstra et al., 1991; Qu et al., 1994; Cheng, 1995; Hornung et al., 1999).

Predicted and measured fluorescence spectra from normal and dysplastic tissue are compared in figure 11.15A. It should be noted that the modeled normal data are scaled to the value of the measured normal data at 475 nm emission. This same scaling factor was then applied to the precancerous data such that the relative magnitude of the normal and dysplastic spectra is predicted rather than the absolute magnitude of each curve. These results show that the Monte Carlo model accurately predicts both the shape and the relative intensity of fluorescence of normal and dysplastic cervix.

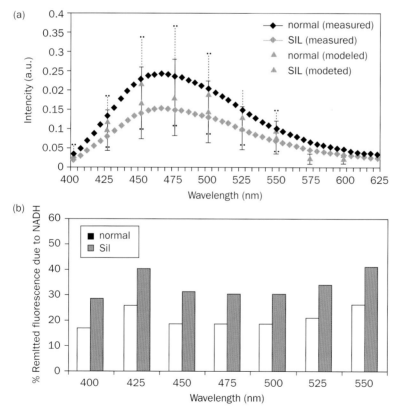

Figure 11.15 (A) Comparison of average measured and modeled spectra. Measured curves come from an in vivo study of 95 women. Modeled curves are generated as described in the text. (B) Percent of remitted fluorescence resulting from NADH and collagen.

In addition, the model is used to indicate how fluorescence from NAD(P)H and collagen add together in normal and abnormal tissue spectra. The predicted contribution of NAD(P)H and collagen to remitted fluorescence at each emission wavelength in the normal and dysplastic case is shown in figure 11.15B. In normal cervix at 380 nm excitation, approximately 20% of remitted tissue fluorescence is the result of epithelial NAD(P)H, whereas the remaining 80% is the result of stromal collagen. In dysplastic cervix, approximately 30% to 40% of remitted fluorescence is the result of epithelial NAD(P)H, whereas 60% to 70% is the result of stromal collagen. This study suggests that at 380 nm excitation, collagen contributes more significantly to the fluorescence signal than NAD(P)H and that the clinically observed decreases in fluorescence intensity with dysplasia are the result of a reduction in the collagen fluorescence, even though the contribution of NADH to measured fluorescence increases with dysplasia. Thus, forward modeling can be applied broadly to data collected in vivo to understand better the biological origins of variations in fluorescence resulting from dysplasia.

In this chapter we reviewed the current use of fluorescence and reflectance spectroscopy to diagnose cervical precancer. We showed that statistical algorithms based on optical spectroscopy maintain the sensitivity of current detection methods while significantly improving specificity. A promising new field is the combined use of reflectance and fluorescence, as in the case of TMS, in which both techniques complement each other to yield an improved diagnostic capability. Another important step toward improving the performance of optical spectroscopy is to relate changes observed in in vivo spectra to specific molecular and morphological alternations that occur with dysplastic conversion. Results from fresh tissue slices and the tissue phantoms models show that there is a distinct biological basis for the spectral differences in cervical tissue. In addition, data measured from tissue slices models can be used as input to forward models that can accurately predict the shape and intensity of clinically measured spectra. Larger clinical trails of these technologies are underway in the developed world to establish the clinical role they can play. Ultimately, inexpensive and portable devices are to be designed for use in the developing world, where 80% of all cervical cancers occur.

REFERENCES

Alfano RR, Tang GC, Pradham A, Lam W, Choy DSJ, Opher E. Fluorescence spectra from cancerous and normal human breast and lung tissues. IEEE J. Quant. Electron. 1987, QE. 23:1806–1811.

Backman V, Wallace MB, Perelman LT, Arendt JT, Gurjar R, Muller MG, Zhang Q, Zonious G, Kline E, McGillican T, Shapshay S, Valdez T, Badizadegan K, Crawford JM, Fitzmaurice M, Kabani S, Levin HS, Seiler M, Dasari RR, Itzkan I, Van Dam J, Feld MS. Detection of preinvasive cancer cells. Nature. 2000, 406(6791):35–36.

Bailey AJ, Paul RG, Knott L. Mechanisms of maturation and ageing of collagen. Mech. Ageing Devel. 1998, 106:1–56.

Balas C. A novel optical imaging method for the early detection, quantitative grading, and mapping of cancerous and precancerous lesions of cervix. IEEE Trans. Biomed. Eng. 2001, 48(1):96–104.

Balas JC, Themelis GC, Prokopakis EP, Orfanudaki I, Koumantakis E, Helidonis E. In vivo detection and staging of epithelial dysplasias and malignancies based on the quantitative assessment of acetic acid tissue interaction kinetics. J. Photochem. Photobiol. B. 1999, 53:153–157.

Baynes JW. Role of oxidative stress in development of complications in diabetes. Diabetes. 1991, 40:405–411.

Blomfield J, Farrar JF. The fluorescent properties of maturing arterial elastin. Cardiovasc. Res. 1969, 3:161–170.

Brookner C, Follen M, Boiko I, Galvan J, Thomsen S, Malpica A, Suzuki S, Lotan R, Richards-Kortum R. Tissue slices autofluorescence patterns in fresh cervical tissue. Photochem. Photobiol. 2000, 71:730–736.

Brookner C, Utzinger U, Staerkel G, Richards-Kortum R, Follen Mitchell M. Cervical fluorescence of normal women. Lasers Surg. Med. 1999, 24:29–37.

Cantor SB, Mitchell MF, Tortolero-Luna G, Bratka CS, Bodurka D, Richards-Kortum R. Cost-effectiveness analysis of the diagnosis and management of cervical intraepithelial neoplasia. Obstet. Gynecol. 1998, 91:270–277.

Chang SK, Dawood MY, Stearkel G, Utzinger U, Richards-Kortum R, Follen M. Fluorescence spectroscopy for cervical precancer detection: Is there variance across the menstrual cycle? J. Biomed. Optics. 2002, 7:595-602 .

Cheng Q-F. In *Summary of Optical Properties in Optical–Thermal Response of Laser Irradiated Tissue*. AJ Welch, M Van Gemert, eds. Plenum Press, New York, 1995.

Coppelson M, Reid BL, Skladnev V, Dalrymple JC. An electronic approach to the detection of pre-cancer and cancer of the uterine cervix: A preliminary evaluation of polar probe. Int. J. Gynecol. Cancer. 1994, 4:79–93.

Cothren RM, Sivak MV, Van Dam J, et al. Detection of dysplasia at colonoscopy using laser-induced fluorescence: A blinded study. Gastrointest. Endosc. 1996, 44:168–176.

Dattamajumdar AK, Parnell J, Wells D, Ganguly D, Wright TC. *Preliminary Results from multi-center Clinical Trials for Detection of Cervical Squamous Intraepithelial Lesions Using a Novel Full Field Evoked Tissue Fluorescence Based Imaging Instrument.* Presented at the European Conference of Biomedical Optics, Munich, Germany, June 2001a.

Dattamajumdar AK, Wells D, Parnell J, Lewis JT, Ganguly D, Wright TC. *Preliminary Experimental Results from Multi-center Clinical Trials for Detection of Cervical Precancerous Lesions Using the Cerviscan System: A Novel Full Field Evoked Tissue Fluorescence Based Imaging Instrument.* Presented at the 23rd annual meeting of IEEE—Engineering in Medicine and Biology, Istanbul, Turkey, October 2001b. Online. www.lifespex.com/1338-Dattamajumdar-EMBS2001.pdf.

Davis RP, Canessa-Fischer M. Spectrofluorometric identification of reduced pyridine nucleotide in the intact isolated urinary bladder of the toad. Anal. Biochem. 1965, 10:325–343.

Deyl Z, Macek K, Adam M, VanCikova O. Studies on the chemical nature of elastin fluorescence. Biochim. Biophys. Acta. 1980, 625:248–254.

Drezek R, Brookner C, Pavlova I, Boiko I, Malpica A, Lotan R, Follen M, Richards-Kortum R. Autofluorescence microscopy of fresh cervical tissue sections reveals alterations in tissue biochemistry with dysplasia. Photochem. Photobiol. 2001a, 73(6):636–641.

Drezek R, Dunn A, Richards-Kortum R. A pulsed FDTD method for calculating light scattering from cells over broad wavelength ranges. Opt. Express. 2000, 6(7):147–157.

Drezek R, Guillaud M, Collier T, Boiko I, Malpica A, Macaulay C, Follen M, Richards-Kortum R. Light scattering from cervical cells throughout neoplastic progression: Influence of nuclear morphology, DNA content, and chromatin texture. J Biomed Opt. 2003, 8:7–16 .

Drezek R, Sokolov K, Utzinger U, Boiko I, Malpica A, Follen M, Richards-Kortum R. Understanding the contributions of NADH and collagen to cervical tissue fluorescence spectra: Modeling, measurements, and implications. J. Biomed. Opt. 2001c, 6:385–396

Dunn A, Richards-Kortum R. Three dimensional computation of light scattering from cells. IEEE J. Sel. Topics Quant. Electr. 1996, 2:889–905.

Dunn A, Smithpeter C, Welch AJ, Richards-Kortum R. FDTD simulation of light scattering from single cells. J. Biomed. Opt. 1997, 2:262–266.

Eyre D, Paz M, Gallop PM. Cross-linking in collagen and elastin. Annu. Rev. Biochem. 1984, 53:717–748.

Ferenczy A, Wright TC. *Anatomy and Histology of the Cervix: Blaustein's Phatology of the Female Genital Tract.* 3rd ed. Springer Verlag, New York, 1994, pp. 185–202.

Ferris DG, Lawhead RA, Dickman ED, Holtzapple N, Miller JA, Grogan S, Bambot S, Agrawal A, Faupel M. Multimodal hyperspectral imaging for the noninvasive diagnosis of cervical neoplasia. J. Lower Genital Tract Dis. 2001, 5:65–72.

Fujimoto D, Akibo KY, Nakamura N. Isolation and characterization of a fluorescent material in bovine achilles-tendon collagen. Biochem. Biophys. Res. Commun. 1977, 76:1124–1129.

Georgakoudi I, Sheets E, Muller M, Backman V, Crum C, Badizadegan K, Dasari R, Feld MS. Trimodal spectroscopy for the detection and characterization of cervical precancer in vivo. Am. J. Obstet. Gynecol. 2002, 186:374–382.

Gorodeski GI, Eckert RL, Utian WH, Sheean L, Rorke EA. Retinoids, sex steroids and glucocorticoids regulate ectocervical cell envelope formation but not the level of the envelope precursor, involucrin. Differentiation. 1989, 42:75–80.

Gurjar R, Backman V, Perelman L, Georgakoudi I, Badizadegan K, Itzkan I, Dasari RR, Feld MS. Imaging human epithelial properties with polarized light scattering spectroscopy. Nat. Med. 2001, 7: 1245–1248.

Harries ML, Lam S, MacAulay C, Qu J. Palcic diagnostic imaging of the larynx: Autofluorescence of laryngeal tumours using the helium-cadmium laser. J. Laryngol. Otol. 1995, 190:108–110.

Hong WK, Sporn MB. Recent advances in chemoprevention of cancer. Science. 1997, 278:1073–1077.

Hornung R, Pham TH, Keefe KA, Berns MW, Tadir Y, Tromberg BJ. Quantitative near-infrared spectroscopy of cervical dysplasia in vivo. Human Reprod. 1999, 14(11):2908–2916.

Hung J, Lam S, LeRiche JC, Palcic B. Autofluorescence of normal and malignant bronchial tissue. Lasers Surg. Med. 1991, 11:99–105.

Kucharz EJ. *The Collagens: Biochemistry and Pathophysiology.* Springer-Verlag, New York, 1992.

Kuntz RM, Saltzman WM. Neutrophil motility in extracellular matrix gels: Mesh size and adhesion affect speed of migration. Biophys. J. 1997, 72:1472–1480.

Lakowicz JR. *Principles ofFluorescence Spectroscopy.* Plenum Press, New York, 1985.

Lam S, Kennedy T, UngerM, MillerYE, Gelmont D, Rusch V, Gipe B, Howard D, LeRiche JC, Coldman A, Gazdar AF. Localization of bronchial intraepithelial neoplastic lesions by fluorescence bronchoscopy. Chest. 1998, 113(3): 696–702.

Lam S, MacAulay C, Hung J, LeRiche J, Profio AE, Palcic B. Detection of dysplasia and carcinoma in situ with a lung imaging fluorescence endoscope device. J. Thoracic Cardiovasc. Surg. 1993a, 105(6):1035–1040.

Lam S, MacAulay C, Palcic B. Detection and localization of early lung cancer by imaging techniques. Chest. 1993b, 103(Suppl.):12S–14S.

LangerR, Vacanti JP. Tissue engineering. Science. 1993, 260:920–926.

Lohmann W, Mussman J, Lohmann C, Kunzel W. Native fluorescence of unstained cryo-sections of the cervix uteri compared with histological observation. Naturwissenschaften. 1989, 96:125–127.

Mahadevan A. *Fluorescence and Raman Spectroscopy for Diagnosis of Cervical Precancers.* Masters thesis, The University of Texas at Austin, Austin, Texas, 1998.

Mahadevan A, Mitchell M, Silva E, Thomsen S, Richards-Kortum R. A study of the fluorescence properties of normal and neoplastic human cervical tissue. Lasers Surg. Med. 1993, 13:647–655.

Masters BR, Chance B. Redox confocal imaging: intrinsic fluorescent probes of cellular metabolism. In *Fluorescent and Luminescent Probes for Biological Activity,* ed. WT Mason. Academic Press, London, 1993.

Mirabal Y, Chang S, Atkinson N, Malpica A, Follen M, Richards-Kortum R. Reflectance spectroscopy for in vivo detection of cervical precancer. J. Biomed. Opt. 2002, 7:587–594.

Mitchell MF, Schottenfeld D, Tortolero-Luna G, Cantor SB, Richards-Kortum R. Colposcopy for the diagnosis of squamous intraepithelial lesions: a meta-analysis. Obstet Gynecol. 1998, 91: 626–631.

Mourant JR, Bigio IJ, Boyer J, et al. Spectroscopic diagnosis of bladder cancer with elastic light scattering. Lasers Surg. Med. 1995, 17(4):350–357.

Mourant J, Bigio I, Boyer J, Johnson T, Lacey J. Detection of GI cancer by elastic scattering spectroscopy. J. Biomed. Opt. 1996, 1:192–199.

Mourant JR, Hielscher AH, Eick AA, Johnson TM, Freyer JP. Evidence of intrinsic differences in the light scattering properties of tumorigenic and nontumorigenic cells. Cancer. 1998, 84(6):366–374.

Nordstrom RJ, Burke L, Niloff JM, Myrtle LF. Identification of cervical intraepithelial neoplasia (CIN) using UV-excited fluorescence and diffuse-reflectance tissue spectroscopy. Lasers Surg. Med. 2001, 29(2):118–127.

Odetti P, Pronzato MA, Noberasck G, Cosso L, Traverso N, Cottalasso D, Marinari UM. Relationships between glycation and oxidation related fluorescences in rat collagen during aging. Lab. Invest. 1994, 70:61–67.

Parkhurst MR, Saltzman WM. Quantification of human neutrophil motility in three-dimensional collagen gels: Effect of collagen concentration. Biophys. J. 1992, 61:306–315.

Parks WC, Mecham RP. *Matrix Metalloproteinases.* Academic Press, San Diego, 1998, pp. 247.

Perelman L, Backman V, Wallace M, Zonios G, Manoharan R, Nusrat A, Shields S, Seiler M, Lima C, Hamano T, Itzkan I, Van Dam J, Crawford JM, Feld MS. Observation of periodic fine structure in reflectance from biological tissue: A new technique formeasur ing nuclearsize distribution. Phys. Rev. Lett. 1998, 80:627–630.

Pogue BW, Burke GC, Weaver J, Harper DM. Development of a spectrally resolved colposcope for early detection of cervical cancer. In *Biomedical Optical Spectroscopy and Diagnostics Technical Digest.* Optical Society of America, Washington, D.C., 1998, p. 87–89.

Pogue BW, Mycek MA, Harper D. Image analysis for discrimination of cervical neoplasia. J. Biomed. Opt. 2000, 5(1):72–82.

Qu J, MacAulay C, Lam S, Palcic B. Optical properties if normal and carcinomatous bronchial tissue. Appl. Opt. 1994, 33(31):7397–7405.

Ramanujam N. Fluorescent spectroscopy of neoplastic and nonneoplastic tissues. Neoplasia. 2000, 2(1–2):89–117.

Ramanujam N, Follen Mitchell M, Mahadevan A, Thomsen S, Malpica A, Wright T, Atkinson N, Richards-Kortum R. Development of a multivariate statistical algorithm to analyze human cervical tissue fluorescence spectra acquired *in vivo.* Lasers Surg. Med. 1996a, 19:46–62.

Ramanujam N, Follen Mitchell M, Mahadevan A, Thomsen S, Malpica A, Wright T, Atkinson N, Richards-Kortum R. Spectroscopic diagnosis of cervical intraepithelial neoplasia (CIN) *in vivo* using laser induced fluorescence spectra at multiple excitation wavelengths. Lasers Surg. Med. 1996b, 19:63–74.

Ramanujam N, Follen Mitchell M, Mahadevan A, Thomsen S, Richards-Kortum R. Fluorescence spectroscopy as a diagnostic tool for cervical intraepithelial neoplasia. Gynecol. Oncol. 1994a, 52:31–38.

Ramanujam N, Follen Mitchell M, Mahadevan-Jansen A, Thomsen SL, Staerkel G, Malpica A, Wright T, Atkinson N, Richards-Kortum R. Cervical pre-cancer detection using a multivariate statistical algorithm based on laser induced fluorescence spectra at multiple excitation wavelengths. Photochem. Photobiol. 1996c, 6:720–735.

Ramanujam N, Mitchell MF, Mahadevan A, Thomsen S, Silva E, Richards-Kortum RR. *In vivo* diagnosis of cervical intraepithelial neoplasia using 337 nm laser induced fluorescence. Proc. Natl. Acad. Sci. USA. 1994b, 91:10193–10197.

Ramanujam N, Richards-Kortum R, Thomsen S, Mahadevan-Jansen A, Follen M, Chance B. Low temperature fluorescence imaging of freeze trapped human cervical biopsies. Opt. Express. 2001, 8(6):335–343.

Richards-Kortum RR. Role of laser induced fluorescence spectroscopy in diagnostic medicine. In *Optical–thermal responses oflaser irradiated tissue*, ed. AJ Welch, M Van Gemert. Plenum Press, New York, 1994.

Richards-Kortum R, Rava RP, Petras RE, Fitzmaurice M, Sivak M, Feld MS. Spectroscopic diagnosis of colonic dysplasia. Photochem. Photobiol. 1991, 53(6):777–786.

Richards-Kortum R, Sevick-Muraca E. Quantitative optical spectroscopy for tissue diagnosis. Ann. Rev. Phys. Chem. 1996, 47:555–606.

Riesle J, Hollander AP, Langer R, Freed LE, Vunjak-Novakovic G. Collagen in tissue-engineered cartilage: Types, structure, and crosslinks. J. Cell. Biochem. 1998, 71:313–327.

Schomacker KT, Frisoli JK, Compton C, et al. Ultraviolet laser-induced fluorescence of colonic tissue: Basic biology and diagnostic potential. Lasers Surg. Med. 1992, 12:63–78.

Sokolov K, Drezek R, Zgossage K, Richards-Kortum R. Reflectance spectroscopy with polarized light: Is it sensitive to cellularand nuclearmor phology? Opt. Express. 1999, 5:302–317.

Sokolov K, Galvan J, Myakov A, Lacy A, Lotan R, Richards-Kortum R. Realistic three-dimensional epithelial tissue phantoms forbiomedical optics. J. Biomed. Opt. 2002, 7;148–156.

Stryer L. *Biochemistry.* 34th ed. WH Freeman, New York, 1988.

Thornhill, DP. Separation of a series of chromophores and fluorophores present in elastin. Biochem. J. 1975, 147:215–219.

Tromberg B. *Optical and Physiological Properties ofTumors.* Presented at the 1998 OSA annual meeting. [CITY, STATE]

Tumer K, Ramanujam N, Ghosh J, Richards-Kortum R. Ensembles of radial basis function networks for spectroscopic detection of cervical precancer. IEEE Trans. Biomed. Eng. 1998, 45:953–961.

Utzinger U, Trujillo V, Atkinson EN, Mitchell MF, Cantor SB, Richards-Kortum R.Performance estimation of diagnostic tests for cervical pre-cancer based on fluorescence spectroscopy: Effects of tissue type, sample size, population and signal to noise ratio. IEEE Trans. Biomed. Eng. 1999, 46:1293–1303.

Vo-Dinh T, Panjehpour M, Overholt BF, Farris C, Buckley FP, Sneed R. *In vivo* cancerdiagnosis of the esophagus using differential normalized fluorescence (DNF) indices. Lasers Surg. Med. 1995, 16:41–47.

Wagnieres GA, Star WM, Wilson BC. In vivo fluorescence spectroscopy and imaging for oncological applications. Photochem. Photobiol. 1998, 68(5):603–632.

Welch AJ, Gardner C, Richards-Kortum R, Chan E, Criswell G, Pfefer J, Warren S. Propagation of fluorescent light. Lasers Surg. Med. 1997, 21:166–178.

Zangaro RA, Silveira L, Manoharan R, Zonios G, Itzkan I, Dasari R, Van Dam J, Feld MS. Rapid multiexcitation fluorescence spectroscopy system for in vivo tissue diagnosis. Appl. Opt. 1997, 35:5211–5219.

Zijlstra WG, Buursma A, Meeuwsen-van der Roest WP. Absorption spectra of human fetal and adult oxyhemoglobin, de-oxyhemoglobin, carboxyhemoglobin, and methemoglobin. Clin Chem. 1991, 37(9):1633–1638.

Zuclich JA, Shimada T, Loree T, Bigio I, Strobl K, Nie S. Rapid noninvasive optical characterization of the human lens. Lasers Life Sci. 1994, 6:39–53.

Zuluaga A, Utzinger U, Durkin A, Fuchs H, Gillenwater A, Jacob R, Kemp B, Fan J, Richards-Kortum R. Fluorescence excitation emission matrices of human tissue: A system for in vivo measurement and data analysis. Appl. Spectrosc. 1999, 53:302–311.

12

Quantitative Absorption and Scattering Spectra in Thick Tissues Using Broadband Diffuse Optical Spectroscopy

DOROTA JAKUBOWSKI, FREDERIC BEVILACQUA,
SEAN MERRITT, ALBERT CERUSSI,
AND BRUCE J. TROMBERG

Near-infrared spectroscopy is a well-established technique that has been applied to a variety of medical diagnostic problems. Recent examples include detecting peripheral vascular disease, compartment syndrome, and various forms of cancer (Fantini et al., 1998; Zonios et al., 1999; Benaron et al., 2000; Giannotti et al., 2000; Tromberg et al., 2000; Wariar et al., 2000; Cerussi 2001). Because near-infrared light penetrates bone, near-infrared spectroscopy is also widely used for noninvasive brain imaging to monitor cortical activation, neonatal cerebral oxygenation, and hematomas (Adcock et al., 1999; Fantini et al., 1999; Benaron et al., 2000; Springett et al., 2000). The use of model-based *photon migration* methods to separate light absorption from scattering quantitatively in multiply scattering tissues is a type of near-infrared spectroscopy broadly referred to as *diffuse optical spectroscopy* or *DOS*.

Diffuse optical spectroscopy methods can be categorized as *time independent* and *time resolved*. Time-independent, steady-state methods use a CW source. They are capable of acquiring a full wavelength-dependent spectrum of diffuse reflectance at one time, but the data cannot be resolved into their scattering and absorbing components without additional information. Research groups using this approach have employed multiple source–detector separations to provide this additional information (Farrell et al., 1992). Short (e.g., millimeter) source–detector separations are more sensitive to scattering properties, whereas far (e.g., centimeter) source–detector separations are sensitive to both scattering and absorption. This method provides fast and accurate results for homogenous systems. However, because two overlapping but different volumes are sampled, accuracy may suffer if the system is heterogenous or layered, as is commonly the case in tissue.

Time-resolved techniques separate absorption from scattering by launching either a single light pulse (time domain) or intensity modulations (frequency domain) into the sample. Time domain techniques measure the delay and spread of a pulse sent through the medium (Patterson et al., 1989). Frequency domain techniques, such as those used in this work, use intensity-modulated light at frequencies between 50 and 1000 MHz, and measure the phase shift and amplitude of the modulated wave as it propagates through the medium (Pham et al., 2000).

Time-resolved techniques generally require sequential scanning over optical wavelengths. This is usually satisfied by a number of diode lasers or a tunable laser. Tunable lasers are capable of acquiring a full near-infrared spectrum (Cubeddu et al., 2000). They are finely tuned and require greater maintenance than laser diodes, posing some difficulties in a clinical setting. In contrast, diode lasers are robust but operate only at discrete wavelengths. Most such frequency domain systems measure optical properties using two to three sources (Franceschini et al., 1997; Moesta et al., 1998); a few operate with up to 7 to 10 sources. In principle, this is well above the four wavelengths necessary to quantify the four major near-infrared chromophores in tissue: deoxygenated hemoglobin, oxygenated hemoglobin, lipid, and water. Nevertheless, our recent experience suggests there is significant inaccuracy in calculating chromophore concentrations even with seven wavelengths (Bevilacqua et al., 2000).

The combined use of both steady-state and frequency domain photon migration (FDPM) techniques can provide complete absorption and scattering spectra with a single source–detector separation. This broadband DOS method expands the optical bandwidth of discrete wavelength FDPM to encompass the entire wavelength spectrum between $\approx$600 and 1000 nm. The steady-state–FDPM combination reduces dependence on the availability of specific diode wavelengths while simultaneously making the system more robust under clinical conditions. Last, it has the potential to characterize absorption backgrounds caused by minor chromophores in such tissues as muscle and brain.

The feasibility of broadband DOS has previously been shown (Bevilacqua et al., 2000). This section validates broadband DOS over a broad range of absorption and scattering values, and compares it with FDPM methods used alone. A series of water-based phantoms with known scattering and absorption properties are measured using broadband DOS and FDPM, and the derived dye concentrations are compared with those expected. As a second step, the limits of both systems are tested on a series of multiple-dye phantoms with absorption properties sufficiently high to challenge the FDPM system. We test the ability of both broadband DOS and FDPM to recover the correct absorption spectrum.

In subsequent sections we describe the application of broadband DOS to measurements of human muscle and breast tissue. These noninvasive in vivo studies highlight the sensitivity of DOS to physiological changes, particularly alterations in water and lipid content that would otherwise be impossible to recover with a standard two- to four-wavelength near-infrared spectroscopy instrument.

THEORY

Broadband DOS combines steady-state and frequency domain photon migration technologies. Combining the two systems is accomplished by joining the theoretical framework on which each is based, as illustrated in figure 12.1. The final result is a full scattering and absorption spectrum at all wavelengths from a single source–detector pair. Here we present a brief overview of this approach. For a thorough discussion and full definitions of reflectance, refer to Bevilacqua et al. (2000).

Optical properties of tissue are typically characterized by μ_a and μ'_s, the absorption and reduced scattering coefficients, respectively. These parameters describe the inverse mean free paths for absorption and reduced scattering events. Reduced scattering is thus named because it attempts to take into account two independent factors: the probability of a scattering event and the angular distribution of such a scattering event. Both are functions of the optical wavelength λ. Foran extensive discussion, referto one of many sources on diffusion theory (Ishimaru, 1978; Farrell et al., 1992; Fishkin, 1994; Haskell et al., 1994).

Frequency domain photon migration methods are capable of measuring μ_a and μ'_s at discrete wavelengths. This is accomplished by recording phase lag and amplitude modulation depth as a function of modulation frequency foreach laserdiode used. The raw phase and amplitude are calibrated to distinguish the tissue signal from the instrument response

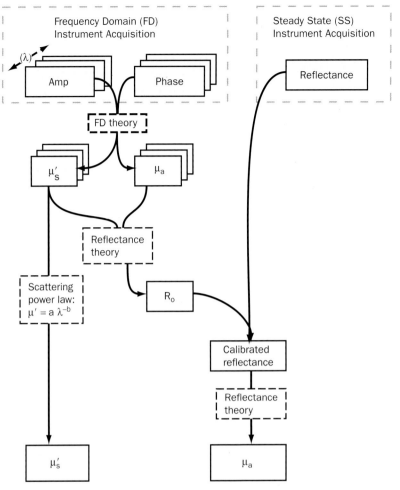

Figure 12.1 The broadband DOS algorithm. It is used for combining the raw data of the FDPM and steady-state systems. In the end, we arrive at continuous μ_a and μ'_s spectra.

using silicon phantoms of known optical properties (Pham et al., 2000). After the amplitude and phase data for a given wavelength are calibrated, it is fit using the P1 diffusion approximation under appropriate semi-infinite boundary conditions, as detailed by Fishkin (1994). The phase and amplitude data are fit simultaneously. Conversely, the real or imaginary components can be fit instead of the original phase and amplitude. Thus, we obtain the free parameters of the fit—in other words, the optical properties $\mu_a(\lambda)$ and $\mu'_s(\lambda)$—of the sample at that wavelength, λ.

In contrast, steady-state methods provide relative diffuse reflectance data for multiple wavelengths over a broad (e.g., $\approx$600–1000 nm) spectral region. Although μ_a and μ'_s cannot be determined from the steady-state spectra, specific pairs can be evaluated for discrete diode wavelengths. To combine steady state with FDPM, the distribution of scatterers in tissue is described by a power law:

$$\mu'_s(\lambda) = a\lambda^{-b}$$

This equation is an approximation based on Mie theory with a wide distribution of scatterers (Graaff et al., 1992; Mourant et al., 1997). It is additionally supported by numerous studies in tissue (Graaff et al., 1992; Mourant et al., 1997, 1998). By applying this equation to the 6 to 10 FDPM-derived μ_s' values, the a and b parameters, and therefore the full μ_s' spectrum, can be determined for our region of interest.

Last, we must consider that the steady-state reflectance measurement is relative. It varies by a constant R_o from the absolute reflectance R. However, the FDPM-derived μ_a and μ_s' pair can be used to calculate the true absolute reflectance R (a function of μ_a and μ_s' at any given wavelength) at 6 to 10 wavelengths. Comparing the measured with the calculated reflectance gives us R_o. With R and μ_s' determined for all wavelengths, μ_a can be calculated at each wavelength. We thus arrive at the full μ_a spectrum for all wavelengths in the 600- to 1000-nm region.

To arrive at chromophore concentrations, the μ_a spectrum is fit using the Beer-Lambert law. For a given chromophore, the μ_a contribution as a function of wavelength would be $\mu_a(\lambda) = C \times \varepsilon(\lambda)$, where C is the concentration and ε is the extinction coefficient of the chromophore at that wavelength. If the chromophores do not interact, the final absorption spectrum is a linear combination of the component spectra, and we may solve the equation by adjusting μ_a and C into the appropriate vectors, ε into a matrix, and solving the matrix equation.

INSTRUMENT

The broadband DOS system consists of two separate components: steady-state spectroscopy and FDPM integrated into a single probe handpiece illustrated in figure 12.2. This probe combines steady-state and FDPM systems by crossing their source–detector paths. Thus, the volume of sample interrogated by each is nearly identical.

Two broadband instruments that differ slightly in their FDPM source components were used in this work. Each FDPM subsystem uses six to nine laser diodes (instrument 1: 672, 780, 807, 852, 896, and 913; instrument 2: 663, 680, 783, 806, 816, 855, 911, 945, and 973). The basic system schematic is shown in figure 12.2.

Diode powers (measured at the source) range between 20 and 100 mW, with the exception of diodes in the upper 900-nm region, which may have powers up to 1 W. Current and temperature control are maintained by a laser power supply. During a measurement, each laserdiode is sequentially selected and intensity modulated. The AC modulation to the laser diodes is provided by a network analyzer, which sweeps through frequencies between 100 to 700 MHz. The network analyzer's radiofrequency (RF) output is coupled to the correct diode via an RF switch. A bias T circuit at each diode mixes the AC signal with DC power provided by the laserpowersupply.

Each diode is pigtailed to a 100-μm graded-index or coupled to 400-μm step-index fibers. Switching between the diodes is accomplished with either a multiplexed optical switch or keeping nonselected diodes below lasing current (instrument 2). Output fibers are bundled together and terminate at the handheld probe, where optical power ranges from 5 to 35 mW. The system characteristics have been described extensively by Pham et al. (2000).

The diffuse optical signal is received from the sample by an APD that is built into the handheld probe and is placed directly on the sample. The source fibers and APD are arranged in reflectance geometry within the probe. The RF output of the APD is directed to the network analyzer. Both amplitude depth and phase shift of the signal are recorded as a function of modulation frequency.

The steady-state system consists of a source and CCD spectrometer. The source is a 100- or 150-W tungsten quartz halogen lamp. Source output is channeled to the sample using a 3-mm fiberbundle of 100-μm fibers (instrument 1) or a 5-mm liquid core light guide (instrument 2). The diffuse reflectance from the tissue is collected by a 1-mm optical fiber that is coupled to a spectrometer. In the case of the first set of experiments, the spectrometer

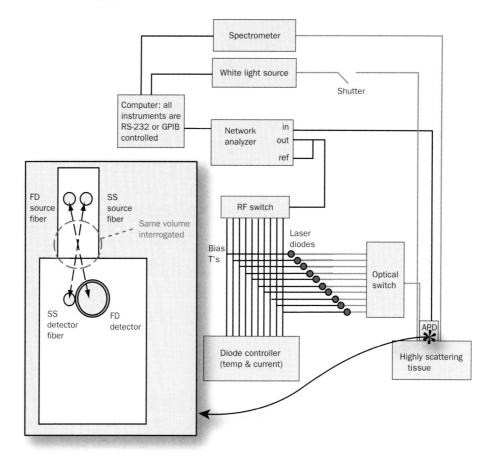

Figure 12.2 Schematic of the broadband DOS system, including the probe design on the left, which combines the steady state (SS) and frequency domain (FD) systems by crossing the interrogating paths.

uses a linear silicon CCD array with a 12-bit dynamic range. In its current configuration, the spectrometer has 5 nm spectral resolution over the 650- to 1000-nm range. Also included in the configuration is a long-pass 515-nm filter to eliminate second harmonic signals up to 1030 nm, and an L2 detectorlens meant to improve light collection efficiency on the CCD. A long-pass filterat 780 nm (Schott RG-780) is partially placed in the light path to flatten the white-light response. The time required for a spectral acquisition is highly dependent on the absorption properties of the sample, varying between 15 ms to minutes. Instrument 2 uses a spectrometer with a higher dynamic range than the previous model. It contains a two-dimensional 1024×256 thermoelectrically cooled silicon CCD with a 16-bit dynamic range. Spectrometer input is provided through an SMA connection on an F/# matching optic used to optimize light collection onto the CCD. The grating chosen is blazed at 500 nm with a line density of 400 lines/mm, and the slit size is 300 µm. Also included is a long-pass glass filter with a cutoff of approximately 550 nm. Wavelength calibration was performed on a mercury–argon lamp.

The network analyzer, laser power supplies, RF switch, optical switch, and spectrometer are controlled with in-house software. Steady-state and frequency domain measurements are performed alternately to minimize the impact of laser diode and CW sources on the CCD and APD detectors, respectively. We perform a weighted, nonlinear least-squares fit using the Levenberg-Marquardt algorithm. All data fitting is done with code written in Matlab.

Materials and Methods

The samples of known scattering and absorption characteristics were made by preparing liquid phantoms from water, intralipid as the scatterer, and dye as the absorber. Intralipid is a commercially available intravenous nutritional supplement with a high-volume fraction of fatty acids (20%). Because of its negligible absorption in the 650- to 1000-nm region, it can be considered solely as a scatterer in these experiments. The scattering properties of Intralipid are published in the literature (Flock et al., 1992). Because of concern over batch-to-batch variations, we performed our own FDPM "gold standard" measurements on the Intralipid in infinite medium geometry using multiple distance calibration. These data matched surprisingly well with original published values for 20% Intralipid by van Staveren et al. (1991) with an error of less than 2% within the 650- to 1000-nm region.

The dye used in the validation measurements was copper phthalocyaninetetrasulfonic acid (CPTA; Aldrich 27,401–1, St. Louis, Missouri) prepared in a stock water-based solution. The molarextinction spectrum of the dye (ε vs. λ) was obtained using a standard spectrophotometer by measuring absorbance, A, at various concentrations, C. Using Beers law, $\mu_a = 2.3 \times \varepsilon \times C$, where ε is the molarextinction coefficient (permole permillimeter). The resulting extinction coefficient spectrum in the near infrared is shown in figure 12.3A.

The second set of experiments used two additional chromophores: naphthol green B (Aldrich 11,991–1) and methylene blue (Aldrich M4,490–7). Peak wavelengths in the 650- to 1000-nm region are 650 nm forCPTA (minorpeak at 690), 710 nm fornapthol green B, and 664 nm formethylene blue. This combination was chosen because a mixture of these probes creates absorption spectra with multiple features in the near-infrared region that add similarly to chromophores in tissue spectra. Extinction coefficient spectra of all three molecules are shown in figure 12.3. These spectrophotometer measurements were taken before the addition of Intralipid.

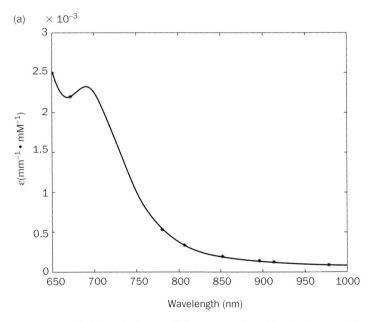

Figure 12.3 (A–C) The extinction coefficient spectra found for the dyes used in the phantoms. Asterisks mark locations of frequency domain diodes. (A) Copper phthalocyaninetetrasulfonic acid (CPTA). (B) Napthol green B. (C) Methylene blue.

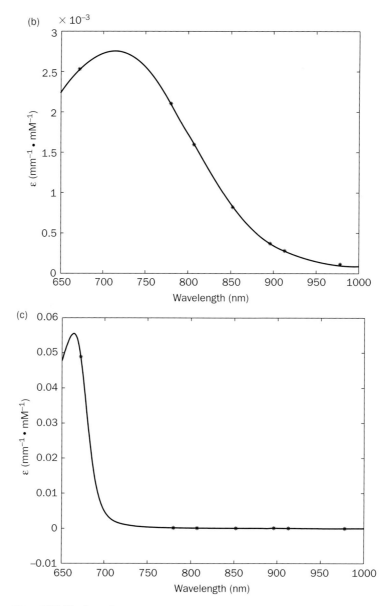

(b)

(c)

Figure 12.3 (*Continued*)

Calibration for the frequency domain system was based on a phantom of known optical properties manufactured by Medlight, SA (Lausanne, Switzerland). The optical properties of the phantom had been previously determined by multidistance, multifrequency measurements. The steady-state system was calibrated by taking source measurements in an integrating sphere coated with a material of known reflectance. Dark measurements and corrections were also made for both sets of experiments.

The liquid phantoms were prepared in a 700-mL glass beaker with a magnetic stir bar used to keep the solution homogenous. A total of 25 phantoms were prepared in the validation study and 29 were prepared in the multidye study. The probe was placed by hand onto each phantom, which was covered with thin plastic film to protect the probe from liquid. We used a source–detector separation of 20 mm in the validation studies. In the multiple-dye studies, we lengthened the separation to 24 mm because of the increased optical power available with this particular system. Integration times were chosen to maximize the signal to at least 60% of the CCD capacity without saturating any part of the spectrum. Each phantom measurement was repeated three times.

In the validation studies, the concentrations of Intralipid and dye were chosen to span physiologically relevant μ_a and μ'_s values derived from human breast tissue measurements. The 672-nm wavelength was chosen as the representative diode. Thus, dye concentrations have a μ_a at 672 nm randomly distributed between 0.00 and 0.03/mm. Intralipid concentrations have a μ'_s at 672 nm randomly distributed between 0.2 and 2/mm.

In the multiple-dye phantoms, each dye had a concentration chosen that would randomly place the μ_a within the range of 0.0001 and 0.025/mm at its peak wavelength. After the phantom was mixed, the sum μ_a at 672 nm in fact spanned the range between 0.01 and 0.055/mm forall phantoms. In comparison, the highest μ_a we encounterin tissue within the 650- to 1000-nm range is the result of water, which at 100% concentration reaches a value of 0.05/mm at its peak wavelength ($\approx$980 nm). Most tissue measurements of μ_a fall well below this absorption.

The value for μ'_s was chosen in a similar way to that described in the validation experiments. However, the lower limit was increased to 0.5/mm to keep measurements within the diffusion approximation. The μ'_s values below the threshold of 0.5/mm had a tendency to have inverse mean free path lengths of less than 10 times the source–detector separation (24 mm) forour μ_a values, and therefore the diffusion approximation was not valid for these conditions. The final μ_a and μ'_s space covered by both sets of phantoms (at the 672-nm wavelength) is shown in figure 12.4, demonstrating a fairly even distribution within the sampled space.

Results and Discussion

Two measurements were removed from the multiple-dye phantom study because frequency domain fits showed the data to be corrupted, likely because of insufficient contact between the probe and phantom. In addition, one phantom was entirely excluded from fits in the multiple-dye study because the raw reflectance spectrum was corrupted by a shutter delay.

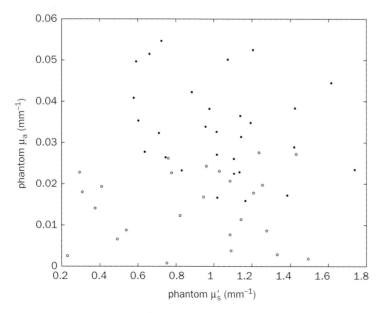

Figure 12.4 Range of μ_a and μ'_s space covered by Intralipid dye liquid Phantoms at a wavelength of 672 nm.

Figure 12.5 shows an example of the fits obtained from the first set of phantoms (single dye). In figure 12.5A, the μ_s' values are shown along with theirpowerlaw fit. The fit matches closely to the original values. The power law fit is used with the steady-state reflectance to determine the μ_a spectrum. This broadband DOS-derived μ_a spectrum, along with the FDPM-derived μ_a spectrum and the dye + water fits to both, are shown in figure 12.5B.

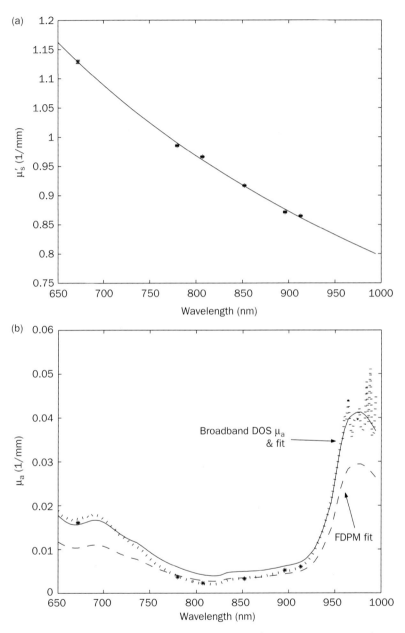

Figure 12.5 Single-dye phantoms. (A) An example of a μ_s' fit (line) to the FDPM μ_s' values (asterisks) for one of the validation phantoms. (B) The broadband DOS-derived μ_a spectrum (dotted line) is shown, along with the FDPM-derived μ_a values (asterisks). The broadband DOS-derived spectrum correlates well with the FDPM-derived μ_a values. However, the fits to the two systems are also shown (broadband DOS, solid line; FDPM. dashed line), and these differ. This example did not add a constant to the chromophore fits.

The DOS-derived spectrum correlates well with the FDPM-derived values; however, the fits to the two systems differ considerably, suggesting discrepancies between the FDPM- and broadband DOS-derived concentrations.

The results of the single-dye, or validation, set of experiments is best shown by comparing final dye concentrations. Obtained versus predicted dye and water concentrations are plotted to establish the accuracy of broadband DOS for a wide range of dye concentrations. Lipid was not fit because of its near-negligible contribution (1%) to the phantoms. Figure 12.6A displays predicted versus obtained CPTA concentrations. We also took standard spectrophotometer readings of all phantoms (sampled before the Intralipid was added). The concentrations determined by the spectrophotometer (vs. predicted values) are also plotted in figure 12.6A. These spectrophotometer values give us some indication of the error inherent in the experimental setup, including the error in mixing in the phantoms. An ideal fit would lie along the 1:1 line. Although the spectrophotometer values have a small and uniform error throughout the CPTA concentration range, both FDPM- and DOS-derived CPTA values begin to deviate considerably at higher concentrations. Frequency domain photon migration-derived values deviate significantly more than those obtained by broadband DOS.

Predicted versus obtained water volume fraction values are shown in figure 12.6B. Because the phantoms are almost completely water, all obtained values should be near 99% to 100%. Although FDPM-derived values deviate by as much as 40% from the true value, broadband DOS confers a modest improvement in accuracy, with a maximum deviation of 30%. Broadband DOS also produces a narrower range of values ($\approx$70%–110%) than FDPM ($\approx$60%–130%).

In figure 12.7, a flat baseline is added as a chromophore in the fit. Both broadband DOS and FDPM methods are able to recover the CPTA concentration and water volume fraction more accurately. Broadband DOS is capable of faithfully quantifying CPTA throughout the entire range of CPTA concentrations tested (fig. 12.7A). The variation seen is very similar to that of the spectrophotometer-derived values. Of note, FDPM-derived CPTA values still deviate at high CPTA concentrations. Most likely, this is the result of the deterioration of the 672-nm diode signal at higherCPTA concentrations.

With the added baseline, the water volume fraction is better recovered by both broadband DOS and FDPM (fig. 12.7B). Diffuse optical spectroscopy-derived values continue to underestimate water in most cases, whereas FDPM now overestimates it. From the raw data (not shown), the part of the μ_a spectrum that drives the water concentration (950–1000 nm) is clearly near the noise floor. Such a problem with water is only seen when there is an extremely high volume fraction, a condition that is not generally observed in tissue measurements.

Results from the second set of phantoms made with multiple dyes are shown in figures 12.8 through 12.11. This set was made to be more challenging for both FPDM and broadband DOS methods. Figure 12.8 shows an example of the fits derived with this set. Because the dye absorption was significantly higher in this set of data, FDPM phase and amplitude data were often unable to provide a satisfactory signal-to-noise ratio. This was especially true of fourof the diodes (663, 680, 945, 973 nm). Because these data were often not salvageable, a condition was written into the fit that would exclude from further fitting those diodes that have modulation amplitudes below a specified level—in this case, –40 dB. Figure 12.8 shows all FDPM- and DOS-derived values, including those FDPM-derived values not used in broadband DOS fitting. In this particular example, the μ_a and μ_s' values of the two highest wavelength diodes were not used in fits, but are shown in the figures as black asterisks. The μ_s' fit in figure 12.8A shows much better correlation to original FDPM-derived μ_s' values without the inclusion of these last two diodes. There is also much better correlation between the DOS-derived μ_a spectrum and the FDPM-derived μ_a values underthe same condition. Notice, however, that the FDPM-derived μ_a fit using only the best FDPM diodes

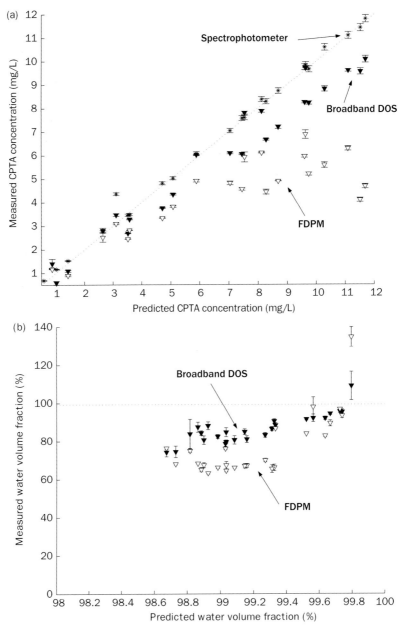

Figure 12.6 (A, B) Single-dye phantom concentrations, no baseline added. Predicted versus measured CPTA concentration (A) and water volume fraction (B) using broadband DOS (filled triangles), FDPM (open triangles), and a standard spectrophotometer (asterisks). Both values deviate from true values (1:1 dotted line). However, FDPM-derived values deviate considerably more than broadband DOS-derived values.

significantly underestimates the water μ_a contribution in the 950- to 1000-nm region and also incorrectly predicts dye μ_a contributions (as predicted by the broadband DOS μ_a spectrum) in the low 600-nm region. This is unsurprising because the lost FDPM data are in important peak regions. Thus, μ_a fits to FDPM data are not tolerant of the removal of diodes if these diodes are in a critical region. Unfortunately, this is frequently the case,

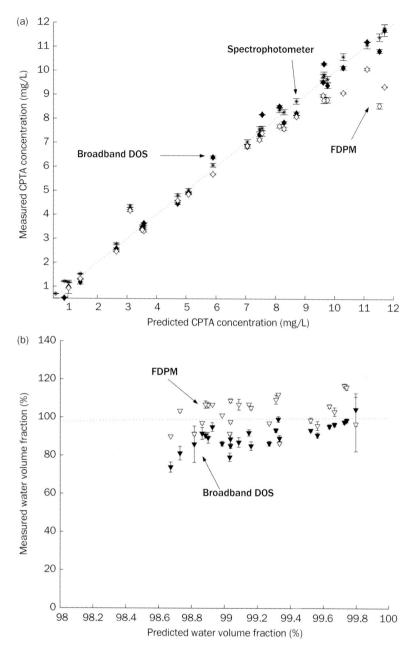

Figure 12.7 (A, B) Single-dye phantom concentrations, with baseline added. Predicted versus measured CPTA concentration (A) and water volume fraction (B) using broadband DOS (filled triangles), FDPM (open triangles), and a standard spectrophotometer (asterisks). A flat baseline was added as a chromophore to the water and CPTA fit. Copper phthalocyaninetetrasulfonic acid is recovered faithfully over the entire range by broadband DOS, whereas FDPM values deviate at higher concentrations. Water volume fraction values still contain a large degree of variation for both methods.

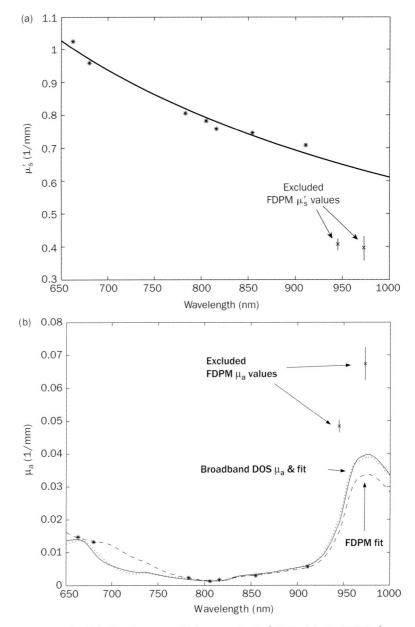

Figure 12.8 Multiple-dye phantoms. (A) An example of μ_s' fit (line) to the FDPM μ_s' values (asterisks) for one of the multiple-dye phantoms. (B) The broadband DOS-derived μ_a spectrum (dashed line) along with the FDPM-derived μ_a values (asterisks). In both views, the crosses mark FDPM-derived μ_a or μ_s' values that were thrown out before fitting. Notice that it is very difficult to get a good μ_a fit to the frequency domain-derived values (dashed line), especially if we disregard key diodes. The broadband DOS μ_a fit, however, does very well (solid line).

because attenuated signal at a particular diode wavelength is usually the result of a high concentration of a relevant chromophore.

In the multiple-dye measurements, absolute dye concentrations could not be accurately determined because the dyes interacted, changing the effective extinction coefficients. Consequently, these results were compared with spectrophotometer results, shown later.

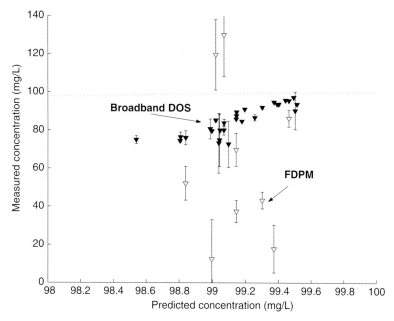

Figure 12.9 Multiple-dye phantom water concentration, with baseline added. Shown are the expected versus obtained water concentrations from the multiple-dye phantoms using broadband DOS (filled triangles), FDPM (open triangles), and a standard spectrophotometer (asterisks). The FDPM system had a difficult time recovering water concentrations under the experimental conditions, whereas the broadband DOS system was able to recover water values with the same accuracy as in the validation phantoms.

It was possible, however, to compare expected versus obtained water concentrations for the multiple dye phantoms (fig. 12.9). Comparison with results in the one-dye phantom experiments using the Ocean Optics spectrometer, shown in figure 12.7B, reveals that water values fall in the same range. At first pass these results are surprising, because the more sensitive Oriel spectrometer used in the second set of phantoms should improve the capability of the system to extract the true water concentrations, even at highly attenuating (100% water fraction) values. However, the second set of phantoms presented a very difficult set of conditions. This is quite apparent if one compares the FDPM-derived water values derived in the multiple-dye phantoms with those in the single-dye phantoms (figs. 12.9 and 12.7B). The FDPM system had a difficult time providing any meaningful data. Frequency domain photon migration-derived water values now range from 10% to 130%, whereas DOS-derived values remain in the 70% to 100% range. The broadband DOS measurement with the Oriel spectrometer was able to provide meaningful data when FDPM alone would have failed. We find this an extremely desirable feature, especially in tissue measurements.

To evaluate the multiple-dye phantom results, we obtained full absorption spectra of each phantom's dye composition (in the absence of intralipid) using the spectrophotometer. Each spectrum was converted to μ_a and corrected for the volume difference of Intralipid, which had been added after the spectrophotometer sample was taken. The known μ_a contribution of the water and lipid spectra were then added to each phantom spectrum. Therefore, each spectrophotometer-derived spectrum is a combination of measured and expected data. To test the ability of broadband DOS to reproduce the μ_a spectrum faithfully in a highly scattering medium, each spectrophotometer-derived μ_a spectrum was compared with that obtained by the broadband DOS method. Two examples of such comparison are shown in figure 12.10. The baseline of the broadband DOS-derived μ_a spectrum was allowed to

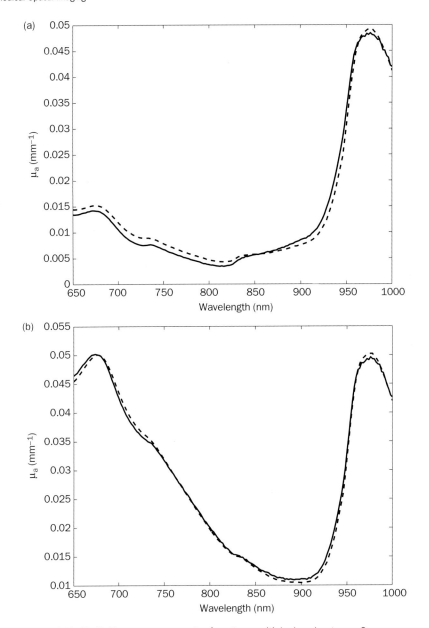

Figure 12.10 (A, B) Shown are μ_a spectra from two multiple-dye phantoms. One spectra was obtained using broadband DOS (solid) and the other was obtained from the spectrophotometer (dashed; with expected water and lipid contributions added).

vary, as was indicated by the results of the one-dye phantoms. The spectrophotometer- and DOS-derived spectra show very good correlation in both panels.

We then subtracted the spectrophotometer spectra from the broadband DOS-derived spectra to obtain residuals for all phantoms. These are shown in the figure 12.11B. In figure 12.11A, we also show all obtained spectra for the multiple-dye phantoms. In no area of the spectrum was the difference between spectra greater than 0.007/mm. This is comparable with the original resolution of the FDPM system, which in our experience has an error of approximately 0.005/mm in μ_a. We conclude that the broadband DOS system is capable

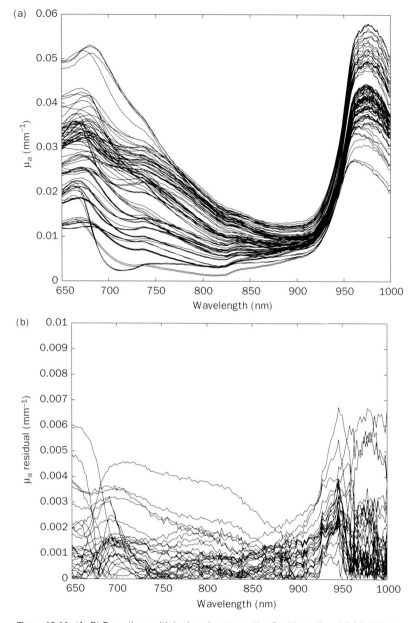

Figure 12.11 (A, B) From the multiple-dye phantoms, the final broadband DOS-derived μ_a spectra and final μ_a residuals obtained from subtracting spectrophotometer-derived spectra from broadband DOS-derived spectra. In no area of the spectrum was the difference between spectra greater than 0.007/mm.

of providing comparable accuracy in μ_a values as the FDPM system, with an added level of robustness in spectral regions of high absorption.

Summary

The combination of steady-state and FDPM is a promising approach for quantifying absorption and scattering in turbid media. The combined method, broadband DOS, consistently performs at least as well as FDPM methods alone. Diffuse optical spectroscopy can enhance the accuracy of FDPM when discrete wavelength methods fail because of high attenuation at key diode wavelengths, (e.g., for high chromophore concentrations). Multiple-dye studies demonstrate that broadband DOS is capable of faithfully reproducing μ_a spectra in highly scattering media to within 0.007/mm if a baseline is introduced. Overall, the broadband DOS technique is capable of meeting orexceeding the accuracy of FDPM in an ideal, single-component system under high-absorption conditions in which frequency domain measurements would normally fail.

BROADBAND DIFFUSE OPTICAL SPECTROSCOPY IN MUSCLE

There have only been a few reports focused on monitoring exercised-induced changes within muscle using quantitative DOS methods (Quaresima et al., 2000; Cooper and Angus, 2003; Miura et al., 2003; Torricelli et al., 2004). In these previous studies, the primary aim was to measure the concentrations of oxyhemoglobin and deoxyhemoglobin using two wavelengths between 650 nm and 850 nm. Although wavelengths in this region are sensitive to blood absorption, they are not spectrally sensitive to water and lipid components in tissue that have high absorption in the 900- to 1000-nm region.

Figure 12.12 shows absorption spectra of oxyhemoglobin, deoxyhemoglobin, water, and lipid—the main chromophores within tissue in the 650- to 1000-nm spectral region. By simply examining the characteristic shapes and the chromophore absorption spectra, it is obvious that water and lipids contribute very little to the tissue absorption below 900 nm. In this section we present results of broadband DOS measurements on human muscle that demonstrate sensitivity to water and lipid that would otherwise be impossible to recover with a standard two- to four-wavelength tissue oximeter. We present these results to highlight two important features: (1) the sensitivity of broadband DOS to subtle physiological changes in muscle and (2) the opportunity for broadband DOS to provide new information on subcutaneous fat layers.

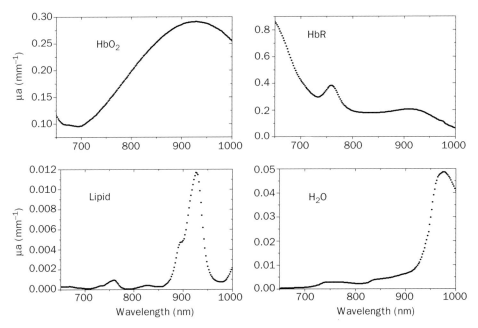

Figure 12.12 Absorption spectra plotted for major absorbers in tissue. Oxyhemoglobin (HbO_2), 1 mM; deoxyhemoglobin (HbR), 1 mM; lipid, 0.9 g/cm^3; H_2O, 1 g/cm^3.

Materials and Methods

The broadband DOS system used for muscle measurements was nearly identical to the system described in the phantom validation section; the only differences were the broadband light source (Ocean Optics LS-1) and the specific laserdiode wavelengths. Forthe muscle measurements, a total of five laser diodes were used at the following wavelengths: 658, 785, 810, 830, and 850 nm.

Measurements were carried out one male volunteer (age, 26 years; weight, 160 lb; height, 188 cm), using approved University of California at Irvine human subjects protocol 2001-1924, who was not participating in any type of strength-training regimen. A single location was chosen at the centerof the subject's two biceps brachii, allowing forr epeat measurements of the same tissue volumes. Ultrasound was used to measure skin/fat layer thicknesses of 0.16 and 0.17 cm for the right and left arms, respectively, at the measurement locations. The source–detector separation used in the study was 2.1 cm, which allowed for sufficient light penetration into the muscle given the relatively thin top layer thickness measured.

Initially, five baseline measurements were taken on each arm, with the axis that bisected the source and detector oriented along the length of the muscle fibers. The subject's non-dominant biceps brachii (left arm) was then exercised by repeatedly curling a 20-lb weight until failure. This was followed by negative curls during which the subject was helped to lift the weight and then asked to apply resistance as it was pulled down by gravity (10 repetitions). The total time elapsed during exercise was 4 minutes and the acqui-sition of postexercise measurements began within 30 s of exercise completion. Three measurements were then taken on the exercised arm followed by a single measurement on the nonexercised arm; this sequence of measurements was repeated until a total of 48 postexercise measurements had been acquired from the exercised arm and 16 from the nonexercised arm. The total time spent on postexercise measurements was just over 1 hour (1:03:20) with an average measurement time of 60 s. The data were processed three separate ways to allow a comparison between broadband DOS and DOS alone. For broadband DOS, the entire absorption spectrum was determined using all five laser diodes plus the broadband information. A chromophore fit was then performed using the extinc-tion coefficients of deoxyhemoglobin, oxyhemoglobin, water, lipid, and a flat baseline. To allow for a comparison with broadband DOS, extinction coefficients of deoxyhe-moglobin, oxyhemoglobin, and a baseline were fit to the absorption coefficients determined at the five laserdiode wavelengths only, whereas waterand lipids were assumed not to contribute significantly to the absorption spectrum in this region. This provided concen-trations of deoxyhemoglobin and oxyhemoglobin determined only from the frequency domain measurements. Last, water, lipid, hemoglobin, oxyhemoglobin, and a baseline were all fit to the absorption coefficients at the five laser diode wavelengths to acquire water and lipid concentrations that could be compared directly with those recovered using broadband DOS.

Results and Discussion

Figure 12.13 is a plot of a typical μ_a spectrum measured on the arm of the volunteer. The fit of the chromophore spectra (deoxyhemoglobin, oxyhemoglobin, water, lipid, and a constant background absorption) to the μ_a spectrum allows for the determination of the concentrations and is described in detail later. The figure shows a qualitatively good fit of the chromophore spectra to the measured μ_a spectrum. The muscle spectra also illustrate how well the endogenous chromophore features stand out, particularly the water peak (978 nm) and the deoxyhemoglobin absorption at lower wavelengths. Notice that the lipid peak is not visually apparent in this measurement because the fat layer thickness of this subject was

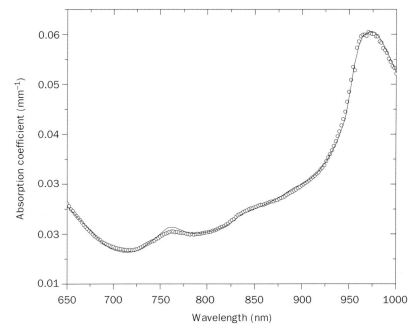

Figure 12.13 Measurements of tissue spectra using DOS. Circles, μ_a muscle spectrum measured with broadband DOS instrument. Solid curve, fit of deoxyhemoglobin, oxyhemoglobin, water, lipid, and baseline absorption spectra to measured μ_a spectrum.

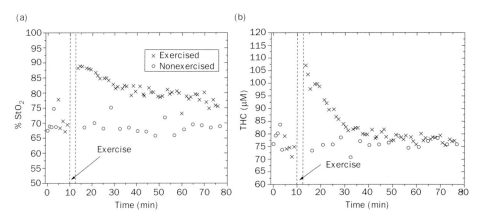

Figure 12.14 (A, B) Tissue oxygen saturation (StO$_2$; A) and total hemoglobin content (THC; B) recovered from muscle measurements on the volunteer's exercised (x) and nonexercised (o) arms. The exercise period is represented by time between dashed vertical lines.

minimal. This lipid peak is obvious in tissues with higherlipid content, such as breast, or in individuals with thickersubcutaneous fat layers.

Figure 12.14 is a plot of the tissue oxygen saturation (StO$_2$) and total hemoglobin content (THC) measured on both the exercised arm and the nonexercised arm over 70 minutes.

Tissue oxygen saturation is an indicator of local tissue metabolism, and is defined as the following:

$$StO_2 = \frac{HbO_2}{HbR + HbO_2}$$

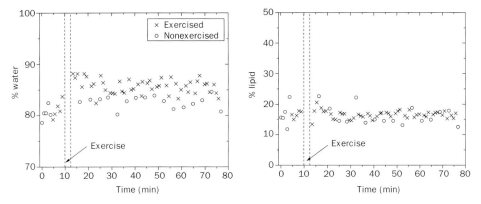

Figure 12.15 (A, B) Water (A) and lipid (B) recovered from muscle measurements on the volunteer's exercised (x) and nonexercised (o) arms. The exercise period is represented by time between dashed vertical lines.

Both StO_2 and THC increase significantly in the exercised arm immediately after exercise, with THC returning close to baseline within 25 minutes. The StO_2 value also demonstrates a fast return to baseline within the first 20 minutes and tapers off to a slower descent, but StO_2 values continue to remain elevated from baseline 60 minutes after exercise. The recovered THC and StO_2 values for the nonexercised arm show no significant changes before and afterexer cise.

Figure 12.15 is a plot of the water content and the lipid content of both the exercised (x) and nonexercised (o) arms. There is a modest increase in the water values measured on the exercised arm immediately after exercise (8%), and these values remain elevated from baseline 60 minutes later. The nonexercised arm also shows a statistically significant increase in water before and after exercise, although not as high (3%) as the increase in the exercised arm. As expected, there are no exercise-induced changes in the lipid content of eitherar m.

The results using only the five laser diodes without the broadband measurement provide similar trends in blood values, with lower contrast between the pre- and postexercise measurements in recovered oxyhemoglobin concentrations. Table 12.1 lists the percentage change between the first three postexercise measurements and the five preexercise measurements. Looking directly at the deoxyhemoglobin and oxyhemoglobin concentrations, it is apparent that the changes in deoxyhemoglobin are nearly the same without the broadband measurement. On the other hand, the contrast between pre- and postexercise measurements of oxyhemoglobin is greatly reduced when the broadband measurement is eliminated. This is a result of two observed differences: The average concentration recovered for oxyhemoglobin at baseline is increased by 30 μM and the absolute change in oxyhemoglobin resulting from exercise is reduced from 40 μM to 20 μM when the broadband measurement is neglected. The absorption spectra plotted in figure 12.12 provide some explanation for these results. The deoxyhemoglobin spectrum is strongly absorbing in the spectral region between 650 and 850 nm, and also has very distinct spectral features, such as the peak at 758 nm. The laser diodes used do a good job of sampling this spectral region. Therefore, the recovered deoxyhemoglobin concentrations preserve the contrast between the pre- and postexercise measurements with and without the broadband measurement, although there are still differences in the absolute concentrations recovered.

In contrast, the absorption spectrum of oxyhemoglobin is much smoother and has a maximum absorption at 929 nm. Because the oxyhemoglobin spectrum lacks distinct features, it may require the increased spectral content from the broadband measurement to fit accurately

Table 12-1. Percent difference between five baseline measurements and three postexercise measurements on muscle.

Chromophore	Percent change, broadband DOS (exercised)	Percent change, broadband DOS (nonexercised)	Percent change, FDPM only (exercised)	Percent change, FDPM only (nonexercised)
Oxyhemoglobin	$+72 \pm 19$	-6 ± 8	$+29 \pm 4$	$+5 \pm 4$
Deoxyhemoglobin	-46 ± 7	-2 ± 7	-45 ± 9	-4 ± 10
Total hemoglobin content	$+37 \pm 8$	-5 ± 5	$+12 \pm 3$	$+2 \pm 2$
Tissue oxygen saturation	$+25 \pm 7$	-1 ± 4	$+15 \pm 3$	$+2 \pm 3$
Water	$+8 \pm 2$	$+3 \pm 2$	$+58 \pm 70$	-6 ± 38
Lipid	$+4 \pm 23$	$+12 \pm 36$	$+485 \pm 1040$	-88 ± 170

The first two columns include the broadband DOS measurements on the exercised and non-exercised arms respectively. The last two columns are the same for values derived from the five wavelength frequency domain data. Errors are represented by the standard deviations determined from repeat measurements.

for this concentration. Ultimately, this reduced sensitivity to oxyhemoglobin decreases the sensitivity of both StO_2 and THC recovered without the broadband measurement.

As expected, the water and lipid concentrations recovered without the broadband measurement have large errors and are meaningless because the wavelengths used are not sensitive to the regions where these chromophores are the dominant absorbers. When water is fit using the broadband measurement, it provides the highest precision of all the chromophores recovered. This is because water is not only highly absorbing, but it is also highly concentrated within tissue. The lipid measurements, on the other hand, have the largest error of all the chromophores, which is a result of their lower concentrations in the subject's arm and lower absorption compared with water and oxyhemoglobin. Despite the higher measurement errors, the recovered lipids are within a range that is anatomically correct and are unchanging during the time course of the measurements.

Summary

These results have been presented to demonstrate the capabilities of broadband DOS to measure physiological changes within muscle after exercise. An important feature of this measurement is the sensitivity of DOS to changes in muscle water content that are difficult to observe. These alterations may provide insight into conditions such as muscle soreness, muscle injury, and compartment syndromes that could follow from tissue damage and edema. We also measure substantially different absolute concentrations of oxyhemoglobin with and without the broadband measurement included (53 ± 3 μM and 84 ± 4 μM, respectively). Although we do not have the absolute chromophore values for a comparison, we can use the phantom measurements presented in the previous section as a validation of broadband DOS accuracy. This improved accuracy has important implications for estimation of commonly used metabolic indices such as THC and StO_2.

BROADBAND DIFFUSE OPTICAL SPECTROSCOPY IN BREAST TISSUE

Near-infrared optical mammography has only recently become clinically feasible as a result of significant improvements in optical detectors, sources, and components, coupled with important advances in the understanding of the interaction between light and tissue. Near-infrared spectra are sensitive to several important physiological components in tissue such as oxyhemoglobin, deoxyhemoglobin, water, and lipids. In the clinical management of breast disease, such functional information suggests a variety of potential medical applications: therapeutic monitoring (angiogenesis, chemotherapy), supplemental lesion characterization (benign vs. malignant), and risk assessment (origins of mammographic breast density). A noninvasive optical imaging technique that provides unique, quantitative physiological information can greatly enhance current screening and diagnostic monitoring for the breast.

Results and Discussion

Many groups have demonstrated that increasing the spectral bandwidth of optical mammography improves tissue functional characterization (Pera et al., 2003; Pifferi et al., 2003; Srinivasan et al., 2003). An example of this is illustrated by point broadband DOS measurements acquired on a 54-year-old human subject with an ≈2-cm-diameteradenocar cinoma ≈1 cm beneath the breast surface (fig. 12.16). Measurements were performed according to University of California at Irvine–approved human subjects protocol 95–563. The tumor site, which represents an average of both normal and diseased tissue, displays increased absorption in the 650- to 850-nm spectral range, corresponding to higher THC. The pronounced peak at 750 nm represents an increase in deoxygenated hemoglobin in the tumor tissue. Additional spectral features are present at 920 nm (lipids) and 980 nm (water). The lipid-to-water ratio is substantially greater for normal tissue, indicating a significant increase in tumorwatercontent.   The components of the tissue may be quantified by fitting these absorption spectra to the assumed known basis spectra (oxyhemoglobin, deoxyhemoglobin, water, and lipids). Overall, the spectral differences between normal and tumor are manifestations of multiple physiological changes associated with increased vascularization, oxygen consumption, cellularity, and edema in the tumor-containing tissue.

However, a practical concern exists. Although the broadband source permitted a detailed tissue functional profile, it is difficult to generate rapid breast images using our current broadband DOS method as a result of (1) slow acquisition times and high number of wavelengths required by FDPM, and (2) long integration times required by steady-state spectroscopy. Technical improvements can significantly reduce FDPM acquisition time, but problem 2 is a signal-to-noise limitation that can only be addressed by better detectors and brighter sources.

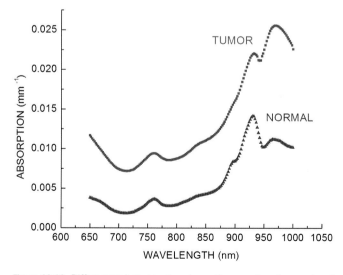

Figure 12.16 Differences between the absorption spectra of normal and tumor-containing breast tissue.

We can minimize the impact of these limitations by using features of broadband DOS to preserve the spectral content, yet decrease measurement time. Figure 12.17 presents measurements of a series of linear point broadband DOS measurements of 24 normal breasts (12 right, 12 left). Each point in figure 12.17 represents a population average of the same data, but has been processed in different ways. Error bars, which represent population variance, are not included because they do not affect the subsequent analysis. The first method (FDPM) uses nine discrete FDPM wavelengths (660, 685, 786, 809, 822, 852, 911, 946, and 973 nm). The second method (SSALL) uses the broadband DOS technique: the same

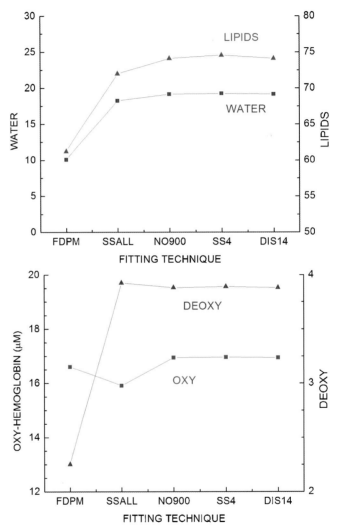

Figure 12.17 Comparison between fitting techniques using both breasts in a population of 12 normal subjects. Each point represents a population average. The FDPM method uses only a nine-wavelength FDPM fit, whereas (SSALL) adds to this the broadband DOS spectra. NO900 is similar to SSALL, but does not use FDPM wavelengths above 900 nm. SS4 is similar to NO900, but it uses only four FDPM wavelengths instead of six. Last, DIS14 uses the same four FDPM wavelengths as SS4, and 14 steady-state wavelengths spread throughout the near infrared. Notice that all the techniques using steady-state wavelengths (SSALL, NO900, SS4, and DIS14) yield similar results.

nine laser diodes plus the broadband light source. We see that the data are consistent between the two cases, but the values of deoxyhemoglobin, water, and lipids are noticeably higher in the broadband DOS measurement. Given the results of the phantom study described earlier, we assume these broadband DOS values are a better characterization of the tissue composition than FDPM alone.

Results similar to SSALL are found by using only the six laserdiodes below 900 nm along with the broadband source (NO900). The results are also seemingly unchanged by using just four common laserdiode wavelengths (660, 685, 786, and 822 nm) along with the broadband source (SS4). These findings are significant because diodes above 900 nm, in our experience, are typically difficult to modulate, sometimes producing inconsistent results because of a lack of modulation depth. Notice that problems with these FDPM wavelengths were minimized by the inclusion of the broadband source in SSALL. By using only laser diodes that are easily modulated, we can drastically improve the signal-to-noise ratio of the system. In addition, we can restrict the modulated laser diodes to wavelengths at which scattering and absorption are easily separated (unlike 980 nm), an approach that has been discussed in a different light by Corlu et al. (2003).

The last case (DIS14) uses only four laserdiode wavelengths (660, 685, 786, and 822 nm) and 14 steady-state discrete wavelengths (634, 652, 686, 762, 812, 822, 832, 852, 872, 882, 902, 920, 952, and 982 nm), which were taken from the broadband measurement. These discrete steady-state wavelengths were selected solely based upon commercial availability as dedicated solid-state sources. On average, the same results are obtained using this "discrete" broadband DOS approach as with any of the previous broadband DOS fitting schemes. This finding is significant because the broadband source may now be replaced by a series of discreet solid-state sources, which can generate far greater amounts of light in the selected wavelength bands than a conventional broadband lamp. Broadband integration times, which ranged anywhere from 1000 to 10,000 ms in this study, could be reduced conservatively by an order of magnitude. Thus, spectral imaging of the breast at a great number of locations becomes a practical endeavor. Image reconstructions using spectral data would only require 14 bands in this case, rather than the 300 we acquired in the full broadband DOS measurements (SSALL, NO900, and SS4).

Summary

Broadband spectroscopy enhances optical mammography by increasing contrast and providing new, physiology-based diagnostic criteria. This improved performance has usually come at great cost because it has been difficult to obtain scatter-corrected data in the 900- to 1000-nm region. However, various approaches can be used to reduce the information content of combined CW and frequency domain measurements while maintaining accuracy. For example, a limited number of discrete-wavelength frequency domain sources in the 650- to 850-nm region can be combined with broadband CW spectra. Similarly, we have seen that using a limited number of steady-state, commercially available, discrete-wavelength laser diodes can provide enough spectral content to maintain the tissue characterization capability of broadband DOS. These observations provide a technical framework that balances the requirements of imaging and spectroscopy in thick tissues.

ACKNOWLEDGMENTS

We acknowledge the National Institutes of Health (LAMMP: P41-RR01192 and NTROI: U54-CA105480), the California Breast Cancer Research Program, the Air Force Office of Scientific Research (MFEL: F49620–00–1–0371), the Beckman Foundation, and the University of California at Irvine Medical Scientist Training Program (D.J.) for support of this research. Use of University of California at Irvine clinical facilities and support in the Chao Family Comprehensive Cancer Center, the General Clinical Resource Center, and the Laser Surgery Clinic is gratefully acknowledged.

REFERENCES

Adcock, L. M., L. S. Wafelman, S. Hegemier, A. A. Moise, M. E. Speer, C. F. Contant, and J. Goddard-Finegold. Neonatal intensive care applications of near-infrared spectroscopy. *Clin Perinatol* 26, no. 4 (1999): 893–903, ix.

Benaron, D. A., S. R. Hintz, A. Villringer, D. Boas, A. Kleinschmidt, J. Frahm, C. Hirth, H. Obrig, J. C. van Houten, E. L. Kermit, W. F. Cheong, and D. K. Stevenson. Noninvasive functional imaging of human brain using light. *J Cereb Blood Flow Metab* 20, no. 3 (2000): 469–477.

Bevilacqua, F., A. J. Berger, A. E. Cerussi, D. Jakubowski, and B. J. Tromberg. Broadband absorption spectroscopy in turbid media by combined frequency-domain and steady-state methods. *Appl Opt* 39, no. 34 (2000): 6498–6507.

Cerussi, A. E., A. J. Berger, F. Bevilacqua, N. Shah, D. Jakubowski, J. Butler, R. F. Holcombe, and B. J. Tromberg. Sources of absorption and scattering contrast for near-infrared optical mammography. *Acad Radiol* 8, no. 3 (2001): 211–218.

Cooper, C. E., and C. Angus. Blood volume changes are controlled centrally not locally: A near-infrared spectroscopy study of one legged aerobic exercise. *Adv Exp Biol* 530 (2003): 627–635.

Corlu, A., T. Durduran, R. Choe, M. Schweiger, E. M. C. Hillman, S. R. Arridge, and A. G. Yodh. Uniqueness and wavelength optimization in continuous-wave multispectral diffuse optical tomography. *Opt Lett* 28, no. 23 (2003): 2339–2341.

Cubeddu, R., C. D'Andrea, A. Pifferi, P. Taroni, A. Torricelli, and G. Valentini. Effects of the menstrual cycle on the red and near-infrared optical properties of the human breast. *Photochem Photobiol* 72, no. 3 (2000): 383–391.

Fantini, S., D. Hueber, M. A. Franceschini, E. Gratton, W. Rosenfeld, P. G. Stubblefield, D. Maulik, and M. R. Stankovic. Non-invasive optical monitoring of the newborn piglet brain using continuous-wave and frequency-domain spectroscopy. *Phys Med Biol* 44, no. 6 (1999): 1543–1563.

Fantini, S., S. A. Walker, M. A. Franceschini, M. Kaschke, P. M. Schlag, and K. T. Moesta. Assessment of the size, position, and optical properties of breast tumors in vivo by noninvasive optical methods. *Appl Opt* 37, no. 10 (1998): 1982–1989.

Farrell, T. J., M. S. Patterson, and B. Wilson. A diffusion theory model of spatially resolved, steady-state diffuse reflectance for the noninvasive determination of tissue optical properties in vivo. *Med Phys* 19, no. 4 (1992): 879–888.

Fishkin, J. B. *Imaging and Spectroscopy of Tissue-Like Phantoms Using Photon Density Waves: Theory and Experiments.* University of Illinois, Urbana-Champaign, 1994.

Flock, S. T., B. C. Jacques, B. C. Wilson, W. M. Star, and M. J. C. van Gemert. Optical properties of intralipid: A phantom medium forlight propagation studies. *Lasers Surg Med* 12, no. 5 (1992): 510–519.

Franceschini, M. A., K. T. Moesta, S. Fantini, G. Gaida, E. Gratton, H. Jess, W. M. Mantulin, M. Seeber, P. M. Schlag, and M. Kaschke. Identification and quantification of intrinsic optical contrast for near-infrared mammography. *Proc Natl Acad Sci USA* 94 (1997): 6468–6473.

Giannotti, G., S. M. Cohn, M. Brown, J. E. Varela, M. G. McKenney, and J. A. Wiseberg. Utility of near-infrared spectroscopy in the diagnosis of lower extremity compartment syndrome. *J Trauma* 48, no. 3 (2000): 396–399; discussion, 399–401.

Graaff, R., J. G. Aarnoudse, J. R. Zijp, P. M. A. Sloot, F. F. M. Demul, J. Greve, and M. H. Koelink. Reduced light-scattering properties for mixtures of spherical particles: A simple approximation derived from Mie calculations. *Appl Opt* 31, no. 10 (1992): 1370–1376.

Haskell, R. C., L. O. Svaasand, T. Tsay, M. S. Feng, M. S. McAdams, and B. J. Tromberg. Boundary conditions for the diffusion equation in radiative transfer. *J Opt Soc Am A* 11, no. 10 (1994): 2727–2741.

Ishimaru, A. *Wave Propagation and Scattering in Random Media.* Vol. 1. Academic Press, New York, 1978.

Miura, H., K. McCully, and B. Chance. Application of multiple NIRS imaging device to the exercising muscle metabolism. *Spectroscopy* 17 (2003): 549–558.

Moesta, K. T., S. Fantini, H. Jess, M. A. Totkas, M. A. Franceschini, M. Kaschke, and P. M. Schlag. Contrast features of breast cancer in frequency-domain laser scanning mammography. *J Biomed Opt* 3, no. 2 (1998): 129–136.

Mourant, J. R., J. P. Freyer, A. H. Hielscher, A. A. Eick, D. Shen, and T. M. Johnson. Mechanisms of light scattering from biological cells relevant to noninvasive optical-tissue diagnostics. *Appl Opt* 37, no. 16 (1998): 3586–3593.

Mourant, J. R., T. Fuselier, J. Boyer, T. M. Johnson, and I. J. Bigio. Predictions and measurements of scattering and absorption over broad wavelength ranges in tissue phantoms. *Appl Opt* 36, no. 4 (1997): 949–957.

Patterson, M. S., B. Chance, and B. C. Wilson. Time resolved reflectance and transmittance for the non-invasive measurement of tissue optical properties. *Appl Opt* 28, no. 12 (1989): 2331–2336.

Pera, V. E., E. L. Heffer, H. Siebold, O. Schutz, S. Heywang-Kobrunner, L. Gotz, A. Heinig, and S. Fantini. Spatial second-derivative image processing: An application to optical mammography to enhance the detection of breast tumors. *J Biomed Opt* 8, no. 3 (2003): 517–524.

Pham, T. H., O. Coquoz, J. B. Fishkin, E. Anderson, and B. J. Tromberg. Broad bandwidth frequency domain instrument forquantitative tissue optical spectroscopy. *Rev Sci Instrum* 71, no. 6 (2000): 2500–2513.

Pifferi, A., P. Taroni, A. Torricelli, F. Messina, R. Cubeddu, and G. Danesini. Four-wavelength time-resolved optical mammography in the 680–980-nm range. *Opt Lett* 28, no. 13 (2003): 1138–1140.

Quaresima, V., S. Homma, K. Azuma, S. Shimizu, F. Chiarotti, M. Ferrari, and A. Kagaya. Calf and shin muscle oxygenation patterns and femoral artery blood flow during dynamic plantar flexion exercised in humans. *Eur J Appl Physiol* 84 (2000): 387–394.

Springett, R., M. Wylezinska, E. B. Cady, M. Cope, and D. T. Delpy. Oxygen dependency of cerebral oxidative phosphorylation in newborn piglets. *J Cereb Blood Flow Metab* 20, no. 2 (2000): 280–289.

Srinivasan, S., B. W. Pogue, S. Jiang, H. Dehghani, C. Kogel, S. Soho, J. J. Gibson, T. D. Tosteson, S. P. Poplack, and K. D. Paulsen. Interpreting hemoglobin and water concentration, oxygen saturation, and scattering measured in vivo by near-infrared breast tomography. *Proc Natl Acad Sci USA* 100, no. 21 (2003): 12349–12354.

Torricellli, A., V. Quaresima, A. Pifferi, G. Biscotti, L. Spinelli, P. Taroni, M. Ferrari, and R. Cubeddu. Mapping of calf muscle oxygenation and haemoglobin content during dynamic plantar flexion exercise by multi-channel time-resolved near-infrared spectroscopy. *Phys Med Biol* 49 (2004): 685–699.

Tromberg, B. J., N. Shah, R. Lanning, A. Cerussi, J. Espinoza, T. Pham, L. Svaasand, and J. Butler. Non-invasive in vivo characterization of breast tumors using photon migration spectroscopy. *Neoplasia* 2, no. 1–2 (2000): 26–40.

van Staveren, H., C. Moes, J. van Marle, S. Prahl, and M. van Gemert. Light scattering in Intralipid-10% in the wavelength range of 400–1100 nm. *Appl Opt* 30, no. 31 (1991): 4507–4514.

Wariar, R., J. N. Gaffke, R. G. Haller, and L. A. Bertocci. A modular NIRS system for clinical measurement of impaired skeletal muscle oxygenation. *J Appl Physiol* 88, no. 1 (2000): 315–325.

Zonios, G., L. T. Perelman, V. M. Backman, R. Manoharan, M. Fitzmaurice, J. Van Dam, and M. S. Feld. Diffuse reflectance spectroscopy of human adenomatous colon polyps in vivo. *Appl Opt* 38, no. 31 (1999): 6628–6637.

13
Detection of Brain Activity by Near-Infrared Light

ENRICO GRATTON, VLAD TORONOV,
URSULA WOLF, AND MARTIN WOLF

There is a series of outstanding questions regarding the detection of brain activity in measurements made at the head surface and regarding the origin of the observed changes in the optical parameters of tissues. In this chapter we review the information generated so far and discuss the evidence available about the origin of the effects observed. Because the field is still controversial and rapidly advancing, we mainly focus on our own work and a few otherstudies that we chose to illustrate ouropinions.

Numerous studies show that neuronal activity can be detected optically. Basically, there are two lines of evidence: one based on optical studies of exposed cortex and another based on measurements through the skull. Although the measurements on exposed cortex directly visualize areas of neuronal activation, the spatial resolution that can be achieved by imaging through the skull is insufficient to delineate the columnar areas of cortical activity. It is also likely that the optical contrast mechanism could be different for the two types of approaches resulting from the different ways light interrogates the tissue. In the direct exposed cortex measurement, the contrast is given by the changes of reflectivity, resulting from either changes in absorption or scattering of the brain surface. In the through-the-skull measurements, we must make some hypothesis about the path that the photons follow through the bone and through the brain and back out. Most of the controversy about the noninvasive optical measurements arises from the particular model used to describe light propagation throughout the head. Although we have quite an accurate theory to describe light propagation in multiple scattering media, the problem of applying this theory to the intact head is that we need to know the details of local anatomy and details of the optical parameters of each tissue element. Probably the most successful approach has been that of considering only average tissue properties, segmenting different tissues with a coarse resolution, on the order of centimeters. The theoretical justification for this approach should be sought in the measured values of the mean free path for photons in the different tissues. This parameter strongly depends on the nature of the tissue. Under the hypothesis that we are in the multiple scattering regime, we can use the value of the reduced scattering coefficient as an indicatorof the mean free path and of the length overwhich we can average the optical properties. For tissues in the head, this value is typically on the order of millimeters. This consideration has led to the hypothesis that we can roughly divide the various tissues in three major segments: the superficial skin layer, the skull, and the brain surface. Clearly, this is an oversimplification, and a lot of effort was spent in validating this assumption.

An important issue is that there are membranes around the brain in which the diffusion regime could hardly be achieved.

There are two different classes of measurements that one can perform on the brain. One attempts to measure the absolute optical properties of the brain tissue. A second approach is only concerned with measuring the variation of optical parameters. Of course, to measure the variations of optical parameters, the knowledge of the absolute value is less crucial. It can be shown, based on a perturbation approach, that the variations of optical parameters can be recovered much more accurately that the absolute values. For most of the brain functional studies, this is the approach that is normally being followed. However, for clinical applications, it is of great interest to determine accurately the absolute values of the tissue's optical parameters.

To determine the absolute values of the optical parameters, there is a general consensus among researchers that it is necessary to measure not only light attenuation, but also the time of flight of the photons in the tissue to recover accurately the absolute values of the optical parameters. This can be achieved by using two alternative methods: the time domain method and the frequency domain method. There are substantial technical differences between the two methods, but from a conceptual point of view, they provide equivalent information, being related through the Fourier transform. In this chapter we discuss the frequency domain approach, because this is the method we use in our lab.

Experiments on exposed cortex have shown that neuronal activity produces measurable changes of optical parameters. For example, in 1973, Cohen showed that the amount of light scattered from a slice of brain tissue rapidly changes after stimulation. Also, reflectivity changes in the near infrared have been measured on exposed cortex (Grinvald et al., 1986; Frostig et al., 1990). The question is: *Can changes be observed through the skull?* Only recently, noninvasive observations of fast changes in optical parameters were reported (Gratton et al., 1994, 1995, 2000). In these pioneering experiments, near-infrared light was brought to the external surface of the head using an optical fiber, and the reflected light was measured at a distance of about 3 to 4 cm using a collection optical fiber. In this configuration, the light must travel deep (1–1.5 cm) into the head to reach the detector fiber. We have shown with experiments on phantoms and with simulations that changes in the optical properties of a small region, and at a depth of about 1.5 cm from the head surface, can affect the amount of light collected. These experiments showed that, in principle, it is possible to measure changes in optical parameters occurring in a deep region. The question that remains is whether we have sufficient sensitivity to measure changes resulting from brain activity and whether the phantoms used to demonstrate sensitivity are realistic models of the brain morphology and optical parameters.

FREQUENCY DOMAIN METHOD TO MEASURE OPTICAL PROPERTIES OF TURBID MEDIA

In 1991, Chance, Patterson, and Wilson (Patterson et al., 1991a,b) showed that the optical parameters of a turbid medium can be obtained from time-resolved measurements of short light pulses propagating in the medium. A fit of the intensity as a function of time, measured at some distance from the source, provides the values of both the absorption and the reduced scattering coefficients of the medium. During the same period, our group developed a method of measurement using intensity-modulated light sources. With this method, the light intensity is modulated at high frequency (in the 100-MHz range), and the DC (mean intensity), AC (modulated amplitude), and phase of the detected light are measured at a distance from the source after the light has traveled in the turbid medium. Because the light pulse used in the first method can be considered the sum of the harmonic signals, both methods are mathematically related by means of the Fourier transform. However, in practice, the frequency domain method has a better sensitivity and is faster than the time domain method. The frequency domain method was adopted by many investigators, including Chance, Patterson, and Wilson (Patterson et al., 1991a,b; O'Leary et al., 1992; Boas et al., 1993; Cui and Ostrander, 1993; Duncan et al., 1993; Hoshi et al. 1993a,b; Kaltenbach and Kaschke, 1993; Maier, et al 1993; Tromberg et al., 1993; Boas et al., 1994). The speed and precision of the frequency domain method are crucial for the detection and measurement of brain activity.

The theory of light propagation in tissue has been extensively described in the context of the transport theory (Case and Zweifel, 1967; Ishimaru, 1978). In the range of source–detectordistances used in ourexper iments (1–5 cm), the physical model based on the diffusion approximation to the Boltzmann transport equation is adequate to describe the propagation of light in highly scattering materials such as tissues. The diffusion approximation is based on using the photon density $U(r, t)$, that has units of photons percubic centimeter and can be interpreted as a quantity proportional to the probability of finding a photon at a distance r from the source at a time t. It was shown (see, forexample, Ishimaru, 1978) that at a large enough distance from the light source and at not very high frequencies (<1 GHz), the photon density obeys the diffusion equation

$$v\mu_a U(r,t) + \frac{\partial U(r,t)}{\partial t} - vD\Delta U(r,t) = Q(r,t) \tag{13.1}$$

where $D = (3\mu_a + 3\mu_s')^{-1}$ is the diffusion coefficient in units of square centimeters per second, μ_s' is the reduced scattering coefficient per centimeter, μ_a is the absorption coefficient percentimeter , v is the velocity of light in the medium in centimeters per second, and r is the distance from the source in centimeters. The source term $S(r, t)$ on the right side of the equation represents a sinusoidal intensity-modulated source at an angular frequency ω. The solution of this equation in an infinite homogeneous medium is

$$-\ln(rU_{DC}) = r\left(\frac{\mu_a}{D}\right)^{\frac{1}{2}} + \ln\left(\frac{Q_0}{4\pi vD}\right) \tag{13.2}$$

$$-\ln(rU_{AC}) = r\left(\frac{v^2\mu_a^2 + \omega^2}{v^2 D^2}\right)^{\frac{1}{4}} \cos\left[\frac{1}{2}\arctan\left(\frac{\omega}{v\mu_a}\right)\right] + \ln\left(\frac{Q_0 A}{4\pi vD}\right) \tag{13.3}$$

$$\phi = r\left(\frac{v^2\mu_a^2 + \omega^2}{v^2 D^2}\right)^{\frac{1}{4}} \sin\left[\frac{1}{2}\arctan\left(\frac{\omega}{v\mu_a}\right)\right] + \varepsilon \tag{13.4}$$

This solution represents a spherical photon density wave that has a constant (DC) component (eq. 13.2) and the oscillatory component (AC) spreading from the light source. Equations 13.3 and 13.4 present the phase and the amplitude of the AC component, respectively. Note that at the frequency of about 100 MHz, the difference between the spatial profiles of AC and DC, is very small. A similar but somewhat more complicated solution was obtained for a semi-infinite medium (Fantini et al., 1994a). Both solutions are very close at distances largerthan 1 cm. Equations 13.2 through 13.4, ortheiranalogues fora semi-infinite medium, can be used for measuring absorption and scattering coefficients of homogeneous media.

Note that equations 13.2 through 13.4 contain the reduced scattering and absorption coefficients, but also some otherunknowns: Q_o, A, and ε, which represent the strength, the modulation, and the phase of the light source, respectively. Unless the source constants are known precisely, they need to be measured. In our work, after the analysis of several possibilities, we have found it more convenient to measure the values of the phase and AC at several distances from the source (Fantini et al., 1994a). We then used only the slope of the plots of phase, $\ln(rU_{DC})$ and $\ln(rU_{AC})$, as a function of the distance to obtain the values of the reduced scattering and absorption coefficients. The details of the method are described in Fantini et al. (1995). The method provides separation of the scattering from the absorption with negligibly small cross-talk.

To implement this idea, an instrument was built that operates simultaneously at two different wavelengths: 758 nm and 830 nm (ISS, Champaign, Illinois). At these wavelengths, the major contributions to absorption in human tissue are the result of deoxyhemoglobin and oxyhemoglobin, respectively. Using this instrument, the light from laser diodes is RF modulated at 110 MHz. Heterodyning at the detectors, using well-established techniques (Alcala et al., 1985; Feddersen et al., 1989), converts this RF frequency to 5 KHz. An analogue-to-digital (A-to-D) converter card in a computer slot simultaneously samples the signal of the four detectors at 160 KHz. A computer program processes the stream of digitized data in real time. Exactly eight points are digitized per each period of the 5-KHz frequency. After a preset number of periods of the 5-KHz frequency has been acquired, an electronic multiplexer circuit turns on the next light source. The multiplexing of the light sources proceeds indefinitely. The fast Fourier transform routine performs the calculation of the DC, AC, and phase of the signal for all four detectors. Because the A-to-D sampling is synchronous with the 5-KHz frequency, the phase of each detected signal has a well-defined relationship with the phase of the light source. The result of data acquisition and preprocessing is a stream of DC, AC, and phase values at a rate as high as 200 Hz per light source, because the data from the four channels are interlaced in our card. In principle, data acquisition can be faster if one uses a larger value of the heterodyning frequency. The obtained values of AC and phase are then used to calculate the optical properties of the tissue. Theoretically, one can use either AC or DC. However, in practice, AC is preferable because, unlike DC, it is not sensitive to ambient light. Because absorption can be measured free of the scattering contribution, an absolute estimation of hemoglobin oxygen saturation and concentration is easily achieved.

NONINVASIVE MEASUREMENT OF BRAIN ACTIVITY: FAST AND SLOW EFFECTS

Several authors have shown that functional changes in brain tissue associated with neuronal activity can be detected by optical methods based on measures of propagation of diffusive light (Chance et al., 1993; Hoshi and Tamura, 1993a,b; Gratton et al., 1995a,b; Maki et al., 1995; Meek et al., 1995; Obrig et al., 1996; Villringer and Chance, 1997). Two types of changes have been reported: *fast effects* (with a 50–500-ms latency) and *slow effects* (with a 2–10-s latency). These two effects probably have different origins that can be distinguished because of their large difference in the timescale. Fast effects are presumably directly related to changes in neuronal tissue. Slow effects have a hemodynamic origin.

Slow and fast effects are measured in different ways. Slow changes in the tissue parameters are observed using a sustained activation task, such as one involving repeated stimulation (or repeated movements). Such an approach is used in optical functional magnetic resonance imaging (Belliveau et al., 1991) and PET studies. Fast effects are detected by time-locked averaging of the changes elicited by many individual stimulations. Apart from studies using optical methods, this approach is also used in electrically recorded potentials and magnetic encephalographic experiments.

Fast and slow effects differ in terms of the optical parameters they affect. Slow effects can be observed using both steady-state and time-resolved methods (Hoshi and Tamura, 1993a,b; Gratton et al., 1995b; Maki et al., 1995; Obrig et al., 1996). In contrast, fast effects are best visualized using time-resolved methods (Gratton et al., 1995a,b; Meek et al., 1995). To a large extent, the slow effects can be attributed to changes in absorption that are mostly produced by oxygenation and hemodynamic changes that occur in active areas of the brain

(Villringer and Dirnagl, 1995; Obrig et al., 1996). They may also be related to the effects observed with PET studies of blood flow.

OPTICAL DETECTION OF NEURONAL ACTIVITY

During the early stage of our studies on the optical detection of fast neuronal effects, to investigate the ability of the optical method to detect fast changes in the brain, we performed experiments using modulated light with a wavelength of 715 nm and a power of less than 1 mW. The source–detector distance was fixed at 3 cm (Gratton et al., 1995b). The phase and the attenuation of the photon density wave were acquired using sampling rates varying between 12.5 and 50 Hz in different experiments. The measurements were obtained using a single-channel system (maps were obtained by repeating measurements at multiple locations). The pulsation of the vascular system, and in particular of large and medium arteries, which is associated with systolic activity of the heart, produces relatively large changes in the transmission of light through tissue (Jennings and Choi, 1983). The signal resulting from pulsation can be quite significant, and is exploited in studies of hemodynamic effects related to brain activity. However, when the interest is in detecting neuronal activity, pulsation may produce substantial artifacts that are difficult to eliminate with simple bandpass filtering approaches. It is, therefore, important to separate the changes in intensity resulting from hemodynamics from the changes resulting from neuronal activity. An algorithm was developed to estimate the pulse artifact (Gratton and Corballis, 1995) based on regression procedures in which the effect of systolic pulsation is estimated from the data. In short, it extracts the mean shape of the pulse by screening each trace separately for pulses. The period of each pulse is adjusted before the averaging. This mean shape corresponds to the best estimate of each pulse and is used to remove each pulse from the data. This requires adjusting the period of the mean shape to the one of each pulse and scaling the shape by linear regression. Tests showed that, when used to estimate and subtract the effects of pulsation from the optical recordings, the procedure substantially reduces the impact of the pulsation artifact.

One of the major problems in the optical detection of neuronal signals is that signals measured using a single source–detector pair are mostly sensitive to the changes occurring very closely to the source or the detector, whereas in the case of the brain, the region of interest is situated about 1.5 cm below the surface. Therefore, special measures should be taken to increase the sensitivity of the optical device to the deep changes and to reduce the effect of biological noise near the surface. Using either equation 13.2 or 13.3, one can plot the density of the photon wave as a function of the distance from the source (note that at a frequency of 110 MHz, both equations provide very similar spatial profiles that are valid both for infinite and semi-infinite media at distances larger than 5 mm). In the following description, we assume that the detector is placed on the surface of the tissue at a distance d from the source. We define the region of sensitivity (ROS) of such a source–detector pair as that region of space that is visited by the photons emitted by the source and collected by the detector. The ROS can be obtained by calculating for each volume element P the product of the probability of a photon emitted by the source to arrive at P times the probability that the detector will measure a photon at P. This is known as the equation of the *light bundle*. The shape of the light bundle depends on the boundary conditions, tissue optical parameters, and, in the frequency domain, on the frequency of light modulation. By considering the effect caused by a small particle at position P on the AC and phase, we can also define the ROS for the AC and phase signal. To this end, we identify with the AC_0 (ϕ_0) and AC_P (ϕ_P), the AC (and phase) measured without the particle and with the particle, respectively. The AC and phase ROS bundles are obtained by plotting the following fractions: $1 - AC_p/AC_0$ and $1 - \phi_p/\phi_0$, respectively. Cross-sections of the ROS of AC and phase bundles by the plane passing through the source–detector line perpendicularly to the surface are shown in figure 13.1. Coordinate x corresponds to the direction along the surface, and coordinate y corresponds to the direction perpendicular to the surface. As seen in figure 13.1A, the AC bundle of the single source–detectorpairhas the highest sensitivity in the region just below the source and the detector. At the same time, the ROS is relatively weak midway between the source and the detector. The phase ROS (fig. 13.1B) extends deeper into the tissue, but it is still quite sensitive in the regions close to the source and the detector. One should note that the phase ROS also depends strongly on whether the signal change arises from changes in the absorption or scattering. We have confirmed these properties of the AC and phase ROS experimentally. For this we inserted a small absorbing object in the medium and measured the difference in the signal (DC, AC, and phase) caused by the introduction of the object (1-mm absorbing sphere) as we moved the object throughout the medium.

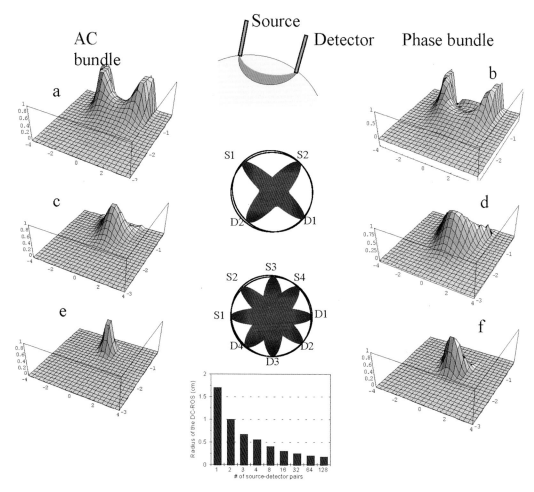

Figure 13.1 (A–F) the AC and phase ROS for different source–detector combinations. The light sources and detectors are on the skin surface. (A, B) The AC and phase ROS for a single source–detector pair. The coordinates x (along the source–detector line) and y (perpendicular to the surface) X are measured in centimeters. The source–detector distance is 4 cm. Note the relatively large density near the surface in A and B. Calculations were done for conditions of 110 MHz, $\mu_a = 0.1/cm$ and $\mu_s' = 10/cm$. The perturbation was an absorbing sphere 1 mm in radius. Locations of individual data points correspond to the positions of the particle in the ROS. The ROS of the pi-sensor are shown in views C through F for the AC and phase ROS (views C and D for two bundles and views E and F for eight bundles). Note that the phase ROS has maximum sensitivity at a deeper position than the AC ROS (compare E and F). The pi-detector geometries for the cases of two and eight source–detector pairs are presented in the center panels.

The measured changes in the AC and phase signals as a function of the object coordinates were in excellent agreement with the ROS shown in figures 13.1A and 13.1B.

To reduce the sensitivity of the optical device to changes in the superficial regions under the source or detector fiber s and to make it more sensitive to changes in the deep region in between, we developed the concept of the *pi-sensor*. In the following we use the term *sensor* to indicate a combination of light sources and detectors that can measure localized changes in the optical parameters of the medium. The basic idea of the pi-sensor is the simultaneous use of two or more source–detector pairs that produce light bundles in a cross-configuration as shown in the center panels of figure 13.1. The electronics and the signal processing are designed to be sensitive only to changes that occur simultaneously in all source–detector pairs. A simple way to obtain such a "correlator" is to cross-correlate numerically the changes measured at different source–detector channels. (We derived the term *pi-sensor* from the type of data processing used to produce the sensor output.) The ROS of the pi-sensor composed by two source–detector pairs is shown in figure 13.1C and 13.1D for the AC and phase signals, respectively. Note that in the AC ROS of the pi-sensor, the sensitivity is larger in the region under the center of the cross rather than immediately under the source or the light detector (compar e fig. 13.1C with fig. 13.1A). An even larger sensitivity to changes at the central region can be achieved by adding more source–detector pairs centered on a common point (fig. 13.1E, F). The AC ROS of the pi-sensor has an average radius of about 3 mm when eight source–detector pairs are used in the semi-infinite geometry. The lower panel of figure 13.1 shows how the radius (at half maximum) of the AC ROS for the semi-infinite medium changes as the number of source–detector pair s is increased. (The ROS has no spherical symmetry. By radius of the ROS we mean the average radius.)

We used the pi-sensor together with a low-noise frequency domain near-infrared spectroscopy instrument, and highly effective filtering algorithms to detect the fast neuronal signals during motor stimulation and to determine the colocation with the slow hemodynamic changes. The protocol was approved by the institutional review board of the University at Urbana-Champaign (IRB no. 94125). Written informed consent was obtained from all subjects prior to the measurements.

The light of four laser diodes at either 758 or 830 nm was combined to achieve a higher light intensity at each source location. To distinguish the light from the four source locations, the laser diodes were multiplexed. The detector fibers had a 3-mm core diameter. The sample rate was 96 Hz. The sensor had four sour ces and four detector s (photo multiplier tube [PMT]) arranged equidistantly along the circular arc of a circle of 3.2 cm in diameter (in total, 16 light bundles) and was placed above the motor cor tex (C3 position). The tapping frequency was set at 2.5 times the heart rate. There was a sequence of alternating periods of 20 s of tapping and 20 s of rest for 5 minutes. In each subject, five measurements were carried out with tapping of different fingers (index, middle, ring, pinkie, all fingers). To provide the synchronization between the stimulus and the optical data, we recorded the tapping signals (the S signals) synchronously with the optical data using the auxiliary input of the data acquisition system. These were the signals resulting from the contact between the metal caps put on the tapping fingertip and the thumb.

To reduce physiological noise from the arterial pulsatility, the AC and phase data were filtered using an adaptive filter (Gratton et al., 1995b). Apart from this, the data were "detrended" by a digital high-pass filter with a cut-off frequency equal to 2.2 times the mean heart rate. This frequency was selected to reduce further the effect of the hemodynamic pulsatility, but still be low enough not to affect the fast neuronal signal at 2.5 times the mean heart rate.

A cross-correlation function (or CCF, as used in the following equation) of the time lag τ between the optical data and the stimulus (S) signal was obtained as well as the cross-correlation function between the optical signals in different source–detector channels of the pi-sensor (pi-correlation). To calculate the cross-correlation function between two signals $f(t)$ and $g(t)$, we used the equation

$$CCF(\tau) = \frac{\langle f(t)g(t+\tau)\rangle}{\langle f\rangle\langle g\rangle} - 1$$

To determine how the level of noise depended on the number of data samples, the S signal was cross-correlated with the optical signals measured on a phantom block. To verify that the observed correlation was not the result of an artifact, we compared the correlation for the data obtained during tapping (after filtering to the noise) with the cross-correlation function for the data acquired on the phantom. The level of random noise reduced as $1/N^{1/2}$, where N was the number of data points. In contrast, if the optical signal contained components that were correlated with the stimulus, this genuine signal would keep its amplitude independent of the numberof data points. Using the measurements on the phantom block, we determined the factor of the noise reduction as a function of N. We also verified that this factorwas the same forphysiological measurements without stimulation.

Filtering the heartbeat was crucial for the detection of the fast signals. Figure 13.2A and 13.2B show the cross-correlation function for the simultaneously recorded stimulus and the unfiltered AC and phase signals, respectively. Comparing these cross-correlation functions with those for the filtered data (black bold curves in fig. 13.2C, D), one can see that the curves corresponding to unfiltered data exhibit no correlation. The black bold curves in figure 13.2C and 13.2D correspond to data acquired at a source–detector distance of 3.2 cm. Figure 13.2C and 13.2D also show the cross-correlation functions for the filtered signals acquired at a 1.3-cm source–detector distance (red hairline curves). One can see that these short-distance signals exhibit a much lower correlation than the signals acquired at the larger distance.

Note that the period of the cross-correlation functions shown in figure 13.2C and 13.2D by the black bold curves is very close to the period of tapping presented using the metronome (about 3 s). This indicates that the correlation is not spurious, but is the result of the coherence between the optical signals and the tapping. Using the Fourier transform, we verified that the mean tapping frequency corresponded to the one of the metronome. However, we found that the tapping signals recorded during different experiments (with the same or different subjects) were uncorrelated. To test whether the cross-correlation technique is sensitive only to the unique tapping signal corresponding to the given data set, we obtained cross-correlation functions between the optical and stimulus signals recorded in different experiments (with the same subject). Such cross-correlation functions for the "wrong" stimulus signal are shown figure 13.2E and 13.2F for AC and phase, respectively. One can see that the cross-correlation functions corresponding to the wrong tapping signals show significantly lower correlation than the cross-correlation functions corresponding to the tapping signal recorded simultaneously with the optical data (bold curves in fig. 13.2C, D).

Figure 13.2G and 13.2H demonstrates the work of the pi-detector. Fine curves in each figure show the cross-correlation functions of the S signal with the signals in the individual source–detector channels (AC in fig. 13.2G, and phase in fig. 13.2F). The bold curves show cross-correlation functions for signals from different channels. One can see that the pi correlation of two single-channel signals shows the fast signal at the tapping frequency as clearly as the cross-correlation functions for single-channel signals and the S signal.

We found no difference in the localization of the hemodynamic signal corresponding to different fingers. On the contrary, the fast signal was highly localized and clearly showed individual spots of activity (fig. 13.3).

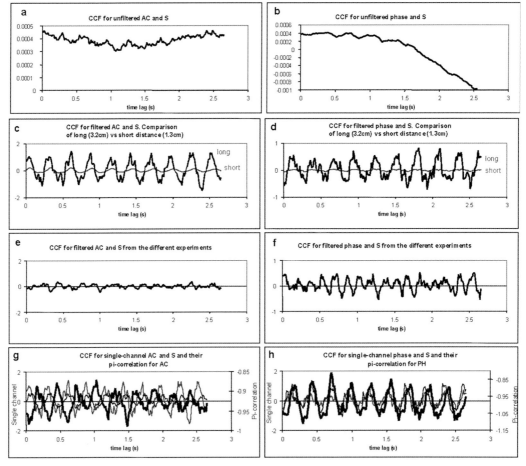

Figure 13.2 (A–H) Cross-correlation analysis for the tapping signal. Total measurement time was 20 minutes. Tapping frequency was approximately 3 Hz. Tapping for 20 s was followed by 20 s of rest. (A, C, E, G) AC signal. (B, D, F, H) Phase signal. Note the improvement in the signal-to-noise ratio after filtering out the pulse (compare view A with view C and view B with view D). Note the different latency of the AC and phase signals (C, D). There is no correlation in case of the wrong tapping signal (E, F). The cross-correlation in views G and H (bold blackcurves) was obtained using two individual bundles in the pi-sensor (thin lines).

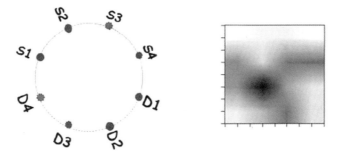

Figure 13.3 (A, B) Back-projection map (A) of the AC cross-correlation of 16 light bundles of the sensor (B) with the finger-tapping signal. The coordinate scales are in centimeters. The resolution of the activation region corresponds to the anticipated resolution using four source–detector pairs.

It is important to outline how we concluded that the signals we observe are "real" and the control experiments we performed.

1. If we do not remove the pulse signal, the cross-correlation function between the raw signal and the tapping channel contains a relatively large contribution resulting from the pulse, but in some cases the neuronal signal can still be distinguished. However, after pulse removal, the signal-to-noise ratio increases dramatically (fig 13.2C, D) and the neuronal activity can be easily detected.

2. If we perform a blank experiment without tapping, there is no cross-correlation signal (of course) and no neuronal activity is detected.

3. If we measure optical signals simultaneously in several locations in the region of the motor cortex, there is a definite spatial dependence of the strength of the cross-correlation signal (fig. 13.3).

4. If we analyze light bundles that are quite superficial (source–detector separation distance, 1.3 cm), the cross-correlation signal is either absent (in the phase signal) or much reduced (by more than a factor of 10) with respect to deeper bundles (fig. 13.2C, D).

5. One possible artifact can result from poor statistics. We analyzed a set of optical data using the S signal from a different experiment (similar in tapping frequency). We found that there was no correlation in this case (fig. 13.2E, F).

6. The measurements were repeated four times on different days and with different tapping frequencies. In all cases, we found very similar results and an excellent signal-to-noise ratio.

INVESTIGATION OF HUMAN BRAIN HEMODYNAMICS BY SIMULTANEOUS NEAR-INFRARED SPECTROSCOPY AND FUNCTIONAL MAGNETIC RESONANCE IMAGING

In recent years, near-infrared spectroscopy has been proposed as a method to study brain hemodynamics, which is simple and inexpensive compared with such "heavy-duty" methods as functional magnetic resonance imaging and PET. Many studies applying near infrared to monitor functional cerebral hemodynamics have been reported (for a review, see Villringer and Chance, 1997). However, the precise location of tissues contributing to the optical signals was still uncertain. This issue could be clarified using coregistration of brain hemodynamics by near-infrared spectroscopy and by functional magnetic resonance imaging, because the blood oxygenation level dependent (BOLD) signal used in functional magnetic resonance imaging is believed to be proportional to the changes in deoxyhemoglobin concentration. We performed a study with the aim of comparing functional cerebral hemodynamic signals measured simultaneously by the optical method and by functional magnetic resonance imaging. The contribution of superficial layers (skin and skull) to the optical signals was also assessed. Both optical and functional magnetic resonance imaging methods were used to generate maps of functional hemodynamics in the motor cortex area during a periodic sequence of stimulation by finger motion.

For near-infrared measurements we used a two-wavelength (758 and 830 nm) frequency domain (110-MHz modulation frequency) oximeter (ISS, Champaign, Illinois), which had 16 laser diodes (eight per each wavelength) and two photomultiplier tube detectors. At a wavelength of 758 nm, light absorption by deoxyhemoglobin substantially exceeded absorption by oxyhemoglobin, whereas at 830 nm oxyhemoglobin absorption prevailed over the deoxyhemoglobin absorption. The laser diodes operated in a sequential multiplexing mode with 10 ms "on" time per each diode. Light emitted by these laser diodes was guided to the

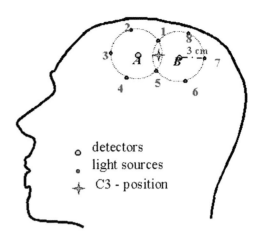

Figure 13.4 Geometric layout of the optical probe and its position on the head.

tissue through the 10-m-long, multimode, silica optical fibers. Two 10-m-long glass fiber bundles collected the scattered light and brought it to the detectors. The paired (758- and 830-nm wavelength) source fibers were attached to the probe at eight positions. Together with two detectors, they provided 10 biwavelength source–detector channels with a source–detector distance of 3 cm. The probe covered an area 9×6 cm^2. The geometric layout of the optical probe and its position on the head are shown in figure 13.4. In this figure, the numbers 1 through 8 correspond to the light sources and the letters A and B correspond to the detectors. Using this notation, we denote the source–detector light channels as A1, A2, and so on through B8. The probe is centered at the measured C3 position according to the International 10–20 System (Jaspers, 1957). Changes in the concentration of oxy-hemoglobin and deoxyhemoglobin were calculated from the changes in the light intensity at two wavelengths using the modified Lambert-Beer law (Villringer and Chance, 1997; Hirth et al., 1997).

Three multimodality radiological markers (IZI Medical Products Corporation, Baltimore, Maryland) were embedded into the optical probe to facilitate correct orientation of the magnetic resonance imaging slices with respect to the probe and to enable recovery of the probe orientation for data analysis.

Magnetic resonance imaging was performed using a 1.5-T whole-body magnetic resonance scanner (Sigma, General Electric Medical Systems, Milwaukee, Wisconsin) equipped with echo-speed gradients and a standard, circularly polarized birdcage head coil. Sagittal T1-weighted localizer scans were used to determine the correct plane for the functional scans. Gradient-echo echo-planar images were acquired using a data matrix of 64×64 complex points, a repetition time of 640 ms, an echo time of 40 ms, a field of view of 240 mm, a slice thickness of 7 mm, no interslice gap, a receiver bandwidth of 62.5 KHz, and a tip angle of 90 deg. Figure 13.5 shows the arrangement of a typical set of five slices denoted as α, β, γ, δ, and ϵ. The slices are parallel to the plane of three radiological markers on the optical probe. The middle slice, γ, was set between the skull and the brain surface at the C3 position. This slice was mostly filled with cerebrospinal fluid, dura mater, arachnoidal tissue, and pia mater, but could also include the cortical tissue depending on the subject's anatomy. Two deeperslices, α and β, mostly contained the brain tissue, whereas two outer slices, δ and ϵ, included the skull, the skin, and the markers. Figure 13.6A through C represents the anatomic images corresponding to the three deepest slices—α, β, and γ—shown in figure 13.5. The numbers 1 through 8 show the source locations and the letters A and B indicate the detectors.

The studies were performed in six healthy right-handed male volunteers, age 18 to 37 years old. Informed consent was obtained from all subjects. Each exercise run consisted

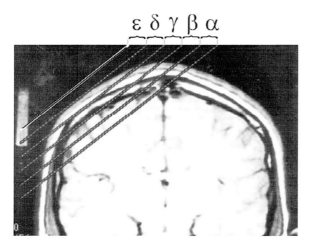

Figure 13.5 Coronal anatomic image of the human head with a set of functional magnetic resonance imaging slices labeled α, β, γ, δ, and ε. The lines show the limits of each slice. Letters L and R indicate orientation and stand for left and right, respectively.

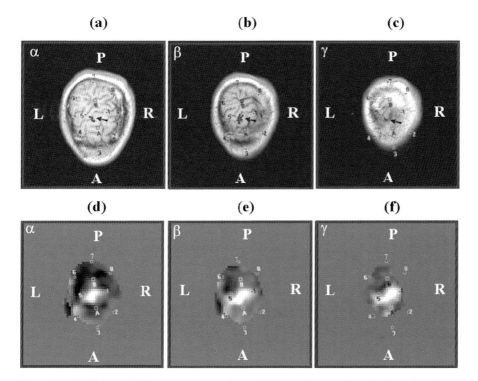

Figure 13.6 (A–C) Anatomic images corresponding to the three deepest slices (α, β, and γ) shown in fig. 13.5. The numbers 1 through 8 correspond to the locations of sources shown by darkdots and the Letters A and B label the detectors depicted as empty squares. (D–F) Correlation coefficient maps of the functional signals for the slices α, β, and γ shown in figure 13.5. A, anterior; L, left; P, posterior; R, right.

of a 30-s preexercise epoch, ten 20-s stimulation epochs, separated by ten 20-s control epochs, and a 50-s postexercise epoch. During stimulation epochs, subjects performed light palm squeezing with the right hand following the rhythm presented by the sound from the vacuum pump of the magnetic resonance imager. The commands to begin and to stop the exercise were presented via a speaker installed in the imaging room and by a light, which was on during the exercise epoch and off during the control epoch. A total of 750 images acquired during each run resulted in an 8-minute imaging time. The synchronization of the exercise sequence with the functional magnetic resonance imaging and optical records was provided by a computer program that generated the commands for the subject and the imaging operator based on the preset command timing.

In all subjects, the analysis of the BOLD signal revealed an area under the optical probe where the signal was highly correlated with the paradigm boxcar function. It was an area in the primary motor cortex with the center close to the central sulcus. The size of this area varied depending on the slice. Figure 13.6D through 13.6F shows the typical correlation coefficient maps of the functional signals for the slices α, β, and γ shown in figure 13.5. White color of the highest intensity corresponds to the voxels, with r greater than .5 (correlation z score > 9.0). Usually, the largest size of the activated area was in the middle slice γ (fig. 13.6F). In the next slice, deeperin the brain (β), which was mostly cortical tissue, the activation area was almost as large as in the middle slice. The deepest slice (α) typically showed a significantly smaller activation area. Using head landmarks for the C3 position, the center of the optical probe was placed very close to the central sulcus so that light channels A1, A5, B1, and B5 were above the activated area (fig. 13.6). In two subjects, the probe center was displaced from the central sulcus by 1.5 to 2 cm toward the frontal lobe, and the major activated area appeared under channels B5, B6, B7, and B8. In two upperslices (δ and ε), the coefficient of correlation between the BOLD signal and the paradigm function was typically less than 0.1 and the corresponding z score was less than 2.0.

Figure 13.7A shows a BOLD signal from the slice β with $r = 0.7$ and a z score of 12.5. Figure 13.7B and 13.7C shows the concentration of deoxyhemoglobin and the concentration of oxyhemoglobin signals, respectively, from the light channel A5 (fig. 13.6) acquired

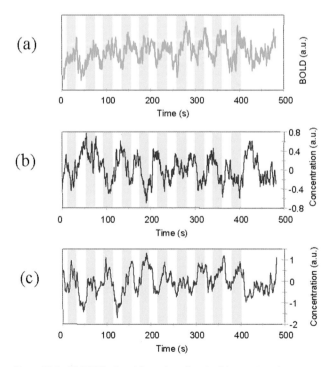

Figure 13.7 (A) BOLD signal from the slice β with $r = .7$ and a z score of 12.5. (B, C) Concentration of deoxyhemoglobin and oxyhemoglobin signals, respectively, from the light channel A5 acquired simultaneously with the BOLD signal shown in A.

simultaneously with the BOLD signal shown in figure 13.7A. One can see that the directions of changes in the BOLD and the concentration of deoxyhemoglobin signals are almost perfectly opposite. As a result of fluctuations in the upper tissue layers corresponding to slices δ and ε, in some cases the concentration of oxyhemoglobin and the concentration of deoxyhemoglobin signals from the activated area were noisier than the signals shown in figure 13.7. However, in all subjects the folding average analysis revealed the same type of concentration of oxyhemoglobin and concentration of deoxyhemoglobin behavior forlight channels situated above the majoractivated area. Figure 13.8 shows a map of the folding average of the concentration of deoxyhemoglobin and the concentration of oxyhemoglobin traces acquired simultaneously with the functional magnetic resonance imaging maps presented in fig. 13.6. For the light channels situated above the activated area (channels A1 and A5), the characteristic feature was a significant decrease in the concentration of deoxyhemoglobin during the stimulation, which was concurrent with a significant increase of the oxyhemoglobin concentration. Typically, rapid changes in the concentration of deoxyhemoglobin and the concentration of oxyhemoglobin began 2 to 3 s after stimulation onset and continued during the next 7 to 15 s. The concentration of deoxyhemoglobin then fluctuated nearits low level forthe rest of the stimulation epoch and the beginning of the resting epoch. During stimulation, the concentration of oxyhemoglobin eitherfluctuated at its high level orexhibited a slight decrease (see channels A1, A5, and B1 in fig. 13.6). A rapid recovery toward the baseline level began 4 to 6 s afterthe onset of the rest epoch in both the concentration of deoxyhemoglobin and the concentration of oxyhemoglobin. Such a behaviorof the concentration of deoxyhemoglobin and the concentration of oxyhemoglobin in the activated area was qualitatively the same in all six subjects.

No significant decrease in the folding average of the concentration of deoxyhemoglobin traces concurrent with the significant concentration of oxyhemoglobin increase was observed during stimulations in the light channels outside the activated area. Usually the

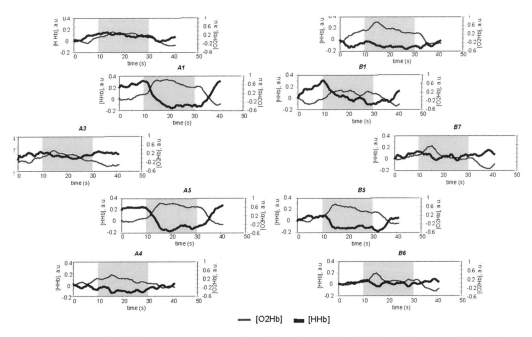

Figure 13.8 A map of the folding average concentration of deoxyhemoglobin ([HHb]) and the concentration of oxyhemoglobin ([O₂Hb]) traces acquired simultaneously with the functional magnetic resonance imaging maps presented in figure 13.7. The title of each panel indicates the corresponding light channel.

concentration of deoxyhemoglobin in such channels (A2, A3, A4, B8, B7, and B6 in fig. 13.6 and the corresponding panels in fig. 13.8) fluctuated without correlation with the paradigm function. In some cases, the concentration of oxyhemoglobin increased (see channel B8 in fig. 13.8), but without a significant change in the concentration of deoxyhemoglobin. In two subjects, both the concentration of deoxyhemoglobin and the concentration of oxyhemoglobin decreased in some channels during the stimulation, indicating a decrease in blood volume in these nonactivated regions.

Significant changes in the folding average traces corresponding to the activated area demonstrate a correlation between the hemodynamic changes and the paradigm function. To determine the location of the tissue contributing to such a correlation, we performed a correlation analysis of the BOLD signals using the sign-inversed concentration of deoxyhemoglobin signals as the reference function. (The concentration of deoxyhemoglobin signal was taken with the opposite sign because an increase in BOLD signal should correspond to a decrease in the concentration of deoxyhemoglobin.) In figure 13.6A through 13.6C, the arrows point at the clusters of voxels where the score for the correlation coefficient between BOLD and the concentration of deoxyhemoglobin signals in channel A5 exceeded 10.0. One can see a good colocation of these clusters with channel A5. Such a direct temporal correlation between optical and BOLD signals in cerebral and near-cerebral tissues was detected in all subjects.

The decrease of the concentration of deoxyhemoglobin concurrent with the increase in the concentration of oxyhemoglobin agrees with the results of the previous simultaneous functional magnetic resonance imaging optical study (Kleinschmidt et al., 1996), BOLD signal theory, and the basic knowledge of brain physiology (Mattay and Weinberger, 1999). Therefore, the colocation of the light channels exhibiting such a behavior in the folding average concentration of deoxyhemoglobin and concentration of oxyhemoglobin traces with the activated area revealed by functional magnetic resonance imaging proves that the near infrared monitors functional brain hemodynamics. Additional support for this conclusion is the temporal correlation between the intracranial BOLD signals and the respective near-infrared signals. We observed such a correlation in several subjects in the deep slices (α, β and γ, δ, ε), which included brain tissue, dura mater, pia mater, and arachnoidal tissue. The area of correlation was close to the central sulcus and was colocated with the corresponding light channels. This result indicates that the task-related hemodynamic changes measured by near-infrared spectroscopy are not the result of artifacts, but have an intracranial origin.

CONCLUSION

Although changes in the optical parameters (near-infrared reflectivity and scattering) have been measured in the exposed brain, the detection of these changes through the skull remained controversial. It was commonly accepted that neuronal signals obtained by currently available techniques are buried in the strong hemodynamic fluctuations and, in general, in the instrument noise. Several researchers have concluded that studies over a population are necessary to extract the signal from the background noise. Our goal was to enhance the quality of the measurement to the level that a measurement on a single subject and during a relatively short time would have a sufficient signal-to-noise ratio to be detectable. We believe that we have now demonstrated that this goal can be achieved, and we understand some of the reasons why earlier studies were affected by large background noise.

In a study of visual cortex activation, the observed fast changes of the optical properties suggested that this signal occurs simultaneously with or before the surface electrical potential (evoked potential or electrically recorded potential). This feature makes the fast optical signal an ideal marker for indexing the occurrence of neuronal activity. The fast optical signal also appears to be very localized, allowing discrimination between different parts of the same neuroanatomic area of the cortex (as in the case of Brodmann area 17). These data indicate that such localization is consistent with that obtained using functional magnetic resonance imaging (Gratton et al., 2000). This may be useful in two ways: First, it may help us to study the relative timing of neuronal activity in different brain areas. This may have profound theoretical implications, because we expect most brain functions (such as vision, memory, attention, language, movement control, and so forth) to depend on the dynamic interactions between different brain areas. Second, the possibility of acquiring data with high resolution both in the temporal and the spatial domains may help to integrate various noninvasive methods for studying brain function and, in particular, electrophysiological and hemodynamic techniques.

We would like to emphasize that if we can routinely obtain the quality of the signal shown in figure 13.2, the optical technique could make a tremendous leap into the methods we use for studying the functioning of the brain. We have an exciting opportunity to integrate the optical method with functional magnetic resonance imaging, which is the long-term goal of our research. Currently we are very active in the development of a multimodal near-infrared functional magnetic resonance imaging instrument. Functional magnetic resonance imaging has the spatial resolution and the penetration to identify the location of the signal. Togetherwith the optically detected hemodynamic changes, the combination of functional magnetic resonance imaging and near-infrared spectroscopy can provide a quantification that is difficult to achieve with functional magnetic resonance imaging alone. If the development that we have initiated is carried out, we can also identify the regions of neuronal activation, enhancing and integrating the functional magnetic resonance imaging maps. This is a truly unique and exciting possibility, and it can lead to advances in our understanding of brain function in general.

REFERENCES

Alcala, R., E. Gratton, and D. M. Jameson. A multifrequency phase fluorometer using the harmonic content of a mode-locked laser. Anal. Inst. 14, 225–250 (1985).

Belliveau, J. W., D. N. Kennedy, R. C. McKinstry, B. R. Buchbinder, R. M. Weisskopf, M. S. Cohen, J. M. Vevea, T. J. Brady, and B. R. Rosen. Functional mapping of the human visual cortex by magnetic resonance imaging. Science 254, 716–719 (1991).

Boas, D. A., M. A. O'Leary, B. Chance, A. G. Yodh. Scattering and wavelength transduction of diffuse photon density waves. Phys. Rev. E 47, R2999–R3002 (1993).

Boas, D. A., M. A. O'Leary, B. Chance, A. G. Yodh. Scattering of diffuse photon density waves by spherical inhomogeneities within turbid media: Analytic solution and applications. Proc. Natl. Acad. Sci. USA 91, 4887–4891 (1994).

Case, K. M., and P. F. Zweifel. *Linear Transport Theory.* Addison-Wesley, Reading, Mass. (1967).

Chance, B., K. L. Kang, L. He, J. Weng, and E. Sevick. Highly-sensitive object location in tissue models with linear in-phase and anti-phase multi-element optical arrays in one and two dimensions. Proc. Natl. Acad. Sci. USA 90, 3423–3427 (1993).

Cohen, L. B. Changes in neuron structure during action potential propagation and synaptic transmission. Physiol. Rev. 53, 373–418 (1973).

Cui, W., and L. E. Ostrander. Effect of local changes on phase shift measurement using phase modulation spectroscopy. SPIE 1888, 289–296 (1993).

Duncan, A., T. L. Whitlock, M. Cope, and D. T. Delpy. A multi-wavelength, wideband, intensity modulated optical spectrometer for near infrared spectroscopy and imaging. SPIE 1888, 248–257 (1993).

Fantini, S., M. A. Franceschini, and E. Gratton. Semi-infinite geometry boundary problem for light migration in highly scattering media: A frequency-domain study in the diffusion approximation. J. Opt. Soc. Am. 11(10), 2128–2138 (1994a).

Fantini, S., M. A. Franceschini, J. S. Maier, S. A. Walker, B. Barbieri, and E. Gratton. Frequency-domain multichannel optical detectorfornoninvasive tissue spectroscopy and oximetry. Opt. Eng. 34, 32–42 (1995).

Feddersen, B. A., D. W. Piston, and E. Gratton. Digital parallel acquisition in frequency domain fluorometry. Rev. Sci. Instrum. 60, 2929–2936 (1989).

Frostig, R. D., E. E. Lieke, D. Y Ts'o, and A. Grinvald. Cortical functional architecture and local coupling between neuronal activity and the microcirculation revealed by in vivo high-resolution optical imaging of intrinsic signals. Proc. Natl. Acad. Sci. USA 87, 6082–6086 (1990).

Graber, H. L., J. Chang, R. Aronson, and R. L. Barbour. Perturbation model for imaging in dense scattering media: Derivation and evaluation of imaging operators. In Medical Optical Tomography: Functional Imaging and Monitoring, Mülleret al. eds. SPIE Vol. IS 11, 121–146 (1993).

Gratton, G., and P. M. Corballis. Removing the heart from the brain: Compensation for the pulse artifact in the photon migration signal. Psychophysiology 32, 292–299 (1995).

Gratton, G., P. M. Corballis, E. Cho, M. Fabiani, and D. C. Hood. Shades of gray matter: Noninvasive optical images of human brain responses during visual stimulation. Psychophysiology 32, 505–509 (1995a).

Gratton, G., M. Fabiani, D. Friedman, M. A. Franceschini, S. Fantini, and E. Gratton. Photon migration correlates of rapid physiological changes in the brain during a tapping task. J. Cogn. Neurosci. 7, 446–456 (1995b).

Gratton, G., J. S. Maier, M. Fabiani, W. W. Mantulin, and E. Gratton. Feasibility of intracranial near-infrared optical scanning. Psychophysiology 31, 211–215 (1994).

Gratton, E., W. W. Mantulin, M. J. vandeVen, J. B. Fishkin, M. B. Maris, and B. Chance. *The Possibility ofa Near-Infrared Optical Imaging System Using Frequency-Domain Methods.* Presented at the Proceedings of the Third International Conference: Peace through Mind/Brain Science, 183–189, Hamamatsu City, Japan (1990).

Gratton, G., A. Sarno, E. Maclin, P. M. Corballis, and M. Fabiani. Toward noninvasive 3-D imaging of the time course of cortical activity: Investigation of the depth of the event-related optical signal. Neuroimage 11, 491–504 (2000).

Grinvald, A., E. Lieke, R. D. Frostig, C. D. Gilbert, and T. N. Wiesel. Functional architecture of cortex revealed by optical imaging of intrinsic signals. Nature 324, 361–364 (1986).

Hirth, C., K. Villringer, A. Thiel, J. Bernarding, W. Muhlnickl, H. Obrig, U. Dirnagl, and A. Villringer. Toward brain mapping combining near-infrared spectroscopy and high resolution 3D MRI. Adv. Exp. Biol. Med. 413, 139–147 (1997).

Hoshi, Y., and M. Tamura. Detection of dynamic changes in cerebral oxygenation coupled to neuronal function during mental work in man. Neurosci. Lett. 150, 5–8 (1993a).

Hoshi, Y., and M. Tamura. Dynamic multichannel near-infrared optical imaging of human brain activity. J Appl. Physiol. 75, 1842–1846 (1993b).

Ishimaru, A. *Wave Propagation and Scattering in Random Media.* Vol. 1. Academic, New York (1978).

Jaspers, H. H. Report of the committee on the methods of clinical examination in electroencephalography. Electroencephalogr. Clin. Neurophysiol. 10, 370–375 (1957).

Jennings, J. R., and S. Choi. Methodology: An arterial to peripheral pulse wave velocity measure. Psychophysiology 20, 410–418 (1983).

Kaltenbach, J., and M. Kaschke. Frequency- and time-domain modeling of light transport in random media. SPIE. IS 11, 65–86 (1993).

Kleinschmidt, A., H. Obrig, M. Requardt, K. D. Merboldt, U. Dirnagl, A. Villringer, and J. Frahm. Simultaneous recording of cerebral blood oxygenation changes during human brain activation by magnetic resonance imaging and near-infrared spectroscopy. J. Cereb. Blood Flow Metab. 16, 817–826 (1996).

Maier, J. S., and E. Gratton. Frequency-domain methods in optical tomography: Detection of localized absorbers and a backscattering reconstruction scheme. SPIE 1888, 440–451 (1993).

Maki, A., Y. Yamashita, Y. Ito, E. Watanabe, Y. Mayanagi, and H. Koizumi. Spatial and temporal analysis of human motoractivity using noninvasive NIR topography. Med. Phys. 22, 1997–2005 (1995).

Mattay, V. S., and D. R. Weinberger. Organization of the human motor system as studied by functional magnetic resonance imaging. Eur. J. Radiol. 30, 105–114 (1999).

Meek, J. H., C. E. Elwell, M. J. Khan, J. Romaya, J. D. Wyatt, D. T. Delpy, and S. Zeki. Regional changes in cerebral haemodynamics as a result of a visual stimulus measured by near infrared spectroscopy. Proc. R. Soc. Lond. B. Biol. Sci. 261, 351–356 (1995).

Obrig, H., C. Hirth, J. G. Junge-Hulsing, C. Doge, T. Wolf, U. Dirnagl, and A. Villringer. Cerebral oxygenation changes in response to motor stimulation. J. Appl. Physiol. 81, 1174–1183 (1996).

O'Leary, M. A., D. A. Boas, B. Chance, and A. G. Yodh. Refraction of diffuse photon density waves. Phys. Rev. Lett. 69, 2658–2661 (1992).

Patterson, M. S., B. Chance, and B. C. Wilson. Time resolved reflectance and transmittance for the noninvasive measurement of tissue optical properties. Appl. Opt. 28, 2231–2236 (1991a).

Patterson, M. S., J. D. Moulton, B. C. Wilson, K. W. Berndt, and J. R. Lakowicz. Frequency-domain reflectance for the determination of the scattering and absorption properties of tissues. Appl. Opt. 30, 4474–4476 (1991b).

Tromberg, B., L. O. Svaasand, T. Tsay, and R. C. Haskell. Properties of photon density waves in multiple-scattering media. Appl. Opt. 32, 607–616 (1993).

Villringer, A., and B. Chance. Non-invasive optical spectroscopy and imaging of human brain function. Trends Neurosci. 20, 435–442 (1997).

Villringer, A., and U. Dirnagl. Coupling of brain activity and cerebral blood flow: Basis of functional neuroimaging. Cerebrovasc. Brain Metab. Rev. 7, 240–276 (1995).

14

In Vivo Optical Imaging of Molecular Function Using Near-Infrared Fluorescent Probes

VASILIS NTZIACHRISTOS AND RALPH WEISSLEDER

Tissue observation with light is probably the most common imaging practice in medicine and biomedical research. The original physicians' visual inspection of their patients has been followed by more elaborate photon technologies applied to biomedical and clinical applications, many of which are described in this book. Light offers unique wavelength-dependent interactions with tissue and has been used to investigate both structural and functional tissue characteristics. A recent development in *optical* applications is the noninvasive imaging of molecular events in tissues. Underpinning these developments are the discovery of biocompatible, target-specific, and activatable fluorescent imaging probes and the development of highly sensitive imaging technologies for in vivo fluorescent detection. Of particular interest are fluorescent probes that emit in the near infrared, a spectral window in which hemoglobin and water photons to penetrate for several centimeters in tissue.

In this chapter we review technological advances that have allowed molecular imaging in the near infrared. First, we outline design strategies for fluorescent probes developed for in vivo applications and focus on the engineering of *molecular beacons,* a particular category of near-infrared fluorescent smart probes that maximize imaging capacity. We describe the chemistry of such probes and delineate current and future molecular targets. Second, we describe optical imaging technologies used for noninvasive imaging of the distribution of such probes. We illuminate the advantages and limitations of simple photographic methods and turn our attention to fluorescence-mediated molecular tomography (FMT), a technique that can resolve fluorescence and quenching/activation in deep tissues in three dimensions. We describe theoretical specifics, and we obtain insight on its in vivo capacity and the sensitivity achieved. Last, we discuss its clinical feasibility.

INTRODUCTION: MOLECULAR IMAGING

Molecular imaging, the noninvasive mapping of cellular and subcellular events in living organisms, represents an exciting field in imaging sciences that is expected to have a major impact in biomedical research and the clinical practice of diagnostic imaging. The significance of imaging molecular events noninvasively lies in the development of new sets of strategies and probes to interrogate targets at subcellular levels (fig. 14.1). Certain targets responsible for pathogenesis also represent key points for therapeutic intervention.

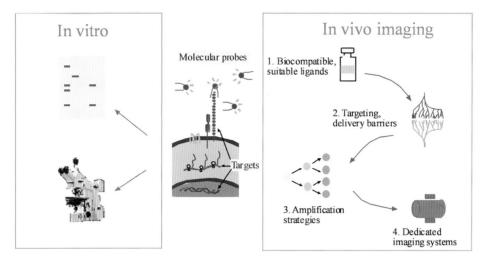

Figure 14.1 Comparison of in vitro and in vivo molecular imaging.

The impact of developing molecular imaging techniques is potentially enormous. First, relatively crude parameters of tumor growth and development (tumor burden, anatomic location, and so forth) could be supplemented by more specific parameters (e.g., growth factor receptors, cell surface markers, transduction signals, cell cycle proteins). Second, this added information is expected to go hand-in-hand with the development of novel targeted therapies and will allow us to assess theirefficacy at a molecularlevel, long before phenotypic changes occur. This, in turn, is expected to have an impact on drug development, drug testing, and choosing appropriate therapies and therapy changes in a given patient. Molecular imaging will be critical in testing novel drug delivery strategies, because there exist significant barriers to the delivery of complex and molecular therapies (Jain, 1996). Understanding these barriers and finding solutions to modulate them will be of the utmost importance in future drug development. Third, molecular imaging potentially allows us to study the genesis of diseases in the intact microenvironment of living systems, which is currently very difficult to do otherwise. The importance of intact micro- and macroenvironments on tumor progression has recently been recognized (Jain, 1996; Koblinski et al., 2000; McCawley and Matrisian, 2000). Last, imaging techniques can accelerate studies of molecular events in tissues compared with conventional techniques that require tissue samples formolecularanalyses.

FLUORESCENT IMAGING PROBES

Molecular imaging using fluorescent probes represents a novel but potentially powerful approach to obtaining molecular information noninvasively in vivo. In general, imaging of specific molecules in vivo requires several key characteristics: (1) availability of high-affinity probes with reasonable pharmacological behavior, (2) the ability of these probes to overcome biological delivery barriers (vascular, interstitial, cell membranes), (3) utilization of amplification strategies (chemical or biological), and (4) fast imaging techniques with high sensitivity in detecting small concentrations of accumulated probes. In a typical scenario, all four prerequisites must be met to successfully image at the molecular level in vivo.

In, general, high-affinity ligands must reach their intended target at sufficient concentrations and for sufficient lengths of time to be detectable in vivo. Rapid excretion,

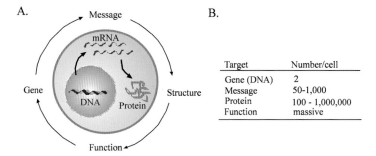

Figure 14.2 Molecular imaging targets. (A) Different intracellular imaging targets and their function are summarized. (B) Numbers of targets per cell.

nonspecific binding, metabolism, and delivery barriers all counteract this process and must be overcome. Delivery barriers are typically the most challenging to deal with, particularly for larger "biotech drugs." A number of strategies have been developed to circumvent existing delivery barriers; current research concentrates on developing even more efficient methods. Examples include the use of peptide-derived membrane translocation signals to shuttle imaging drugs actively into cells (Lewin et al., 2000), "pegylation" to decrease both immunogenicity and rapid recognition (Bogdanov et al., 1993), use of long-circulating drugs to achieve more homogenous distribution (Bogdanov et al., 1996), and/or local delivery combined with pharmacological or physical methods to improve targeting (Neuwelt et al., 1994; Muldoon et al., 1995). Utilization and future developments of efficient amplification strategies (increasing the signal imaged) and appropriate cellular and subcellular targeting that yields sufficient contrast is a critical component of much of molecular imaging research. Figure 14.2 summarizes generic cellular targets including DNA, mRNA, and proteins. Although the number of DNA and mRNA targets per cell is limited (requiring extreme levels of signal amplification for visualization), imaging proteins and/or protein function is much more feasible (referred to as *imaging downstream*).

Targeted Fluorescent Probes

A number of different approaches to image molecular targets using fluorescent probes have been recently advocated; most of them rely on targeting specific molecules (affinity ligands) and imaging them after intravenous injection when a fraction of an agent has bound to its target, and the remainder of unbound agent has been cleared (Achilefu et al., 2000; Becker et al., 2001). These approaches have been particularly useful for imaging receptors and cell surface-expressed molecules. To achieve higher penetration depths, the probes are engineered to fluoresce in the near-infrared region, where minimal photon absorption occurs by tissue. The strategy of biocompatible, near-infrared fluorochromes coupled to peptide ligands (rather than monoclonal antibodies), may well represent a step toward smaller, more penetrable reporter probes and may have particular applications in unique clinical situations in which nuclear imaging is not an option (e.g., for reasons of resolution, during endoscopy or in surgery). The peptide-coupled near-infrared fluorochromes are also expected to have significant advantages over nonspecific fluorochromes such as indocyanine green, because the latter primarily reflects initial vascular distribution (through binding to plasma proteins) and subsequent hepatobiliary excretion (Licha et al., 2000). The concept of tagging near-infrared fluorochromes can potentially be extended to a myriad of other peptide/small-molecule receptorsystems. The technology may also be useful forin vivo screening of limited peptide libraries and/or for identifying structure–activity relationships. A number of toxicology and safety issues with these probes remain to be addressed prior to their clinical use. Some of the near-infrared fluorochromes such as indocyanine green, however, have been used in tens of thousands of patients with reported side effects of less than 0.15 %, an extremely favorable index when compared with other reported agents (Hope-Ross et al., 1994).

One potential caveat, however, of using targeted conjugates is the fact that target-to-background ratios can be limited by receptor density/availability, limited clearance kinetics from the interstitial space, and/ornonspecific cellularuptake oradhesion of certain fluorescent probes. In particular, it may be difficult to differentiate specifically the bound from unbound ligands, and this is the reason why imaging is usually performed after nonspecifically distributed surplus has cleared.

Activatable Fluorescent Probes

Fluorescent probes offer several pathways for signal amplification and suppression of signals emanating from nonspecific uptake, by utilizing activation techniques (quenching/dequenching) and wavelength-shifting techniques, and by loading multiple fluorochromes on the same delivery molecule (backbone). Besides signal amplification, activation offers the ability to interrogate highly specific molecular function or structure resulting in "smart" probes. A number of different designs of optical smart agents have been described, two of which we describe briefly here (fig. 14.3). The first agent has pioneered detection of complementary DNA strands (Tyagi and Kramer, 1996). The molecular beacon contains a stem-loop structure with a fluorochrome at one terminus and DABCYL (4-[4′dimethylaminophenylazo] benzoic acid) as universal quencher at the other. The probe DNA sequence is in the center of the molecule, and the bases at both termini are self-complementary, yielding a five- to eight-nucleotide stem with a loop. After duplex formation occurs, the fluorochrome and quencher become spatially separated and FRET and/or fluorescence direct energy transfer is no longer possible. The fluorochrome then

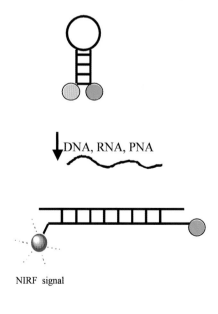

Figure 14.3 Examples of activatable enzyme-specific (left) and DNA/RNA-specific (right) imaging probes. Note that the fluorochromes are excited in the near infrared.

emits light of the appropriate wavelength when excited. Several investigators have demonstrated that molecular beacons can be used to assess DNA duplex formation directly in living cells. More recently, wavelength-shifting molecular beacons that fluoresce in a variety of different colors, yet are excited by a common monochromatic light source, have been developed. These probes contain a harvester fluorophore that absorbs strongly in the wavelength range of the monochromatic light source, an emitter fluorophore of the desired emission color, and a nonfluorescent quencher. In the absence of complementary nucleic acid targets, the probes do not emit light; in the presence of targets, they fluoresce in the emission wavelength of the emitter fluorophore and not in the emission wavelength of the harvester fluorophore that absorbs the light. This shift in emission spectrum is a result of the transfer of the absorbed energy from the harvester fluorophore to the emitter fluorophore by FRET, and it only takes place in probes that are bound to targets. Wavelength-shifting molecular beacons are reported to be substantially brighter than conventional molecular beacons that contain a fluorophore that cannot efficiently absorb energy from the available monochromatic light source (Tyagi et al., 2000).

The second type of probe described here can be designed to report on specific enzyme activities. The probes typically consist of three building blocks: (1) near-infrared fluorochromes, (2) peptide substrates, and (3) a delivery vehicle. The design of the probes has been described in detail elsewhere (Weissleder et al., 1999; Tung et al., 2000) and basically uses certain peptide substrates (table 14.1) attached to a delivery molecule to impart molecular specificity. When a given target protease cleaves the substrate, the fluorochromes are released and fluoresce (dequenching). Specific examples are tumor lysosomal endopeptidases that recognize and cleave free lysine residues (cathepsin B) (Weissleder et al., 1999); cathepsin D, which cleaves two phenylalanine peptides (Tung et al., 1999); matrix metalloproteinase 2-specific probes (Bremer et a., 2001); caspase 3-specific probes; and others as summarized in table 14.1. Although it is conceivable to design alternative small-molecular weight probes in which a fluorescence donor and a quencher are directly attached to a substrate peptide (Gulnik et al., 1997; Zlokarnik et al., 1998), such compounds are typically subject to fast excretion in vivo. For this reason and to improve tumoral delivery of the near-infrared fluorescent probes, a long circulating synthetic graft copolymer poly-L-lysine/methoxypolyethylene glycol graft copolymer (PGC) is utilized. This copolymer has been tested in clinical trials (Callahan et al., 1998) and accumulates in tumors by slow extravasation through permeable neovasculature, reaching up to 2% to 6% of injected dose pergr am of tissue in mice within 24 to 48 hours afterinjection in some tumormodels (Marecos et al., 1998). Uptake of the polymerinto tumorcells occurs by pinocytosis and is comparable in magnitude with that of tumor-specific internalizing monoclonal antibodies. The intracellular release of near-infrared fluorescent probes results in a fluorescent signal that can be detected in vivo at depths sufficient forexper imental orclinical imaging, depending on the near-infrared fluorescent image acquisition technique (Weissleder et al., 1999).

Table 14-1. Applications and peptide substrates used for near-infrared fluorescent probes

Protease target	Disease	Substrate
Cathepsin B/H	Cancer, inflammation	KK
Cathepsin D	Breast cancer > others	PIC(Et)FF
Cathepsin KOsteoporosis		GGPRGLPG
PSA	Prostate cancer	HSSKLQG
Matrix metalloproteinase 2	Metastases	P(L/Q)G(I/L)AG
CMV protease	Viral infection	GVVQASCRLA
HIV protease	HIV	GVSQNYPIV
Thrombin	Cardiovascular	dFPipR
Caspase 3	Apoptosis	DEVD

This approach has fourmajoradvantages overothermethods that use single fluorochromes attached to affinity molecules: (1) a single enzyme can cleave multiple fluorochromes, thus resulting in one form of signal amplification; (2) reduction of background "noise" by several orders of magnitude is possible; (3) very specific enzyme activities can potentially be interrogated; and (4) multiple probes can be arranged on delivery systems to probe simultaneously for a spectrum of enzymes. These probes are currently being used to detect disease at their earliest stage (e.g., small cancers, vulnerable plaque, rheumatoid arthritis), to image transgenes, or to test the in vivo efficacy of enzyme inhibitors within hours after administration (table 14.1).

OPTICAL IMAGING TECHNOLOGIES OF MOLECULAR FUNCTION

A number of different optical imaging approaches can be used for imaging fluorescence in vivo. Traditionally, optical methods have been used to look at surface and subsurface fluorescent events using confocal imaging (Rajadhyaksha et al., 1995; Gonzalez et al., 1999; Korlach et al., 1999), multiphoton imaging (Masters et al., 1997; So et al., 1998; Buehler et al., 1998; Konig, 2000), microscopic imaging by intravital microscopy (Monsky et al., 1999; Dellian et al., 2000), or total internal reflection fluorescence microscopy (Toomre and Manstein, 2001). Recently, however, light has been used for in vivo interrogations deeper into tissue using photographic systems with continuous light (Mahmood et al., 1999; Beckeret al., 2001) orwith intensity-modulated light (Reynolds et al., 1999) and tomographic systems (Ntziachristos et al., 2000). Potentially, phase-array detection (Chance et al., 1993) can be also applied. Next, we discuss imaging techniques that use the diffuse component of light forpr obing molecularevents deep in tissue. Specifically, we focus on reflectance imaging and FMT, because these approaches are more commonly used currently for imaging near-infrared fluorescent probes in deep tissues. We further predict the capacity of near-infrared fluorescent signals to propagate through human tissue for noninvasive medical imaging and address feasibility issues for clinical studies.

Reflectance Imaging

Simple "photographic methods" of tissue, in which the light source and the detector reside on the same side of the animal imaged, are generally referred to by using the term *reflectance imaging*. Reflectance imaging is the typical method of choice foraccessing the distribution of fluorescent probes in vivo, but the method can be applied more generally to imaging fluorescent proteins or even bioluminescence, even if in the latter case no excitation light is used.

Near-infrared fluorescence reflectance imaging in particular operates on light with a defined bandwidth as a source of photons that encounters a fluorescent molecule (*optical contrast agent*), which emits a signal with different spectral characteristics that can be resolved with an emission filter and captured by a high-sensitivity CCD camera.

A typical reflectance imaging system is shown in figure 14.4. The light source can be either a laser at an appropriate wavelength for the fluorochrome targeted or white-light sources using appropriate low-pass filters. Generally, laser sources are preferable because they offerhigherpowerdeliver y at narrowerand betterdefined spectral windows (typically ±3 nm forlaserdiodes vs. ±10 nm or more for filtered white-light sources). The laser beam is expanded on the animal surface with an optical system of lenses (not shown). Narrow wavelength selection is important, especially in the near infrared in which the excitation and emission spectra overlap, and it is likely that excitation photons can propagate into the fluorescent images. The CCD camera is usually a high-sensitivity camera because fluorescent signals are of low strength. On the other hand, because the targeted measurement

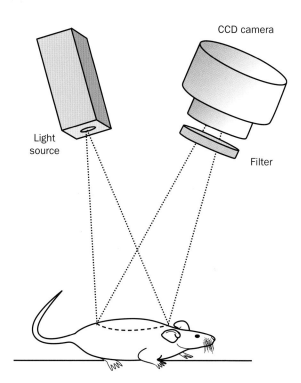

CCD camera

Light
source

Filter

Figure 14.4 A typical reflectance
imaging system. The construction is
usually encased in a photon-sealing
box (blackbox).

is a diffuse light measurement, emanating from a virtually flat surface, CCD chip resolution and dynamic range are not crucial factors in these types of systems. Also, the noise requirements of a CCD camera for fluorescence imaging are not as stringent because, in general, the detection sensitivity limit is set by the background tissue fluorescence (background probe distribution and autofluorescence), and fluorescence strength can be adjusted above the camera noise floor by optimizing illumination strength and acquisition times. These features are in contrast to bioluminescence measurements, in which the detection sensitivity is usually set by the CCD camera noise floor.

Typically the fluorescence image is accompanied by a second image that is acquired without the fluorescence filter to obtain a "photograph" of the animal for registration purposes at the excitation wavelength. This intrinsic light image also serves as a calibration measurement because it records the exact spatial distribution of the excitation light strength forlatercor rection of excitation field inhomogeneities.

Figure 14.5 depicts such measurement from a nude mouse implanted with a cathepsin B-rich HT1080 fibrosarcoma, which had been implanted (10^5 cells) into the mammary fat pad of the mouse 7 to 10 days prior to the experiment. The animal received an intravenous injection of a cathepsin B-sensitive imaging probe (Weissleder et al., 1999) 24 hours prior to the imaging experiments. The animal was anesthetized with an intraperitoneal injection of 90 mg/kg ketamine and 9 mg/kg xylazine and was imaged by a surface imaging system (Mahmood et al., 1999). The acquisition time was 0.5 s forthe intrinsic light image and 10 s for the fluorescence image, which demonstrates significant fluorescence activation from the site of implantation. Immunohistochemistry and Western blotting results are usually used to verify and correlate the imaging findings with the underlying biology. It should be noted that the intrinsic light image records the structure on the animal surface and can yield high-resolution images after appropriate focusing, whereas the fluorescence (or luminescence) image records fluorescence signals emanating from under the surface and is, by nature, a low-resolution image that uses mainly the diffuse component of light. The side-by-side

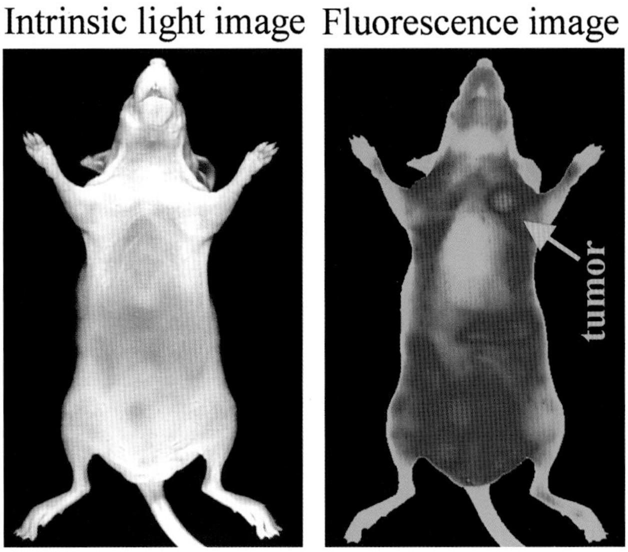

Intrinsic light image Fluorescence image

Figure 14.5 Reflectance fluorescence imaging of cathepsin B expression in vivo.

comparison or the superposition of the two images into a single image yields a nice visual effect because high resolution is retained.

Such approaches have been used to evaluate activatable probes (Weissleder et al., 1999; Bremer et al., 2001) or peptide–near-infrared dye conjugates (Achilefu et al., 2000; Becker et al., 2001). Reflectance imaging could be also used to assess endogenous gene products macroscopically by using fluorescent proteins (Wouters et al., 2001). Specifically, imaging of florescent proteins can be done in analogy to near-infrared fluorescent imaging with the exception that (1) absorption/excitation are shifted usually to the visible light range and that (2) no exogenous fluorochromes have to be administered to visualize fluorescent protein expression. In a similar paradigm that is presented analytically elsewhere in this volume, bioluminescence imaging (Contag et al., 1997) is based on the emission of visible photons at specific wavelengths based upon energy-dependent reactions catalyzed by luciferases. Luciferase genes have been cloned from a large number of organisms, including bacteria, firefly (*Photinus pyralis*), coral (*Renilla*), jellyfish (*Aequorea*), and dinoflagellates (*Gonyaulax*). In the firefly, luciferase utilizes energy from adenosine triphosphate to convert its substrate luciferin to oxyluciferin, with the emission of a detectable photon. Sensitive imaging systems have been built to detect quantitatively small numberof cells oror ganism expressing luciferase as a transgene (Sweeney et al., 1999).

Advantages and Limitations

Reflectance imaging is an ideal tool for high-throughput imaging and the screening of animals and excised tissues. It offers simplicity of operation and high sensitivity for molecular events that are close to the surface. Typical acquisition times range from a few seconds to a few minutes. For laboratory settings, multiple animals can be imaged simultaneously. Hardware development and implementation are also straightforward and relatively inexpensive. Reflectance systems do not use ionizing radiation, use safe laser powers, can be made portable, and have small space requirements that are ideal for a laboratory bench.

On the other hand, reflectance imaging has fundamental limitations as a research or a clinical tool. The technique attains only small penetration depths (a few millimeters or less) and lacks quantification. The latter is illustrated in figure 14.6. A small structure of high-fluorochrome concentration that is deeper into tissue could yield the same appearance

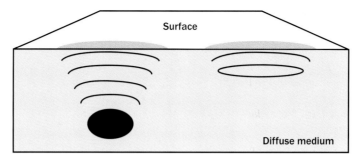

Figure 14.6 Different fluorescent lesions embedded in a diffuse medium can have the same appearance in the surface, impeding quantification.

on the surface as a larger structure of low-fluorochrome concentration that is closer to the surface. This is because of the nature of propagation of diffuse photon density waves into tissue. In an animal, the photon count reading at the surface resulting from a lesion on or under the surface depends on the lesion depth, the lesion volume, and the optical properties of both the lesion and the surrounding tissue. Therefore, images from different animals, or from the same animal at different time points, are generally insufficient to yield quantitative insights.

Fluorescence-Mediated Molecular Tomography

To resolve and quantify fluorochromes deep in tissue, tomographic approaches are necessary. The general framework of reconstruction techniques using diffuse light has been developed during the past decade. Rigorous mathematical modeling of light propagation in tissue, combined with technological advancements in photon sources and detection techniques, have made possible the application of tomographic principles (Barbour et al., 1995; Yodh and Chance, 1995; Arridge, 1999; Colak et al., 1999; Hielscher et al., 1999; Schotland, 1997) for imaging with diffuse light. The technique, generally termed *diffuse optical tomography* (DOT), uses multiple projections and measures light around the boundary of the illuminated body. It then effectively combines all measurements into an inversion scheme that takes into account the highly scattered photon propagation to deconvolve the effect of tissue on the propagating wave, even though high-frequency components are generally significantly attenuated. Diffuse optical tomography has been used for quantitative imaging of absorption and scattering (Oleary et al., 1995; Change et al., 1997b; Jiang et al., 1998; Ntziachristos et al., 1998), as well as fluorochrome lifetime and concentration measurements (Oleary et al., 1996; Chang et al., 1997a; Paithankar et al., 1997; Sevick-Muraca et al., 1997; Chernomordik et al., 1999; Ntziachristos and Weissleder, 2001). Recently, DOT has been applied clinically for imaging tissue oxy- and deoxyhemoglobin concentration and blood saturation (Benaron et al., 2000; Pogue et al., 2000; Hillman et al., 2001), and contrast agent uptake (Ntziachristos, 2000).

A particular class of these techniques was developed specifically for molecular interrogations of tissue in vivo and is termed *fluorescence-mediated molecular tomography* (or FMT). The method is, by definition, a volumetric method that produces quantified three-dimensional reconstructions of fluorescence concentration. In its optimal implementation, FMT uses measurements at both emission and excitation wavelengths to offer significant advantages for in vivo imaging. The basic concept has been described for a normalized Born expansion (Ntziachristos and Weissleder, 2001), but the method can be applied to different mathematical constructions of the forward problem, including numerical approaches. The main advantage of using both intrinsic and fluorescence contrast lies in that no absolute photon-field measurements are required, yet absolute fluorochrome concentrations can be reconstructed. Furthermore, the method does not require any measurements obtained before contrast agent administration, which is a crucial parameter for in vivo imaging of probes that require long circulation times to achieve sufficient accumulation in their intended targets. In the following paragraphs we describe in more detail these unique characteristics and present key features of FMT performance and imaging examples.

Fluorescence-Mediated Molecular Tomography Theoretical Specifics: Forward Problem

To enable a better understanding of FMT, we present in this section a simplified review of the theoretical specifics of the technique. In general, the tomographic approaches for fluorescence imaging in the near infrared that have been developed in the past are straightforward expansions of algorithms originally developed for DOT of absorption and scattering (Oleary et al., 1996; Paithankar et al., 1997). To obtain accurate reconstructions of tissue in vivo, DOT requires *differential measurements* (Ntziachristos et al., 1999a). In the case of vascular or micromolecular contrast agents, such differential measurements can be constructed using pre- and postcontrast agent administration optical recordings (Ntziachristos, 2000). Conversely, in vivo imaging of long-circulating probes does not allow preadministration recordings because such probes need to circulate in the vascular system for several hours oreven days to allow sufficient accumulation and activation ornonspecific clearance. This imposes certain limitations in conventional DOT methods.

To facilitate molecularimaging on the basis of imaging and quantifying the distribution of long-circulating near-infrared fluorescent probes, differential measurements can be constructed by using intrinsic and fluorescence measurements. Assuming an intensity-modulated source at position $\vec{r}_s$, a photon wave detected in a unbounded diffuse medium zat position $\vec{r}$ can be written as $U_0(\vec{r}_s, \vec{r}, \omega) = \Theta_s(\vec{r}_s)\exp(ik\vec{r})/4\pi D\vec{r}$, where $k = ((-\upsilon\mu_a + i\omega)/D)^{1/2}$ is the photon wave propagation vector, υ is the speed of light into the medium, ω is the angularmodulation frequency, μ_a is the absorption coefficient, $D = \upsilon/3\mu'_s$ is the diffusion coefficient, μ'_s is the reduced scattering coefficient, and $\Theta_s(\vec{r}_s)$ is a gain factorassociated with light source strength, fibercoupling losses, and otherattenuation in the optical system. If $\Theta_s(\vec{r}_s)$ accounts forthe detectorgain and the detection fiber losses, and $QE^{\lambda 1}$ is the detectorquantum efficiency at wavelength λ_1, then the incident photon field detected at position $\vec{r}_d$ can be written as

$$U_{inc}(\vec{r}_s, \vec{r}_d) = QE^{\lambda 1} \times \Theta_s(\vec{r}_s) \times \Theta_d(\vec{r}_d) \times U_0(\vec{r}_s - \vec{r}_d, k^{\lambda 1}), \qquad (14.1)$$

where $k^{\lambda 1}$ denotes the wave propagation vector for the optical properties of the medium of investigation at wavelength λ_2, and frequency dependence is omitted for simplicity.

When fluorescent objects exist in the medium, the field $U_{fl}(\vec{r}_s, \vec{r}_d)$ detected at position $\vec{r}_d$ at the emission wavelength can be written as (Oleary et al., 1996):

$$U_{fl}(\vec{r}_s, \vec{r}_d) = \int d^3r \times \Theta_s(\vec{r}_s) \times \Theta_f \times QE^{\lambda 2} \times \Theta_d(\vec{r}_d) \times U_0(\vec{r}_s - \vec{r}, k^{\lambda 1}) \times \frac{n(\vec{r})}{1 - i\omega\tau(\vec{r})}$$

$$\times \frac{\upsilon}{D^{\lambda 2}} \times G(\vec{r}_d - \vec{r}, k^{\lambda 2}) \qquad (14.2)$$

where $n(\vec{r})$ is the product of the fluorochrome absorption coefficient and fluorescence quantum yield, is the $\tau(\vec{r})$ fluorescence lifetime, $k^{\lambda 2}$ is the wave propagation vector at wavelength λ_2, Θ_f is the attenuation of the filterthat is used to collect the fluorescent field, $QE^{\lambda 2}$ and is the detectorquantum efficiency at the emission wavelength. $G(\vec{r}_d - \vec{r}, k^{\lambda 2})$ is the Green's function solution to the diffusion equation and describes the propagation of the emission photon wave from the fluorochrome to the detector. Equation 14.2 treats fluorochromes as two-level quantum systems and ignores fluorescence saturation effects and high-order scattering interactions. These approximations hold for the weak fluorescent concentrations expected in many biological systems.

Finding solutions for equation 14.2 requires determination of the factors $\Theta_s(\vec{r}_s)$, $\Theta_d(\vec{r}_d)$ for each source–detector pair $\vec{r}_s, \vec{r}_d$. To reach more manageable expressions for the fluorescent field, we can divide equation 14.2 by equation 14.1 and obtain the field U^{nB}:

$$
\begin{aligned}
U^{nB}(\vec{r}_s, \vec{r}_d) &= \frac{1}{\Theta_f} \times \frac{U_{fl}(\vec{r}_s, \vec{r}_d)}{U_{inc}(\vec{r}_s, \vec{r}_d)} \times \frac{QE^{\lambda 1}}{QE^{\lambda 2}} \\
&= \frac{1}{U_0(\vec{r}_s, \vec{r}_d, k^{\lambda 1})} \times \int d^3 r \times (U_0(\vec{r}_s - \vec{r}, k^{\lambda 1}) \frac{n(\vec{r})}{1 - i\omega\tau(\vec{r})} \frac{u}{D^{\lambda 2}} G(\vec{r}_d - \vec{r}, k^{\lambda 2})
\end{aligned}
$$

(14.3)

Equation 14.3 is similar to expressions obtained for the linear perturbation DOT problem using the Rytov approximation (Kak and Slaney, 1988) and offers experimental advantages because all the position-dependent gain factors are canceled out. The terms Θ_f and $QE^{\lambda 2}/QE^{\lambda 1}$ can be determined experimentally and are virtually independent of position and time. Usually, $QE^{\lambda 2}/QE^{\lambda 1}$ is approximately one, as a result of the proximity of λ_1, λ_2.

Equation 14.3 assumes a measurement of the incident field U_{inc} (i.e., a measurement through a homogeneous medium). This measurement is not practical when probing biological media. However, it has been shown (Ntziachristos et al., 2001b) that U_{inc} can be substituted by a measurement U'_{inc}, obtained through the heterogeneous medium if U_0 at the right part of equation 14.3 is computed for the average optical properties of the medium at λ_1 (assuming a random, weak optical background heterogeneity). It is important to note that U'_{inc} can be obtained even if the fluorochrome is present during the measurement. The weak sensitivity of the algorithm to the selection of U_{inc} versus U'_{inc} has been shown experimentally (Ntziachristos and Weissleder, 2001) and is a significant and unique advantage of FMT when compared with differential absorption/scattering DOT.

Fluorescence-Mediated Molecular Tomography Theoretical Specifics: Inverse Problem

For reconstruction purposes, equations such as equation 14.3 are "discretized" into a number of volume elements (voxels), which yield a set of coupled linear equations (Kak and Slaney, 1988). The terms U_0 and G can be calculated numerically or analytically for the appropriate boundary conditions and background optical properties at λ_1, λ_2, respectively. The resulting system of equations can then be inverted (Kak and Slaney, 1988) for the unknown quantities $n(\vec{r})$ and $\tau(\vec{r})$. There are several ways to solve the inverse problem, such as direct inversion, χ^2-based fits, or algebraic reconstruction techniques (Kak and Slaney, 1988; Schotland, 1997; Arridge, 1999). Higher order solutions can be obtained when a solution is fed back into the forward problem to produce more accurate forward propagation models, and this process can be repeated iteratively (Kak and Slaney, 1988). It should be noted that the proposed inversion methods are applied in the linear approximate solution given in equation 14.3. If the differential scheme was to be written in its complete nonlinear form (without using the Born approximation) it could be solved by approximating it numerically, using, for example, finite differences or finite element methods.

Diffuse Optical Tomographic Imaging Systems

To perform tomography, tissue has to be illuminated at different projections, and multiple measurements should be collected from the boundary of the tissue under investigation. Fluorescence image reconstruction benefits significantly from constant-intensity laser illumination and detection systems that are linear in intensity. Fluorescence-mediated molecular tomography, in its simplest implementation, needs only measurements of fluorescence and intrinsic light strength (amplitude) to resolve the single quantity targeted—namely, fluorochrome concentration. This is in contrast with DOT that, in its general form, requires decomposition of absorption from scattering and therefore requires more elaborate photon technologies such as intensity-modulated sources or short photon pulses and appropriate detection systems. Fluorescence-mediated molecular tomography is in need of an estimate of the average background absorption and reduced scattering coefficient, but it is less sensitive to theirexact knowledge ortheirbackgr ound heterogeneity, which is similar to differential DOT (Ntziachristos et al., 2001b). The ability of FMT to operate with constant-wave systems allows for a practical and relatively inexpensive implementation of multiple detection channels using CCD cameras. In CW mode, CCD cameras offer very high sensitivity and low noise levels. A sample CCD system is proposed in figure 14.7.

The light source (a, in fig. 14.7) is generally a laser diode at the appropriate wavelength to excite the targeted fluorochrome. Multiple laser diodes at different wavelengths could be combined in the same system (either by time sharing or spectral separation) to excite multiple fluorochromes simultaneously. The light from the laser diode can be directed to an optical switch (d, in fig. 14.7) for time sharing one input to many outputs and directed with optical fibers (e, in fig. 14.7) at different points around the body of investigation ora specially designed *optical bore* (f, in fig. 14.7). The optical bore would contain the body of examination similarto a computed tomographic scannerora   magnetic resonance imager. Typically, such an implementation would require the animal to be immersed into a "matching fluid," such as a watersolution of a scatterer(e.g., TiO_2 particles) and an absorber (e.g., India ink or something similar) that matches the optical properties of the tissue being investigated. The matching fluid serves virtually the same function as gel is to ultrasound. Fiber bundles (g, in fig. 14.7) can be used to collect photons through the turbid medium and direct them onto the CCD chip (j, in fig. 14.7) either by direct fiber coupling or by an appropriate positioning arrangement (h, in fig. 14.7) so that the fiber output can be imaged via a lens system as depicted in figure 14.7. Appropriate filters (i, in fig. 14.7) are necessary to reject intrinsic or fluorescence light according to the measurement performed and the background ambient light. A reference measurement could also be introduced to account for temporal variations in laser intensity. Figure 14.7 shows that such approach has been realized using a beam splitter (b, in fig 14.7) to direct part of the source light directly onto the CCD chip via an optical fiber(c, in fig. 14.7). The use of an optical bore and matching fluid is mainly for illustration purposes and is by no means limiting. Different schemes, including direct fiber positioning on the tissue surface, at arbitrary geometries, including geometries that could be used by endoscopic probes, could be devised.

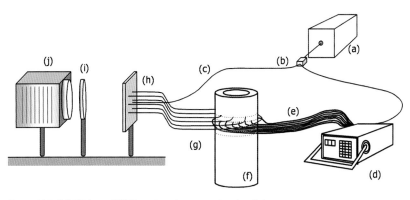

Figure 14.7 A CCD-based FMT system (see text for details).

Generally, the use of CCD cameras can allow the parallel detection of multiple detection channels. In principle, the number of channels that can be implemented equals the number of pixels of the CCD chip and can easily exceed vectors of 1000×1000 pixels ormor e. However, it would be advantageous to bin pixels together to improve the signal-to-noise ratio, because the diffuse nature of the signals does not generally require detection with high spatial sampling (Culveret al., 2000). A potential disadvantage of CCD cameras is dynamic range, which rarely exceeds three orders of magnitude (although the dynamic range of a measurement can be improved by summing multiple frames). Generally, for measurements through tissues, an order of magnitude drop is expected every ≈ 2 cm ormor e of photon propagation (for organs with high absorption such as brain or liver, an order of magnitude drop is expected every 1–1.5 cm). Therefore, for measurement schemes that may include measurements more than 4 to 6 cm apart, different detection technology would be required, such as independent silicon photon detectors that attain dynamic ranges that may exceed 10 orders of magnitude (Schmitz et al., 2000). Obviously, it is more complicated to increase the numberof channels forsystems that have independent parallel photon detectors. Such systems are generally prone to more experimental errors because each independent detector has much larger gain and noise variation than the spatial variation encountered at the surface of a common CCD chip.

In Vivo Imaging

Fluorescence-mediated molecular tomography has been recently used for the imaging of cathepsin B activity from within deep structures. Figure 14.8 shows a representative experiment of the results obtained from 9L gliosarcomas stereotactically implanted into unilateral brain hemispheres of nude mice, as cathepsin B activity had been implicated in glioma invasion (Rempel et al., 1994; Yan et al., 1998; Demchik et al., 1999). Correlative magnetic resonance imaging was performed to determine the presence and location of tumors prior to the FMT imaging studies. Figures 14.8A and 14.8B depicts the gadolinium-enhanced tumor (enhancement is shown in a green color map superimposed onto a T1-weighted image) on axial (fig. 14.8A) and sagittal (fig. 14.8B) slices. Figures 14.8C, E, and F depict the three consecutive FMT slices obtained from top to bottom of the volume of interest. The location and volume covered by the three slices is indicated in figure 14.8B by thin white horizontal lines. Figure 14.8C shows marked local probe activation relative to adjacent slices, which is congruent with the location of the tumor identified on the magnetic resonance images. Figure 14.8D shows a superposition of the magnetic resonance axial slice passing through the tumor (fig. 14.8A) onto the corresponding FMT slice (fig. 14.8C) after appropriately translating the magnetic resonance image to the actual dimensions of the FMT image. For coregistration purposes, we used special water-containing fiducials and body marks that facilitated matching of the magnetic resonance and FMT orientation and animal positioning. The in vivo imaging data correlated well with surface-weighted reflectance imaging of the excised brain. Figure 14.8G depicts the axial brain section through the 9L tumor examined with white light by using a CCD camera mounted onto a dissecting microscope. Figure 14.8H shows the same brain section imaged using a previously developed reflectance imaging system (Mahmood et al., 1999) at the excitation wavelength (675 nm), and figure 14.8I is the fluorescence image obtained at the emission light wavelength using appropriate three-cavity cut-off filters that demonstrate marked fluorescent probe activation, congruent with the tumor position identified by gadolinium-enhanced magnetic resonance imaging and FMT. These results confirm that cathepsin B can be used as an imaging marker (Weissleder et al., 1999), because the protease is produced in considerable amounts by tumor cells and by recruited host cells (Koblinski et al., 2000). Cathepsin B expression in the tumors was further confirmed by immunohistochemistry, Western blotting, and reverse transcription–polymerase chain reaction.

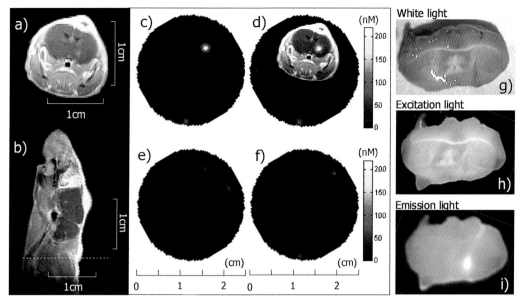

Figure 14.8 (A–I) In vivo FMT of cathepsin B expression levels in 9L gliosarcomas stereotactically implanted into unilateral brain hemispheres of nude mice. (A, B) Axial (A) and sagittal (B) magnetic resonance slices of an animal implanted with a tumor, which is shown in green after gadolinium enhancement. (C, E, F) Consecutive FMT slices obtained from top to bottom from the volume of interest shown in B by thin white horizontal lines. Superposition of the magnetic resonance axial slice (D) passing through the tumor (A) onto the corresponding FMT slice (C) after appropriately translating the magnetic resonance image to the actual dimensions of the FMT image. (G, H) Axial brain section through the 9L tumor imaged with white light (G) and with monochromatic light (H) at the excitation wavelength (675 nm). (I) Fluorescence image of the same axial brain section demonstrating marked fluorescent probe activation, congruent with the tumor position identified by gadolinium-enhanced magnetic resonance imaging and FMT.

Sensitivity

A significant advantage of fluorescent measurements is their high sensitivity, which allows detection of pico- to femtomole concentrations of fluorescent probes. The detection sensitivity of FMT depends on the optical properties of the tissue investigated and the fluorochrome depth. A second important parameter is the emission signal response as a function of light intensity and fluorochrome concentration. Deviation from linear behavior indicates photobleaching or quenching phenomena that could result in quantification errors and data misinterpretation. In the following paragraphs we show experimental measurement of the fluorescence response for the common fluorescent dye Cy5.5 (Amersham Pharmacia Biotech, Piscataway, New Jersey). In the next section, we use such measurements to predict the efficiency of fluorescence to propagate in the breast, lung, brain, and muscle.

The fluorescence counts collected as a function of light power and Cy5.5 concentration are shown in figures 14.9A and 14.9B in linear and logarithmic plots, respectively. The measurements were obtained using a diffuse medium with geometry shown in figure 14.10. A fused quartz tube (Wilmad Glass, Buena, New Jersey) 4 mm in diameter and 8 mm in length, containing the fluorochrome at different concentrations, was immersed into a medium with breastlike optical properties (absorption coefficient, $\alpha_a = 0.04$/cm and reduced scattering coefficient $\mu_s' = 10$/cm). The concentration range examined was 1 nM to 800 nM and it represents a wide span of biologically expected fluorochrome accumulations. The measured fluorescence count was obtained for 5 s of exposure time using the detector marked in figure 14.10 with the letter Z as a result of a source at the position marked with the letter A. The signal plotted is the actual fluorescence measurement minus the background signal

(a)

(b)

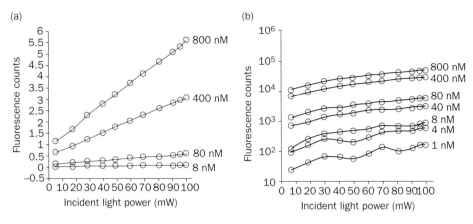

Figure 14.9 (A, B) Fluorescence responses from the experimental setup plotted in figure 14.10. The plots depict, in linear and semilogarithmic plots, the responses detected by detector Y (fig. 14.10).

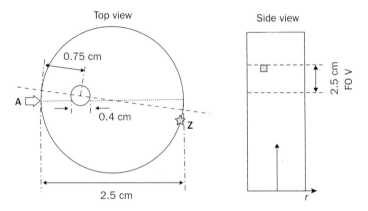

Figure 14.10 (A, B) Experimental setup used in the measurement of figure 14.9. FOV, field of view.

recorded by the same detector at 0 nM fluorochrome concentration. Figures 14.9A and 14.9B is the linear plot and semilogarithmic plot, respectively, of the fluorescence counts obtained.

Figure 14.9 demonstrates a linear response for the entire range of light powers and concentrations. This finding indicates that no self-quenching effects or photobleaching occur with the investigated indocyanine dye. Furthermore, concentrations of 1 nM (100 fM Cy5.5 dye) are clearly detectable, even when using low light power. This is particularly evident on the semilogarithmic plots. By appropriately selecting the light power and acquisition times, subnanomole concentrations could be also detected forsmallervolume fractions.

Quantification

The feasibility to reconstruct in three dimensions and to quantify fluorochromes embedded in diffuse media with the reported instrument has been demonstrated (Ntziachristos and Weissleder, 2001). Herein we investigate another aspect of the reconstruction performance: the linearity of the reconstructed fluorochrome concentration as a function of the real fluorochrome concentration. Although a linear relation between fluorescence signals and fluorochrome concentrations is shown in figure 14.9, tomographic techniques based on perturbation methods have been reported to give nonlinear responses (Kak and Slaney, 1988) and underestimate the perturbation magnitude as this magnitude increases. This is

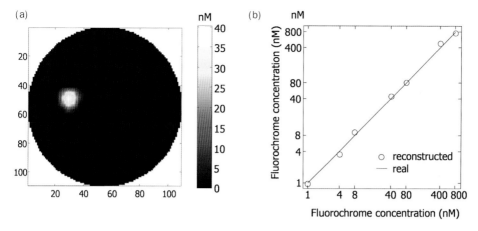

Figure 14.11 (A, B) Reconstructions of a single fluorochrome for different fluorochrome concentrations.

because perturbation methods operate optimally on the approximation that the perturbation signal detected is very small when compared with the intrinsic signal (i.e., the signal at the excitation wavelength).

To examine tomographic performance, we imaged the medium shown in figure 14.10 for the range of concentrations examined in figure 14.9 (1–800 nM). The laser power used in all reconstructions was 30 mW at the fiber tip and the acquisition time (integration time) was 10 s per source, resulting in a 4-minute acquisition time for a full three-dimensional reconstruction. The reconstruction segmented the field of view in three slices with 700 voxels perslice. The voxel size was $\approx 1 \times 1 \times 8.3$ mm^3. The inversion technique used was an iterative algebraic reconstruction technique that was run until 100 more iterative steps yielded a change that was less than 1%. The typical numberof iterative steps was ≈ 2000.

A typical reconstructed image is seen in figure 14.11A. The image shown is interpolated to a 4×4 finer grid than the original. Figure 14.11B plots the reconstructed chromophore concentration as a function of the expected concentration. The reconstructed value demonstrates a remarkably linear response for the whole range of examined concentrations. It is possible that perturbation methods applied in the fluorescence regime can reconstruct linearly in a larger concentration range when compared with absorption/scattering DOT reconstructions. This could be partly the result of the low quantum yield that produces fluorescent perturbation signals much smaller than the corresponding absorption or scattering perturbation signals, and partly the result of the differential nature of the forward problem.

Molecular Quantification and Specificity

One of the majorcontributions of FMT is the ability to obtain quantitative insights of enzyme concentration and to differentiate between different enzymes. The FMT capacity to associate fluorochrome quantification with enzyme concentration is shown in figure 14.12. The experimental setup, shown in figure 14.12A, was such that four capillaries contained 200 µL of a 1-µM cathepsin B-sensitive imaging probe (Weissleder et al., 1999) to which was added either0, 5, 25, or50 µg purified cathepsin B (Calbiochem, La Jolla, California). The capillaries were enclosed in turbid media (resin insert) as shown in figure 14.12A and were imaged using an FMT imager (c.f. figure fig. 14.7) over time, with each acquisition taking 2 minutes. The surrounding turbid media was constructed out of polyester resin to which TiO$_2$ particles and India ink had been added to simulate tissue scattering and absorption similar to breast tissue. The reconstructed values shown in figure 14.12B demonstrate

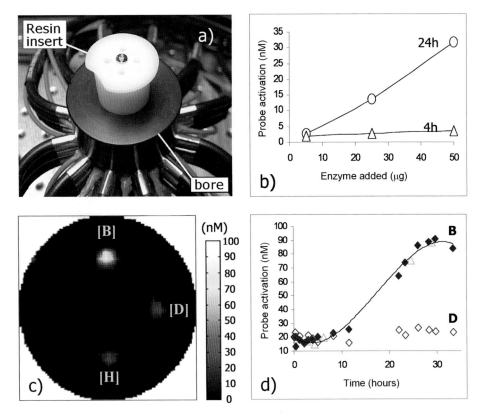

Figure 14.12 (A) Photograph of the turbid resin tube (shown inserted into the optical chamber), used for two in vitro experiments summarized here. The four capillaries shown were filled with 1 μM of a cathepsin B-sensitive near-infrared activatable probe for both experiments. (B) Probe activation as a function of cathepsin B concentration obtained with FMT at 4 and 24 hours after enzyme addition. (C) Fluorescence-mediated molecular tomographic reconstruction obtained from the middle slice of the resin tube 24 hours after 25 μg cathepsins (B, D, H) were added in different capillaries, as marked on the image. The fourth capillary did not contain any enzyme. (D) Activation kinetics obtained with FMT from the capillaries containing cathepsins B and D. The orange triangles indicate measurements obtained in parallel with a fluorometer in the absence of the turbid medium.

a linear relation between enzymatic concentration and fluorochrome activation covering a range of physiologically encountered enzyme levels obtained 4 and 24 hours after enzyme administration.

Fluorescence-mediated molecular tomography has also been shown to be capable of resolving specific enzymatic activity by injecting the same cathepsin B-sensitive probe at environments rich in different enzymes. Cathepsins B, D, and H were added in different capillaries of the same experimental setup shown in figure 14.2A. The enzyme concentrations were kept at 25 μg/capillary diluted in appropriate buffers. All capillaries contained the cathepsin B-sensitive near-infrared fluorescent probe (1 μM). The fourth capillary was absent an enzyme for control purposes. Figure 14.12C shows the reconstructed middle slice 24 hours after substrate/enzyme interaction. Activation of the cathepsin B-sensitive probe by cathepsin B is clearly evident, whereas it was lacking for the other cathepsins. The kinetics of probe activation resulting from cathepsin B versus cathepsin D is shown in figure 14.12D. Cathepsin H demonstrated a very similar activation pattern with cathepsin D, whereas the fourth tube demonstrated minimal background fluorescence. Measurements performed under identical conditions in nonturbid media using a fluorometer (F4500, Hitachi, Chula

Vista, California) are plotted in figure 14.12D as triangles and they verify the activation rate calculated by FMT.

In summary, these results show that (1) the technique is specific in differentiating fluorescence activation imparted by different enzymes, that (2) there is a predictable relation between enzyme activity and generated fluorescence under the tested conditions, and that (3) measurements obtained by FMT imaging from within turbid media closely reflect those obtained using an analytical fluorometerundernontur bid conditions.

Resolution and Coregistration

The resolution limits of diffuse light imaging techniques have been studied theoretically, but there are limited experimental demonstrations on resolution limits, mainly because of the lack of high source detection systems. Similar to DOT, increasing the number of sources and detectors improves resolution not only as a result of the higher spatial sampling, but also as a result of the improvements in the signal-to-noise ratio achieved in increased data sets (Culver et al., 2000). Naturally, improvements of source and detector density do not scale linearly with resolution improvements. It is expected that the resolution for small-animal imaging would be on the order of 1 to 2 mm, whereas for larger tissues it would approach 4 to 5 mm (Culveret al., 2000).

Fluorescence-mediated molecular tomography can be combined with high-resolution reflectance imaging in superimposing surface architectural features and markers on the underlying three-dimensional reconstruction to correlate the tomographic results with anatomic features, especially in animal imaging. Furthermore, the combination of FMT with high-resolution imaging methods that primarily target structure, such as X-ray computed tomography or magnetic resonance imaging, could yield a superior hybrid imaging modality in which highly sensitive molecular information and high resolution is achieved. Such an approach is feasible as a result of the optical component compatibility with most other medical imaging modalities and has been recently applied to performing clinical investigations of the breast using a combined magnetic resonance imaging and DOT system (Ntziachristos et al., 2000).

CLINICAL FEASIBILITY

Clinical imaging of fluorescent probes would require detection of fluorescent signals that have propagated for several centimeters into tissue. Figure 14.13 demonstrates the predicted fluorescence photons detected at the boundary of different tissues containing a 100-⎡L tumorlike fluorescent volume, for different tissue diameters. This calculation is based on experimental measurements obtained from a diffuse volume (similar to the one shown in fig. 14.10). The experimental measurements obtained were similar to the ones shown in figure 14.9, and used a 100-μL fluorescent volume containing 100 nM (10 pM) Cy5.5 dye, which is a typical uptake of macromolecules into malignant tumors, as previously determined by biodistribution studies (Marecos et al., 1998) 24 hours after administration. The FMT imagerused forthese measurements has been briefly discussed in the past (Ntziachristos and Weissleder, 2001). These experimental measurements were used to calibrate an analytical solution of the diffusion equation, based on the Kirchhoff approximation (Ripoll et al., 2001) for cylindrical geometries. Using this experimentally calibrated solution, we computed the fluorescence signals expected from the human breast, lung, and brain, as a function of diameter (Ntziachristos et al., 2001). The simulated geometry assumed is similar to the experimental one shown in figure 14.10. The position and size of the "tumor" structure was kept constant and only the outer diameter of the surrounding diffuse medium

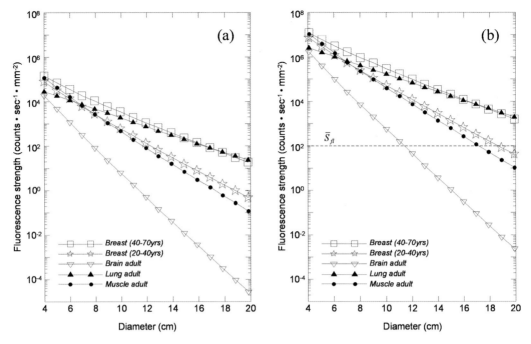

Figure 14.13 (A) Average fluorescent photon counts expected at the periphery of different organs resulting from the fluorochrome shown in figure 14.1, as a function of diameter. (B) Average fluorescent photon counts predicted for a two orders of magnitude improvement in detection technology. The $\bar{S}_{fl}$ horizontal line indicates a signal-to-noise ratio of 20 dB, assuming shot noise-limited systems.

Table 14-2. Optical properties used in the simulations

Tissue type (780 nm)	Absorption coefficient μ_a (cm^{-1})	Reduced scattering coefficient μ_s' (cm^{-1})	Literature source
Breast (ages 40–70 y)	0.03	9	Ntziachristos (2000)
Breast (ages 20–40 y)	0.05	11	Ntziachristos (2000)
Lung	0.01	20	Beek et al. (1997)
Brain (adult)	0.15	12	Ntziachristos et al. (1999b)
Muscle	0.15	9	Torricelli et al. (2001)

was varied from 4 to 20 cm. The optical properties used for the surrounding medium simulating different tissues are summarized in table 14.2 (Beek et al., 1997; Ntziachristos et al., 1999b; Torricelli et al., 2001).

Figure 14.13A indicates that the attenuation rate of breast tissue and the adult lung in the near infrared is one order of magnitude every 4 cm, the attenuation rate in denser breast and muscle approaches one order of magnitude every 3 cm, and the attenuation rate in brain tissue is approximately one order of magnitude every 1.5 cm. In absolute terms, figure 14.13A denotes that fluorescent photons from small cancerlike volumes can be detected in most tissues examined, even after propagating for several centimeters. This estimate is conservative because technological advances could currently lead to significant detection improvements, relative to the experimental setup used in this study (Ntziachristos and Weissleder, 2001). More efficient photon detection can readily yield a 40-fold detection improvement. The use of longeracquisition times and higherpowerlaser s could significantly increase the expected fluorescence photon strength, especially because we have found that no saturation orphotobleaching effects occurforlaserpower s up to 100 mW and fluorochrome concentrations that exceed 800 nM (c.f. fig. 14.9). For example, a twofold increase in laser strength and a fivefold increase in exposure time could gain one more magnitude in detection capacity.

To indicate how technological improvements could improve fluorescent detection, we plotted figure 14.13B, which predicts the photon counts expected for a projected two-order-of-magnitude increase in detection capacity relative to figure 14.13A. The $\overline{S}_{fl}$ horizontal line marks the 20-dB signal-to-noise ratio level, assuming shot noise-limited detection. In the absence of background signals, this 20-dB signal-to-noise ratio level could also represent a typical reconstruction limit for fluorescence-mediated reconstructions of single fluorochromes in tissuelike media. Background signals resulting from nonspecific dye distribution in the vascular system could, however, become a limiting factor in reconstructions and lesion detection. After appropriate subtraction schemes, background fluorescence can be seen as "biological noise" that raises the $\overline{S}_{fl}$ bar(Wu et al., 1997). On the otherhand, optimizing the optode arrangement and density using higher information data sets, and using algorithms that more efficiently account for measurement noise (Eppstein et al., 1999) could be used toward compensating for this increase. Considerations like this could be addressed on a case-to-case scenario. These findings, however, demonstrate that in vivo fluorescence imaging of human tissues becomes an issue of contrast rather than of propagation feasibility. Most important to the latter argument is that fluorescence offers mechanisms for multiple-fold background suppression, as described earlier. These technologies could significantly minimize background signals in vivo and, combined with instrumental and algorithmic improvements, could potentially allow fluorescence imaging of even larger diameters than the ones predicted in figure 14.12.

SUMMARY

Despite advances in medical imaging technologies during the past two decades, we are still limited in ourability to detect tumors orotherdiseases in theirear liest stage of formation, phenotype tumors during complex cycles of growth, and invasion and metastases; to use imaging techniques to accelerate drug testing; or to use imaging as an objective end-point for tailoring therapies in a given individual. Similar limitations also exist in neuro-degenerative, cardiovascular, and immunological diseases. Traditional cross-sectional imaging techniques such as magnetic resonance, computed tomography, and ultrasound primarily rely on physical (e.g., proton density, relaxation times, absorption, scattering) and/or occasionally physiological parameters as the main source of image signal. We described a new set of technologies that could allow imaging of specific molecular markers in vivo. This has been enabled by several developments: (1) the development of "smart" and targeted fluorescent molecular probes developed for an increasing number of targets identified by gene array profiles, (2) miniaturization and development of highly sensitive imaging equipment that allows recording of fluorescent signals at pico- to femtomole accumulations, and (3) the ability to produce tomographic images of near-infrared fluorescent probe distribution in whole bodies. The technology has been applied to imaging small animals, but it could be expanded to clinical use as well, as a result of the penetration of near-infrared fluorescent signals for several centimeters in tissues. Furthermore, the optical technology is relatively inexpensive, can be made portable, and uses nonionizing radiation and stable molecular markers. These features can allow easy laboratory and benchtop use, and enable monitoring of molecularevents repeatedly and overtime. The combination of molecularbeacons and optical imaging technologies is expected to play a fundamental role in biomedicine during the next decade, and continued developments in probe design and imaging technologies will undoubtedly further expand current capabilities.

REFERENCES

Achilefu, S., et al. Novel receptor-targeted fluorescent contrast agents for in vivo tumor imaging [in-process citation]. Invest. Radiol., 2000, 35(8):479–485.

Arridge, S. R. Optical tomography in medical imaging. Inverse Problems, 1999, 15(2):R41–R93.

Barbour, R., et al. MRI-guided optical tomography: Prospects and computation for a new imaging method. IEEE Comp. Sci. Eng., 1995, 2(4):63–77.

Becker, A., et al. Receptor-targeted optical imaging of tumors with near-infrared fluorescent ligands. Nat. Biotechnol., 2001, 19(4):327–331.

Beek, J. F., et al. The optical properties of lung as a function of respiration. Phys. Med. Biol., 1997, 42:2263–2272.

Benaron, D.A., et al. Noninvasive functional imaging of human brain using light. J. Cereb. Blood Flow Metab., 2000, 20(3):469–477.

Bogdanov, A. A., Jr., et al. A new macromolecule as a contrast agent for MR angiography: Preparation, properties, and animal studies. Radiology, 1993, 187(3):701–706.

Bogdanov, A. A., Jr., et al. An adduct of cis-diaminedichloroplatinum(II) and poly(ethylene glycol)poly(L-lysine)-succinate: Synthesis and cytotoxic properties. Bioconjug. Chem., 1996, 7(1):144–1449.

Bremer, C., C. Tung, and R. Weissleder. Imaging of metalloproteinase2 inhibition in vivo. Nat. Med., 2001, 7(6):743–748.

Buehler, C., et al. Innovations in two-photon deep tissue microscopy. IEEE Eng. Med. Biol. Mag., 1999, 18(5):23–30.

Callahan, R. J., et al. Preclinical evaluation and phase I clinical trial of a 99mTc-labeled synthetic polymer used in blood pool imaging. AJR Am. J. Roentgenol., 1998, 171(1):137–143.

Chance, B., et al. Highly sensitive object location in tissue models with linearin-phase and anti-phase multi-element optical arrays in one and two dimensions. Proc. Natl. Acad. Sci. USA, 1993, 90(8):3423–3427.

Chang, J. H., H. L. Graber, and R. L. Barbour. Imaging of fluorescence in highly scattering media. IEEE Trans. Biomed. Eng., 1997a, 44(9):810–822.

Chang, J., et al. Optical imaging of anatomical maps derived from magnetic resonance images using time-independent optical sources. IEEE Trans. Med. Imaging, 1997b, 16(1):68–77.

Chernomordik, V., et al. Inverse method 3-D reconstruction of localized in vivo fluorescence: Application to Sjögren syndrome. IEEE J. Sel. Top. Quant. Electr., 1999, 5(4):930–935.

Colak, S., et al. Clinical optical tomography and NIR spectroscopy for breast cancer detection. IEEE J. Sel. Top. Quant. Electr., 1999, 5(4):1143–1158.

Contag, C. H., et al. Visualizing gene expression in living mammals using a bioluminescent reporter. Photochem. Photobiol., 1997, 66(4):523–531.

Culver, J., et al. Optimization of optode arrangements for diffuse optical tomography: A singular value analysis. Opt. Lett., 2000, 26(10):701–703.

Dellian, M., et al. Vascular permeability in a human tumour xenograft: Molecular charge dependence. Br. J. Cancer, 2000, 82(9):1513–1518.

Demchik, L. L., et al. Cathepsin B and glioma invasion. Int. J. Dev. Neurosci., 1999, 17(5–6):483–494.

Eppstein, M. J., et al. Biomedical optical tomography using dynamic parameterization and Bayesian conditioning on photon migration measurements. Appl. Opt., 1999, 38(10):2138–2150.

Gonzalez, S., et al. Characterization of psoriasis in vivo by reflectance confocal microscopy. J. Med., 1999, 30(5–6):337–356.

Gulnik, S., et al. Design of sensitive fluorogenic substrates for human cathepsin D. FEBS Lett., 1997, 413:379–384.

Hielscher, A., A. Klose, and K. Hanson. Gradient-based iterative image reconstruction scheme for time-resolved optical tomography. IEEE Trans. Med. Imaging, 1999, 18(3):262–271.

Hillman, E. M. C., et al. Time resolved optical tomography of the human forearm. Phys. Med. Biol., 2001, 46(4):1117–1130.

Hope-Ross, M., et al. Adverse reactions due to indocyanine green. Ophthalmology, 1994, 101(3):529–533.

Jain, R. Delivery of molecular medicine to solid tumors. Science, 1996, 271:1079–1080.

Jiang, H., et al. Improved continuous light diffusion imaging in single- and multi-target tissue-like phantoms. Phys. Med. Biol., 1998, 43(3):675–693.

Kak, A., and M. Slaney. Principles ofComputerized Tomographic Imaging. 1988, New York: IEEE Press.

Koblinski, J. E., M. Ahram, and B. F. Sloane. Unraveling the role of proteases in cancer. Clin. Chim. Acta, 2000, 291(2):113–135.

Konig, K. Multiphoton microscopy in life sciences. J. Microsc., 2000, 200(Pt 2):83–104.

Korlach, J., et al. Characterization of lipid bilayer phases by confocal microscopy and fluorescence correlation spectroscopy. Proc. Natl. Acad. Sci. USA, 1999, 96(15):8461–8466.

Lewin, M., et al. Tat peptide-derivatized magnetic nanoparticles allow in vivo tracking and recovery of progenitor cells. Nat. Biotechnol., 2000, 18(4):410–414.

Licha, K., et al. Hydrophilic cyanine dyes as contrast agents for near-infrared tumor imaging: Synthesis, photophysical properties and spectroscopic in vivo characterization. Photochem. Photobiol., 2000, 72(3):392–398.

Mahmood, U., et al. Near infrared optical imaging system to detect tumor protease activity. Radiology, 1999, 213: 866–870.

Marecos, E., R. Weissleder, and A. Bogdanov, Jr. Antibody-mediated versus nontargeted delivery in a human small cell lung carcinoma model. Bioconjug. Chem., 1998, 9:184–191.

Masters, B. R., P. T. So, and E. Gratton. Multiphoton excitation fluorescence microscopy and spectroscopy of in vivo human skin. Biophys. J., 1997, 72(6):2405–2412.

McCawley, L. J., and L. M. Matrisian. Matrix metalloproteinases: Multifunctional contributors to tumor progression. Mol. Med. Today, 2000, 6(4):149–156.

Monsky, W. L., et al. Augmentation of transvascular transport of macromolecules and nanoparticles in tumors using vascular endothelial growth factor. Cancer Res., 1999, 59(16):4129–4135.

Muldoon, L. L., et al. Comparison of intracerebral inoculation and osmotic blood–brain barrier disruption for delivery of adenovirus, herpesvirus, and iron oxide particles to normal rat brain. Am. J. Pathol., 1995, 147(6):1840–1851.

Neuwelt, E. A., et al. Delivery of virus-sized iron oxide particles to rodent CNS neurons. Neurosurgery, 1994, 34(4): 777–784.

Ntziachristos, V. Concurrent diffuse optical tomography, spectroscopy and magnetic resonance of breast cancer. In *Bioengineering,* Ph.D. dissertation. 2000, Philadelphia, Pa.: University of Pennsylvania.

Ntziachristos, V., B. Chance, and A. G. Yodh. Differential diffuse optical tomography. Opt. Express, 1999a, 5(10): 230–242.

Ntziachristos, V., X. H. Ma, and B. Chance. Time-correlated single photon counting imager for simultaneous magnetic resonance and near-infrared mammography. Rev. Sci. Instrum., 1998, 69(12):4221–4233.

Ntziachristos, V., J. Ripoll, and R. Weissleder. Would near-infrared fluorescence signals propagate through large human organs for clinical studies? Opt. Lett. 2001a, 27:333–335

Ntziachristos, V., and R. Weissleder. Experimental three-dimensional fluorescence reconstruction of diffuse media using a normalized Born approximation. Opt. Lett., 2001, 26(12):893–895.

Ntziachristos, V., et al. Multichannel photon counting instrument for spatially resolved near infrared spectroscopy. Rev. Sci. Instrum., 1999b, 70(1):193–201.

Ntziachristos, V., et al. Concurrent MRI and diffuse optical tomography of breast after indocyanine green enhancement. Proc. Natl. Acad. Sci. USA, 2000, 97(6):2767–2772.

Ntziachristos, V., et al. Diffuse optical tomography of highly heterogeneous media. IEEE Trans. Med. Imaging, 2001b, 20(6):470–478.

Oleary, M. A., et al. Experimental images of heterogeneous turbid media by frequency-domain diffusing-photon tomography. Opt. Lett., 1995, 20(5):426–428.

Oleary, M. A., et al. Fluorescence lifetime imaging in turbid media. Opt. Lett., 1996, 21(2):158–160.

Paithankar, D. Y., et al. Imaging of fluorescent yield and lifetime from multiply scattered light reemitted from random media. Appl. Opt., 1997, 36(10):2260–2272.

Pogue, B., et al. Hemoglobin imaging of breast tumors with near-infrared tomography. Radiology, 2000, 214(2):G05H.

Rajadhyaksha, M., et al. In vivo confocal scanning laser microscopy of human skin: Melanin provides strong contrast. J. Invest. Dermatol., 1995, 104(6):946–952.

Rempel, S. A., et al. Cathepsin B expression and localization in glioma progression and invasion. Cancer Res., 1994, 54(23):6027–6031.

Reynolds, J. S., et al. Imaging of spontaneous canine mammary tumors using fluorescent contrast agents. Photochem. Photobiol., 1999, 70(1):87–94.

Ripoll, J., et al. The Kirchhoff approximation for diffusive waves. Phys. Rev. E, 2001, 64(5):051917-1-8.

Schmitz, C., H. Graber, and R. Barbour. A fast versatile instrument for dynamic optical tomography, 94–96. In *Advances in Optical Imaging and Photon Migration, OSA.* Technical digest. Ed. Fujimoto, J., and Patterson, M. Washington, D.C., Optical Society of America.

Schotland, J.- C. Continuous wave diffusion imaging. J Opt. Soc. Am., 1997, A-14(1):275–279.

Sevick-Muraca, E. M., et al. Fluorescence and absorption contrast mechanisms for biomedical optical imaging using frequency-domain techniques. Photochem. Photobiol., 1997, 66(1):55–64.

So, P. T., et al. New time-resolved techniques in two-photon microscopy. Cell. Mol. Biol., 1998, 44(5):771–793.

Sweeney, T. J., et al. Visualizing the kinetics of tumor-cell clearance in living animals. Proc. Natl. Acad. Sci. USA, 1999, 96(21):12044–12049.

Toomre, D., and D. J. Manstein. Lighting up the cell surface with evanescent wave microscopy. Trends Cell Biol., 2001, 11(7):298–303.

Torricelli, A., et al. In vivo optical characterization of human tissues from 610 to 1010 nm by time-resolved reflectance spectroscopy. Phys. Med. Biol., 2001, 46(8):2227–2237

Tung, C. H., et al. Preparation of a cathepsin D sensitive near infrared fluorescence probe for imaging. Bioconjug. Chem., 1999, 10(5):892–896.

Tung, C., et al. In vivo imaging of proteolytic enzyme activity using a novel molecular reporter. Cancer Res., 2000, (60):4953–4958.

Tyagi, S. ,and F. R. Kramer. Molecular beacons: Probes that fluoresce upon hybridization. Nat. Biotechnol., 1996, 14(3):303–308.

Tyagi, S., S. A. E. Marras, and F. R. Kramer. Wavelength-shifting molecular beacons. Nat. Biotechnol., 2000, 18(11):1191–1196.

Weissleder, R., et al. In vivo imaging of tumors with protease-activated near-infrared fluorescent probes. Nat. Biotechnol., 1999, 17(4):375–378.

Wouters, F. S., P. J. Verveer, and P. I. Bastiaens. Imaging biochemistry inside cells. Trends Cell Biol., 2001, 11(5):203–211.

Wu, J., et al. Fluorescence tomographic imaging in turbid media using early-arriving photons and Laplace transforms. Proc. Natl. Acad. Sci. USA, 1997, 94(16):8783–8788.

Yan, S., M. Sameni, and B. F. Sloane. Cathepsin B and human tumor progression. Biol. Chem., 1998, 379(2):113–123.

Yodh, A. G., and B. Chance. Spectroscopy and imaging with diffusing light. Phys. Today, 1995, 48(3):34–40.

Zlokarnik, G., et al. Quantitation of transcription and clonal selection of single living cells with beta-lactamase as reporter. Science, 1998, 279(5347):84–88.

15

Revealing the Subtleties of Disease and the Nuances of the Therapeutic Response with Optical Reporter Genes

CHRISTOPHER H. CONTAG

The use of optical imaging to study biology in animal models and to detect markers of disease in humans provides a window through which we can view features of disease that are either not detectable by conventional approaches or that can be detected more efficiently with optics. Moreover, with optics we can often appreciate patterns in biological processes that would be otherwise overlooked or missed. Because the nuances of disease mechanisms and the subtleties of the response to therapy are key to understanding and resolving disease, imaging has become an essential tool forr evealing pathogenic mechanisms and for developing effective therapeutic strategies. Optical imaging tools comprise a key set of approaches for the study of biology in animal models, and are emerging as powerful methods for early disease detection, staging of progression, and assessing response to therapy in the clinic. Sensitive and rapid imaging methods based on the optical properties of tissues or the detection of reporter proteins and dyes provide real-time information that will contribute to our ability to unravel the mechanisms of disease, enable us to intervene with treatment strategies, and allow us to assess therapeutic outcomes.

Because optical imaging methods cover a range of scales, with resolutions from microns to centimeters, the preclinical and clinical applications are far-reaching, with great potential for changing the way these studies are performed. The clinical utility of optical imaging, however, is limited to a narrow set of applications in which the depth of penetration of light is sufficient to interrogate the intended target in mammalian tissues. These intended targets could be biological structures or molecules that are intrinsic to the tissue, or exogenous reagents that can label cells orstain tissues. The small numberof approved optical contrast agents for human use further confines optical imaging in the clinic to specific application areas. In contrast, in preclinical studies the applications are limitless because the study subjects are typically small laboratory rodents in which many tissues and organs are accessible to optical tools. In addition, the cells and tissues in laboratory animals are amenable to staining with a range of exogenous dyes and may even be manipulated genetically to express optical reporters. Perhaps the greatest impact that optical imaging will have on human health is by providing a greater understanding of mammalian biology and more rapid evaluation of new therapies in laboratory animals.

Because laboratory mice are well suited to evaluation by optical imaging (as a result of theirsmall size), the study of mammalian biology in these models has been greatly facilitated by optical imaging, and these tools have contributed immensely to our understanding of mammalian biology. The increasing number of reagents and reporter genes with optical signatures that can be "tuned" to reveal specific biological process will continue to uncover biological mechanisms in rodent models, and these models will accelerate drug development by providing rapid in vivo assays that reveal drug efficacy and enable target evaluation. Genetically manipulating transplanted cells, or those of the host animal, can be used to incorporate optical labels that are linked to cellular physiology, with the optical signals perpetuated through cell division. Reporter genes can be incorporated into animal models with both specificity for a unique cell or tissue type, and exquisite genetic control such that they can be turned on and off at will.

Among the large number of tools that are based on optical reporter genes, in vivo bioluminescence imaging (BLI) has had, and will continue to have, a significant role in optical imaging (Contag and Bachmann, 2002; Contag and Ross, 2002; Negrin and Contag, 2006). Bioluminescence imaging is a cornerstone technology in the emerging field of molecular imaging. This field is comprised of a set of tools that have molecular specificity and that can be used to assess biological function in living subjects. As a molecular imaging tool, BLI is based on the incorporation of light-emitting enzymes, called *luciferases*, into cellular physiology, and this is typically done by transferring the genes that encode these enzymes into cells (Hastings, 1996; Hastings and Morin, 2006; Wilson and Hastings, 1998; Wood, 1990). Alternatively, some luciferases can be conjugated to molecules that target cell surface markers or sense enzymatic activity, and they can be injected into the animal to reveal biological change (So et al., 2006a,b; Venisnik et al., 2006, 2007). Bioluminescence imaging is a versatile modality with tremendous signal-to-noise ratios (SNRs) and offers a large number of opportunities for examining molecular features of mammalian biology (Olafsen et al., 2004; Venisnik et al., 2006, 2007). In vivo fluorescent imaging (FLI) is analogous to BLI with some distinct differences, and this modality offers unique opportunities for understanding biology in animal models with the potential for clinical applications (Hsiung et al., 2008). Bioluminescence imaging and FLI can be incorporated into studies with single molecular tags or can be used in combinations of imaging approaches to create opportunities that are not possible with single modalities. The genes that encode luciferases or fluorescent proteins can be fused to the coding sequences of each other or to other reporter genes that can be detected by other modalities (Edinger et al., 2002; Ray et al., 2004). The resulting multi-functional reporters can be used for cross-validation or increased functionality (Ponomarev et al., 2004; Ray et al., 2003, 2004, 2007). The properties of the various reporter proteins and the parameters that control transmission of light through mammalian tissues determine the capabilities of each reporter; understanding these parameters is important for selecting the appropriate reporter for a given application.

OPTICAL IMAGING IN PRECLINICAL STUDIES

Optical imaging has led to the creation of visible animal models of human biology and disease, which has enabled the study of biology in the context of living tissues when the biology is intact and dynamic. Although many of the optical imaging approaches that are used with great efficiency in small laboratory animals may not translate easily to the clinic, the information that is generated by refining and accelerating preclinical studies can be used to inform to clinical studies by providing insights that can not be obtained with out imaging, or can be used to develop new therapeutic strategies. In this manner, the experimental design for the clinical study can be more effective, because more is known about the mammalian response from the preclinical studies. Optical imaging is extremely sensitive for detecting disease states in animal models. Figure 15.1 demonstrates detection of the response to therapy and, perhaps more importantly, optical imaging can be used to localize cancer cells when the disease is remission. The cells that persist after therapy are those that give

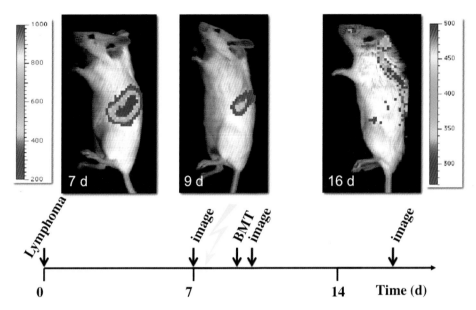

Figure 15.1 Demonstrating the response to conventional therapy in a mouse model of lymphoma. Radiation therapy (yellow arrow) and bone marrow transplantation (BMT) were used to treat an orthotopic mouse model of lymphoma (Edinger et al., 2003a) 7 days after initiation of the cancer. The lymphoma cells (Bcl1) constitutively expressed luciferase as a means of localizing the cancer cells to the spleen (days 7 and 9) and to the spinal column (day 16) using BLI. Despite significant reduction of the signal in the spleen after therapy, residual disease could be detected at day 16. Relapse from states of minimal residual disease such as demonstrated at day 16 remains a major treatment challenge that can be addressed using optical imaging.

rise to relapse and are the source of significant morbidity and mortality in cancer patients. The ability to study the small numberof cells that persist afterther apy will enable the development of new approaches to eliminating these cells and preventing relapse.

An example of the development of a therapy that may target the small numbers of persistent cancer cells is a combination therapy comprised of an immune cell population and an oncolytic virus (Thorne and Contag, 2007; Thorne et al., 2006). Using this dual biotherapy, a synergistic effect between the immune cell and the oncolytic virus was demonstrated with optical imaging, and this effect can be exploited foran effective means of eliminating small numbers of cancer cells (Thorne and Contag, 2007; Thorne et al., 2006). Combination therapies such as this dual biotherapy have a number of parameters that can dramatically affect efficacy, and they need to be tested if such therapies are to be optimized. This can best be done with optical imaging, because groups of animals can be monitored over time both to save effort and to reduce the numbers of animals that are required. Thus, optical imaging can have an impact on human health by improving the studies of animal models for increased understanding and more sophisticated testing of treatment strategies.

REPORTER GENES THAT ENCODE PROTEINS WITH DISTINCT OPTICAL SIGNATURES

The characteristics of reporter genes that make them well suited for in vivo optical imaging studies are well-defined. They should encode well-characterized gene products that have optical signatures that can be detected using deeply penetrating light and excitation and/or

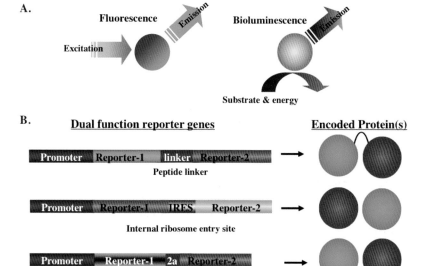

Figure 15.2 (A, B) Multifunctional reporter gene strategies. The benefits of fluorescent and bioluminescent reporter genes (A) can be combined using peptide linkers to make a single protein with two functional components, or the coding sequences can be combined to make a single mRNA that produces two proteins (B). A single message making two proteins is called *polycistronic*, and there are several genetic tools to create these genes. One strategy uses an internal ribosome entry site (IRES) and the other uses a ribosome slippage site from the second gene (2a) in the genome of the foot-and-mouth disease virus.

emission, and have the potential for a high SNR. With optical imaging, the SNR is determined by the levels and location of the reporter in the body, and the optical properties of mammalian tissues (Benaron et al., 1997; Jobsis, 1977; Rice et al., 2001; Troy et al., 2004). There are several types of reporter genes with optical signatures that have been used in vivo, and they include bioluminescent and fluorescent reporter genes as well as combinations of genes when two or more are coexpressed (figure 15.2). The coexpressed reporters take advantage of the strengths of each reporter such that multiple assays can be performed after transfer of a single genetic construct. For example, coexpression of luciferase and a fluorescent protein enables tracking of labeled cells in vivo using luciferase. After the cells are removed from the animal, they can be analyzed via fluorescence microscopy and/or flow cytometry (Edinger et al., 2002, 2003a).

Luciferases

Light-emitting enzymes, luciferases, have played a significant role in in vitro and cell culture, and thus they have become critical tools of discovery in biology. The transition to imaging in live animals using these reporters was first demonstrated with thin, transparent organisms (Millar et al., 1995; Plautz et al., 1997). Their use as a readout for biological function in mammals was first reported by Contag et al. in 1995. This report led to the use of luciferases as in vivo reporters of a wide range of biological processes in living mammals, including in vivo measures of gene expression (Contag et al., 1997, 1998), and the use of luciferase as a reporter of tumor burden and growth in laboratory rodents (Edinger et al., 1999; Sweeney et al., 1999). The ease of use of BLI has led to widespread use of this tool in biomedical research and drug discovery/development that spans the areas of research from stem cell biology to gene therapy. Data generated by BLI have been used to support applications to the Food and Drug Administration for many drugs as a result of the ability of imaging to accelerate preclinical studies and to provide additional data that cannot be obtained using otherbiological assays.

Luciferases used in BLI are derived from a variety of different nonmammalian species obtained from marine and terrestrial environments. To adapt these genes for use as reporters in mammalian cells and tissues, they have been modified extensively with changes that include, optimization of codon usage and removal of cryptic mammalian regulatory sequences that affect expression in mammalian cells (Hastings, 1996; Shapiro et al., 2005; Wood, 1990; Wood et al., 1989). The modifications also include removal of sequences that direct the encoding enzyme to subcellular compartments (peroxisomes [de Wet et al., 1987] orto the extracellularspace [Michelini et al., 2008; Shao and Bock, 2008; Tannous et al., 2005]). Most luciferases in nature emit with relatively short wavelengths and, thus, attempts have been made to improve their utility in mammals by altering the wavelength of emission. This has only been moderately successful, with a 615-nm peak for the longest emitting luciferases (both the click beetle [CBLuc] and the firefly [FLuc]) (Doyle et al., 2004; Loening et al., 2007; Stolz et al., 2003; Wood, 1990). All luciferases have broad emission spectra and many of these spectra are overlapping, which restricts the development of multiplexed bioluminescence assays. However, there are several unique chemistries that enable the use of two reporters in one animal (Bhaumik and Gambhir, 2002; Bhaumik et al., 2004), and spectral resolution of two luciferases is possible (Kadurugamuwa et al., 2003, 2005).

Because the transmission of light through mammalian tissue is most efficient at longer wavelengths as a result of the diminished absorbance of light by hemoglobin when wavelengths longer than 600 nm are used, the luciferases with the longest emission (FLuc and the red click beetle luciferase [CBLuc^{red}]) are most often used in vivo (Zhao et al., 2005). Up to 60% of the emission from these two enzymes is in the region of the spectrum above 600 nm when measured at 37°C (Zhao et al., 2005). The broad emission spectra of luciferases offer a solution to the problem of determining the depth of the source within the body. Because these broad spectra span the region that is differentially affected by hemoglobin absorption, a ratio of short to long wavelengths can be assessed, and the ratio indicates the depth of the signal within the body.

Bioluminescence imaging in its most common form is requires the incorporation of the reporter gene into the genome of the target cells. This can be a significant advantage for some studies because it allows both specificity and selectivity to be built into the animal model. Of course, this approach is limited to animal models and is one reason why both FLI and BLI will have more applications in preclinical imaging than clinical imaging. The versatility of imaging using optical reporter genes is extreme, given that any number of genes can be labeled. Relative to other reporter gene imaging modalities, BLI is sensitive, rapid, and relatively inexpensive. When applied to the study of laboratory animals, BLI has a relatively high throughput. Bioluminescence imaging and FLI are inherently functional imaging tools; however, as a result of the scattering nature of mammalian tissue, they are low-resolution modalities unless the overlying tissue is removed. Because luciferase reactions require oxygen and an energy source, the signals from these enzymes are often linked to the metabolic activity of the labeled cells. Although a disadvantage when the absolute number of reporter molecules needs to be determined, this requirement can be adapted foruse as a markerof cellularfunction ortissue physiology (Cecic et al., 2007). The use of optical reporter genes enables functional measurements of a wide range of biological processes.

Fluorescent Proteins

The first autocatalytic fluorescent protein that was used as a reporter gene to mark cellular function was GFP and, like luciferases, it was first used in small, transparent organisms

(Chalfie et al., 1994). The short excitation and emission wavelengths of GFP limit its ability to be detected in mammalian tissues (Tsien, 1998). Longerwavelengths of excitation and emission of fluorescent proteins is desirable, because they would allow for improved depth of imaging. Proteins with excitation and emission wavelengths in the range of 590 to 680 nm have been described orar e underdevelopment (Giepmans et al., 2006; Shaneret al., 2004, 2005; Shu et al., 2006). Microscopic detection of cells labeled with fluorescent proteins can be performed after removing the overlying tissues and placing a traditional fluorescent or confocal microscope on the target tissue (Izzo et al., 2001). Multiphoton excitation using ultrashort, pulsed infrared light improves high-resolution imaging in scattering tissue, and labeled cells at depths of up to approximately 1 mm can be studied in vivo (Helmchen and Denk, 2005; Zipfel et al., 2003). Alternatively, imaging of fluorescent markers at greater depths is possible using serial reconstruction techniques in fixed tissue (Tsai et al., 2003) or through tomographic imaging in live tissues (Keller et al., 2006; Ntziachristos et al., 2005).

The intrinsic brightness of fluorescent proteins is controlled by characteristics shared with otherfluor s such as the extinction coefficient, quantum yield, and photostability, but in addition it is also determined by the rate and efficiency of protein maturation (Shaner et al., 2005). Forphotostability, it is essential to have the optimal excitation filtersets to obtain the best performance from the fluorescent protein. Fluorescent proteins are stable in general and can accumulate over time, which on one hand compromises the quantitative assessment of dynamic processes, but on the other hand the accumulation of reporters increases the signal from a given cell. This stability makes it possible to assay reporter levels in removed cells and tissues and, therefore, offers the benefit of analyses after biopsy ornecr opsy (Contag et al., 2000; Rice et al., 2001; Shaneret al., 2005; Shu et al., 2006). In contrast, the activity of luciferases is inherently short-lived, enabling dynamic measures of biological changes in cells and in the tissues of live animals (Contag and Bachmann, 2002; Contag et al., 2000; Rice et al., 2001). Bioluminescence also differs from fluorescence in that the signal is generated by a chemical reaction, rather than the absorption of excitation of light (figure 15.2). Because excitation is not required for bioluminescence, the background signals are dramatically less with bioluminescence than with fluorescence, and this significantly improves SNR.

New Fluorescent Proteins

The need to improve in vivo assays based on fluorescent proteins, increase the capability of multiplexing, and develop additional partners for FRET has driven the development of new monomeric red, orange, and yellow fluorescent proteins, which are largely derived from the red fluorescent protein from *Discomsoma sp*. The longerexcitation and emission wavelengths of these proteins range from 550 to 600 nm, and thus have reduced autofluorescence and increased depth of penetration (Shaner et al., 2004; Shu et al., 2006). Several new proteins have been reported that have increased tolerance to fusion with other proteins at both the N- and C-termini of the fluorescent protein (Shaner et al., 2004, 2005; Shu et al., 2006). Proteins with improvements in quantum yields, extinction coefficients, and photostability have also been described (Shaner et al., 2004, 2005; Shu et al., 2006). These proteins are increasing the versatility of bioassays and leading to advances in preclinical optical imaging of biological function.

Multifunctional Reporter Genes

Combinations of reporter genes can be generated by creating gene fusions during which multiple proteins are made from a single mRNA (polycistronic message) (Cao et al., 2004, 2005; O'Connell-Rodwell et al., 2004). Gene fusions can also be generated such that the reporter genes are comprised of multiple protein domains, within single large reporter

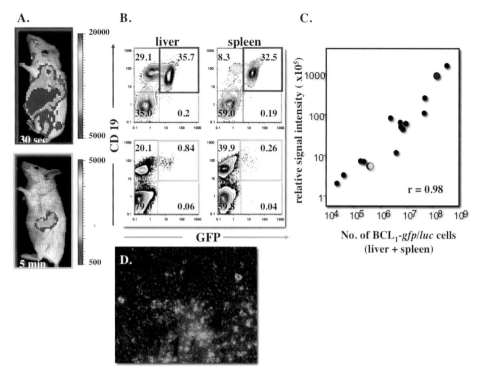

Figure 15.3 Dual-function reporter genes serve to linkin vivo and in vitro assays. In this lymphoma model, the tumor cells are labeled through the expression of luciferase and GFP as a fusion protein with a peptide linker (figure 15.2). (A, B) Mice at different stages of disease progression were imaged using the luciferase signal (A), the animals were euthanized, and then the tumor burden in the liver and spleen was measured by flow cytometry (B). (C) The data obtained in vivo correlated well with the cytometric data obtained on excised tissues using the GFP signal in 15 animals. Complete data sets are shown for only two animals; the data are boxed in red or blue and the data points are shown in red and blue. (D) The GFP signal also enables fluorescent microscopy on the excised tissues. Dual or multifunctional reporters can be used to enhance a preclinical study through validation and providing additional data.

protein (Ray et al., 2004, 2007). These multifunctional reporter genes (figure 15.2) can be used to link in vitro and in vivo assays, or to enable both macro- and microscopic imaging within a single study. They can also be used to connect BLI measurements in mice based on the luciferase signal to ex vivo assays, such as flow cytometry (figure 15.3), and fluorescence microscopy, which use the fluorescent signal (Edinger et al., 2003a). Multireporter gene fusions were first described using luciferase and GFP for use in cells and flies (Day et al., 1998). The ease of linking genes through genetic tools has led to the development of a number of these reporters, which can be detected by two or more modalities (Edinger et al., 2003a; Ponomarev et al., 2004; Ray et al., 2004). Triple gene fusions have been created and serve to link optical imaging to other modalities such as PET (Ponomarev et al., 2004; Ray et al., 2004, 2007). These multifunctional reporter genes can greatly increase the flexibility of an approach relative to a single reporter gene for in vivo assays (Edinger et al., 2003a, b; Mandl et al., 2003).

   Bioluminescent and fluorescent reporters often have complementary functions, including providing optimal macro- and microscopic detection in vivo, and linked in vivo and ex vivo assays. Macroscopic imaging with bioluminescence can provide information relative to when and where to look via microscopic detection, and can guide the sampling time and tissue type for analyses by procedures that are conducted after biopsy or necropsy (figure 15.3). Image guidance greatly enhances animal models and optimizes the data that can be obtained during preclinical studies. Such approaches will continue to increase our understanding of mammalian biology, refine preclinical studies, and accelerate the development of new therapeutic approaches. Optical imaging has become an essential tool for predictive and informative studies of small-animal models of human biology and disease.

Expression of Reporter Genes

How a reporter gene is to be used determines how it is designed and assembled for use in vivo. If optical reporter genes are to be used as an indication of tumor burden or to assess the trafficking patterns of a cell population, it is necessary to have the reporter constitutively on in the target cell, and thus the reporter is designed to be expressed from a strong, constitutively expressed promoter (such as those from the β actin, orubiquitin c genes, or from viruses such as cytomegalovirus) (Cao et al., 2004, 2005). Integration of these genetic constructs into the genome of the target cell permanently labels the cell, which can be accomplished using a variety of genetic tools including mammalian transposons and viral vectors, or can occur through random integration into the genome. Integration of the reporter gene into the genome of the cells obviates the problem of losing the optical signal through dilution as the cell divides, and eliminates the confounding signal resulting from dissociation of the optical signal from the viable target cell population. Reporter genes designed in this manner can, under rare circumstances, be silenced by chromosomal modifications. Viral promoters are prone to silencing, and promoters derived from mammalian cells are less likely to be silenced.

There are a number of promoters that are regulated in specific cell or tissue types and they can be used to create reporters that reveal the level of expression of the selected gene. Using this approach, the specific genetic regulatory elements (i.e., promoters) are fused to the appropriate reporter gene. Such reporters have been used in cultured cells as a primary means of assessing dynamic changes in gene expression, and these reporters are now being applied to similar research in living animal models. The transition from cell-based assays to in vivo assays has been enabled by advances in imaging instrumentation specifically designed for the study of small animals. A variety of reporter genes has also been developed that report the steps that follow transcription-in other words, protein–protein interaction (Contag and Ross, 2002; Laxman et al., 2002; Luker and Piwnica-Worms, 2004; Paulmurugan and Gambhir, 2003; Paulmurugan et al., 2002; Ray et al., 2002; Wang et al., 2001; Zhang and Kaelin, 2005). These reporters can be expressed either constitutively or from cell-specific promoters.

USES OF SMALL-ANIMAL IMAGING

Optical imaging can serve to validate innovations in other imaging modalities by refining the preclinical development of these approaches. By linking preclinical modalities that are based on clinical imaging such as PET, SPECT, and MRI with imaging tools that are best suited foranimal models, like BLI and FLI, we can betterdevelop strategies for imaging humans. In these cases, the optical imaging modality is often easierto use and can provide additional information that can support and guide the other modality. The use of gene fusions comprised of those encoding bioluminescent reporters and those that encode proteins that concentrate radiolabeled compounds such as thymidine kinase and its substrates (Ponomarev et al., 2004; Ray et al., 2004, 2007) can be used to develop new gene transfer techniques for clinical use, and enables more frequent imaging at a reduced cost.

The vast majority of preclinical BLI studies are not aimed at developing tools that translate to the clinic; rather, these studies are designed to validate potential therapeutic targets, test new compounds that target the basis of disease, and develop delivery tools that can carry established or experimental compounds to the target site. The use of BLI to evaluate compounds and methods that can be translated to the clinic comprises the largest number of the studies using luciferases to report biological functions. In a growing number of studies, BLI is used to reveal basic features of mammalian biology and biological responses to insult. Preclinical imaging using bioluminescent and other reporters has been used to accelerate drug development and to test new delivery schemes for therapy.

In cancerstudies, the sensitivity of BLI enables the study of early disease and minimal residual disease (figure 15.1), and, therefore, the initiating events or causes of relapse from states of remission can be investigated (Edinger et al., 2003a, 2003b; Shachaf et al., 2004). Most cancer patients respond to the initial chemo- and radiotherapies, and go into remission and even have extended disease-free periods, nonetheless, most will relapse. Relapse results in significant morbidity and mortality, and thus we should be directing the development of new therapies that effectively eliminate cancers cells that lead to states of minimal disease and control relapse. The sensitivity of BLI and the combination of fluorescence-based tools with BLI will enable the effective study of minimal disease states (Dickson et al., 2007; Nathan et al., 2007; Thorne and Contag, 2008; Thorne et al., 2006; Wetterwald et al., 2002). As such, the use of optical imaging tools will play a significance role in the development of new treatment paradigms forcancerand otherdiseases.

Regenerative medicine is a research area that benefits from new optical imaging tools. This largely results from the fact that the regeneration of tissue must either occur in the context of the living body, or the survival of tissues regenerated in vitro need to be studied in vivo. Reporter genes have been built into mouse models such that regenerated tissue and its survival can be assessed (BitMansour et al., 2002; Cao et al., 2004, 2005; Chan et al., 2007; Degano et al., 2008; Heckl, 2007; Li et al., 2007; Okada et al., 2005; Olivo et al., 2008; Sheikh et al., 2007; Tolaret al., 2005). This includes determining the location and number of transplanted stem cells and the maturation of these cells into tissues (Cao et al., 2004, 2005; Li et al., 2007; Tanaka et al., 2005; Wu et al., 2002). This approach is also being used to assess the duration of graft survival after transplantation (Cao et al., 2005; Li et al., 2007). The ability to monitor engraftment and to assess graft survival presents new opportunities for discovery that have not previously been possible. The development of transgenic mice with strong constitutive promoters driving expression of the reporter gene has been useful in transplantation studies (Cao et al., 2004, 2005), and has also been found to be useful fordeveloping new tools forsmall-molecule delivery (Kim et al., 2007).

SUMMARY

Optical imaging modalities have utility in the clinic forthe study of superficial tissue sites but can be limited by poor tissue penetration. However, these methods excel in the study of laboratory rodents, and as such are being used in a variety of preclinical studies to reveal mammalian biology and to develop novel therapies. The data from these studies are being used to direct clinical studies as well as advance a variety of scientific fields by accelerating and refining the study of rodents as models of human biology and disease. Because optical reporters can be built into animals and linked to target genes and cells, the opportunities for developing new, robust, and informative models are significant. Because the use of optical imaging can be used in conjunction with existing tools and established models, these modalities greatly enhance the study of well-developed mouse and rat models.

Established and emerging optical imaging tools are enabling dynamic cellular and molecularanalyses of disease mechanisms in living animal models and in humans. These advances are serving to inform and guide clinical trials, and to affect clinical studies and medical practice with a greater understanding of mammalian biology. These tools have the potential to refine a number of fields of study and are being applied to the study of infection, cancer, stem cell biology, novel therapies, and immunology. As new disciplines emerge, the adoption of optical imaging tools will rapidly impact these fields due to noninvasive assays that accelerate the study and create new insights. Evidence of this is seen in area regenerative medicine and stem cell therapies. The new tools of optical imaging, which have already had a tremendous impact on preclinical studies, hold great promise for bringing important and novel information to the clinician and the patient. These approaches are likely to enable early diagnosis, rapid typing of molecular markers, immediate assessment of therapeutic outcome, and ready measures of the extent of tissue regeneration after damage. However, the full impact of these new techniques will be determined by our ability to translate them to the clinic, and to develop a general strategy that integrates them with other tools in the areas of moleculardiagnostics and molecularmedicine.

REFERENCES

Benaron, D. A., Cheong, W. F., and Stevenson, D. K. (1997). Tissue optics. Science *276*, 2002–2003.

Bhaumik, S., and Gambhir, S. S. (2002). Optical imaging of *Renilla* luciferase reporter gene expression in living mice. Proc Natl Acad Sci USA *99*, 377–382.

Bhaumik, S., Lewis, X. Z., and Gambhir, S. S. (2004). Optical imaging of *Renilla* luciferase, synthetic *Renilla* luciferase, and firefly luciferase reporter gene expression in living mice. J Biomed Opt *9*, 578–586.

BitMansour, A., Burns, S. M., Traver, D., Akashi, K., Contag, C. H., Weissman, I. L., and Brown, J. M. (2002). Myeloid progenitors protect against invasive aspergillosis and *Pseudomonas aeruginosa* infection following hematopoietic stem cell transplantation. Blood *100*, 4660–4667.

Cao, Y. A., Bachmann, M. H., Beilhack, A., Yang, Y., Tanaka, M., Swijnenburg, R. J., Reeves, R., Taylor-Edwards, C., Schulz, S., Doyle, T. C., Fathman, C. G., Robbins, R. C., Herzenberg, L. A., Negrin, R. S., and Contag, C. H. (2005). Molecular imaging using labeled donor tissues reveals patterns of engraftment, rejection, and survival in transplantation. Transplantation *80*, 134–139.

Cao, Y. A., Wagers, A. J., Beilhack, A., Dusich, J., Bachmann, M. H., Negrin, R. S., Weissman, I. L., and Contag, C. H. (2004). Shifting foci of hematopoiesis during reconstitution from single stem cells. Proc Natl Acad Sci USA *101*, 221–226.

Cecic, I., Chan, D. A., Sutphin, P. D., Ray, P., Gambhir, S. S., Giaccia, A. J., and Graves, E. E. (2007). Oxygen sensitivity of reporter genes: Implications for preclinical imaging of tumor hypoxia. Mol Imaging *6*, 219–228.

Chalfie, M., Tu, Y., Euskirchen, G., Ward, W. W., and Prasher, D. C. (1994). Green fluorescent protein as a marker for gene expression. Science *263*, 802–805.

Chan, K. M., Raikwar, S. P., and Zavazava, N. (2007). Strategies for differentiating embryonic stem cells (ESC) into insulin-producing cells and development of non-invasive imaging techniques using bioluminescence. Immunol Res *39*, 261–270.

Contag, C. H., and Bachmann, M. H. (2002). Advances in in vivo bioluminescence imaging of gene expression. Annu Rev Biomed Eng *4*, 235–260.

Contag, C. H., Contag, P. R., Mullins, J. I., Spilman, S. D., Stevenson, D. K., and Benaron, D. A. (1995). Photonic detection of bacterial pathogens in living hosts. Mol Microbiol *18*, 593–603.

Contag, C. H., Jenkins, D., Contag, F. R., and Negrin, R. S. (2000). Use of reporter genes for optical measurements of neoplastic disease in vivo. Neoplasia *2*, 41–52.

Contag, C. H., and Ross, B. D. (2002). It's not just about anatomy: In vivo bioluminescence imaging as an eyepiece into biology. J Magn Reson Imaging *16*, 378–387.

Contag, C. H., Spilman, S. D., Contag, P. R., Oshiro, M., Eames, B., Dennery, P., Stevenson, D. K., and Benaron, D. A. (1997). Visualizing gene expression in living mammals using a bioluminescent reporter. Photochem Photobiol *66*, 523–531.

Contag, P. R., Olomu, I. N., Stevenson, D. K., and Contag, C. H. (1998). Bioluminescent indicators in living mammals. Nat Med *4*, 245–247.

Day, R. N., Kawecki, M., and Berry, D. (1998). Dual-function reporter protein for analysis of gene expression in living cells. Biotechniques *25*, 848–850, 852–854, 856.

Degano, I. R., Vilalta, M., Bago, J. R., Matthies, A. M., Hubbell, J. A., Dimitriou, H., Bianco, P., Rubio, N., and Blanco, J. (2008). Bioluminescence imaging of calvarial bone repair using bone marrow and adipose tissue-derived mesenchymal stem cells. Biomaterials *29*, 427–437.

de Wet, J. R., Wood, K. V., DeLuca, M., Helinski, D. R., and Subramani, S. (1987). Firefly luciferase gene: Structure and expression in mammalian cells. Mol Cell Biol *7*, 725–737.

Dickson, P. V., Hamner, B., Ng, C. Y., Hall, M. M., Zhou, J., Hargrove, P. W., McCarville, M. B., and Davidoff, A. M. (2007). In vivo bioluminescence imaging for early detection and monitoring of disease progression in a murine model of neuroblastoma. J Pediatr Surg *42*, 1172–1179.

Doyle, T. C., Burns, S. M., and Contag, C. H. (2004). In vivo bioluminescence imaging for integrated studies of infection. Cell Microbiol *6*, 303–317.

Edinger, M., Cao, Y. A., Hornig, Y. S., Jenkins, D. E., Verneris, M. R., Bachmann, M. H., Negrin, R. S., and Contag, C. H. (2002). Advancing animal models of neoplasia through in vivo bioluminescence imaging. Eur J Cancer *38*, 2128–2136.

Edinger, M., Cao, Y. A., Verneris, M. R., Bachmann, M. H., Contag, C. H., and Negrin, R. S. (2003a). Revealing lymphoma growth and the efficacy of immune cell therapies using in vivo bioluminescence imaging. Blood *101*, 640–648.

Edinger, M., Hoffmann, P., Contag, C. H., and Negrin, R. S. (2003b). Evaluation of effector cell fate and function by in vivo bioluminescence imaging. Methods *31*, 172–179.

Edinger, M., Sweeney, T. J., Tucker, A. A., Olomu, A. B., Negrin, R. S., and Contag, C. H. (1999). Noninvasive assessment of tumor cell proliferation in animal models. Neoplasia *1*, 303–310.

Giepmans, B. N. G., Adams, S. R., Ellisman, M. H., and Tsien, R. Y. (2006). Review: The fluorescent toolbox for assessing protein location and function. Science *312*, 217–224.

Hastings, J. W. (1996). Chemistries and colors of bioluminescent reactions: A review. Gene *173*, 5–11.

Hastings, J. W., and Morin, J. G. (2006). Photons for reporting molecular events: Green fluorescent protein and four luciferase systems. Methods Biochem Anal *47*, 15–38.

Heckl, S. (2007). Future contrast agents for molecular imaging in stroke. Curr Med Chem *14*, 1713–1728.

Helmchen, F., and Denk, W. (2005). Deep tissue two-photon microscopy. Nat Methods *2*, 932–940.

Hsiung, P. L., Hardy, J., Friedland, S., Soetikno, R., Du, C. B., Wu, A. P., Sahbaie, P., Crawford, J. M., Lowe, A. W., Contag, C. H., and Wang, T. D. (2008). Detection of colonic dysplasia in vivo using a targeted heptapeptide and confocal microendoscopy. Nat Med *14*, 454–458.

Izzo, A. D., Mackanos, M. A., Beckham, J. T., and Jansen, E. D. (2001). In vivo optical imaging of expression of vascular endothelial growth factor following laser incision in skin. Lasers Surg Med *29*, 343–350.

Jobsis, F. F. (1977). Noninvasive, infrared monitoring of cerebral and myocardial oxygen sufficiency and circulatory parameters. Science *198*, 1264–1267.

Kadurugamuwa, J. L., Modi, K., Coquoz, O., Rice, B., Smith, S., Contag, P. R., and Purchio, T. (2005). Reduction of astrogliosis by early treatment of pneumococcal meningitis measured by simultaneous imaging, in vivo, of the pathogen and host response. Infect Immunol *73*, 7836–7843.

Kadurugamuwa, J. L., Sin, L., Albert, E., Yu, J., Francis, K., DeBoer, M., Rubin, M., Bellinger-Kawahara, C., Parr, T. R., Jr., and Contag, P. R. (2003). Direct continuous method for monitoring biofilm infection in a mouse model. Infect Immunol *71*, 882–890.

Keller, P. J., Pampaloni, F., and Stelzer, E. H. K. (2006). Life sciences require the third dimension. Curr Opin Cell Biol *18*, 117–124.

Kim, J. B., Leucht, P., Morrell, N. T., Schwettman, H. A., and Helms, J. A. (2007). Visualizing in vivo liposomal drug delivery in real-time. J Drug Target *15*, 632–639.

Laxman, B., Hall, D. E., Bhojani, M. S., Hamstra, D. A., Chenevert, T. L., Ross, B. D., and Rehemtulla, A. (2002). Noninvasive real-time imaging of apoptosis. Proc Natl Acad Sci USA *99*, 16551–16555.

Li, Z., Wu, J. C., Sheikh, A. Y., Kraft, D., Cao, F., Xie, X., Patel, M., Gambhir, S. S., Robbins, R. C., and Cooke, J. P. (2007). Differentiation, survival, and function of embryonic stem cell derived endothelial cells for ischemic heart disease. Circulation *116* (supplement), I46–54.

Loening, A. M., Wu, A. M., and Gambhir, S. S. (2007). Red-shifted *Renilla reniformis* luciferase variants for imaging in living subjects. Nat Methods *4*, 641–643.

Luker, K. E., and Piwnica-Worms, D. (2004). Optimizing luciferase protein fragment complementation for bioluminescent imaging of protein–protein interactions in live cells and animals. Methods Enzymol *385*, 349–360.

Mandl, S., Mari, C., Edinger, M., Negrin, R. S., Tait, J. F., Contag, C. H., and Blankenberg, F. (2003). Multi-modality imaging identifies key times for annexin V imaging as an early predictor of therapeutic outcome. Molec Imaging *3*(1), 1-8.

Michelini, E., Cevenini, L., Mezzanotte, L., Ablamsky, D., Southworth, T., Branchini, B. R., and Roda, A. (2008). Combining intracellular and secreted bioluminescent reporter proteins for multicolor cell-based assays. Photochem Photobiol Sci *7*, 212–217.

Millar, A. J., Carre, I. A., Strayer, C. A., Chua, N. H., and Kay, S. A. (1995). Circadian clock mutants in *Arabidopsis* identified by luciferase imaging. Science *267*, 1161–1163.

Nathan, C. O., Amirghahari, N., Rong, X., Giordano, T., Sibley, D., Nordberg, M., Glass, J., Agarwal, A., and Caldito, G. (2007). Mammalian target of rapamycin inhibitors as possible adjuvant therapy for microscopic residual disease in head and neck squamous cell cancer. Cancer Res *67*, 2160–2168.

Negrin, R. S., and Contag, C. H. (2006). In vivo imaging using bioluminescence: A tool for probing graft-versus-host disease. Nat Rev Immunol *6*, 484–490.

Ntziachristos, V., Ripoll, J., Wang, L. H. V., and Weissleder, R. (2005). Looking and listening to light: The evolution of whole-body photonic imaging. Nat Biotechnol *23*, 313–320.

O'Connell-Rodwell, C. E., Shriver, D., Simanovskii, D. M., McClure, C., Cao, Y. A., Zhang, W., Bachmann, M. H., Beckham, J. T., Jansen, E. D., Palanker, D., Schwettman, H. A., and Contag, C. H. (2004). A genetic reporter of thermal stress defines physiologic zones over a defined temperature range. FASEB J *18*, 264–271.

Okada, S., Ishii, K., Yamane, J., Iwanami, A., Ikegami, T., Katoh, H., Iwamoto, Y., Nakamura, M., Miyoshi, H., Okano, H. J., Contag, C. H., Toyama, Y., and Okano, H. (2005). In vivo imaging of engrafted neural stem cells: Its application in evaluating the optimal timing of transplantation for spinal cord injury. FASEB J *19*, 1839–1841.

Olafsen, T., Cheung, C. W., Yazaki, P. J., Li, L., Sundaresan, G., Gambhir, S. S., Sherman, M. A., Williams, L. E., Shively, J. E., Raubitschek, A. A., and Wu, A. M. (2004). Covalent disulfide-linked anti-CEA diabody allows site-specific conjugation and radiolabeling for tumor targeting applications. Protein Eng Des Sel *17*, 21–27.

Olivo, C., Alblas, J., Verweij, V., Van Zonneveld, A. J., Dhert, W. J., and Martens, A. C. (2008). In vivo bioluminescence imaging study to monitor ectopic bone formation by luciferase gene marked mesenchymal stem cells. J Orthop Res *26*(7): 901-9

Paulmurugan, R., and Gambhir, S. S. (2003). Monitoring protein–protein interactions using split synthetic *Renilla* luciferase protein-fragment-assisted complementation. Anal Chem *75*, 1584–1589.

Paulmurugan, R., Umezawa, Y., and Gambhir, S. S. (2002). Noninvasive imaging of protein–protein interactions in living subjects by using reporter protein complementation and reconstitution strategies. Proc Natl Acad Sci USA *99*, 15608–15613.

Plautz, J. D., Kaneko, M., Hall, J. C., and Kay, S. A. (1997). Independent photoreceptive circadian clocks throughout *Drosophila*. Science *278*, 1632–1635.

Ponomarev, V., Doubrovin, M., Serganova, I., Vider, J., Shavrin, A., Beresten, T., Ivanova, A., Ageyeva, L., Tourkova, V., Balatoni, J., Bornmann, W., Blasberg, R., and Gelovani Tjuvajev, J. (2004). A novel triple-modality reporter gene forwhole-body fluorescent, bioluminescent, and nuclearnoninvasive imaging. EurJ Nucl Med Mol Imaging *31*, 740–751.

Ray, P., De, A., Min, J. J., Tsien, R. Y., and Gambhir, S. S. (2004). Imaging tri-fusion multimodality reporter gene expression in living subjects. Cancer Res *64*, 1323–1330.

Ray, P., Pimenta, H., Paulmurugan, R., Berger, F., Phelps, M. E., Iyer, M., and Gambhir, S. S. (2002). Noninvasive quantitative imaging of protein–protein interactions in living subjects. Proc Natl Acad Sci USA *99*, 3105–3110.

Ray, P., Tsien, R., and Gambhir, S. S. (2007). Construction and validation of improved triple fusion reporter gene vectors formolecularimaging of living subjects. CancerRes *67*, 3085–3093.

Ray, P., Wu, A. M., and Gambhir, S. S. (2003). Optical bioluminescence and positron emission tomography imaging of a novel fusion reporter gene in tumor xenografts of living mice. Cancer Res *63*, 1160–1165.

Rice, B. W., Cable, M. D., and Nelson, M. B. (2001). In vivo imaging of light-emitting probes. J Biomed Optics *6*, 432–440.

Shachaf, C. M., Kopelman, A. M., Arvanitis, C., Karlsson, A., Beer, S., Mandl, S., Bachmann, M. H., Borowsky, A. D., Ruebner, B., Cardiff, R. D., Yang, Q., Bishop, J. M., Contag, C. H., and Felsher, D. W. (2004). MYC inactivation uncovers pluripotent differentiation and tumour dormancy in hepatocellular cancer. Nature *431*, 1112–1117.

Shaner, N. C., Campbell, R. E., Steinbach, P. A., Giepmans, B. N. G., Palmer, A. E., and Tsien, R. Y. (2004). Improved monomeric red, orange and yellow fluorescent proteins derived from *Discosoma sp.* red fluorescent protein. Nat Biotechnol *22*, 1567–1572.

Shaner, N. C., Steinbach, P. A., and Tsien, R. Y. (2005). A guide to choosing fluorescent proteins. Nat Methods *2*, 905–909.

Shao, N., and Bock, R. (2008). A codon-optimized luciferase from *Gaussia princeps* facilitates the in vivo monitoring of gene expression in the model alga *Chlamydomonas reinhardtii*. Curr Genet *53*, 381–388.

Shapiro, E., Lu, C., and Baneyx, F. (2005). A set of multicolored *Photinus pyralis* luciferase mutants for in vivo bioluminescence applications. Protein Eng Des Sel *18*, 581–587.

Sheikh, A. Y., Lin, S. A., Cao, F., Cao, Y., van derBogt, K. E., Chu, P., Chang, C. P., Contag, C. H., Robbins, R. C., and Wu, J. C. (2007). Molecular imaging of bone marrow mononuclear cell homing and engraftment in ischemic myocardium. Stem Cells *25*, 2677–2684.

Shu, X. K., Shaner, N. C., Yarbrough, C. A., Tsien, R. Y., and Remington, S. J. (2006). Novel chromophores and buried charges control color in mFruits. Biochemistry *45*, 9639–9647.

So, M. K., Loening, A. M., Gambhir, S. S., and Rao, J. (2006a). Creating self-illuminating quantum dot conjugates. Nat Protoc *1*, 1160–1164.

So, M. K., Xu, C., Loening, A. M., Gambhir, S. S., and Rao, J. (2006b). Self-illuminating quantum dot conjugates for in vivo imaging. Nat Biotechnol *24*, 339–343.

Stolz, U., Velez, S., Wood, K. V., Wood, M., and Feder, J. L. (2003). Darwinian natural selection for orange bioluminescent color in a Jamaican click beetle. Proc Natl Acad Sci USA *100*, 14955–14959.

Sweeney, T. J., Mailander, V., Tucker, A. A., Olomu, A. B., Zhang, W., Cao, Y., Negrin, R. S., and Contag, C. H. (1999). Visualizing the kinetics of tumor-cell clearance in living animals. Proc Natl Acad Sci USA *96*, 12044–12049.

Tanaka, M., Swijnenburg, R. J., Gunawan, F., Cao, Y. A., Yang, Y., Caffarelli, A. D., de Bruin, J. L., Contag, C. H., and Robbins, R. C. (2005). In vivo visualization of cardiac allograft rejection and trafficking passenger leukocytes using bioluminescence imaging. Circulation *112*, I105–110.

Tannous, B. A., Kim, D. E., Fernandez, J. L., Weissleder, R., and Breakefield, X. O. (2005). Codon-optimized *Gaussia* luciferase cDNA for mammalian gene expression in culture and in vivo. Mol Ther *11*, 435–443.

Thorne, S. H., and Contag, C. H. (2007). Combining immune cell and viral therapy for the treatment of cancer. Cell Mol Life Sci *64*, 1449–1451.

Thorne, S. H., and Contag, C. H. (2008). Integrating the biological characteristics of oncolytic viruses and immune cells can optimize therapeutic benefits of cell-based delivery. Gene Ther *15*, 753–758.

Thorne, S. H., Negrin, R. S., and Contag, C. H. (2006). Synergistic antitumor effects of immune cell–viral biotherapy. Science *311*, 1780–1784.

Tolar, J., Osborn, M., Bell, S., McElmurry, R., Xia, L., Riddle, M., Panoskaltsis-Mortari, A., Jiang, Y., McIvor, R. S., Contag, C. H., Yant, S. R., Kay, M. A., Verfaillie, C. M., and Blazar, B. R. (2005). Real-time in vivo imaging of stem cells following transgenesis by transposition. Mol Ther *12*, 42–48.

Troy, T., Jekic-McMullen, D., Sambucetti, L., and Rice, B. (2004). Quantitative comparison of the sensitivity of detection of fluorescent and bioluminescent reporters in animal models. Mol Imaging *3*, 9–23.

Tsai, P. S., Friedman, B., Ifarraguerri, A. I., Thompson, B. D., Lev-Ram, V., Schaffer, C. B., Xiong, C., Tsien, R. Y., Squier, J. A., and Kleinfeld, D. (2003). All-optical histology using ultrashort laser pulses. Neuron *39*, 27–41.

Tsien, R. Y. (1998). The green fluorescent protein. Annu Rev Biochem *67*, 509–544.

Venisnik, K. M., Olafsen, T., Gambhir, S. S., and Wu, A. M. (2007). Fusion of *Gaussia* luciferase to an engineered anti-carcinoembryonic antigen (CEA) antibody for in vivo optical imaging. Mol Imaging Biol *9*, 267–277.

Venisnik, K. M., Olafsen, T., Loening, A. M., Iyer, M., Gambhir, S. S., and Wu, A. M. (2006). Bifunctional antibody–*Renilla* luciferase fusion protein for in vivo optical detection of tumors. Protein Eng Des Sel *19*, 453–460.

Wang, Y., Wang, G., O'Kane, D. J., and Szalay, A. A. (2001). A study of protein–protein interactions in living cells using luminescence resonance energy transfer (LRET) from *Renilla* luciferase to *Aequorea* GFP. Mol Gen Genet *264*, 578–587.

Wetterwald, A., van der Pluijm, G., Que, I., Sijmons, B., Buijs, J., Karperien, M., Lowik, C. W., Gautschi, E., Thalmann, G. N., and Cecchini, M. G. (2002). Optical imaging of cancer metastasis to bone marrow: A mouse model of minimal residual disease. Am J Pathol *160*, 1143–1153.

Wilson, T., and Hastings, J. W. (1998). Bioluminescence. Annu Rev Cell Dev Biol *14*, 197–230.

Wood, K. V. (1990). Luc genes: Introduction of colour into bioluminescence assays. J Biolumin Chemilumin *5*, 107–114.

Wood, K. V., Lam, Y. A., Seliger, H. H., and McElroy, W. D. (1989). Complementary DNA coding click beetle luciferases can elicit bioluminescence of different colors. Science *244*, 700–702.

Wu, J. C., Inubushi, M., Sundaresan, G., Schelbert, H. R., and Gambhir, S. S. (2002). Optical imaging of cardiac reporter gene expression in living rats. Circulation *105*, 1631–1634.

Zhang, G. J., and Kaelin, W. G., Jr. (2005). Bioluminescent imaging of ubiquitin ligase activity: Measuring Cdk2 activity in vivo through changes in p27 turnover. Methods Enzymol *399*, 530–549.

Zhao, H., Doyle, T. C., Coquoz, O., Kalish, F., Rice, B. W., and Contag, C. H. (2005). Emission spectra of bioluminescent reporters and interaction with mammalian tissue determine the sensitivity of detection in vivo. J Biomed Opt *10*, 41210.

Zipfel, W. R., Williams, R. M., and Webb, W. W. (2003). Nonlinearmagic: Multiphoton microscopy in the biosciences. Nat Biotechnol *21*, 1368–1376.

Index

A2RE, 114

Aberration correction, in confocal microscopy, 16–19

Absorption
  chromophores responsible for, in tissues, 316
  described, 308

Acousto-optic beam deflectors, 3

Acousto-optic tunable filters (AOTFs), 35–36

β-Actin messengerRNA, 114, 115–116

Action potential initiation and propagation
  in *Aplysia* ganglion, 134–135, 139–143
  in invertebrate neurons, 135–137
  in vertebrate neurons, 137–139

Activatable fluorescent probes, 377–379

Active transport, cellular, fluorescence correlation spectroscopy and, 224–225

Adaptive aberration correction, in confocal microscopy, 17

Adenocarcinoma
  broadband diffuse optical spectroscopy of, 351
  optical coherence tomography of, 182–183

Afterpulsing, 208

Aggregation measurements, in fluorescence correlation spectroscopy, 221–222

ALA. *See* Aminolevulinic acid (ALA)

Alexa dye family, in fluorescence correlation spectroscopy, 218, 220

Aminolevulinic acid (ALA), 39, 275, 284, 287, 293, 297

Animal models. *See* Preclinical studies

Anomalous subdiffusion, 222–224

Aperture mask system, in confocal microscopy, 11–13

Aplysia ganglion action potentials, membrane potential imaging of, 134–135, 139–143

Arc lamps, in membrane potential imaging, 150

Arrays, silicon diode, 154–155

Ash1 messengerRNA, 114–115

AOTFs (acousto-optic tunable filters), 35–36

Auditory epithelium, multiphoton microscopy of, 88

Autocorrelation analysis, in two-photon fluorescence correlation spectroscopy, 201–207

Autofluorescence
  in medical diagnostic imaging, 273
  signal to background and, 122

Autofocus image, confocal microscopy, 8

Automated end-member determination, in spectral optical imaging data processing, 52–53

Axial resolution, improvement with two lenses, 239–241

Background rejection, in medical diagnostic imaging, 278

Ballistic fraction, 78

Band-sequential spectrometers, 32–36

Barrett's esophagus, optical coherence tomography of, 183–184

beta-Globulin premessenger RNAs, human, 111

bicold messengerRNA, 115

Binding and unbinding studies, fluorescence correlation spectroscopy and, 225

Biographical covariant effects, in spectroscopic detection of cervical cancer, 312–313

Biological applications
  of optical coherence tomographic three-dimensional imaging, 185–186
  of spectral imaging, 56–61

Bioluminescence imaging (BLI), 398, 400–401, 404–405

Biomolecules, individual, analysis of, 197–198. *See also* Fluorescence correlation spectroscopy (FCS)

Biopsy
  optical. *See* Optical biopsy
  optical coherence tomographic guidance of, 182–185
  spectroscopically guided, 268–269

BLI. *See* Bioluminescence imaging (BLI)

Blood imaging, spectral analysis and, 62

Blood oxygenation level dependent (BOLD) signal, 365, 368–370

Brain activity detection, 356–371
  brain hemodynamics investigations, 365–370
  fast effects and, 359, 360–363, 370–373
  frequency domain method for turbid media, 357–359
  future potential of, 370–373
  instrument for noninvasive measurements, 359
  noninvasive measurements: slow vs. fast effects, 359–360
  optical, 360–365
  overview, 356–357
  slow effects and, 359–360

Brain slices, living, multiphoton microscopy of, 85–86, 88–89

Breast tissue, broadband diffuse optical spectroscopy of, 350–353

Broadband diffuse optical spectroscopy. *See* Diffuse optical spectroscopy (DOS)

Bronchus, medical diagnostic fluorescence imaging of, 271

Caged calcium, multiphoton photochemistry and, 93

Caged compounds, 87

Caged fluorophores, in messenger RNA imaging, 106

Cajal bodies, 113

Calmodulin-dependent protein kinase II, 114
Cameras
    charge-coupled device, 137, 149, 150, 151–152,
        154–155
    complementary metal-oxide semiconductor, 155
    vacuum photocathode, 155
Camera systems
    comparison of, 149
    in membrane potential imaging, 154–155
CaMKII messengerRNA, 114
Cancerdiagnosis and treatment. *See also* Cervical
    neoplasia; Medical diagnostic fluorescence
    imaging
    bioluminescence imaging and, 405
    epithelial, three-dimensional tissue phantom
        models of, 323–324
    molecularimaging and, 375
    optical coherence tomography and, 182–185
    spectral imaging and, 63–66
    techniques for, reviewed, 266
CARS (Coherent Anti-Stokes Raman
    Spectroscopy), 61
Catheter/endoscope devices, in optical coherence
    tomography, 176–178
CCD cameras. *See* Charge-coupled device (CCD)
    cameras
Cell labeling techniques, in neuroscience, 76–77
Cell permeant dyes, in messenger RNA imaging,
    107–108
Cells, active transport in, fluorescence correlation
    spectroscopy and, 224–225
Cellularmyelocytomatosis protein (c-myc) messenger
    RNA, 112
Cervical dysplasia
    clinical perspective on detection of, 306–308
    model system for spectrographic detection of,
        320–324
Cervical neoplasia, 306–326
    chromophores in epithelial tissue, 308, 315–319,
        321–323
    clinical applications
        biographical and patient-to-patient covariant
            effects, 312–313
        fluorescence and reflectance spectroscopy,
            311–312
        fluorescence spectroscopy, 308–309
        menstrual cycle effects, 313
        multispectral optical imaging, 313–315
        reflectance spectroscopy, 309–311
    clinical perspective on detection of, 306–308
    mathematical model of spectral data, 324–326
    model system, modulation of chromophore
        concentration and optical properties in,
        320–324
Cervix, medical diagnostic fluorescence imaging of,
    271, 295
c-fos messengerRNA, 115
Charge-coupled device (CCD) cameras
    characteristics of, 154–155
    noise limitations, 149, 150
    rat olfactory bulb imaging example, 137
    two-photon fluorescence microscopy and, 151–152

Chemical two-photon uncaging, 87–88
Chromophores. *See also* Dyes; *names ofspecific
    chromophores*
    absorbers, 308, 316
    core structure of, 125, 126–127
    cyanine dyes as, 125–128
    fluorophores, 54, 106, 308, 316–319
    intensity modulations of, 205–206, 228–229
    limitations of, 124–125
    for membrane potential monitoring, 132–134
    photobleaching of. *See* Photobleaching
    scatterers, 308, 319
    in tissues, 308, 315–319, 321–323
Classification imaging, in spectral data visualization,
    42, 43
Classification-quantitative hybrid imaging
    quantitative imaging compared to, 47–49
    in spectral data visualization, 43, 44, 45–47
Clinical applications, of spectral imaging, 61–66
Clustering algorithms, in automated end-member
    determination, 52–53
CMOS (complementary metal-oxide semiconductor)
    cameras, 155
c-myc messengerRNA, 112
Coherence length, 166
Coherent Anti-Stokes Raman Spectroscopy
    (CARS), 61
Collagen autofluorescence, 316–319, 322, 325–326
Colon
    neoplasia of, medical diagnostic fluorescence
        imaging and, 272, 287
    optical coherence tomographic three-dimensional
        imaging of, 185
Colorspace schemes, 44–45
Complementary metal-oxide semiconductor (CMOS)
    cameras, 155
Computed tomography imaging spectrometer (CTIS),
    37–38
Concentration measurements, fluorescence correlation
    spectroscopy and, 220–222
Confocal microscopy, 3–27. *See also* 4Pi-confocal
    microscopy
    aberration correction, 16–19
    depth discrimination applications, 7–9
    fluorescence microscopy, 9–11
    image formation, 3–7
    limitations of, 124
    membrane potential imaging and, 151–152
    optical architectures
        aperture mask system, 11–13
        beam scanning configuration, 3
        structured illumination vs., 13–16
    optical coherence tomography
        compared to, 164–165
    optical field generation and focusing, 19–27
    scanning methods, 3
    summary, 26–27
Core chromophore structure, 125, 126–127
Coronary artery, optical coherence
    tomography of, 162–163
Corralled diffusion, 114
Correlation, as spectral similarity measure, 40–41

Correlation coefficient. *See* Correlation
Cortical blood flow, multiphoton microscopy and, 91
Cross-correlation analysis, in two-photon
fluorescence correlation spectroscopy, 207–213
CTIS (computed tomography imaging spectrometer), 37–38
Cyanine dyes, as chromophores, 125–128
Cytopathology, spectral imaging for, 61
Cytoplasmic mRNA studies, 113–116

Dark noise, in membrane potential imaging, 150
Data cube, defined, 40
Data processing, in medical diagnostic imaging, 276–278
Dendritic excitation in vivo, 91
Dendritic membranes, membrane potential
imaging of, 135
Dendritic spine physiology, multiphoton
microscopy and, 89
Deoxyhemoglobin
concentration of
blood oxygenation level dependent (BOLD)
signal correlation, 365
in breast tissue, 352
in muscle, 346–350
as tissue chromophore, 316
Depth discrimination in confocal microscopy
applications of, 7–9
image formation and, 4, 5–7
Depth of focus, in membrane potential imaging, 151
Detection systems, in multiphoton microscopy, 83–84
Developmental biology specimens, optical coherence
tomographic three-dimensional imaging of,
185–186
Diabetic retinopathy, optical coherence tomography
and, 180
Diffraction barrier, 237–239, 253–256
Diffraction-unlimited resolution in fluorescence
imaging, 253–256
Diffuse optical spectroscopy (DOS), 330–353
in breast tissue, 350–353
instrumentation, 333–346
in muscle, 346–350
overview, 330–331
theory, 331–333
Diffusion
anomalous, fluorescence correlation spectroscopy
and, 222–224
in solution and cellular systems, fluorescence
correlation spectroscopy and, 222
three-dimensional
of multiple species, 206–207
of single species, 203–205
Dimensionality reduction, in spectral optical
imaging, 50–52
Disease mechanisms
messengerRNA imaging and, 102–103
molecularimaging in investigation of, 375
optical reporter genes in study of, 397, 405–406
DNA probes, fluorescent
molecularimaging and, 377–378
spectral imaging and, 61–62

DOS. *See* Diffuse optical spectroscopy (DOS)
Drosophila oocytes, messengerRNA studies in, 115
Drug delivery strategies, molecular imaging and, 375
Dual-band fluorescence imaging, 278–279, 287–290
Dual-color cross-correlation
in enzyme kinetics studies, 226–227
in two-photon fluorescence correlation
spectroscopy, 208, 212–213, 215–216
Dye aggregation, fluorescent probes and, 128
Dyes. *See also* Chromophores
cell permeant, in messenger RNA imaging,
107–108
Dye systems, for fluorescence correlation
spectroscopy
one-photon, 217–218
two-photon, 218–220

Earth science, spectral optical imaging in, 39–40
Echo time delay of light, 165
interferometric measurement of, 165–166,
186–190
EGFP. *See* Enhanced GFP (EGFP)
Elastin autofluorescence, 316
Embryogenesis, multiphoton microscopy and, 89
End member
automated determination of, 52–53
defined, 49
Endogenous markers, spectral analysis of, 62–63, 66
Endoscopy
optical coherence tomographic three-dimensional
imaging and, 186
in optical coherence tomography, 176–178
short-wavelength excitation fluorescence, 286–287
spectral analysis and, 62–63
Enhanced GFP (EGFP)
pH-dependent emission characteristics of, 228
photobleaching of, 230–232
Enzyme activity monitoring, fluorescent probes for,
378–379
Enzyme kinetics, fluorescence correlation
spectroscopy and, 225–227
Epiglottis ex vivo, optical coherence tomography of,
168–169
Epithelial cancer, three-dimensional tissue phantom
models of, 323–324
Epithelial tissues
chromophores present in, 315–319, 321–323
three-dimensional tissue phantom model of,
323–324
Esophagus, optical coherence tomography of,
183–184
Euclidean distance, as spectral similarity
measure, 40–41
Excitation sources, for medical diagnostic
imaging, 285
Exogenous markers, spectral imaging and, 63–66
Extended-focus image, confocal microscopy, 7–8
Extraneous noise, in membrane potential imaging,
149–150

Fast effects, brain activity and, 359, 360–363,
370–373

FCS. *See* Fluorescence correlation spectroscopy (FCS)

FDPM. *See* Frequency domain photon migration (FDPM)

Feature space, in computational processing of spectral data, 39–40, 52

Fellgett advantage, 34

Female genital tract, medical diagnostic fluorescence imaging of, 271, 295

Femtosecond lasers, as optical coherence tomography light sources, 171–174

Fiber optics, in optical coherence tomography, 172–174

FIDA (Fluorescence Intensity Distribution Analysis), 221–222

Field of view, lateral resolution versus, 4

Filterwheels, 32–33

FlAsH (Fluorescein-based Arsenical Hairpin binder), 121

Flavin adenine dinucleotide, as tissue chromophore, 316, 322

Flexible endoscope devices, in optical coherence tomography, 176–178

Flickering, 205–206, 228–229

Fluorescein, in fluorescence correlation spectroscopy, 217–218

Fluorescein with As(III) substituents (FlAsH) technology, 121

Fluorescence
chromophores responsible for, in tissues, 316–319
described, 308
light-induced, 268

Fluorescence correlation spectroscopy (FCS)
dye systems for, 217–220
history of, 198–201
in messengerRNA imaging, 109, 111
principles of, 196–198
stimulation emission depletion and, 258
two-photon, 196–233
  advantages of, 199–201
  autocorrelation analysis in, 201–207
  commercial instruments, 217
  concentration and aggregation measurements with, 220–222
  cross-correlation analysis in, 207–213
  dye systems for, 218–220
  internal dynamics studies, 227–229
  in intracellular observations, 198, 199–201
  in microfluidic fields, 229
  mobility analysis applications, 222–225
  molecularinter action studies, 225–227
  photobleaching in, 199, 229–233
  standard confocal configurations, 196–197, 213–216
  two-photon auto- and cross-correlation setup, 216–217

Fluorescence emission spectrum, 307

Fluorescence excitation spectrum, 307

Fluorescence imaging modes, in medical diagnostic imaging, 278–283

Fluorescence Intensity Distribution Analysis (FIDA), 221–222

Fluorescence-mediated tomography (FMT), 382–391

diffuse optical tomographic imaging systems, 385–386
in vivo imaging, 386–387
molecularquantification and specificity, 389–391
sensitivity, 387–389
theoretical specifics: forward problem, 383–384
theoretical specifics: inverse problem, 384–385

Fluorescence microscopy/imaging. *See also specific methods, e.g.,* Confocal microscopy
illumination systems, 13–16
in medical diagnostics. *See* Medical diagnostic fluorescence imaging
radially polarized illumination and, 24–26
random access, membrane potential imaging and, 152–154
spectral unmixing and, 49–50, 54, 59–60, 65

Fluorescence nanoscopy, 253–261
breaking the diffraction barrier, 253–256
stimulated emission depletion microscopy, 253–254, 256–260, 261
variations of reversible saturable optically linear transition between two molecular states, 260–261

Fluorescence recovery after photobleaching (FRAP)
in messengerRNA imaging, 109, 111
in multiphoton microscopy, 92–93

Fluorescence resonance energy transfer (FRET), in intramolecular dynamics studies, 228

Fluorescence spectroscopy. *See also* Spectral optical imaging
in detection of cervical cancer, 308–309
model system for, 320–324
with reflectance spectroscopy, 311–312
of tissues, overview, 307–308

Fluorescent in vivo hybridization, in messenger RNA imaging, 104

Fluorescent probes, 120–128. *See also* Molecular beacons
activatable, 377–379
core chromophore of, 125–127
current state-of-the-art, 120–121
development goals for, 121–122
DNA, 61–62, 377–378
forenzyme activity monitoring, 378–379
fundamental components of, 125–126
in molecularfunction imaging, 375–379
multiple, difficulties in use of, 124
near-infrared
  applications of, 122–124
  design of new, 125–127
  limitations of current, 124–125
  optimization of, 128
oligonucleotide, in messengerRNA imaging, 104–107
R-groups of, 125–126, 128
targeted, 376–377

Fluorescent proteins
in fluorescence correlation spectroscopy, 218
labeling with, 77
in live-cell imaging, 260–261
as optical reporter genes, 401–402
spectral unmixing and, 59–60, 65

as tissue chromophores, 316–317
Fluorescent RNA, microinjection of, 104
Fluorettes, 121
Fluorophores. *See* Chromophores; *names of specific fluorophores*
FMT. *See* Fluorescence-mediated tomography (FMT)
Fourier domain detection, 166, 171, 186–190
    spectral/Fourier domain detection, 187–188
    swept-source/Fourier domain detection, 187, 188–190
Fourier transform imaging spectrometers, 33–34
Fourier transform spectroscopic imaging, in medical diagnostics, 281–282, 294
4Pi-confocal microscopy
    axial resolution and, 240
    live mammalian cell imaging with, 253
    optical transfer functions of, 245–246, 250–251
    point-spread function of, 241–245, 249–250
    and stimulated emission depletion, 258, 261
Fragile X syndrome, 102–103
FRAP. *See* Fluorescence recovery after photobleaching (FRAP)
Free diffusion model of nuclear organization, 110–111
Frequency domain interferometry. *See* Fourier domain detection
Frequency domain method for turbid media, brain activity detection and, 357–359
Frequency domain photon migration (FDPM), in diffuse optical spectroscopy, 331, 333
FRET. *See* Fluorescence resonance energy transfer (FRET)
Full-widths-at-half-maximum (FWHM), 237
Functional magnetic resonance imaging, brain hemodynamics measurements by, 365–370, 371
Fuzzy C-means algorithms, 52
FWHM (full-widths-at-half-maximum), 237

Galvanometer-type mirrors, vibrating, scanning method in confocal microscopy, 3
Ganglia, *Aplysia*, membrane potential imaging of, 134–135, 139–143
Gastrointestinal tract. *See also* Colon
    esophagus, optical coherence tomography of, 183–184
    medical diagnostic fluorescence imaging of, 272
Gated detection, 278
Genital tract, female, medical diagnostic fluorescence imaging of, 271, 295
Geology, spectral optical imaging in, 39–40
Green fluorescent protein (GFP). *See also* Enhanced GFP (EGFP)
    current research on, 120
    in messengerRNA imaging, 108, 113
    as optical reporter gene, 401–402
    photobleaching of, 232
    spectral unmixing and, 59
Growth hormone (hGH) messenger RNA, human, 116

Hadamard masks, 31–32
Handheld imaging probes, in optical coherence tomography, 176

Hartmann-Shack wavefront sensors, 17
Head, medical diagnostic fluorescence imaging of, 272
Hematoporphyrin derivative (HpD), 284
Hemodynamic changes, brain activity detection and, 360, 362
Hemodynamic investigations, brain, near-infrared spectroscopy and, 365–370
Hemoglobin, as tissue chromophore, 316
Hemoglobin content, total, in exercised and nonexercised muscle, 348–350
hGH (human growth hormone) messenger RNA, 116
Histopathology, spectral imaging for, 61
hnRNP A2 response element, 114
Holographic spectral imaging, 36–37
HpD (hematoporphyrin derivative), 284
Hue, saturation, brightness (HSB) color space, 44–45
Human *beta*-globulin premessenger RNAs, 111
Human growth hormone (hGH) messenger RNA, 116
Hybrid imaging
    in medical diagnostics, 283, 297
    quantitative imaging compared to, 47–49
    in spectral data visualization, 43, 44, 45–47
Hyperspectral imaging., 30. *See also* Multispectral imaging

I^5M microscopy
    axial resolution and, 240
    imaging in aqueous media with, 252–253
    optical transfer functions of, 245–246, 250–251
    point-spread function of, 241–245, 374
Image space, in computational processing of spectral data, 39–40
Image space models, in spectral data visualization, 44–45
Imaging devices
    in membrane potential imaging, 154–155
    in optical coherence tomography, 170–171, 174–178
Infrared fluorescence probes. *See* Near-infrared fluorescent probes
Intensity point-spread function, in fluorescence microscopy, 9–10
Interferometric measurement of echo time delay of light, 165–166, 170, 186–190
Intersystem crossing, 228
Intracellular observations. *See also* Live-cell imaging
    fluorescence correlation spectroscopy and, 198, 199–201
    mobility analysis, fluorescence correlation spectroscopy and, 222
Intracellular transport, multiphoton photochemistry and, 92–93
Intramolecular dynamics, reversible, fluorescence correlation spectroscopy and, 228
Intrinsic optical signals, spectral analysis of, 62–63, 66
Invertebrate neurons, action potential initiation and propagation in, 135–137
In vivo hybridization, fluorescent, in messenger RNA imaging, 104

In vivo imaging. *See also* Live-cell imaging
  multiphoton microscopy, 86, 89–91, 94
  spectral methods, 54–56, 62–63
Ion concentrations, fluorescence correlation
  spectroscopy and, 228

Known confounding fluorophore (KCF), 54

Labeling techniques, in neuroscience, 76–77
Lamps, as excitation sources for medical diagnostic
  imaging, 285
Laparoscopes, in optical coherence tomography, 176
Laser-induced fluorescence (LIF), 268
Laserlight sources
  femtosecond, in optical coherence tomography,
    171–174
  formedical diagnostic imaging, 285
  in membrane potential imaging, 150
Lasers, ultrafast, in multiphoton microscopy, 74
Lateral resolution, vs. field of view, 4
LCTFs (liquid crystal tunable filters), 34–35
LEDs (light-emitting diodes), 285
Lenses, dual, axial resolution improvement with,
  239–241
LIF (light-induced fluorescence), 268
LIFE system, 289–290
Light
  back-reflected and backscattered, in optical
    coherence tomography, 165
  echo time delays of. *See* Echo time delay of light
Light bundle, equation of, 360–361
Light-emitting diodes (LEDs), 285
Light-induced fluorescence (LIF), 268
Light propagation theory, in tissues, 358–359
Light-scattering spectroscopy, in detection of
  cervical cancer, 310–311
Light-sensitive proteins, coupling of, to effector ion
  channels, 94
Light sources
  membrane potential imaging, 150
  optical coherence tomography, 171–174
  spectral optical imaging, 30
Linear2 ′ OMe RNA probes, in messenger RNA
  imaging, 106
Liquid crystal tunable filters (LCTFs), 34–35
Live-cell imaging. *See also* Intracellular observations
  4Pi-confocal microscopy and, 253
  fluorescent proteins in, 260–261
Living animals, imaging in. *See* In vivo imaging
Living tissues
  messengerRNA imaging of. *See* MessengerRNA
    imaging
  neural, multiphoton microscopy of, 75, 85–86,
    89–91, 94
Low-coherence interferometry, 165–166
Lower urinary tract, medical diagnostic fluorescence
  imaging of, 270
Luciferases, 398, 400–401

Macular diseases, optical coherence tomography and,
  179–182
Malignancies. *See* Cancerdiagnosis and treatment

Mammalian cell lines, messengerRNA studies in,
  115–116
Mammalian preparations, voltage-sensitive
  population signal measurements in, 146
Mammography, near-infrared optical, 350–353
MCFM (multichannel fusion method), 46–47
Medical diagnostic fluorescence imaging, 265–299
  basic concepts, 272–276
  clinical challenges in, 269–272
  data processing, 276–278
  future directions, 297–299
  imaging modes, 278–283
  imaging systems, 283–297
    current status, 284–286
    history of, 283–284
    hybrid imaging mode, 297
    Mode A: single-band, 286–287
    Mode B: dual-band imaging, 287–290
    Mode C: multicolorimaging, 290–293
    Mode E: Fourier transform spectroscopic
      imaging, 293
    Mode F: time-gated imaging, 294–297
  malignancy detection techniques, 266
  spectroscopy-based optical detection, 268–269
  statistical concepts in, 267
Membrane-bound dyes, membrane potential sensitive
  optical properties of, 132
Membrane-bound receptors, anomalous diffusion of,
  222–224
Membrane potential imaging, 132–156
  amplitude of voltage change, 146–147
  Aplysia ganglion action potentials, 134–135,
    139–143
  future directions, 154–155
  individual neuron processes, 134, 135–139
  measuring technology, 147–155
    imaging devices, 154–155
    light sources, 150
    noise, 147–150
    optics, 150–154
    photodetectors, 154
    silicon diode arrays, 154–155
  methods, 146
  overview, 132–135
  signal type, 146
  turtle olfactory bulb population signals, 134, 135,
    143–146
MEMS (Micro-Electro-Mechanical Systems)
  scanning devices, 176
Menstrual cycle, and spectroscopic detection of
  cervical cancer, 313
MessengerRNA imaging, 102–117
  cytoplasmic RNA studies, 113–116
  future directions, 116–117
  nuclearRNA studies, 110–113
  rationale for, 102–103
  RNA dynamics analysis, 108–110
  technologies for, 104–108
Metastasis, mechanism of, messengerRNA imaging
  studies of, 102

Micro-Electro-Mechanical Systems (MEMS) scanning devices, 176

Microfluidic systems, fluorescence correlation spectroscopy in, 229

Microinjection of fluorescent RNA, in messenger RNA imaging, 104

Microscopy, optical coherence tomography compared to, 164–165

Minimum noise fraction (MNF), in spectral optical imaging data processing, 51–52

MMM-4Pi microscopy, 253

MNF. *See* Minimum noise fraction (MNF)

Mobility analysis, fluorescence correlation spectroscopy and, 222–225

Molecularbeacons. *See also* Fluorescent probes
    in messengerRNA imaging, 107
    formolecularfunction   optical imaging, 374, 377–379

Molecularbr ightness, 207, 221

Molecularfunction optical imaging, 374–393
    clinical feasibility, 391–393
    fluorescence-mediated tomography for, 382–391
        diffuse optical tomographic imaging systems, 385–386
        in vivo imaging, 386–387
        molecularquantification and specificity, 389–391
        sensitivity, 387–389
        theoretical specifics: forward problem, 383–384
        theoretical specifics: inverse problem, 384–385
    fluorescent imaging probes for, 375–379
    future potential of, 374–375
    molecularbeacons and, 374, 377–379
    reflectance imaging for, 379–382

Molecular interaction studies, fluorescence correlation spectroscopy and, 225–227

Molecularpr ocesses, analysis of, 196–198. *See also* Fluorescence correlation spectroscopy (FCS)

Monte Carlo modeling of spectral data, 324–326

Morphological plasticity, multiphoton microscopy and, 89

Moving slit spectrometers, 31

MPCM (multipeak contrast method), 45–46

MPE (multiphoton excitation), 73–75, 77–78

MPM. *See* Multiphoton microscopy (MPM)

mRNA imaging. *See* MessengerRNA imaging

MS2-GFP fusion protein, in messenger RNA imaging, 108, 113

Multichannel fusion method (MCFM), 46–47

Multicolorimaging. *See* Multispectral imaging

Multifunctional reporter genes, 402–404

Multiparameter imaging. *See* Multispectral imaging

Multipeak contrast method (MPCM), 45–46

Multiphoton concepts, superresolution and, 238

Multiphoton excitation (MPE), 73–75, 77–78

Multiphoton fluorescence recovery after photobleaching (FRAP), 92–93

Multiphoton microscopy (MPM) in neuroscience, 72–94
    advantages of, 73–76
    applications of, 88–93
    fluorescence imaging, 76, 88–91

future potential of, 93–94
    limitations of, 124
    photochemistry, 76, 86–88, 92–93
    technical issues, 76–88
        data acquisition, 84–85
        detection, 83–84
        excitation, 77–78
        labeling, 76–77
        tissue penetration, 78–83
        working with living brain tissue, 85–86

Multiphoton uncaging, 87, 92

Multislit spectrometers, 31–32

Multispectral imaging
    described, 30, 124
    in detection of cervical cancer, 313–315
    in medical diagnostics, 273, 277, 279–281, 290–293

Multitemporal imaging, in medical diagnostics, 277

Multivariate techniques, in multispectral imaging, 273

Muscle, broadband diffuse optical spectroscopy of, 346–350

Myelin basic protein messenger RNA, 114

Myotonic dystrophy, messenger RNA imaging studies of, 102

NA. *See* Numerical aperture (NA)

Nanoscopy, 237–261
    applications of, 253
    axial resolution improvement, 239–241
    deconvolution in frequency domain, 248–249
    deconvolution in spatial domain, 247–248
    diffraction barrier and, 237–239, 253–256
    optical transfer functions, comparison and significance of, 241, 245–246, 250–253
    point-spread function comparison, 241–245, 249–250
    stimulated emission depletion microscopy, 253–254, 256–260, 261

Near-infrared fluorescent probes, 122–128
    applications of, 122–124
    design of new, 125–127
    limitations of current, 124–125
    optimization of, 128

Near-infrared spectroscopy
    in brain activity detection, 356–371
        brain hemodynamics investigations, 365–370
        fast effects and, 359, 360–363, 370–373
        frequency domain method for turbid media, 357–359
        future potential of, 370–373
        instrument for noninvasive measurements, 359
        noninvasive measurements: slow vs. fast effects, 359–360
        optical detection of, 360–365
        overview, 356–357
        slow effects and, 359–360
    in brain hemodynamics investigations, 365–370
    of in vivo molecularfunction. *See* Molecular function optical imaging
    medical applications of, 333
    in thick tissues. *See* Diffuse optical spectroscopy (DOS)

Neck, medical diagnostic fluorescence imaging of, 272

Needle imaging devices, in optical coherence tomography, 178

Negative predictive value, defined, 267

Neoplasia. *See* Cancerdiagnosis and treatment

Neural network studies, multiphoton microscopy and, 94

Neural tissues
  factors influencing penetration depth in, 79–80
  light propagation theory in, 358
  living, multiphoton microscopy of, 75, 85–86, 89–91, 94

NeuroCCD camera. *See* Charge-coupled device (CCD) cameras

Neuronal activity
  fast and slow effects in, 359–360
  optical detection of, 356–357, 360–365

Neuronal mRNA binding proteins, 103

Neurons
  individual processes of, membrane potential imaging of, 134, 135–139
  messengerRNA studies in, 113–114

Neuron-type specific staining, for membrane potential imaging, 156

Neuroreceptor mapping, multiphoton photochemistry and, 93

Neuroscience. *See also* Brain activity detection
  cell labeling techniques in, 76–77
  intact tissue studies and, 75
  membrane potential imaging in
    Aplysia ganglion action potentials, 134–135, 139–143
    individual neuron processes, 134, 135–139
    turtle olfactory bulb population signals, 134, 135, 143–146
  multiphoton microscopy in, 72–94
    advantages of, 73–76
    applications of, 88–93
    data acquisition, 84–85
    detection, 83–84
    excitation, 77–78
    fluorescence imaging, 76, 88–91
    future potential of, 93–94
    labeling, 76–77
    photochemistry, 76, 86–88, 92–93
    technical issues, 76–88
    tissue penetration, 78–83
    working with living brain tissue, 85–86

Nicotinamide adenine dinucleotide, as tissue chromophore, 316, 321–323, 325–326

Noise, in membrane potential imaging, 147–150

N protein messenger RNA, vesicular stomatitis virus, 112

NuclearRNA binding proteins, 112–113

NuclearRNA studies, 110–113

Nucleic acid aptamers, 121

Numerical aperture (NA), in membrane potential imaging, 150–151

Object cube, defined, 40

OCT. *See* Optical coherence tomography (OCT)

ODNs. *See* Oligodeoxynucleotide probes (ODNs)

Odor-induced neural responses, membrane potential imaging of, 143–146

Olfactory bulb
  multiphoton microscopy of, 88
  rat, membrane potential imaging of, 137–139
  turtle, population signals, membrane potential imaging of, 134, 135

Oligo(dT) probes, 106, 110–111

Oligodeoxynucleotide probes (ODNs), in messenger RNA imaging, 104, 106

Oligonucleotide probes, fluorescent, in messenger RNA imaging, 104–107

Oligonucleotides, with nonconstant fluorescent properties, 107

2′ OMe RNA probes, linear, in messenger RNA imaging, 106

One-photon fluorescence microscopy, 10–11

Ophthalmology
  multiphoton microscopy in, 88
  optical coherence tomography in, 161–162, 168, 174–175, 179–182

Optical biopsy, 161, 178–185, 186

Optical coherence tomography (OCT), 161–190
  basic principles of, 161
  in cancerdiagnosis, 182–185
  examples of, 161–164
  image resolution, detection sensitivity, and image generation, 167–169
  interferometric measurement of echo time delay of light, 165–166, 170, 186–190
  in ophthalmology, 179–182
  optical biopsy, 161, 178–185, 186
  otherimaging technologies compared to, 164–165
  potential applications of, 186
  technology and systems for, 169–178
    imaging instruments and probes, 174–178
    light sources and ultrahigh-resolution imaging, 171–174
    overview, 169–171
  three-dimensional imaging, 185–186

Optical fields
  optimum resolution and, 19–20
  radially polarized, generation and focusing of, 20–26

Optical frequency domain imaging. *See* Swept-source/Fourier domain detection

Optical nanoscopy. *See* Nanoscopy

Optical reporter genes, 397–406
  characteristics of, 399–400
  expression of, 404
  fluorescent proteins, 401–402
  luciferases, 398, 400–401
  multifunctional reporter genes, 402–404
  overview, 397–398
  potential applications of, 405–406
  preclinical studies, 397, 398–399, 404–405
  small animal imaging applications, 404–405

Optical sectioning. *See also* Three-dimensional imaging
  confocal microscopy and, 7
  structured illumination and, 13–16

Optical signal, in membrane potential imaging, 146

Optical transfer functions (OTFs), 238, 241
    comparison and significance of, 245–246, 250–253

Optics, in membrane potential imaging, 150–154

Optimized contrast function, in multispectral
    imaging, 273

Optophysiology in brain slices, 88–89

Oral cancer, medical diagnostic fluorescence
    imaging of, 272

Organic infrared dyes, limitations of, 124–125

Orthogonal polarization spectral imaging, 62

OTFs. *See* Optical transfer functions (OTFs)

Out-of-focus light, in membrane potential
    imaging, 151

Oxyhemoglobin concentrations, in muscle,
    346–350

Pair-rule transcripts, injection into *Drosophila*
    oocytes, 115

Parallel readout arrays, in membrane potential
    imaging, 154

Particle tracking, in messenger RNA imaging,
    109–110

Patient variation, in spectroscopic detection of
    cervical cancer, 312–313

PCA. *See* Principal component analysis (PCA)

PDT (photodynamic therapy), 284

Peptide nucleic acid probes, in messenger RNA
    imaging, 106

Peptide substrates, for near-infrared fluorescent
    probes, 378–379

pH-dependent emission characteristics, 228

Photobleaching. *See also* Fluorescence recovery after
        photobleaching (FRAP)
    in messengerRNA imaging, 108–109, 112
    two-photon fluorescence correlation spectroscopy
        and, 199, 229–233

Photochemistry, multiphoton microscopy, 76, 86–88,
    92–93

Photodamage, avoidance of, in multiphoton
    microscopy, 86

Photodetectors, in membrane potential imaging,
    152–154

Photodiode array, noise limitations, 149, 150

Photodynamic therapy (PDT), 284

Photosensitizers, 274

Photostability
    of cyanine dyes, 127
    of infrared chromophores, 125

Photostimulation, multiphoton microscopy and, 94

pi-sensor, 361

Point-scanning spectrometers, 31

Point-spread function
    defined, 237
    in fluorescence microscopy, 9–10
    image deconvolution and, 252
    SWM, 4Pi, and I^{5}M compared, 241–245,
        249–250

Poly(A) RNA studies, 110–111, 113

Polycyclic chromophores, limitations of, 125

Polyene dyes, 155–156

Polymethine fluorophores, limitations of,
    124–125

Polypoid adenoma of colon, optical coherence
        tomographic three-dimensional imaging of, 185

Positive predictive value, defined, 267

Potential-dependent optical properties, molecular
    mechanisms of, 133–134

PpIX (protoporphyrin IX), 275, 287

Preclinical studies, optical imaging in, 397, 398–399,
    404–405

Premessenger RNAs
    human *beta*-globulin, 111
    rat prokephalin, 111

Primary cells, messenger RNA studies in, 116

Principal component analysis (PCA), in spectral
    optical imaging data processing, 51

Protein fluorescence. *See* Fluorescent proteins

Protein trafficking, multiphoton microscopy and, 94

Protoporphyrin IX (PpIX), 275, 287

Pushbroom hyperspectral imaging, 281, 294

Pushbroom imaging spectrometers, 31

Pyramidal neuron, membrane potential
    imaging of, 137

Quantitative imaging
    hybrid imaging compared to, 47–49
    in spectral data visualization, 42–44

Quantum dots, 120–121

Radially polarized optical fields, generation and
    focusing of, 20–26

Raman spectral imaging, 31, 43, 58–59, 60–61

Random access fluorescence microscopy, membrane
    potential imaging and, 152–154

Rat olfactory bulb, membrane potential imaging of,
    137–139

Rat prokephalin premessenger RNAs, 111

Red, green, blue (RGB) color space, 44

Reference spectral signature selection, 52

Reflectance imaging
    advantages and limitations of, 381–382
    in detection of cervical cancer, 309–311
        with fluorescence spectroscopy, 311–312
    formolecularfunction studies, 379–382
    of tissues, overview, 307–308

Regenerative medicine, optical imaging in, 405

Reporter genes. *See* Optical reporter genes

Representation space, in computational processing of
    spectral data, 39–40

RESOLFT (reversible saturable optically linear
    fluorescence transition), 239, 254, 260–261

Resolution improvement, in confocal microscopy, 4–5

Resolution limits, 237–239

Retina
    multiphoton microscopy of, 88
    optical coherence tomography of, 161–162, 168,
        174–175, 179–182

Retinopathy, diabetic, optical coherence tomography
    and, 180

Reversible saturable optically linear (fluorescence)
    transition between two molecular states
    (RESOLFT), 239, 254, 260–261

RGB (red, green, blue) color space, 44

R-groups, of fluorescent probes, 125–126, 128

Rhodamine dyes, in fluorescence correlation
  spectroscopy, 218

RLS (RNA localization region), 114

RNA, fluorescent, microinjection of, 104

RNA binding proteins, nuclear, 112–113

RNA dynamics in living cells, analysis of, 108–110

RNA imaging. *See* MessengerRNA imaging

RNA localization region (RLS), 114

RNA probes, linear 2<′> OMe, in messengerRNA
  imaging, 106

RNA tagged with sequence-specific green fluorescent
  protein, 108

RNA transport sequence (RTS), 114

RNA transport tracking, 109–110

SAM (spectral angle measure), 42

Saturation factor, 255

Scanning methods
  in confocal microscopy, 3
  in spectral optical imaging, 30
    band-sequential spectrometers, 32–36
    single-shot spectrometers, 37–38
    slit spectrometers, 30–32
    volume holographic spectral imaging, 36–37

Scattering
  chromophores responsible for, in tissues, 319
  described, 308
  in membrane potential imaging, 151
  tissue penetration and, 78–80, 122–124

SCS (spectral correlation similarity), 41

Sensitivity, defined, 267

Serial readout arrays, in membrane potential imaging,
  154–155

Shot noise, in membrane potential imaging, 147–149

Signal-to-noise ratios
  in medical diagnostic imaging, 278
  in membrane potential imaging, 147–149

Signature contribution, 55–56

Signature specificity degree, 55

Silicon diode arrays, in membrane potential imaging,
  154–155

Single-band fluorescence imaging, 278, 286–287

Single-shot spectrometers, 37–38

Single-slit spectrometers, 31

Skin, medical diagnostic fluorescence imaging of,
  271, 286, 293

SLDs. *See* Superluminescent diodes (SLDs)

Slit spectrometers, 30–32

Slow effects, brain activity and, 359–360

Small animal imaging, 65, 404–405. *See also*
  Preclinical studies

Small molecularweight indicators, labeling with,
  76–77

Small nuclearRNA (snRNA), 112

Smart biopsy, 268–269

Smart probes, 377–379

snRNA (small nuclearRNA), 112

Solid phase model of nuclearor ganization, 110–111

Solubility, of fluorescent probes, 128

Spatial cross-correlation, in two-photon fluorescence
  correlation spectroscopy, 208–212

Specificity, defined, 267

Spectral angle measure (SAM), 42

Spectral correlation similarity (SCS), 41

Spectral endoscopy, 62–63

Spectral/Fourier domain detection, 187–188

Spectral imaging, defined, 29–30

Spectrally resolved fluorescence imaging, 275

Spectral optical coherence tomography.
  See Spectral/Fourier domain detection

Spectral optical imaging, 29–67
  biological applications, 56–61
  clinical applications, 61–66
  data processing, 38–56
    automated end-memberdeter mination, 52–53
    dimensionality reduction, 50–52
    examples using in vivo data, 54–56
    spectral signature and spectral similarity
      measures, 40–42
    spectral unmixing, 49–50, 54, 59–60, 65
    three views of spectral data and, 39–40
    visualization. *See below at* visualization methods
  future potential of, 66–67
  light sources, 30
  multiplexing strategies, 30
  spectral selection methods, 29–38
    band-sequential spectrometers, 32–36
    single-shot spectrometers, 37–38
    slit spectrometers, 30–32
    volume holographic spectral imaging, 36–37
  terminology, 29–30
  visualization methods, 42–49
    classification imaging, 42, 43
    classification-quantitative hybrid imaging, 43,
      44, 45–49
    image space models and, 44–45
    quantitative imaging, 42–44, 47–49
    strategies for, 42–43

Spectral signature
  defined, 39
  versus end member, 49
  measures of, 40–42

Spectral signature display algorithm, 54–56

Spectral similarity, measures of, 40–42

Spectral similarity value (SSV), 41–42

Spectral space, in computational processing of
  spectral data, 39–40

Spectral subtraction, 54

Spectral unmixing
  in fluorescence imaging, 58, 59–60
  in spectral optical imaging, 49–50
  spectral subtraction and, 54
  forwhole-animal imaging, 65

SSV (spectral similarity value), 41–42

Standing wave microscopy (SWM)
  axial resolution and, 239
  optical transfer functions of, 245–246, 250–251
  point-spread function of, 241–245, 249–250

Statistical concepts in medical diagnosis, 267

Stimulated emission depletion (STED) microscopy,
  253–254, 256–260, 261

Storz system of fluorescence imaging, 287
Stroma
    fluorescence of, 318–319, 320–321, 322, 325–326
    neoplastic changes and, 306–307, 319
    three-dimensional tissue phantom model of,
        323–324
Structured illumination microscopy, 13–16
Substage detection, in multiphoton microscopy, 84
Superluminescent diodes (SLDs), as optical coherence
        tomography light sources, 171, 172
Superresolution, 238
Supervised end-member determination, 52
Support vector machines (SVMs), 52
SV40 messengerRNA, 116
SVMs (support vector machines), 52
Swept-source/Fourier domain detection, 187,
        188–190
SWM. *See* Standing wave microscopy (SWM)
Synapses, optophysiology of, 89
SYTO-14, 107–108, 113

Targeted fluorescent probes, 376–377
Targeted therapies, molecular imaging and, 375
Technical noise, in membrane potential imaging,
        149–150
Temporal decay time, 277
Temporally resolved fluorescence imaging,
        275–276
Tetramethylrhodamine (TMR), photobleaching of,
        230–232
Three-dimensional diffusion
    of multiple species, 206–207
    of single species, 203–205
Three-dimensional imaging. *See also* Optical
        sectioning
    confocal microscopy and, 4, 8 (*see also* Depth
        discrimination in confocal microscopy)
    optical coherence tomography and, 185–186
Three-dimensional tissue phantom models, for
        simulation of epithelial cancer, 323–324
Three-photon excitation, 78
Time domain detection, 166, 186
Time-gated imaging, in medical diagnostics, 282,
        295–297
Time-independent diffuse optical spectroscopy, 330
Time-lapse microscopy, in messenger RNA imaging,
        109–110, 112–113
Time-resolved diffuse optical spectroscopy,
        330–331
Tissue oxygen saturation, in exercised and
        nonexercised muscle, 348–350
Tissue penetration
    in multiphoton microscopy, 78–83
    wavelength and, 122–124
Tissue phantom models, three-dimensional, for
        simulation of epithelial cancer, 323–324
Tissues
    chromophores present in, 308, 315–319,
        321–323
    breast, 350
    muscle, 346

fluorescence and reflectance spectroscopy of,
        307–308
    light propagation theory in, 358
    optical properties of, 331
TMR (tetramethylrhodamine), photobleaching of,
        230–232
TMS (trimodal spectroscopy), in detection of cervical
        cancer, 312
Total hemoglobin content, in exercised and
        nonexercised muscle, 348–350
Trimodal spectroscopy (TMS), in detection of cervical
        cancer, 312
Tumor markers, fluorescent, 284
Tungsten filament lamps, in membrane potential
        imaging, 150
Turtle olfactory bulb population signals, 134, 135,
        143–146
Two-photon fluorescence correlation spectroscopy.
        See Fluorescence correlation spectroscopy
        (FCS), two-photon
Two-photon fluorescence microscopy, 10–11. *See
        also* Multiphoton microscopy (MPM)
    membrane potential imaging and, 151–152

Ultrafast lasers, in multiphoton microscopy, 74
Ultrahigh-resolution imaging, in optical coherence
        tomography, 172–174
Ultrasound, optical coherence tomography compared
        to, 164–165
Uncaging
    chemical two-photon, 87–88
    multiphoton, 87, 92
Unsupervised end-member determination, 52
Urinary tract, lower, medical diagnostic fluorescence
        imaging of, 270

Vacuum photocathode cameras, 155
Vascularimaging, 62
vav protooncogene, messenger RNA, 115–116
Vertebrate neurons, membrane potential imaging of,
        137–139
Vesicular stomatitis virus N protein
        messengerRNA, 112
Virology, messenger RNA imaging
        applications in, 103
Visualization methods forspectral data, 42–49
    classification imaging, 42, 43
    classification-quantitative hybrid imaging, 43, 44,
        45–49
    image space models and, 44–45
    quantitative imaging, 42–44, 47–49
    strategies for, 42–43
Vocal cords, medical diagnostic fluorescence
        imaging of, 272, 287, 291
Voltage-sensitive dyes. *See also* Membrane potential
        imaging
    potential future developments in, 155–156
    properties of, 132

Volume holographic spectral imaging, 36–37

Walsh functions, 32
Water, light absorption in near-infrared, 122–123
Wavefront biasing, in aberration correction, 17
Wavefront generation, 20
Wavefront sensors, in aberration correction, 17–19
Wavelength, tissue penetration and, 122–124
White-light interferometry, 165
Whole-animal imaging, spectral unmixing for, 65
wingless messengerRNA, injection into *Drosophila* oocytes, 115

Xenopus laevis tadpole, optical coherence tomographic three-dimensional imaging of, 185–186
XFPs, 77

Yeast, messengerRNA studies in, 114–115
Yellow fluorescent protein (YFP), pH-dependent emission characteristics of, 228

Zebra fish egg, optical coherence tomography of, 163–164
Zernike circle polynomials, in aberration correction, 17